SECOND EDITION

Biochemistry and Physiology of Protozoa

Volume 2

CONTRIBUTORS

Robert A. Bloodgood
P. F. L. Boreham
Marvin H. Cantor
G. S. Coleman
Donald L. Cronkite
B. M. Honigberg
S. H. Hutner
Joanne K. Kelleher

Theo M. Konijn
M. Levandowsky
David Lloyd
José M. Mato
Jytte R. Nilsson
Yoshinori Nozawa
Guy A. Thompson, Jr.
John J. Wille, Jr.

SECOND EDITION

Biochemistry and Physiology of Protozoa

Volume 2

Edited by

M. LEVANDOWSKY

S. H. HUTNER
Haskins Laboratories of Pace University
New York, New York

Consulting Editor

LUIGI PROVASOLI
Department of Biology
Yale University
New Haven, Connecticut

1979

ACADEMIC PRESS

A Subsidiary of Harcourt Brace Jovanovich, Publishers

New York London Toronto Sidney San Francisco

ACADEMIC PRESS, INC.
111 Fifth Avenue, New York, New York 10003

United Kingdom Edition published by
ACADEMIC PRESS, INC. (LONDON) LTD.
24/28 Oval Road, London NW1 7DX

Library of Congress Cataloging in Publication Data

Main entry under title :

Biochemistry and physiology of protozoa.

First ed. published in 1951–64, entered under
A. Lwoff.
Includes bibliographies and indexes.
1. Protozoa––Physiology. I. Levandowsky, Michael.
II. Hutner, Seymour Herbert, Date. III. Lwoff,
Andre, Date   ed. Biochemistry and physiology of
protozoa.
QL369.2.L87 1979   593'.1'041   78–20045
ISBN 0–12–444602–7 (v. 2)

Contents

4 Microtubules
Joanne K. Kelleher and Robert A. Bloodgood

5 Chemosensory Transduction in *Dictyostelium discoideum*
José M. Mato and Theo M. Konijn

6 The Genetics of Swimming and Mating Behavior in *Paramecium*
Donald L. Cronkite

List of Contributors

Numbers in parentheses indicate the pages on which the authors' contributions begin.

Robert A. Bloodgood (151), Department of Anatomy, Albert Einstein College of Medicine, Bronx, New York 10461

P. F. L. Boreham (429), Imperial College Field Station, Silwood Park—Ascot, Berkshire, SL5 7PY England

Marvin H. Cantor (9), Department of Biology, California State University, Northridge, California 91330

G. S. Coleman (381), Biochemistry Department, Agricultural Research Council, Institute of Animal Physiology, Babraham, Cambridge, CB2 4AT England

Donald L. Cronkite (221), Biology Department, Hope College, Holland, Michigan 49423

B. M. Honigberg (409), Department of Zoology, University of Massachusetts, Amherst, Massachusetts 01003

S. H. Hutner (1), Haskins Laboratories of Pace University, New York, New York 10038

Joanne K. Kelleher* (151), Department of Developmental Biology, Boston Biomedical Research Institute, Boston, Massachusetts 02114

Theo M. Konijn (181), Cell Biology and Morphogenesis Unit, Zoological Laboratory, University of Leiden, Leiden, The Netherlands

M. Levandowsky (1), Haskins Laboratories of Pace University, New York, New York 10038

David Lloyd (9), Department of Microbiology, University College, Newport Road, Cardiff, Wales, United Kingdom

José M. Mato (181), Cell Biology and Morphogenesis Unit, Zoological Laboratory, University of Leiden, Leiden, The Netherlands

---

* Present address: Department of Physiology, George Washington University Medical Center, Washington, D.C. 20037.

Jytte R. Nilsson (339), Institute of General Zoology, University of Copenhagan, Copenhagen, Denmark

Yoshinori Nozawa (275), Department of Biochemistry, Gifu University School of Medicine, Gifu, Japan

Guy A. Thompson, Jr. (275), Department of Botany, University of Texas at Austin, Austin, Texas 78712

John W. Wille, Jr. (67), Department of Zoology and Physiology, Louisiana State University, Baton Rouge, Louisiana 70803

Preface to the Second Edition

This inaugurates, some 15 years after its predecessor, a multivolume second edition of "Biochemistry and Physiology of Protozoa." In this sense, in retrospect the preceding volumes (three in all) constitute a first edition, but as the intervals between the new volumes will be measured in months and a year or two rather than decades, and the new volumes have been planned as a whole, "second edition" seems fitting, and emphasizes that protozoology has vastly expanded in recent years and, by most evidence, will continue expanding.

The causes of this expansion are easily detected. That the gulf separating prokaryotes and eukaryotes seems evolutionarily the widest among extant organisms is unchallenged. Kluyverian unity of biochemistry remains firmly established, but is perceived in a perspective at once deeper and more practical. How easy it was to find drugs against prokaryotes, how cursedly hard and expensive to find them against the eukaryotic parasites—protozoa, fungi, and helminths!

The World Health Organization designates malaria, leishmaniasis, and trypanosomiasis as three of the six infectious diseases posing the most important global challenges. In the developed countries recognition of the grudging pace of progress with chronic diseases and aging is widening the demand for eukaryotic "models" which will be easier to handle than conventional laboratory animals. The more conspicuously animal protozoa are increasingly meeting this hunger for expeditious approaches to eukaryotic fundamentals; they will not be neglected in this new series nor will the pathogens (taking into account that several other volumes on parasitic protozoa have recently been published or are listed in press).

Advances in identifying molecular kinships have attracted biochemists (and other gentry not formally protozoologists) into the enterprise of building abutments for bridges between eukaryotes and prokaryotes. Fittingly, therefore, this edition leads off with an overview of phytoflagellate phylogeny.

The increase in knowledge of metabolic pathways and descriptive biochemistry permits more penetrating analyses of the protozoan equivalents of endocrinology, neurology, especially as manifest in behavior. I therefore am delighted to welcome as senior editor for this edition my colleague Dr. Michael Levandowsky. In doing so I follow the precedent set up by the founder of this enterprise, Andre Lwoff, when he invited me to serve as senior editor with him for Volume II of what was, in retrospect, a three-volume first edition, with long intervals between Volumes II and III. The pace has quickened; the old verities need new kinds of substantiations.

S. H. Hutner

Preface to Volume 2

As noted in the introduction to Volume 1 of this edition, we seek to define the advancing paths of research, to predict the shape of things to come in protozoology. In this, we willfully, even passionately, eschew standard categories and instead follow a broad story line.

In Volume 1 we began with biochemical views of current issues in protozoan phylogeny and evolution, and related areas of physiological ecology. In this volume we indulge our preoccupation with the interactions of structure and function; here our scope has ranged from the level of subcellular organization to that of host–parasite interactions.

To be sure, the logical trail from chapter to chapter is again a winding one, even doubling back on itself at times, for we were concerned to construct the most scenic route through a vast territory. With this in mind, then, we have tried to provide a Baedeker in the introduction for travelers from other scientific realms; it is our fond hope that this will prove useful to the casual tourist and the seasoned voyager, and, perhaps, to some local commuters as well.

We wish to thank L. G. Goodwin for advice on parasitological topics. We are also greatly indebted to the editors of Academic Press for their amiable persistence and patience.

M. Levandowsky

Contents of Other Volumes

* In preparation.

Introduction

1

M. LEVANDOWSKY AND S. H. HUTNER

Protozoan biochemistry had its origins in three themes whose reverberations can still be clearly felt: nutrition of *Tetrahymena,* a ciliate, of *Crithidia,* a kinetoplastid, and of the acetate flagellates, with *Euglena, Polytomella,* and *Chlamydomonas* as prime examples (Pringsheim, 1921, 1937; Lwoff, 1932, 1943). We begin Volume 2 of this treatise with the last of these classical themes.

I. ACETATE FLAGELLATES AND THE EVOLUTION OF HETEROTROPHY

As a group, these species fit very well into the general viewpoint of this treatise outlined in our introduction to Volume 1 (Levandowsky and Hutner, 1979). We deal here with a polyphyletic group that includes euglenids, cryptomonads, and several chlorophytes. Some of these are facultative photoautotrophs and others are colorless heterotrophs; all have acquired the ability to thrive on the short-chain fatty acids and alcohols

BIOCHEMISTRY AND PHYSIOLOGY OF PROTOZOA
SECOND EDITION, VOL. 2

produced in abundance in certain common natural fermentative processes, such as the anaerobic decomposition of cellulose. They are often found in such pungent man-made habitats as sewage ponds and barnyard wastes. There is a flickering borderline between autotrophy and heterotrophy in the group which retains its deep fascination for students of biochemical evolution, but modern biochemistry is preoccupied with the spatiotemporal deployment of metabolic systems—the cell is no longer a bag of enzymes. Lloyd and Cantor explore this architectural side of biochemistry in the present volume.

II. TIME AND SPACE: THE BIOCHEMISTRY OF EUKARYOTIC CELL ARCHITECTURE

Contemporary biochemistry is looking for the control systems of metabolism. The first successes were the analyses of inducible enzyme systems in prokaryotes; the demonstration by genetic analysis of negative feedback mechanisms involved in selective inhibition of parts of the prokaryote genome not only put an end to a rather sterile, somewhat vitalistic tradition of theorizing on this subject, but also suggested new experimental approaches to eukaryotic problems such as cell differentiation in embryology.

The biological clock, ubiquitous in eukaryotes from man to *Euglena* but apparently absent from prokaryotes, may turn out to be a fundamental control system defining eukaryotic architecture. As an object of study, it has always been rather shadowy—though mirrored by many rhythmic phenomena, its biochemical basis proved singularly elusive despite many attempts to pinpoint it. The accomplishment of Wille in his exhaustive review here of protozoan rhythms is to distill a useful new theoretical suggestion from the mass of experiments. This is an elaboration of an abstract theme introduced by Winfree (1973), who suggested that the biological clock may reside in a system of reactions forming a coupled nonlinear system oscillating in a stable limit cycle. Wille now suggests that there are actually *two* such limit cycles: one, the circadian (or ultradian), is viewed as ticking over continually with a low amplitude in the plateau phase of growth, while the other, the infradian, is evoked under the optimal conditions of exponential growth and represents a bifurcation in the model's solution, when some parameter (perhaps a storage pool of a metabolite) attains a critical value in a bifurcation diagram. The evidence for some such mechanism now seems very strong (Kauffman, 1977), and biochemists must seriously address the problem of fleshing out these abstractions. What, biochemically, is the oscillating system and where is it in the cell? Wille has some suggestions. For those who are tempted to enter further into these mysteries, a handy glossary to the lingo is provided: from *acrophase* to *zeitgeber*.

So much for the temporal architecture. Spatial organization in eukaryotes is intimately associated with the deployment of microtubules. With the exception of certain spirochetes and blue-green bacteria (Margulis, 1978; Bisaluptra *et al.*, 1975), tubulin appears to be restricted to the eukaryotes, where it is ubiquitous. Like some other architecturally important molecules (histones, actin), tubulin appears to be extremely conservative phylogenetically—a curious, unexpected finding of modern comparative biochemistry. This being so, the review of Kelleher and Bloodgood is not restricted to protozoa, but considers evidence from all eukaryotes. The presumption here is that, since certain kinds of tubulin in, for example, vertebrate brain tissue and *Chlamydomonas,* have virtually the same primary structure (Binder *et al.*, 1975), control of microtubule construction is probably a fundamental, phylogenetically constant piece of eukaryotic equipment as well. Within a given organism, however, there are different tubulins with different functions, some of which (e.g. membrane-bound, nonmicrotubular tubulin) are as yet very little understood. This chapter deals with the problem of initiation and control of polymerization and microtubule formation—a very central problem in protozoan ontogeny and differentiation.

III. INTERCELLULAR SIGNALS

Ontogeny, in the sense of well-defined successive stages of the cell cycle, can no doubt be seen in all eukaryotes, and the appearance from time to time of some major, noncyclical transformation (spore or cyst formation, induction of sexuality, metamorphosis of parasites on entering or leaving a host) which can be truly labeled differentiation may also be universal. Certain cases are particularly striking, such as the ameba-to-flagellate transformation of *Naegleria,* reviewed in Volume 1 (Schuster, 1979), and the aggregation of cellular slime molds, treated here by Mato and Konijn.

An important difference between rhythmic ontogenetic changes—the mitotic cycle, circadian phenomena—and such arrhythmic metamorphoses is that in the latter the initiating signal(s) are external. In slime molds chemical signals (acrasins) from other cells are involved, and we must recognize this intercellular communication as at least a metaphor for developmental processes in metazoa and metaphyta, as well as an incipient sociality.

It may be more than a metaphor. By now it is fairly clear that enough similarity, and no doubt also evolutionary homology, exists between the physiology of chemoreception in protozoan and metazoan cells to make a detailed study of chemosensory transduction in the former fruitful of insights regarding the latter. Recent progress has been such that Mato and Konijn are able to present a biochemical model of initial events in transduction of the external cAMP signal by *Dictyostelium discoideum*. In this model, internal Ca^{2+} levels modulate short-term responses such as pseudopod formation and contraction, whereas internal cGMP levels trigger long-term responses—i.e., cell differentiation.

Current work on ciliate behavior, treated here by Cronkite, reveals a very similar picture. The short-term control of *Paramecium* behavior—swimming speed, avoiding reactions—is determined by internal Ca^{2+} levels, and the electrophysiology of this will be reviewed in a later volume. Induction of mating behavior is less well understood however; strictly speaking, this is a long-term change, but it perhaps should not be viewed as arrhythmic since it involves a slow cycle of maturation, through many asexual mitotic divisions after the previous mating event. Chemical mating signals are involved, and specific membrane-bound receptor molecules, but so far rather little is known about the intracellular transduction process triggering the transformation to mating behavior in this species. The availability of the potentially powerful tool of genetic analysis suggests a cautious optimism, however.

It is obvious that membrane structure and physiology are central aspects of both reception and transduction of chemical signals. However, most of the information on membranes comes from work with *Tetrahymena*, reviewed here by Nozawa and Thompson. This is now a very exciting, fast-moving area. Recent studies have revealed control systems that modify lipid composition in response to temperature in such a way as to maintain membrane fluidity. As regards receptors, a membrane-bound adenylate cyclase appears to be an epinephrine receptor, raising the question of the biologic function of responses to this and other biogenic amines in protozoa. Guanylate cyclase is also found in *Tetrahymena* membranes.

Besides its role in transport and recognition of signals, the ciliate membrane and its receptors are central to studies of phagocytosis, to which we now turn.

IV. PHAGOCYTOSIS AND PARASITISM

As noted, these two themes are also classic in the history of our subject. Somewhat paradoxically, most studies of the nutritional biochemistry of phagotrophs use nonparticulate media. The fact that *Tetrahymena* grows axenically in autoclaved peptone media, or in chemically defined media containing an appropriate mix of salts, amino acids, and water-soluble growth factors, has made it the tool of choice. It is only recently that we are in a position to analyze the subcellular events leading to uptake of particulate food, and the as yet exploratory nature of the field comes through clearly in the chapter by Nilsson. Many interacting factors are no doubt involved in controlling phagocytosis. The studies of Csaba and Lantos (1973, 1975, 1976, 1977), in which phagocytosis by *Tetrahymena gelei* was stimulated by low levels of epinephrine, serotonin, and a number of other metazoan hormones are of great interest.

It seems probable that many protozoan parasites evolved from free-living phagotrophs that took to histophagy, a tendency cropping up in a number of *Tetrahymena* species that parasitize snails and other invertebrates.

In rumen ciliates, reviewed here by Coleman, a curious intermediate situation is seen. It is not quite clear whether the host ruminant is benefited nutritionally by these organisms; most of them can only be grown for limited periods outside of the rumen, however, and require periodic passages through the host. They are true phagotrophs requiring bacterial food with organic supplements. Relatively little is known in detail of their anaerobic metabolism, most work having focused on characterization of nutritional requirements and waste products.

With the final two chapters by Honigberg and Boreham, we reach the true parasites. Emphasis, appropriately, is on pathogenesis and clinical phenomena. Later volumes will discuss the underlying biochemistry, especially as it relates to possible chemotherapeutic targets and the development of new drugs. In both trichomonads and trypanosomes, a variety of secreted cytotoxic substances as well as immune mechanisms appear to be responsible for the various pathologic effects, but in general very little is known yet about the nature of these substances. Thus, these two chapters serve as a challenge for clinical biochemists.

V. SUMMARY

A. Nutritional Physiology

In our view, which clearly stems from that of Lwoff (1932, 1943), certain evolutionary trends appeared independently in phylogenetically distinct groups, as follows:

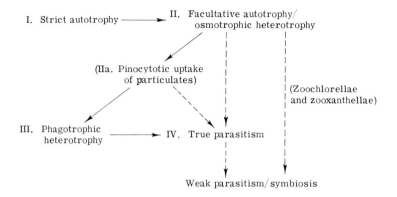

The organisms in group I of this scheme tend to inhabit relatively pure, even oligotrophic, waters with few dissolved organics; their membranes may be impermeable to most organic molecules. Metabolism is aerobic, and they may be poisoned by moderate levels of fatty acids or alcohols. An example is *Dunaliella,* treated in Volume 1 (Brown and Borowitzka, 1979).

Group II contains the acetate flagellates, and many others such as *Ochromonas* (cf. Pringsheim's *Zuckerflagellaten*). These inhabit waters high in organics—smelly habitats, teeming with bacteria and particulates. They have great tolerance for such conditions and a tendency to facultative anaerobiosis; transport systems and permeability to organic substrates are well developed, and they are usually quite flexible nutritionally, being able to subsist on any one of a large variety of reduced carbon and nitrogen sources; auxotrophic tendencies—nutritional dependency on the abundant bacteria or other sympatric organisms for synthesis of certain molecules—are common, however.

The hypothetical IIa group represents the plausible assumption that the flagellates in II must often take up proteins and perhaps other macromolecules pinocytotically, since these may abound in their environment.

Group III has achieved an evolutionarily irreversible stage; these organisms can take up entire cells (usually prokaryote) as their entire supply of reduced substrate. Not only do they become nutritionally dependent on their prey in the sense of auxotrophy, with absolute requirements for a variety of growth factors, but they also become dependent on an array of specific major substrates such as amino acids, and furthermore, these must be available in the diet in specific proportions, which are characteristic of the usual food cells. Food selection by chemosensory mechanisms becomes important. Many of these organisms (*Tetrahymena* is an exception) inhabit relatively clean waters, and are able by one means or another (mucoid secretion, feeding currents) to concentrate particulates; some are poisoned by high levels of dissolved organics.

Finally, such trends reach their extreme in the completely dependent parasites and symbionts. These are nutritionally fastidious, being specialized for a specific mix of nutrients available in a particular kind of tissue. Sensory mechanisms come into play in finding the host and its appropriate tissue, and sophisticated recognition machinery must be required to combat the host's defense mechanisms.

There are some interesting loose ends to our scheme. We have not discussed the dinoflagellates, which as a group straddle all the categories; these will appear in a later volume. Euglenids, found in groups II and III, are possibly ancestral to or in some sense distantly related to the trypanosomes (group IV) via *Bodo* (group III). The ciliates (groups III and IV) are a gigantic puzzle: Who were their ancestors in groups I and II?

B. Biochemical Organization of Eukaryotic Cells in Space and Time

This topic is the subdominant theme of Volume 2. In it we have a glimpse of future research: the biological clock quietly ticking in the background (but where, precisely?), the chemical signals from cell to lonely cell—a primitive language leading ultimately to multicellularity, hormones, and highly specialized neuromuscular systems—and the sometimes startling metamorphic events that they trigger, including in ciliates an intimation of metazoan senescence and mortality; these are, we presume, the primitive eukaryotic features which were created in the proposed symbiotic origin of eukaryotes (Margulis, 1970). They will be further explored in later volumes.

REFERENCES

Binder, L. I., Dentler, W. L., and Rosenbaum, J. L. (1975). *Proc. Natl. Acad. Sci. U.S.A.* **72**, 1122–1126.

Bisalputra, T., Oakley, B. R., Walker, D. C., and Shields, C. M. (1975). *Protoplasma* **86**, 2–28.

Brown, A., and Borowitzka, L. J. (1979). *In* "Biochemistry and Physiology of Protozoa" (M. Levandowsky and S. H. Hutner, eds.), 2nd ed., Vol. 1, pp. 139–190. Academic Press, New York.

Csaba, G., and Lantos, T. (1973). *Cytobiologie Z. Exp. Zellforsch.* **7**, 361.

Csaba, G., and Lantos, T. (1975). *Acta Protozool.* **13**, 409.

Csaba, G., and Lantos, T. (1976). *Endokrinologie* **68**, 239.

Csaba, G., and Lantos, T. (1977). *Differentiation* **8**, 57.

Kauffman, S. A. (1977). *In* ''Mathematical Models in Biological Discovery'' (D. L. Solomon and C. Walter, eds.), pp. 95–131. Springer-Verlag, Berlin and New York.

Levandowsky, M., and Hutner, S. H. (1979). *In* ''Biochemistry and Physiology of Protozoa'' (M. Levandowsky and S. H. Hutner, eds.), 2nd ed., Vol. 1, pp. 1–5. Academic Press, New York.

Lwoff, A. (1932). ''Recherches Biochimiques sur la Nutrition des Protozooaires.'' Masson, Paris.

Lwoff, A. (1943). ''L'Evolution Physiologique. Étude des Pertes de Fonction Chez les Microorganisms.'' Hermann, Paris.

Margulis, L. (1970). ''Origin of Eukaryotic Cells.'' Yale Univ. Press, New Haven, Connecticut.

Margulis, L., To, L., and Chase, D. (1978). *Science* **200,** 1118–1124.

Pringsheim, E. G. (1921). *Beitr. Allg. Bot.* **11,** 88.

Pringsheim, E. G. (1927). *Planta* **26,** 631.

Schuster, F. L. (1979). *In* ''Biochemistry and Physiology of Protozoa.'' (M. Levandowsky and S. H. Hutner, eds.), 2nd ed., Vol. 1, pp. 215–285. Academic Press, New York.

Winfree, A. T. (1973). *In* ''Biological and Biochemical Oscillators'' (B. Chance, E. K. Pye, A. K. Ghosh, and B. Hess, ed.), pp. 461–502. Academic Press, New York.

Subcellular Structure and Function in Acetate Flagellates

2

DAVID LLOYD AND MARVIN H. CANTOR

I. INTRODUCTION

The ease with which many acetate flagellates can be grown in defined media to high cell densities has made them hardly less popular as subjects

BIOCHEMISTRY AND PHYSIOLOGY OF PROTOZOA
SECOND EDITION, VOL. 2

for biochemical study than *Escherichia coli* or yeast. For many purposes
they have few rivals; topics most extensively studied in recent years
include bioenergetics, the genetics and development of chloroplasts, the
organization of microtubules and flagella, the physiology of motility, and
biological rhythms and associated phenomena. All these topics are covered by specialized contributions elsewhere in these volumes; here we
deal with other more general aspects of structure and function which
continue to receive experimental attention. Nor do we reiterate much of
the information contained in the excellent earlier reviews of Hutner and
Provasoli (1951), Danforth (1967), and Droop (1974). Instead we emphasize recent achievements in the areas of intermediary metabolism,
subcellular fractionation, phenotypic modification, and the temporal organization of cell cycles and life cycles.

II. SUBCELLULAR STRUCTURE

Biochemical studies of the acetate flagellates have often been accompanied by or resulted in a closer examination of the ultrastructure of many
members of this group. Specific studies have been made on *Astasia longa*
(Ringo, 1963), *Chilomonas paramecium* (Anderson, 1962), *Euglena* spp.
(Leedale, 1967), *Khawkinea quartana* (Schuster and Hershenov, 1974),
Polytoma uvella (Lang, 1963; Siu *et al.*, 1976a), and *Polytomella* spp. (Gittleson *et al.*, 1969; Moore *et al.*, 1970; Brown *et al.*, 1976a) (Figure 1). The
ultrastructure of *Chlamydomonas reinhardii* has also been examined (Ohad
et al., 1967; Johnson and Porter, 1968). Although a largely photoautotrophic organism, *Chlamydomonas* is included among the acetate flagellates as a borderline example, since its growth in the dark centers on
acetate. Much of this information is reviewed elsewhere in these volumes.

III. METABOLIC STUDIES

A. Carbon and Nitrogen Metabolism

The term "acetate flagellate" was introduced by Pringsheim (1921,
1937); however, their chemotrophy is not as limited as this name suggests.
Thus, Lwoff (1932, 1944) showed that several species could also utilize
pyruvic and lactic acids. Droop (1974) has grouped these organisms by
nutritional characteristics. There are species which exhibit dark growth
on acetate but not on glucose, or vice versa; others grow heterotrophically
on either carbon source; and others can utilize an amino acid as sole

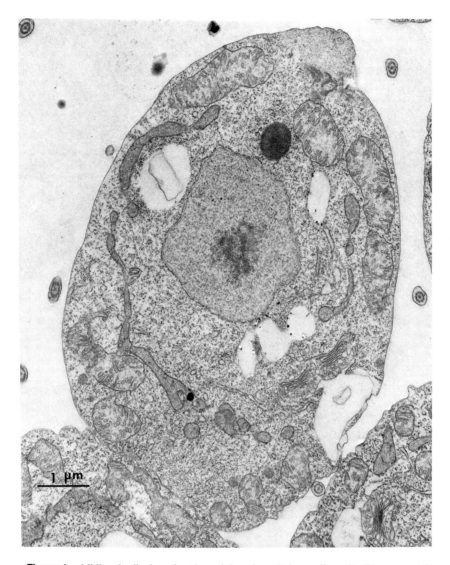

Figure 1. Midlongitudinal section through log-phase *Polytomella agilis*. The concentric arrangement of the internal organelles may be seen. Magnification: ×29,000. (Courtesy of Dr. Joseph Moore.)

carbon source. The carbon and energy metabolism of the acetate flagellates has been extensively studied, because it is generally believed that these organisms represent some of the most primitive protozoa which bridge the gap between photosynthetic and heterotrophic modes of energy production (Danforth, 1967). Acetate supports the heterotrophic growth of all strains of *Euglena gracilis* and *Astasia* tested (Cramer and Myers, 1952; Danforth, 1953; Hutner and Lee, 1962). The range of other carbon sources utilized for growth varies from species to species, and closely related strains may vary enormously in versatility. Short-chain fatty acids, alcohols, and tricarboxylic acids (TCAs) stimulate the respiration of *E. gracilis* var. *bacillaris,* whereas the Mainx and Vischer strains use only acetate and butyrate. In general, short straight-chain fatty acids and alcohols with an even number of carbon atoms are the most often used [e.g., in the nonphotosynthetic cryptomonad *C. paramecium* (Cosgrove and Swanson, 1952)]. However, odd-numbered fatty acids fail to support the growth of *Polytomella caeca* (Wise, 1955, 1959). Whereas *Chlamydomonas pseudagloe* resembles many strains of *Euglena* and *Astasia* in that it grows in the dark with glucose or acetate (Lucksch, 1933), *C. reinhardii* is limited to acetate utilization (Sager and Granick, 1953), and *C. moewusii* and *C. eugametos* are obligate phototrophs (Lewin, 1950; Wetherell, 1958). Phagotrophy is common only in colourless members of the Euglenophyceae, Dinophyceae, and Chrysophyceae, and only three genera have been examined in any detail in this respect (Droop, 1974).

The acetate flagellates show an unusual tolerance to highly acid growth conditions and to high concentrations of fatty acids and alcohols (Hutner and Provasoli, 1951). These peculiarities result from their relative impermeability to substrates; undissociated acids penetrate the limiting membranes most easily, and the optimum pH for acid utilization decreases with the strength of the acid (Holz, 1954; Wise, 1959). Fatty acids are toxic at lower concentrations in media of low pH. Thus it has been suggested that absorbability limits nutritional diversity rather than metabolic deficiency, and the relative impermeability of these organisms may be an advantage in the polluted environments they inhabit. It seems that these flagellates might provide interesting experimental material for future studies of transport processes. It is thought that succinate transport in *E. gracilis* involves a specialized transport system, as nonpermeant inhibitors block uptake (Levedahl, 1965). The inability of many species to use glucose has been discussed by Danforth (1967).

The unconventionally high resistance to ordinarily inhibitory or even lethal concentrations of fatty acids and alcohols, and the extraordinarily rapid and efficient utilization of some of them, notably acetate, butyrate, and ethanol, is in a biochemical sense a superficial characteristic. Early work indicated that the central pathways of carbon metabolism in acetate flagellates were conventional and that glucose utilization involved the Embden–Meyerhof pathway of glycolysis and the oxidative pentose phosphate pathway. Hurlbert and Rittenberg (1962) demonstrated glucose-6-phosphate dehydrogenase and 6-phosphogluconate dehydrogenase in *Euglena,* but tracer studies suggested that glycolysis predominates in glucose utilization in extracts of *A. longa* (Barry, 1962). Two triosephosphate isomerases and two aldolases have been separated from both phototrophically and heterotrophically grown *E. gracilis* (Mo *et al.,* 1973). The cytoplasmic enzymes predominate over the chloroplast enzymes in heterotrophically grown cells, and vice versa, in light-grown organisms. Changes in the activity of glucose-6-phosphate dehydrogenase under different growth conditions in *E. gracilis* (Huth, 1967; Ohmann *et al.,* 1969) indicate changes in the net flux of carbon entering the TCA cycle by alternative routes.

The control mechanisms for the changing enzyme levels in light- and dark-grown flagellates are far from clear. Enzymes for both formation and destruction of adenosine 3',5'-cyclic monophosphate (cAMP) have been detected in phototrophically grown E. *gracilis* (Keirns *et al.,* 1973), and cAMP receptor proteins or chloroplast and nonchloroplast origin have been demonstrated (Nicholas and Nigon, 1974). Cyclic nucleotide phosphodiesterase in *Chlamydomonas* hydrolyses cytosine 3', 5'-cyclic monophosphate (cCMP) three times more rapidly than cAMP, does not act on $N^6,O^{2'}$-dibutyryl cAMP, and in many respects resembles the enzyme from other microorganisms but is quite different from that of green plants.

Suprisingly, even the cardinal attribute of acetate flagellates, rapid utilization of short-chain fatty acids, shows at least in some forms a definite lability. Adaptation to fatty acids by *P. uvella,* originally described by Provasoli (1938), depends on induced synthesis of enzymes of fatty acid oxidation and requires a nitrogen source (Cirillo, 1956, 1957). Adaptation of acetate-grown *Polytomella agilis* to propionate or butyrate shows a similar lability (Cantor and James, 1965). Pathways of oxidation of higher fatty acids in these organisms await study, but butyryl–coenzyme A (CoA) formation in *Polytoma obtusum* involves butyryl–CoA kinase (Chapman *et al.,* 1965).

The alcohol dehydrogenases of *E. gracilis* grown heterotrophically on ethanol as sole carbon source have been investigated (Mego and Farb, 1974). Isopropanol and cinnamyl alcohol were oxidized more rapidly than ethanol; mercaptoethanol inhibited isopropanol oxidation but not that of ethanol. A cytosolic alcohol dehydrogenase reached elevated activities when *A. longa* was grown under hyperbaric oxygen (Morosoli and Bégin-Heick, 1974). Such organisms no longer grew on ethanol as sole carbon source.

As expected, the glyoxylate cycle provides a high-capacity mechanism for the growth of flagellates on acetate as the sole carbon and energy source. Isocitrate lyase has been demonstrated in cell-free extracts of acetate-grown *Chlamydomonas dysosmos* and *P. agilis* (Haigh and Beevers, 1964a), and malate synthase has also been detected in *P. caeca* (Haigh and Beevers, 1964b); this investigation confirmed the *in vivo* function of the cycle, as malate was the most highly labeled compound detected after a brief exposure to [2-^{14}C]acetate. The glyoxylate cycle is also necessary for the growth of *P. caeca* on propionate, as the product of β oxidation of the C_3 acid is acetyl-CoA (Lloyd *et al.*, 1968).

Heterotrophic carbon dioxide fixation in *Euglena* (reviewed in Levedahl, 1968) may serve as an anaplerotic pathway when glucose is the carbon source (Peak and Peak, 1976). The large transient elevations in dark carbon dioxide fixation do not, however, occur when acetate or ethanol replaces glucose in the growth medium, suggesting that induction of the glyoxylate cycle obviates the need to replenish TCA cycle intermediates by carbon dioxide fixation reactions. Another factor regulating the level of heterotrophic carbon dioxide fixation is the concentration of exogenous ammonium ions (Peak and Peak, 1977). Two separate phosphoenolpyruvate carboxylating systems have been demonstrated: one with higher activity in light-grown organisms and the other most active in dark-grown cells (Perl, 1974). TCA cycle intermediates and 3-phosphoglycerate inhibit the carboxylation reactions.

Synthesis of fatty acids in *E. gracilis* occurs on microsomally bound fatty acid synthetase (Khan and Kolattukudy, 1973a,b, 1975); this system is also implicated in the initial steps in the synthesis of wax-ester storage products. The multienzyme fatty acid synthetase complex activity is halved after the exposure of organisms to light for 24 hr and does not show a requirement for acyl carrier protein *in vitro* (Ernst-Fonberg *et al.*, 1974). A second system of fatty acid synthesis consists of discrete enzymes, is acyl carrier protein-dependent, and increases from low levels on illumination of dark-grown organisms but decreases in several chloroplast mutants. The purification method for acyl carrier protein has been detailed (Dinello and Ernst-Fonberg, 1975).

A question arises about how far the high-capacity substrate carbon pathways, postulated as underlying the definition of acetate flagellates as a group, extend to amino acid metabolism. The amino acid pool of *E. gracilis* stays constant under widely differing growth conditions (Kempner and Miller, 1974). The major amino acid was glutamate, not only when organisms were grown with this carbon source but also when acetate was provided heterotrophically, or under phototrophic conditions with carbon dioxide. Hydrolysis of the soluble pool gave larger amounts of aspartate, glycine, and arginine, as well as increasing free glutamate, indicating that small peptides make a significant contribution to this fraction. An unusual peptide was L-arginyl-L-glutamine. Glutamate metabolism has also been studied by Tokunaga *et al.* (1976), who showed the presence of two succinyl-semialdehyde dehydrogenases.

The archetypal photosynthetic acetate flagellate, *E. gracilis*, does not use nitrate as sole nitrogen source (Wolken, 1961) and, as noted earlier, *C. reinhardii* may be regarded as a borderline acetate flagellate. Hence the divergence between *C. reinhardii* and *E. gracilis* is also evident in that *C. reinhardii* does not assimilate nitrate or ammonium unless a suitable carbon source (carbon dioxide in light or, in darkness, acetate or carbon reserves accumulated in nitrogen-starved cells) is present (Thacker and Syrett, 1972a,b). 3'-(3,4-Dichlorophenyl)1',1'-dimethylurea (DCMU)-inhibited phototrophically grown cells did not assimilate nitrate or ammonium unless acetate was added. Assimilation of nitrate was inhibited by ammonium or nitrite. The nitrite reductase system as assayed in extracts was inducible; maintenance of its activity was dependent on the availability of a suitable carbon source. Barea and Cardenas (1975) purified nitrate and nitrite reductases from this system and demonstrated involvement of a ferredoxin in the six-electron reduction step from nitrite to ammonium.

Pathways of porphyrin synthesis in *E. gracilis* have been investigated by Rossetti *et al.*, (1977).

B. Endogenous Storage Products: Their Synthesis and Mobilization

The amylose and amylopectin reserve materials of *P. caeca* were characterized by Barker and Bourne (1955), and the distribution of isotopic carbon incorporated from [1-^{14}C]- and [2-^{14}C]acetate fitted the pattern predicted: synthesis via pyruvate and a reversal of glycolysis. *P. agilis* forms more amylose at 18°C than at 9°C (Sheeler *et al.*, 1968a). The effects of agitation, temperature, and culture age on starch composition and enzymes involved in starch synthesis in *P. uvella* have been studied by McCracken and Badenhuizen (1970) and Mangat and Badenhuizen (1970, 1971). After the onset of the stationary phase of growth, ADP glucose and UDP glucose concentrations and ADP glucose transferase and phosphorylase activities became maximal. The ratio of amylose to amylopectin declined later in the stationary phase as the activity of the Q enzyme increased. A lowered content of amylose in cells grown at 30°C reflects changes in the phosphorylase/Q enzyme ratio.

The paramylum of euglenids consists solely of β-(1 $\rightarrow$ 3)-linked D-glucose residues (Barras and Stone, 1968). Extracts of *Euglena* contain enzymes with β-(1 $\rightarrow$ 3)-glucosyltransferase, hydrolase, and phosphorylase activities (Smillie, 1968), and an exohydrolase (which releases glucose residues sequentially from the polymer) increases during the stationary phase of growth (Vogel and Barber, 1968). An endohydrolase is also present and is partly bound to the paramylum granules (Barras and Stone, 1969a,b). Paramylum synthesis ceases as the stationary phase of growth is attained. The utilization of reserve glucan is accelerated by illumination, even in permanently bleached strains of *Euglena* (Mitchell, 1971). A theoretical discussion of the kinetics of deposition and utilization of storage materials has been presented by Cohen and Parnas (1976). Several end products of metabolism are excreted into the growth medium. Examples of compounds released from *C. reinhardii* include glycolate and an unidentified compound which stimulates the induced synthesis of a specific bacterial protein in mixed cultures (Stageman and Hoober, 1975).

IV. SUBCELLULAR FRACTIONATION: PROPERTIES OF ISOLATED ORGANELLES

The most important step in the isolation of metabolically functional organelles is the first step: disruption of the organism. In the acetate flagellates we see a wide range of susceptibilities to mechanical breakage from the most fragile, *P. caeca,* to the most resistant, *Chlamydomonas.* A variety of disruption procedures have been employed; these include gentle hand homogenization, grinding or shaking with small glass beads, or recourse to a French or Yeda pressure cell. Theoretical and practical aspects which govern the choice of method have been described (Hughes *et al.,* 1971; Coakley *et al.,* 1977); these considerations are all-important where analytical subcellular fractionation studies are undertaken with organisms not easily giving high yields of spheroplasts (Lloyd, 1974). Many acetate flagellates fall into this category, and the special difficulties encountered, for instance, with the glycoprotein-containing cell walls of *Chlamydomonas* (Miller *et al.,* 1974), account for the lack of information available for these species; the wall-defective mutants introduced by Davies and Plaskitt (1971), and the autolysin (Schlosser *et al.,* 1976) and bacterial lytic systems (Gunnison and Alexander, 1975) may alleviate the problem. Few of the studies undertaken have been based on the philosophy of analytical subcellular fractionation as detailed by deDuve (1971); most have instead pursued the less rigorous goal of particle-rich fractions with little regard for the overview of enzyme distributions which only comes from complete balance sheets and calculation of recoveries. Nevertheless many methods allow the properties of isolated nuclei (Buetow, 1974), mitochondria (Buetow, 1970; Lloyd, 1974), peroxisomes (Müller, 1975), chloroplasts (Graham and Smillie, 1971), and other organelles to be defined, albeit with a degree of caution as to the deficiencies of function and its control which inevitably accompany their extraction.

A. Nuclei

Nuclei have been isolated from *Euglena* by several different procedures. Freezing and thawing followed by Triton X-100 extraction or treatment with pancreatic protease were used by Parenti *et al.* (1969), whereas Aprille and Buetow (1973, 1974) replaced the protease with pepsin. The method of choice which gives structurally intact nuclei is that of Lynch and Buetow (1975); organisms are frozen under carefully controlled conditions of pH in solid carbon dioxide–acetone, thawed, washed, and incubated in a buffer containing Triton X-100 and then sonicated to about 75% cell breakage. Extracts are then centrifuged through 10% dextran and 2.4 *M* sucrose; the final pellet contains 22–37% of the nuclei (only occasionally contaminated with paramylum granules). These nuclei are still rich in acid-soluble nuclear proteins. They resolve into nine bands during electrophoresis on polyacrylamide gels; five of the bands correspond to the histones of chromatin from higher eukaryotes (Lynch *et al.*, 1975). Whole chromatin showed the presence of condensed fibres (Haapala and Soyer, 1975), even from interphase nuclei, and was subfractionated into a euchromatic fraction (containing 14% of the nuclear DNA and more than 80% of the RNA polymerase) and a heterochromatic fraction (Lynch *et al.*, 1975).

The use of a cell wall-defective mutant of *C. reinhardii* has facilitated the extraction of nuclei (Robreau and LeGal, 1975). Failure to sediment chromatin in high yield led to the suggestion that this chromatin was low in basic proteins.

Nucleoli of *A. longa* isolated after disruption with a French press retained their morphology in that they exhibited distinct granular and fibrillar zones (Morosoli and LaFontaine, 1976). Nucleolar DNA showed a buoyant density of 1.708 gm/cm^3 in CsCl. Sheared nucleolar DNA was separated into two fractions in Cs_2SO_4 containing Ag^+ ions; one of these hybridized with ^{125}I-labelled RNA and therefore corresponds to ribosomal DNA. Ribosomal RNA cistrons of nuclear and chloroplast origin have been characterized in DNA extracted from *E. gracilis,* and they are not related to one another (Gruel *et al.*, 1975).

The informational complexities of nuclear and chloroplast genomes of *C. reinhardii* were analysed by Howell and Walker (1976). They concluded from DNA–DNA reassociation experiments that the nuclear genome was 24-fold more complex than the *E. coli* genome, whereas the chloroplast genome was 13-fold less complex than that of the bacterium. The number of DNA copies per genome was calculated to be 1 for the nuclear DNA, but approximately 50 for the chloroplast DNA. The only reiterated DNA in the *Chlamydomonas* nuclear genome is the γ-component satellite, which consists of about 200 copies of rRNA genes (Howell, 1972). In a different strain, Wells and Sager (1971) found that nearly 30% of the nuclear DNA renatured with a mean reiteration frequency of 1000-fold, but Chiang *et al.* (1975) found no such rapidly renaturing nuclear DNA fraction. Howell and Walker (1976) discuss the problem of the nuclear ploidy of *Chlamydomonas;* their results favour the view that the haploid nuclear genome is a single unreplicated copy in most vegetative organisms but, as most cells in an asynchronous vegetative culture have replicated their chloroplast DNA, the unreplicated chloroplast genome would contain 12 or 13 unique copies. Reassociation kinetics for *E. gracilis* DNA has shown that this organism is also not polyploid (Rawson, 1975). The heterogeneity and complexity of the nuclear genome of *P. obtusum* have also been studied (Siu *et al.,* 1974).

Subcellular fractionation of *E. gracilis* strain Z and a bleached derivative of this strain have allowed localization of several DNA polymerases (McLennan and Keir, 1975a,b,c,d). Polymerase A (MW, 18,500) has a predominantly nuclear location, and polymerase B (MW, 240,000) is predominantly cytoplasmic; their activities decline 4- and 15-fold, respectively, when organisms enter the stationary phase of growth. The activity of mitochondrial DNA polymerase (MW, 170,000) increased threefold during this transition, whereas the chloroplast enzyme was more active in growing organisms. It was proposed that both polymerases A and B are oligomers of 3.0 S subunits together with other possibly dissimilar subunits. After purification the two enzymes were further differentiated on the basis of associated enzymic activities, although both had similar requirements for K^+ and Mg^{2+} when assayed with an "activated" DNA primer template, and both were inhibited by N-ethylmaleimide, novobiocin, and o-phenanthroline. They differed in chromatographic behaviour on DEAE-cellulose and in electrophoretic mobility on polyacrylamide gels; neither possessed endodeoxyribonuclease activity, but polymerase B exhibited another nuclease activity and nucleoside diphosphokinase activity, and their primer template specificities were different. Exonuclease I has been purified from *C. reinhardii,* and elevated levels of this enzyme suggest that it may have a repair function (Tait and Harris, 1977a,b).

B. Mitochondria

Mitochondria from a steptomycin-bleached strain of *E. gracilis* were investigated by Buetow and Buchanan (1964, 1965). Oxidation of NADH, succinate, L-malate, L-glutamate, α-ketoglutarate, lactate, malate, and malate plus pyruvate was shown, and although the preparations were capable of oxidative phosphorylation, coupling to electron transport was not adequate for demonstration of respiratory control. NAD-Linked substrates gave P/O ratios of between 1.6 and 2.0, lactate and succinate 0.9 to 1.0, and NADH 0.4. The suggestion that one of the three classical sites of phosphorylation was missing has now been shown not to be valid by assays specific for each of the phosphorylation sites (Sharpless and Butow, 1970a). Buetow and Buchanan (1965) showed that amytal, antimycin A, and cyanide all inhibited oxidation and phosphorylation and that 2,4-dinitrophenol inhibited both oxidation and coupled phosphorylation with malate as substrate.

Extracts contained NADH and NADPH diaphorases, NADH– and NADPH–cytochrome c reductases, NADH and NADPH oxidases, and NADH-lipoyl dehydrogenase (Mohanty et al., 1977). The NADPH-linked enzyme systems had lower activities and were less sensitive to N-ethylmaleimide and p-hydroxymercuribenzoate than their NADH-linked counterparts.

Whereas light-grown wild-type E. gracilis var. bacillaris possessed at least four cytochromes, Perini et al. (1964) found only cytochromes a_{605} and c_{556} in the dark-grown wild type, in the albino mutants (W_3 and W_8) which lack chloroplasts, and in the yellow mutant (Y_3) which has only a rudimentary plastid. Experiments with particulate preparations demonstrated the respiratory roles of both these components, and an interesting observation was that respiration was 95% inhibited by 1.0 mM cyanide but only 15% inhibited by a carbon monoxide/oxygen ratio of 19 : 1.

That unusual electron transport components are present in Euglena mitochondria was confirmed by Raison and Smillie (1969); both cytochrome c and the a-type cytochromes had unusual α-absorption maxima (at 556 and 609 nm, respectively). This investigation also revealed the presence of a cyanide- and antimycin A-insensitive component of respiration which accounted for as much as 40% of the overall oxygen consumption rate. A permanently bleached strain Z grown with glutamate plus malate also gave mitochondrial preparations which oxidized NADH and D- and L-lactate by an antimycin A-insensitive route which bypassed one coupling site (Sharpless and Butow, 1970a). This pathway was, however, sensitive to inhibition by 2 mM cyanide. Cytochromes present included a_{607} (γ band at 453 nm), a_{593} (γ band at 444 nm), b_{568}, b_{561}, c_{555} (soluble), and c_{551}. The carbon monoxide-reacting species included cytochrome a_{593} and a b-type cytochrome.

An exhaustive investigation of methods of isolation of mitochondria from E. gracilis strain Z has also failed to yield preparations showing respiratory control (Datta and Khan, 1977). The method finally adopted involved breakage at 11 MN/m² in a French press followed by purification by centrifugation into a sucrose cushion. Nupercaine (0.5 nM) in the isolation buffer helped stabilize the phosphorylating capacity of the mitochondria; phosphorylation was inhibited by 2,4-dinitrophenol (DNP) and carbonyl cyanide m-chlorophenylhydrazone (CCCP), and valinomycin, atractyloside, and carboxyatractyloside gave about 60% inhibition. Both phosphorylation and oxidation showed discontinuities in Arrhenius plots at 34° and 18°C. The relative activities of the Mg^{2+}-dependent ATPases with different nucleoside triphosphates were as follows: ATP and GTP, 1; ITP, 0.6; GTP and UTP, 0.15. ATPase activity was stimulated by DNP and CCCP up to 200%. In the presence of uncouplers the optimum pH of mitochondrial ATPase was shifted from 7 to 8.

These observations cannot account for the ability of *E. gracilis* to show reversible adaptation to growth in the presence of uncouplers (or of the energy transfer inhibitors oligomycin and tri-n-butylchlorotin) (Kahn, 1973, 1974); the adaptation involved a response to the high intracellular levels of ADP produced by these compounds. Because no evidence for uncoupler binding proteins or modification of properties of isolated mitochondria was detected, it was suggested that adapted cells might utilize a modified or altered mode of energy coupling.

Isolation of mitochondria from phototrophically grown *E. gracilis* strain Z has been described (Collins and Merrett, 1975; Collins *et al.*, 1975). These mitochondria oxidize glycolate by an antimycin A-sensitive route; the partial reactions of glycolate–cytochrome *c* oxidoreductase and cytochrome *c* oxidase were demonstrated using purified *Euglena* cytochrome c_{558}. The mitochondria did not show respiratory control, but a P/O ratio of 1.7 was obtained for glycolate oxidation. Rotenone was hardly inhibitory. All these results are consistent with the suggestion that electrons from glycolate enter the respiratory chain at the flavoprotein level. Unlike higher plant mitochondria, those from phototrophically grown *Euglena* contain a glycolate dehydrogenase and a glyoxylate-glutamate aminotransferase (Collins and Merrett, 1975); sequential action of these two enzymes yields glycine, which also serves as a respiratory substrate (P/O ratio = 0.9). The mitochondrial location of glycolate dehydrogenase has been confirmed cytochemically (Beezley *et al.*, 1976). Glycolate dehydrogenase is also present in the peroxisomes in this organism.

Increasing the time of exposure to light of heterotrophically grown *E. gracilis* leads to repression of mitochondrial enzymes such as succinate dehydrogenase and a lowered buoyant density of the organelles in sucrose gradients (Brown and Preston, 1975).

Adaptation of *E. gracilis* to growth on succinate in the presence of antimycin A involves a novel succinoxidase mechanism (Sharpless and Butow, 1970b); a similar electron transport pathway is also present in ethanol-grown cells. This cyanide-insensitive pathway is AMP-stimulated, does not produce detectable hydrogen peroxide, and can also serve as a route for the oxidation of NADH.

There is now widespread agreement that the number of mitochondria in *E. gracilis* is lower than surmised from casual examination of thin sections. A single reticular mitochondrial structure occurs at least at some stages of the cell cycle (Buetow, 1968; Leedale and Buetow, 1970; Osafune *et al.*, 1975; Pellegrini and Pellegrini, 1976), and giant mitochondria have been described in antimycin A-adapted organisms (Calvayrac and Butow 1971; Calvayrac *et al.*, 1971). Thus it is probably almost impossible to isolate intact mitochondria from these organisms, and whatever functional integrity survives the isolation procedures inevitably reflects the properties of submitochondrial particles. The measurement of a respiratory control ratio of 3.5 *in vivo* by Kahn (1973) confirms that loose coupling is an artefact of isolation rather than an inherent lack of coupling in an organism which may use a nonphosphorylating bypass of the main respiratory chain.

Mitochondria from *A. longa* did not oxidize reduced mammalian cytochrome *c*, suggesting the presence of an unusual cytochrome oxidase (Webster and Hackett, 1965). They contained two *c*-type and three *b*-type cytochromes and a carbon monoxide-reacting cytochrome a_3 which showed an unusual α-band at 605 nm at 77°K. The oxidative and phosphorylation abilities of these mitochondria are low (Kahn and Blum, 1967; Buetow and Buchanan, 1969).

The presence of cytochromes *b*, *c*, and $a + a_3$ in *P. uvella* and *P. caeca* was reported by Webster and Hackett (1965). In *Polytoma* mitochondria NADH oxidase was sensitive to inhibition by rotenone, amytal, diphenylamine, antimycin A, and potassium cyanide. An unidentified carbon monoxide-binding pigment may have been due to a denatured cytochrome *b*.

Mitochondria showing respiratory control have been isolated from *P. caeca* (Lloyd *et al.*, 1968). P/O ratios for α-ketoglutarate, succinate, and NADH-linked substrates were in the ranges 3.1–3.4, 1.6–1.8, and 2–2.9, respectively. Evidence for the integrity of inner and outer membranes was obtained in electron micrographs and was also indicated by observation of the latency of some TCA cycle enzymes (D. A. Evans and D. Lloyd, unpublished observations) and by the clear differences between responses to inhibition of internally generated and externally added NADH (Lloyd and Chance, 1968). Several different flavoproteins were distinguished by simultaneously recording absorption and fluorescence responses to substrates, and an iron–sulfur signal at $g = 1.94$ was detected in electron paramagnetic resonance (epr) spectra on the reduction of mitochondrial components with succinate, ethanol, and, in damaged mitochondria, by

NADH. When anaerobic mitochondrial suspensions were rapidly mixed with oxygenated buffer, the determination of half-times for the reoxidation of cytochromes and flavoproteins showed an anomalously slow reaction for cytochrome c (600 msec). However, measurement of the steady state redox levels of electron transport components (in the absence of ADP) clearly indicates the functional sequence: oxygen, cytochrome a_3, a, c, b, flavoprotein; flavoprotein reduction is the rate-limiting step in electron transport (Lloyd et al., 1970).

Adaptation of P. caeca to propionate (after growth on acetate as sole carbon source) involves the formation of a β-oxidation pathway which is predominantly mitochondrial (Lloyd et al., 1968). Thus mitochondria either from propionate-grown cells or from propionate-adapted cells oxidized succinate, α-ketoglutarate, β-hydroxypropionate, and malonic semialdehyde and showed respiratory control with these substrates. Mitochondria from acetate-grown cells also exhibited ADP-dependent oxidation of succinate and α-ketoglutarate, but did not oxidize β-hydroxypropionate or malonic semialdehyde. The progressive development of fully coupled respiratory chains accepting electrons from these substrates was followed during adaptation to propionate and was the first clear demonstration of enzymes inducible in preexisting mitochondria. That adaptation proceeds in the presence of chloramphenicol suggests that it involves extramitochondrial enzyme synthesis and then transport and integration of enzymes of the new pathway into the inner membranes of preexisting mitochondria.

The mitochondria of C. reinhardii have not been extensively studied, despite early reports that the pale-green mutant has hardly any detectable a-type cytochromes (Chance and Sager, 1957; Hiyama et al., 1969), and the remarkable transition from cyanide- and carbon dioxide-sensitive respiration to an alternative inhibitor-resistant state which accompanies the germination of zygospores (Hommersand and Thimann, 1965). The observation of Lang (1963) suggested that the mitochondria of Chlamydomonas consist of long, sinuous elements, forming a network, and the elegant scale models constructed on the basis of serial sectioning show that only a small proportion of the organelles have a simple elliptical or spherical shape (Arnold et al., 1972). Most of the mitochondria are elongated and show extensive branching (Schötz et al., 1972; Osafune et al., 1972a,b, 1975; Grobe and Arnold, 1975). Similar findings on careful electron microscope analysis of P. agilis (Burton and Moore, 1974) and Polytoma papillatum (Gaffal and Kreutzer, 1977) underline the difficulty of isolating intact mitochondria from flagellated protozoa. The mutants of Chlamydomonas with defective cell walls obtained by Hyams and Davies (1972) have, however, given new hope to those wrestling with problems of subcellular fractionation of this species.

Mitochondrial DNA from several strains of *E. gracilis* has been characterized as being double-stranded and having a buoyant density of 1.698–1.691 gm/cm³ (Edelman *et al.*, 1966; Krawiec and Eisenstadt, 1970a,b; Manning *et al.*, 1971; Talen *et al.*, 1974; Fonty *et al.*, 1975). Only linear molecules were isolated, but in the report of Nass *et al.* (1974) a small proportion of membrane-associated small circles of about 1 μm contour length were released from osmotically lysed mitochondria. This mitochondrial DNA (MW, about 2×10^6) is considerably smaller than other known animal and plant mitochondrial DNAs.

Cycloheximide (15 μg/ml) significantly reduced the synthesis of nuclear DNA in *E. gracilis* but allowed continued synthesis of mitochondrial DNA (Richards *et al.*, 1971; Calvayrac *et al.*, 1972). The synthesis and turnover of mitochondrial DNA have been further studied by Richards and Ryan (1974).

RNA synthesis in mitochondrial suspensions from a streptomycin-bleached strain of *E. gracilis* (SM-L1) has been studied by Olson and Buetow (1977). Incorporation of [³H]GTP into an RNase-sensitive product required Mg^{2+} or Mn^{2+} and the presence of all four riboside triphosphates for maximum reaction rates. The RNA polymerase was strongly inhibited by actinomycin D, acriflavin, ethidium bromide, and DNase, whereas rifamycin SV and α-amantin (specific inhibitors of bacterial and nuclear DNA-dependent RNA polymerases, respectively) hardly inhibited. RNase also prevented the reaction, but atractyloside (a specific inhibitor of the adenine nucleotide translocase) did not prevent incorporation of [³H]ATP into the RNA of isolated mitochondria. Thus in these experiments, permeability barriers to ribonucleoside triphosphates had been lost, and the low level of inhibition by α-amanitin gives a true measure of contamination by nuclear fragments and confirms the differences in inhibition sensitivities of the two systems of RNA synthesis.

Mitochondrial ribosomes and RNAs have been isolated from *E. gracilis* var. *bacillaris* and strain Z. The sedimentation coefficients of ribosomes from the mitochondria, cytoplasm, and chloroplasts are 71–72 S, 86–87 S, and 69 S, respectively (Avadhani and Buetow, 1972). Mitochondrial polyribosomes sedimented in 9 or 10 peaks in gradients and were disaggregated by DNase or EDTA to give monomers. The RNA of the 50 and 32 S ribosome subunits is 21.4 S and 16 S. The G + C content of mitochondrial rRNA is 29.8%, whereas that of chloroplast and cytoplasmic species is 50.0 and 55.7%, respectively. The mitochondrial polyribosomes were active in protein synthesis without the addition of supplementary mRNA, and were sensitive to inhibition by chloramphenicol but not cycloheximide. Mitochondrial RNA synthetases of *E. gracilis* were quite different from those obtained from the chloroplasts (Koslev and Eisenstadt, 1972).

C. Peroxisomes

Peroxisomes have been isolated from several acetate flagellates (Müller, 1969, 1975). A clear separation of mitochondria and peroxisomes has been achieved in both rate separations and on equilibrium density centrifugation of homogenates of *P. caeca* (Cooper and Lloyd, 1972). The median equilibrium densities of mitochondria and peroxisomes in aqueous sucrose were 1.20 and 1.25 gm/cm^3, respectively, and median $s_{20,w}$ values were 24.4 × 10^3 and 4.4 × 10^3 S, respectively. Enzymes present in peroxisomes of other protozoa but not detectable in *P. caeca* included urate oxidase, D- or L-amino acid oxidases, L-hydroxyacid oxidase, glycolate reductase, and glycolate oxidase. Isocitrate lyase was present and nearly all (>90%) nonsedimentable. This observation confirmed the findings of Gerhardt (1971), although Haigh and Beevers (1964b) had earlier detected some association of the key enzymes of the glyoxylate cycle with a particular fraction of *P. caeca*. Peroxisomal urate oxidase has been demonstrated in extracts of this organism after growth on a complex medium (Gerhardt, 1971). Attempts at cytochemical localization of catalase using the 3,3′-diaminobenzidine–hydrogen peroxide reaction proved unsuccessful, both with intact organisms and with isolated particles, although the organelles in the fractions rich in catalase showed the same fine structural characteristics as microbodies *in situ* (Gerhardt and Berger, 1971). Peroxisomes of *C. reinhardii* and *Chlorogonium elongatum* have been separated from mitochondria on sucrose gradients (Stabenau, 1974; Stabenau and Beevers, 1974), but similar experiments with *A. longa* have proved unsuccessful (Bégin-Heick, 1973).

Peroxisomes from *E. gracilis* have a diversity of enzymes, including the key enzymes of the glyoxylate cycle (isocitrate lyase, malate synthease, malate dehydrogenase, and citrate synthase) (Graves *et al.*, 1972; White and Brody, 1974), the enzymes of β-oxidation of fatty acids (Graves and Becker, 1974), and the enzymes of the glycolate pathway (glycolate dehydrogenase, glutamate-glyoxylate aminotransferase, serine-glyoxylate aminotransferase, aspartate-α-ketoglutarate aminotransferase, and hydroxypyruvate reductase) (Graves *et al.*, 1971, 1972; Codd *et al.*, 1969; White and Brody, 1974; Collins and Merrett, 1975). The activities of all these enzymes are quite variable and depend on the growth conditions; thus glucose represses the glyoxylate cycle enzymes, and cells grown in the dark with acetate have low activities of the glycolate pathway enzymes (Brody and White, 1972, 1973, 1974; White and Brody, 1974). Exposure of these heterotrophically grown cells to light leads to increased activity of the glycolate pathway enzymes in parallel with increased photosynthetic ability, whereas the glyoxylate cycle enzymes and catalase show only a small increase. White and Brody (1974) suggest that the existence of two distinct populations of peroxisomes, each specialized for a particular metabolic role, could explain these results. Catalase is not detectable in ethanol-grown streptomycin-bleached *E. gracilis* var. *bacillaris* (SM-L1) (Graves *et al.*, 1972). Photorespiration and its relationship to glycolate metabolism have been studied in *Chlamydomonas* (Cheng and Colman, 1974; Nelson and Surzycki, 1976).

D. Acid Hydrolase-Containing Organelles

Histochemical ultrastructural localization studies on acid phosphatase in growing and starved *E. gracilis* (Brandes *et al.*, 1964) have not been paralleled by exhaustive subcellular fractionation studies, although the streptomycin-bleached strain (SM-L1) used in this early study has recently been used in a study of acid hydrolases (Baker and Buetow, 1976). Acid phosphatase, β-galactosidase, β-glucosidase, β-fucosidase, cathepsin D, RNase, DNase, and esterase were all active in homogenates. Amylase was hardly detectable, and β-glucuronidase, arylsulfatase, β-N-acetylglucosaminidase, α-fucosidase, and α- and β-mannosidase were not detected. Pinocytotic uptake of protein from the reservoir in *E. gracilis* has been followed by Kivic and Vesk (1974).

The only acid hydrolase highly active in extracts of *P. caeca* was acid phosphatase (Cooper and Lloyd, 1972). The distribution of *p*-nitrophenyl phosphatase and 4'-chloro-3-hydroxy-2-naphtho-*o*-toluidide phosphatase was similar in sucrose density gradient fractionations; the latter activity was located histochemically in vacuoles, in areas of focal degradation, and in the Golgi apparatus, and was distributed throughout the cytosol (Cooper *et al.*, 1974).

Phosphatases of *C. reinhardii* have been characterized. Two constitutive acid phosphatases and three derepressible phosphatases (one neutral and two alkaline) are present (Matagne *et al.*, 1976). Mutants deficient in some of these enzymes have been isolated (Matagne and Loppes, 1975); those lacking alkaline phosphatase have cell walls which are leaky to this enzyme (Loppes and Deltour, 1975).

E. Leucoplasts

Double-membrane leucoplasts were first observed in *P. obtusum* by Lang (1963). These organelles are regarded to have evolved from chloroplasts and to be sites of amylogenesis, as they contain starch grains; they also contain DNA ($\rho = 1.682$ gm/cm^3) and 73 S ribosomes (Siu *et al.*, 1976a,b). The plasmid ribosomes sediment faster than *Chlamydomonas* chloroplast ribosomes (Siu *et al.*, 1976b), probably because their RNAs are larger than those of the green alga (25 and 18 S as against 23 and 16 S). Only one copy of the RNA cistron is present per leucoplast genome (Siu *et al.*, 1976c), and there are an estimated 56 to 65 repetitions of this genome per organism. Renaturation kinetics reveals inversely repeated, repetitive, and unique sequences in the leucoplast DNA, and a total of 32% of the leucoplast unique sequences hybridize with *Chlamydomonas* chloroplast DNA, indicating that affinity still exists between the two species (Siu *et al.*, 1975). Plastids and their RNAs have also been described in bleached mutants of *E. gracilis* (Heizmann *et al.*, 1976).

V. ENVIRONMENTAL MODIFICATIONS

The regulation of metabolism, development, and structure may be conveniently studied by altering the environment. In the acetate flagellates such studies have emphasized the effects of light (including irradiation), temperature, medium changes, and various inhibitors.

A. Light and Radiation

Among the acetate flagellates light is known to affect growth, metabolism, chloroplast development, motility, and circadian rhythms. The effect of light on growth rate has been studied extensively in *Euglena*. Growth of *E. gracilis* on acetate in continuous light is slower than heterotrophic growth in the dark, the population doubling times being 45 and 32 hr, respectively, in the light and dark (Padilla and Cook, 1964).

Photoinhibition of cell division has been demonstrated in *Euglena*, the prophase of mitosis being the most sensitive. When cells were transferred from dark to light (300 foot-candles, fluorescent lamps), division was inhibited, followed by slower growth. No inhibition was noted when illumination was provided by incandescent lamps, indicating a greater sensitivity to shorter wavelengths. Photoinhibition was markedly reduced in complex media and at elevated temperatures. No significant differences in dry mass, cell volume, protein, and RNA concentrations were noted in light- and dark-grown cells. In continuous culture (doubling time, 30 hr), there was a transitory decline in cell number upon illumination, followed by recovery, regardless of carbon source. No recovery occurred at faster dilution rates when acetate was the carbon source, suggesting a light-sensitive acetate metabolism pathway. Protection against inhibition is afforded by chlorophyll. Wild-type *E. gracilis* strain Z and *E. gracilis* var.

bacillaris are less inhibited than bleached mutants (SM-L1). *Astasia longa* is very much inhibited by light, but *C. paramecium* is unaffected (Cook, 1968).

No significant changes in the DNA content of light- or dark-grown cells have been noted. When the doubling time of autotrophically grown cells was shortened, DNA increased (Brawerman, 1968). This may be related to changes in ploidy.

In *Euglena,* light represses the synthesis of an isozyme of the malic enzyme (L-malate : NADP oxidoreductase). Three electrophoretically distinguishable bands are observed in heterotrophically grown cells in the dark. The slow band is absent in light-grown cells. Light-grown cells also have four to six times less activity than dark-grown cells (Karn and Hudock, 1973).

Two electrophoretically distinguishable aldolases have been demonstrated in *Euglena* strains B and Z. Class I (MW, ~146,000) is predominant autotrophically but is not produced in the dark or in strains W_3BUL or W_3BHL in the light. Class II (MW, ~70,000) is found in all these strains except autotrophically grown strain Z (Karlan and Russell, 1976).

Chloroplast structure, function, and development are affected by light. When light-grown cells are transferred to the dark, chlorophyll is diluted among the progeny in growing cells; stationary-phase cells retain their chlorophyll. The decrease in chlorophyll is accompanied by a reduction in the number of chloroplast lamellae and thylakoids. Dividing cells also show a dilution of the rate of oxygen evolution after the removal of illumination. Although photosystem I activity increases during chloroplast division in the dark, photosystem II activity increases for four divisions in the dark and declines after six divisions. When dark-grown cells are illuminated, chlorophyll synthesis in exponential-phase cells is faster, even exceeding the levels of stationary-phase cells. Photosystem I activity is the same in dividing and nondividing cells, whereas photosystem II activity is proportional to chlorophyll accumulation and linked to that of photosystem I (Ophir *et al.,* 1975).

Normally there are changes in chloroplast number per cell and amount of chlorophyll per cell. These features are constant during logarithmic growth. In late exponential phase the amount of chlorophyll per cell increases, followed by an increase in plastid number. The exponential increase in chloroplasts continues after onset of the stationary phase. The qualitative pattern does not change in reduced light, although there is an increase in the number of plastids per cell from lower limits of 7 to 10 up to 15 per cell (Cook, 1973).

Dark-grown cells have plastid precursors, if not discrete structures. Such cells regreen within 1 day in the light. The presence of paramylum

Table I Effect of Growth Conditions on Total Cellular DNA and Chloroplast DNA in *Euglena*

Origin of DNA	Chloroplast DNA/ total cell DNA	Chloroplast DNA molecules per cell
Heterotrophy in the dark	1.446×10^{-4}	217
Heterotrophy in the light	1.581×10^{-3}	530
Autotrophy in the light	1.280×10^{-4}	1014

[a] From Rawson and Boerma (1976).

in these cells suggests plastid activity. Additional evidence is the presence of pale-fluorescing bodies in dark-grown cells (Gibor and Granick, 1962a). More recently proplastids have been described in *Euglena:* irregular, flattened, ellipsoid structures 1.6 μm long and 0.6 μm wide. Two to five girdlelike thylakoids were seen in them, as well as one prolamellarlike body connecting with the thylakoids (Klein *et al.*, 1972). Chlorophyll synthesis and chloroplast development began after 12 hr of illumination. This lag period could be shortened by preillumination for 90 min, followed by 12 hr of darkness prior to continuous illumination. Blue and red light were the most effective during preillumination, and oxygen was necessary (Holowinsky and Schiff, 1970). Additional evidence for the presence of chloroplast precursors in dark-grown cells has been the demonstration of chloroplast DNA molecules in these cells. *Euglena* chloroplast DNA consists of supercoiled closed circles. When the chloroplast DNA is isolated, fragmented by sonication, and reannealed with total DNA, both the fraction of cell DNA and the number of molecules per cell can be determined (Table I). In a heat-bleached mutant no complementary sequences have been found (Rawson and Boerma, 1976). Chloroplast RNA genes correspond to 1.9% of the chloroplast DNA (Rawson and Haselkorn, 1973). The transcription products of both 16 and 23 S rRNA have been demonstrated in dark-grown cells, although in a lesser concentration than in light-grown cells (Heizmann *et al.*, 1976). Finally, ribulose-1,5-diphosphate carboxylase activity has been demonstrated in dark-grown cells. Although it was an order of magnitude less than in light-grown cells, it was 5 to 10 times more active in dark-grown cells than in chloroplast mutants (Evers and Ernst-Fonberg, 1974).

The photoreceptor apparatus is affected by light. In the dark, the carotenoid spheroid is drawn back from the edge and appears fragmented. The volume of the organelle decreases by approximately half. Although the carotenoid content of dark-grown cells is 20% of that of green cells, the former are still positively phototactic (Ferrara and Banchetti, 1976).

Light may affect mitochondrial development also. When *Euglena* was grown phototrophically, no succinic dehydrogenase activity was measurable in the mitochondrial fraction, but dark-grown cells had considerable activity. The mitochondrial fractions consisted of material sedimenting at densities of 1.22 and 1.119 gm/cm³ in dark-grown cells. In mixotropic cells or illuminated cultures activity was noted only in the fraction with a density of 1.119 gm/cm³ (Brown and Preston, 1975).

Growth, chloroplast structure, and differentiation of *Euglena* are sensitive to ultraviolet radiation. Blue light can counteract the effects of irradiation. Normal and streptomycin-, heat-, and uv-bleached cells have slower dry mass accumulation and cell division. Inhibition is proportional to the energy of radiation. Cells grown in the light are more sensitive to growth suppression, and dark-grown cells are inhibited at higher intensities. Ultraviolet sensitivity is independent of the presence or absence of the photosynthetic machinery. The optimum wavelengths for overcoming this inhibition are 365–375 nm. Wavelengths above 660 nm are ineffective in photoreactivation (Michaels and Gibor, 1973). Dark-grown cultures respond more favourably to photorestoration (Klein *et al.*, 1963; Lyman *et al.*, 1959). A uv microbeam apparatus was used to localize the uv-sensitive components in the cytoplasm. Irradiation of the nucleus alone never bleached, although high doses killed. When whole cells or only the cytoplasm was irradiated, at least 50% of the irradiated organisms gave rise to bleached or mixed (bleached and green) colonies. Irradiation of five or more cytoplasmic areas produced mixed colonies (Gibor and Granick, 1962b). Irradiation of stationary-phase cells inhibited $^{32}P_i$ incorporation into DNA. The inhibition was not overcome by photoreactivation with red or blue light. Bleached mutants were also inhibited. This metabolically active DNA was considered to be involved in a repair process (Gibor, 1969).

Irradiated cells showed no differences from normal light-grown cells for three divisions after treatment. Plastids were present after four divisions but had fewer lamellae, and pyrenoids were absent. The thylakoids became vesiculated during further degeneration. All other organelles had normal structures (Michaels and Gibor, 1973).

In cells synchronized by repetitive light–dark cycles, sensitivity to ultraviolet light was greatest just after cell division. During the first 4 hr of the nondividing portion of the cell cycle, cells became increasingly resistant. If the uv-sensitive target is the chloroplast DNA, this suggests that it is synthesized at the time of cytokinesis and not concomitantly with nuclear DNA (Cook and Hunt, 1965).

Adaptation for utilization of butyrate in *P. obtusum* was inhibited by ultraviolet irradiation (Cirillo, 1955). Glutamate dehydrogenase activity in

C. reinhardii is sensitive to both light and nutritional conditions. Wild-type cells maintained on a light–dark cycle of 12 hr of each showed increased activity in the dark. A mutant (y-2) incapable of chlorophyll synthesis or survival under organotrophic conditions had reduced activity, which was reversed upon illumination. Activity was associated with two isoenzymes distinguished by starch gel electrophoresis. The fast component copurified with chloroplasts and was absent in chlorotic cells. The slow, soluble isoenzyme was found in organotrophic cultures, but only in trace amounts in mixotrophic cultures (Kivic *et al.,* 1969).

B. Nutrition

1. Carbon Sources

The composition of the medium has a significant effect on the growth rate, metabolism and, in some cases, the structure of cells. In *Euglena,* the concentrations of various small molecules and macromolecules are a function of the growth medium (Brawerman, 1968; Kempner and Miller, 1965a, 1972b, 1974; Miller and Kempner, 1976). Although the population doubling time is approximately the same in complex and defined media, the variation in individual cell generation times is significantly less in complex, as compared with defined, medium (Cook and Cook, 1962).

The presence of the Embden–Meyerhof pathway in euglenids and the formation of starch in *Polytomella* spp. indicates glucose metabolism in the acetate flagellates (Sheeler *et al.,* 1968b). *Euglena gracilis* strain Z (Cook and Heinrich, 1965) and *A. longa* (Barry, 1962) grew on glucose as sole carbon source. In *A. longa,* the parent strain J had a lower growth rate and cell yield than the glucose-utilizing strain 460-17. In strain J, glucose did not stimulate the rate of respiration above the endogenous rate in whole cells, but homogenates showed a twofold increase. Whole cells and homogenates of strain 460-17 consumed two to three times more oxygen above the endogenous levels. Both strains had hexokinase, lactic dehydrogenase, and phosphoglucomutase activities in homogenates. The specific activity of hexokinase in whole cells of strain J was one-eighth that of strain 460-17, but homogenates of strain J had a higher specific activity. Thus a permeability difference was assumed to account for the variation (Barry, 1962). Lack of hexokinase activity in *Euglena* was thought to be the likely explanation for its inability to utilize glucose; however, in *E. gracilis* var. *bacillaris,* grown in the dark on 1% glucose, hexokinase was found in the postmitochondrial supernatant. Enzyme activity showed a sigmoidal relationship to glucose concentration, suggesting cooperativity between subunits. The apparent K_m for glucose was 0.5

mM, which is similar to that for yeast and other plants. ADP was inhibitory, suggesting that it may be a negative effector (Belsky and Schultz, 1962). The rate of glucose uptake was the same in glucose- and ethanol-grown cells, but very low in acetate-grown cells, yet hexokinase activity was demonstrated in all three cases. The activity in glucose-grown cells was three times that in the others (Graves, 1971). Homogenates of *P. obtusum* have measurable glucokinase activity, although the organism cannot grow on glucose. Like that for *Euglena*, the pH optimum for the enzyme is 8–8.5. Its activity is unaffected by fructose, mannose, and sorbose, but is reduced in the presence of galactose. The cells are impermeable to hexoses, lactose, and raffinose. Thus, inability of cells to grow on glucose is due to a permeability barrier (Chapman *et al.*, 1965).

The effects of glucose on cell metabolism in *Euglena* have been much studied. Although growth rates were essentially the same on glucose and acetate (doubling times of 13–17 and 12–15 hr, respectively), cell yield was significantly higher with glucose ($3.6–5.1 \times 10^6$ and 10^5 cells/ml on glucose and acetate, respectively). Differences in cell mass were attributable to paramylum. Acetate-grown cells contained more RNA (32 and 21 μg/10^6 cells on acetate and glucose, respectively). There were no differences in the endogenous Q_{O_2} between the two groups. As in *A. longa*, glucose did not stimulate cell respiration whatever the carbon source for growth. Acetate stimulated respiration fourfold in acetate-grown cells (Cook and Heinrich, 1965).

Besides glucose, succinate, malate, and fumarate support growth but not respiration in *Euglena*. Ethanol stimulates both activities. Acetate and other C_2 compounds are known to induce synthesis of the glyoxylate bypass enzymes malate synthase and isocitrate lyase. In synchronized populations the induction occurs continuously over the cell cycle (Woodward and Merrett, 1975).

Induction of the glyoxylate bypass was not prevented completely by glucose. Malate synthase was three times more active in acetate-grown cells. Its activity in homogenates was not diminished by glucose, nor was it lower in cells grown on acetate plus glucose. Isocitrate lyase activity was also higher in acetate-grown cells. The activities of isocitrate dehydrogenase and citrate synthase were independent of the carbon source for growth, but malate dehydrogenase and the malic enzyme were repressed by glucose (Heinrich and Cook, 1967).

Glucose-grown *Euglena* fixes carbon dioxide in the dark at twice the rate in acetate-grown cells. Initially it is incorporated into the ethanol-soluble and protein fractions; later it is found in the paramylum (Heinrich and Cook, 1967). Acetate inhibits this fixation, suggesting the replenishment of TCA cycle intermediates by the glyoxylate bypass (Peak and Peak, 1976).

Growth of *Euglena* on succinate, fumarate, and malate was pH-dependent, suggesting a differential permeability for the un-ionized species of these organic acids (Wilson *et al.*, 1959). When *Euglena* was grown on succinate at an initial pH of 6.9, the cell yield was lower than at pH 3.5. When the pH was lowered at the beginning of the stationary phase, growth resumed, suggesting that this phase of growth depended on the concentration of the un-ionized species (Votta *et al.*, 1971). Glutamate was also used as a carbon and nitrogen source, yielding higher cell densities that acetate (Kempner and Miller, 1972a). Ethanol-grown cells had a lower Q_{O_2} than acetate-grown cells. Ethanol-grown cells, respiring on acetate, adaptively increased their Q_{O_2} after 40 min; transferred to acetate, such cells oxidized ethanol faster than acetate-grown cells (Danforth and Wilson, 1957). Whatever the carbon source for growth, the extent of oxidation of ethanol and acetate was the same, i.e., 42% (Wilson and Danforth, 1958).

Euglena can utilize glycolate, glycine, and serine as sole carbon source when grown in the light, but not in the dark. Glycolate utilization depends on photosynthetic reactions—perhaps ATP synthesis—since it is inhibited by DCMU. The role of light in these processes is not the promotion of photosynthetic carbon fixation, for cells growing on glycolate, serine, or glycine contain less chlorophyll than photoautotrophically grown cells. Furthermore, carbon dioxide fixation is reduced by 30% in the presence of glycolate and by 70% in the presence of serine. Growth on glycine or serine is accompanied by the release of ammonia into the medium. Glycolate utilization is accompanied by an uptake of phosphate from the medium (Murray *et al.*, 1970).

Astasia longa and *E. gracilis* both utilize ethanol. The population doubling time of *Astasia* was ~10 hr at concentrations of ethanol between 1 m*M* and 0.3 *M*. At 0.2 *M,* peak populations of 6.6×10^6 cells/ml were obtained. This result is in contrast to growth on acetate, where the rates were lower above 60 m*M* and the maximum population density was 6.8×10^5 cells/ml at 60 m*M*. The dry weight of acetate-grown cells was greater because of increased carbohydrate (Buetow and Padilla, 1963). The Q_{O_2} of ethanol-grown cells was 17% higher than for acetate-grown cells (Danforth and Wilson, 1957). In *Euglena,* cells behaved in essentially the same fashion. The respiratory quotient (0.3) and the oxygen/ethanol ratio (0.966) were constant with time, suggesting no accumulation of intermediates (Eshelman and Danforth, 1964).

Polytomella caeca grows on propionate, butyrate, valerate, butanol, and amyl alcohol, but not on caproate, isobutyrate, propanol, isopropanol, or isobutanol (Wise, 1955). Growth on propionate and butyrate involves, as noted earlier, adaptive responses in *P. agilis* and *P. caeca*. In *P. agilis*, there were lag periods of 34 and 18 hr prior to growth on propionate and butyrate, respectively. Acetate-grown cells did not oxidize these carbon sources. Propionate-grown cells had a higher Q_{O_2} on acetate, propionate, and butyrate than acetate-grown cells had on acetate alone. Butyrate-grown cells oxidized substrates at half the rate of propionate-grown cells. The extent of oxidation of these substrates (when it occurred) was independent of the carbon source for growth and was always ~50%. Azide inhibited propionate oxidation after a lag period but did not inhibit acetate oxidation in propionate-grown cells (Cantor and James, 1965). Nongrowing, acetate-grown *P. caeca* cells oxidized propionate after a lag period of 1 hr (Lloyd *et al.*, 1968; Cartledge *et al.*, 1971).

Polytomella caeca can also grow on acetaldehyde and propionaldehyde at low concentrations (1 m*M* or less). Higher concentrations of acetaldehyde supported growth only in the presence of uridine. Other nucleosides and nucleotides could not be substituted. Acetaldehyde inhibited the uptake of [14C]acetate into RNA (Wise, 1968). The effects of acetaldehyde were the same in succinate-grown cells (Wise, 1970).

Metabolic adaptation for utilization of butyrate and caproate occurs in *P. obtusum* (incorrectly referred to as *P. uvella*). The adaptations were characterized by lag periods before growth, diauxic growth on acetate and butyrate, sensitivity to ultraviolet radiation, and increased oxidative activity (Cirillo, 1955, 1956, 1957). Butyrate adaptation could not be explained in terms of permeability, since uptake occurred over a wide pH range. Butyrate–Co A kinase activity could be found in butyrate- but not acetate-grown cells, indicating the induced synthesis of specific enzymes (Chapman *et al.*, 1965).

Chilomonas paramecium also undergoes adaptation for utilization of butyrate and hexanoate. The lag period before substrate oxidation was 4 hr. Butyrate- and hexanoate-grown cells oxidized acetate, butyrate, and hexanoate. Fatty acid-activating enzyme activity appeared in acetate-grown cells 3 hr after transfer to butyrate. When transferred to acetate, butyrate-grown cells lost their activity only after 12 hr. Utilization of butyrate involves formation of an acyl adenylate as shown by the production of 1 mole of pyrophosphate per mole of reduced Co A. This enzyme activity is localized in the soluble fraction of the cell (Kramer and Hutchens, 1969).

2. Carbon Starvation

Carbon deprivation affects the growth, metabolism, and structure of these organisms. In *Euglena* the cells enter a prolonged stationary phase, although they remain viable and motile for approximately 13 days (Malkoff and Buetow, 1964). During starvation there is a decline in dry weight mainly because of the loss of paramylum, which is complete by ~7 days (Malkoff and Buetow, 1964; Brandes et al., 1964). A decline in cellular protein could be demonstrated by following the loss of ^{14}C or ^{35}S from protein and its appearance in the cold TCA-soluble fraction (which contains amino acids and nucleotides) during starvation (Kempner and Miller, 1965b). A 50% decrease in total cell RNA occurred (Blum and Buetow, 1963; Brawerman, 1968). Guanine was seen to be metabolized in both the cytoplasm and nucleus. Over a 6-day period of starvation, labelling in the nucleic acid fraction declined from 30 to 18% and increased from 17 to 30% in the nucleotide fraction. Carbon-limited cells did not incorporate guanine into the nucleus. RNase treatment markedly reduced nuclear label, showing that RNA is degraded during starvation (Schuit and Buetow, 1968).

Changes occur in the structure of carbon-starved cells. Besides a loss of paramylum during the first 4 days, swelling and vacuolation of the cisternae of the endoplasmic reticulum occurred. An increase in the size of these vacuoles continued during the next few days. Large regions of normal cytoplasm were segregated from the rest of the cell by a limiting membrane; with the subsequent degeneration of the contents, vacuolation and fusion of cytoplasmic membranes into myelin figures were observed in this region (Malkoff and Buetow, 1964). In the initial stages of starvation, mitochondrial profiles became constricted and elongated. In later stages a few extremely long mitochondria were visible in the periphery of the cell. After 2 days, the Golgi body hypertrophied and was dispersed in the cytoplasm. The cisternae displayed saccular distensions (Brandes et al., 1964).

Cytolysosomes—cytoplasmic membrane-bound polymorphic inclusions—appeared in starved Euglena. Within these bodies membranellar structures were seen: components of the Golgi body, mitochondria, bleached and replicating chloroplasts, and other vesicles. Cytolysosomes were noted in association with the endoplasmic reticulum and the nucleus (Brandes et al., 1964; Leedale and Buetow, 1976). Acid hydrolases have been localized in these structures. Acid phosphatase was shown cytochemically to be in the mitochondrial matrix and cristae and in the Golgi body during the first week of starvation. The mitochondrial activity may have been a reflection of organelle turnover. Other hydrolases identified included cathepsin D, β-glucosidase, β-naphtholacetate esterase, and acid DNase. It has been suggested that Golgi vacuoles give rise to cytolysosomes. Interestingly, hydrolytic enzyme activity was maximal at 4–7 days of starvation, although the cells remained viable for an additional week (Brandes et al., 1964; Baker and Buetow, 1976).

After acetate replenishment in *Euglena,* the disappearance of vacuoles was observed, although lysosomal activity continued. Long mitochondrial profiles and lysosomes were seen up to 23 hr after replenishment. Normal mitochondrial, nuclear, and endoplasmic reticulum structure, and paramylum granules, were seen after 2 days, although lysosomes were still present (Malkoff and Buetow, 1964). The accumulation of cell protein, RNA, and DNA depended on the duration of starvation, there being a lag period in cells starved for long periods. The duration of starvation had no effect on the rate of resynthesis (Blum and Buetow, 1963).

Little work has been done on the effects of amino acid starvation, though it could be expected to affect cell growth and metabolism significantly. Cysteine and methionine promoted cell division in *Euglena* and *Astasia,* although cell growth was not impaired in their absence (James, 1964). In acetate-grown *C. reinhardii,* arginine starvation inhibited cell division, although the cells remained viable for more than four generations. [3H]Leucine incorporation into protein continued for 15 hr of starvation, but adenine incorporation into RNA and DNA stopped within 1 hr. During starvation there was no change in the nucleotide pool. The appearance of autophagic vacuoles was also noted. The structural analogue of arginine—canavanine—also inhibited growth and resulted in cellular vacuolation. Canavanine resulted in decreased synthesis and accumulation of RNA. Cycloheximide inhibited its incorporation into protein and protected against cell death (McMahon and Langstroth, 1972).

3. Inorganic Ions

Although defined media for acetate flagellates usually include numbers of ions, few studies have actually been done to determine the specific ion requirements. Thus, while one high-salts medium for *Euglena* contains (besides phosphate, magnesium, sulfate, calcium, potassium, manganese, zinc, and chloride) molybdate, iron, borate, cobalt, iodide, and copper, the latter could be omitted without a significant change in the growth rate or cell yield. The latter parameters are functions mainly of carbon, sulfur, and phosphorus concentrations. A low-salts medium results in decreases in dry mass, protein, lipid, chlorophyll, and carotenoids (Kempner and Miller, 1965a). It must contain sufficient amounts of iron and copper in the water or as contaminants in other reagents, for synthesis of heme-containing or copper-dependent proteins of electron transport, as well as others proteins, is inconceivable in their absence.

4. Sulfur Deprivation

In acetate-grown *Euglena,* the final cell yield was proportional to the sulfur concentration, although at high concentrations (0.68 mg/ml) there was a slower rate and a lower cell yield. Homocysteine and cysteine could not replace sulfate, and cell viability declined after 1.5 days. After 15 days, only 25% of the cells were viable. When ^{35}S-labelled cells were transferred to sulfur-deficient medium, there was an immediate drop in the labelling of the ethanol-soluble and cold and hot TCA-soluble fractions, although the residual protein retained label for several hours. Thus it was suggested that sulfur was redistributed to maintain "structural proteins" during starvation (Buetow, 1965). A sulfur limitation in cells in continuous culture resulted in fewer plastids per cell and slower DNA synthesis (Epstein and Allaway, 1967).

In *A. longa*, grown in continuous culture with a sulfate limitation, cellular carbohydrate remained constant at relatively short generation times and increased dramatically at doubling times longer than 60 hr. The total cell protein was constant regardless of the growth rate. With increasing doubling times there were linear decreases in RNA and DNA content per cell. The DNA content of cells in balanced growth was always less than in the logarithmic phase of batch cultures. The changes in DNA content may have been due to changes in cell ploidy or polyteny (Morimoto and James, 1969a).

5. Phosphate Deprivation

In *Euglena*, phosphate deprivation resulted in lower growth rates and peak populations (Kempner and Miller, 1972a; Buetow and Schuitt, 1968). Survival was up to 7 days. At 1/50 the normal phosphate concentration RNA markedly decreased (80%), as did DNA (60%). The rate of plastid replication did not appear to be coupled to that of cell division. Prior to the deceleration of cell division the plastid number per cell declined from 7 to 9 to 4 to 6, and during the stationary phase the plastid number increased to more than 10 per cell. Chlorophyll synthesis, however, continued at a constant exponential rate even into the stationary phase (Parenti *et al.*, 1972).

Changes in cell ploidy have been noted. Normal cells have 92 ± 2 chromosomes. In medium containing 1/500 normal phosphate there are 48 ± 3 chromosomes. Based on the kinetics of bleaching by uv radiation, a decrease in ploidy of plastids from 38 to 18 was demonstrated (Epstein and Allaway, 1967).

A phosphate transport system, which is saturated at very low concentrations ($10^{-9} M$), has been described in *Euglena*. Uptake was inhibited at low temperatures and by DNP, K^+, Li^+, and Na^+. Arsenate was a competitive inhibitor. Although much of the phosphate taken up was esterified into organic forms, enough was present as orthophosphate to suggest that an active transport mechanism was operative (Blum, 1966).

In *Euglena* at least two acid phosphatases have been demonstrated cytochemically. In cells grown without substrate, enzymatic activity was localized in the lamellae and vesicles of the Golgi body as well as in the vesicles and vacuoles around the reservoir, and also near portions of the endoplasmic reticulum at the corner of the pellicle. An additional inducible enzyme was localized in the pellicle. This enzyme was not detected in uninduced cells (Sommer and Blum, 1965). Arsenate competitively inhibited the induced enzyme, suggesting that it may be involved in phosphate uptake (Blum, 1966).

Astasia longa has only one phosphatase in both induced and uninduced cells. No enzyme was demonstrable in the pellicle of these cells (Sommer and Blum, 1965).

In *P. caeca,* grown in phosphate-limited continuous culture, the RNA/protein ratio increased with shorter doubling times (Jeener, 1953). In *P. agilis* glucose-6-phosphatase was localized in the proplastidlike body and starch bodies. ATPase was found in the endoplasmic reticulum, mitochondria, and nuclear membranes (Senko *et al.,* 1971).

Phosphate limitation in a chemostat did not induce chlorosis in a mutant of *C. reinhardii* incapable of organotrophic growth in the dark. The chlorophyll concentration remained constant and somewhat elevated over a 25-day period. Alkaline phosphatase activity also remained constant over an extended duration. Just as dark growth did not inhibit chlorophyll synthesis, so nutritional limitation also did not induce chlorosis (Hudock *et al.,* 1971). Phosphate concentration did not affect the concentration of acid and alkaline phosphatases in synchronous populations of *Chlamydomonas* (Lien and Knutsen, 1972, 1973a,b).

6. Zinc Deprivation

Zinc is important for the growth of *Euglena*. It is a component of several dehydrogenases, aldolases, peptidases, and phosphatases. It is bound to phosphates in DNA and RNA, and is indispensable for DNA and RNA polymerases. RNA polymerase II (an α-amanitin-insensitive enzyme) contains 2.2 gm-atoms of zinc per mole of enzyme. Its activity is inhibited by various chelating agents (Falchuk *et al.,* 1976). Zinc deficiency results in slower growth but an increase in cell volume and slower RNA and protein synthesis. It has little effect on cell ultrastructure, except that there seem to be more paramylum granules. Zinc-deficient cells accumulate Mn^{2+}, Ca^{2+}, and Fe^{2+} ions (Falchuk *et al.,* 1975a). Other workers using cells in steady state growth conditions have found that a 1000-fold change in the Zn^{2+} concentration of the medium results in only slight differences in growth rate, mean cell volume, and peak population levels (Shehata and Kempner, 1977).

A cytofluorometric study of zinc-deficient cells from the stationary phase showed that most cells were in the S or G_2 phase of the cell cycle (see p. 48), but that no division occurred. Thus, the increased DNA content of these cells can be explained by their failure to progress from S to G_2. Evidence has been obtained showing that the transition from G_1 to S was also blocked by a Zn^{2+} deficiency. When stationary-phase cells were transferred from complete to zinc-deficient medium, only 25% of the cells divided; most cells were in the G_1 phase. The readdition of zinc resulted in cell division (Falchuk et al., 1975b).

Stationary-phase cells in zinc-limited medium lack ribosomes. Replacement of Zn^{2+} after 6 days resulted in the reappearance of ribosomes within 3 days. During the first 4–5 days of deficiency, there was a heterogeneous distribution of ribosomal material rather than a single peak at 87 S. Thus, it is likely that Zn^{2+} deficiency results in general ribosomal disintegration and not merely dissociation into subunits (Prask and Plocke, 1971).

7. Vitamins

All acetate flagellates need exogenous thiamine; the medium can be autoclaved with this vitamin. Thiamine deprivation in P. agilis did not change the logarithmic growth rate, but peak populations were 50% of normal. Electron microscopy of thiamine-deprived cells revealed a changed mitochondrial morphology; profiles appeared smaller, and the inner membranes were more tightly packed and organized in a regular parallel array. Few if any a- or b-type cytochromes were detectable spectrophotometrically, and the concentration of c-type cytochromes was markedly reduced. These effects were reversible within one generation after vitamin replacement (Cantor and Burton, 1975). Thiamine-deprived cells had no detectable pyruvic, α-ketoglutarate, or succinic dehydrogenase activity, and there were three- to fourfold increases in the activity of pyruvic decarboxylase and mitochondrial ATPase (Cantor and Burton, 1974; Cantor, unpublished observations).

Vitamin B_{12} deprivation has been studied extensively in Euglena. The population doubling time and peak population density are proportional to the vitamin concentration. Mean cell size is inversely proportional to concentration (Epstein et al., 1962). Exponentially growing populations showed a greater increase in cell size when deprived of the vitamin compared to stationary-phase cells (Bertaux and Valencia, 1973). After 14 days of deprivation (6 days after the cessation of cell division) chromosomes were no longer visible, the chloroplasts were abnormal, and the mitochondria were very much smaller (Bertaux and Valencia, 1973; Bré and Lefort-Tran, 1974).

The growth of green cells is inhibited in vitamin-free medium when illuminated constantly or periodically, but not in total darkness (using lactate as a carbon source). When vitamin B_{12}-deprived cells were transferred from the dark to white light, the growth rate was inversely proportional to the light intensity. Blue and green light were noninhibitory even at a high intensity. Since an albino mutant also has an absolute requirement for the vitamin, it has been suggested that the ability to divide in the dark in the absence of the vitamin is related to the ability to make a chloroplast. An increased pO_2 or a decreased pCO_2 has also been shown to inhibit cell division in the absence of the vitamin (Bré et al., 1975).

Vitamin-deficient cells incorporated [^{14}C]formate into DNA at slow but continuous rates, suggesting that such cells are in an extended S phase in the cell cycle. The addition of cycloheximide at the same time or prior to that of the vitamin resulted in a transitory rise in the incorporation of formate, indicating that protein synthesis was required for the completion of DNA synthesis in these cells (Goetz et al., 1974).

Chloroplast development and division have been associated with the vitamin B_{12} requirement. Vitamin-deficient cells show increases in both cellular protein and chlorophyll concentrations. They also have more chloroplasts per cell (Carell, 1969), although the chlorophyll/chloroplast ratio remains constant (Christopher et al., 1974), suggesting an uncoupling of chloroplast and cell division. During chloroplast development synthesis of a chloroplast-associated alkaline DNase occurs, which is inhibited by cycloheximide but not by streptomycin or chloramphenicol. During vitamin deprivation the activity of this enzyme increases (Egan and Carell, 1972).

C. Oxygen and Carbon Dioxide

The growth of A. longa, P. caeca, and P. agilis is inhibited in the absence of carbon dioxide (Rahn, 1941; Cantor and James, 1965). The generation time of acetate-grown A. longa is doubled when oxygen plus carbon dioxide (95 : 5) is bubbled through the medium. On exposure to high oxygen tension, the Q_{O_2} decreased as did the activities of succinic dehydrogenase, succinic oxidase, succinate–cytochrome c oxidoreductase, and NADH–cytochrome c oxidoreductase; cytochrome oxidase activity did not decline (Bégin-Heick and Blum, 1967). Oxygen lowered mitochondrial enzyme activity in Euglena grown only in low-phosphate medium (20 mM). Even in the low-phosphate medium, air did not affect these enzymes except after prolonged exposure (Blum and Bégin-Heick, 1967).

Chloroplast development in Euglena may be potentiated by preillumination, provided oxygen is present. Oxygen did not support this potentiation when provided in the dark (Klein et al., 1972).

D. Temperature

Within the range of tolerable temperatures, population doubling times decrease with increasing temperature in all species examined [e.g., in *A. longa* (James and Padilla, 1959; Padilla and James, 1960)]. Different strains of *Euglena* adapt differently to various temperatures. Thus, *E. gracilis* var. *bacillaris* has a shorter doubling time, less protein per cell, and higher exogenous respiration rates at all temperatures examined than *E. gracilis* strain Z (Cook, 1966b). There is, in general, a direct relationship between temperature and Q_{O_2} (Cook, 1966b; Padilla and James, 1960). However, the proportion of cells in fission and the ratio of dry weight to cell volume were constant at all temperatures in *A. longa* (Padilla and James, 1960; James and Padilla, 1959). In *C. paramecium* the Q_{O_2} was independent of the incubation temperature (Johnson, 1962). In *P. agilis* there appeared to be a direct relationship between population growth rate and cellular starch and protein concentrations at lower incubation temperatures (9° and 18°C) but not at 25°C, suggesting that more balanced growth occurs at lower temperatures (Sheeler *et al.*, 1968a, 1970). However, at 9°C there was less starch per cell at any time than at higher incubation temperatures (Sheeler *et al.*, 1968a,b). In *C. paramecium,* no differences in dry weight, DNA per cell, or protein per cell were noted as a function of temperature, although significant differences in cellular RNA were found (Johnson and James, 1960).

Elevation of the temperature above 32°C is an effective method of bleaching *Euglena*. The loss of chlorophyll at elevated temperatures is accompanied by structural changes in the cell. Lipid granules accumulate, and chloroplast structure is disrupted (Wolken, 1967). Normal green strains of *E. gracilis* var. *bacillaris* contain carotenoids, β-carotene, lutein, and neoxanthin; heat-bleached cells contain echinenone (4-keto-β-carotene), zeaxanthin (3,3'-dihydroxy-β-carotene), and lutein (Goodwin and Gross, 1958). Cell gigantism occurs in this strain at 35°C. Heat-bleached cells grow better in the light than in the dark at the elevated temperature (Gross, 1962). There are changes in the c-type cytochromes in heat-bleached cells. Absorption peaks for normal green strain Z cells occur at 552 and 416 nm for photosynthesizing cells, and at 556, 525, and 421 nm for dark-grown cells. Heat-bleached cells lack a peak 552 nm but have one at 556 nm (Wolken and Gross, 1963).

Temperature-sensitive mutants of *C. reinhardii* have been isolated. These mutants grow normally at 23°C, but not at 34°C. One such mutant was resistant to colchicine. At 2 mM colchicine, normal flagellar regeneration occurred; at 5 mM colchicine inhibition was only 40–60% (Sato,

1976). Other mutants which are blocked at specific points in the cell cycle have proven extremely valuable for analysis of the cell cycle (Howell, 1974).

E. Inhibitors

Inhibitors of various sorts have been effectively used as tools for the study of growth, metabolism, and differentiation. These have included inhibitors of protein and nucleic acid synthesis, amino acid analogs, metabolic inhibitors, and uncoupling agents.

In *Euglena*, many of these inhibitors have been used as bleaching agents or in studies on chloroplast structure, function, and differentiation. Since 1948 it has been known that streptomycin and its analogs (which inhibit ribosome function in protein synthesis) induce bleaching (Provasoli *et al.*, 1948). Streptomycin blocks the synthesis of chlorophyll, as dark-grown cells (plus streptomycin) do not regreen when transferred to the light. The bleaching effect is antagonized by divalent ions, especially Ca^{2+} and Mn^{2+} (Rosen and Gawlick, 1961). Ultrastructurally these cells have normal mitochondria but lack chloroplasts (Moriber *et al.*, 1963). The presence of fluorescing plastidlike structures has been noted, indicating the presence of protoporphyrin (Gibor and Granick, 1962a). In contrast to earlier reports (Provasoli *et al.*, 1948), recent studies showed that streptomycin decreased lipid density and vesicle diameter of the stigma after 3 weeks of treatment. By 7 weeks the stigma region contained barely recognizable vesicles. These structural changes were accompanied by declines in carotenoid content and phototactic response (Ferrara and Banchetti, 1976). In bleached cells there is a qualitative change in the carotenoids; in addition to β-carotene, phytofluene and a ζ-carotene are found. No major xanthophylls are present (Goodwin and Gross, 1958; Gross, 1962). Whereas bleached cells do not synthesize carotenoids during exponential growth, stationary-phase cells produce these substances, and such cultures appear yellow and bright orange. There is also an increase in cellular lipid at the same time. These data suggest that nondividing cells do not use carotenoids as sources of carbon and energy (Smillie and Rigopoulos, 1962).

Streptomycin and its analogs vary in toxicity. Streptomycin and kanamycin are not toxic, whereas neomycin and paramycin result in cell death (Aaronson and Scher, 1960; Zahalsky *et al.*, 1962; Ebringer, 1964). Indeed, the cell yield of streptomycin-bleached cells is greater than that of normal green cells, but individual cells are smaller (Tong *et al.*, 1965).

Other effective bleaching agents in *Euglena* include the following:

1. DNP results in the conversion of chlorophyll *a* to pheaophytin *a* accompanied by a loss of lamellar structure. Bleaching is more pronounced in acid pH, suggesting that DNP induces an accumulation of hydrogen ions which are not pumped out, presumably because of inhibition of the contractile vacuole. Other uncouplers or inhibitors of oxidative metabolism and glycolysis which are effective include pentachlorophenol, iodoacetamide, fluoride, and azide. Fluoroacetate is effective, but only in dark-adapted, aerated cells (Greenblatt and Sharpless, 1959). *Euglena* can, however, adapt to grow in the presence of uncouplers (see Section IIB).

2. A variety of antihistamines are effective bleaching agents. These include benadryl, pyribenzamine, decapryn, theophorin, tegathen, and pyronil. Benadryl bleaching is reversible, but that of pyribenzamine is not. Unlike streptomycin, pyribenzamine bleaching results in a loss of carotenoids (Gross *et al.*, 1955).

3. Aminotriazole is a nontoxic, reversible bleaching agent (Aaronson and Scher, 1960).

4. Pigment formation is irreversibly inhibited by *O*-methylthreonine. The inhibition is overcome competitively by isoleucine, homoserine, α-aminobutyrate, and α-ketobutyrate. The supernatant fluid of bleached cultures contains large quantities of free amino acids, but no porphyrins. This contrasts with the medium for other bleached organisms (Aaronson and Bensky, 1962).

5. Erythromycin is an irreversible bleaching agent; only dividing cells are sensitive. Leucoplasts have been described in the bleached cells (Ebringer, 1962, 1964). Related to erythromycin are spiramycin, carbomycin, lincomycin, and triacetyloleandomycin. These substances, as well as tetracycline, inhibit chlorophyll synthesis and growth (Linnane and Stewart, 1967).

6. Hydroxyurea inhibits both cell division and chlorophyll synthesis when added to cultures placed in the light after 5 days of dark growth, or after 3 days of continuous illumination. Heat-bleached cells continue to divide in the presence of hydroxyurea, although reduced peak populations are obtained; this is more likely an effect on the cell cycle than on the synthesis of chlorophyll. Normal dark-grown cells (plus hydroxyurea) do not divide when returned to light. The effect of hydroxyurea is reversible (Buetow and Mego, 1967).

7. Hadacin partially inhibits chlorophyll synthesis, the degree of inhibition being concentration dependent. Growth of the cells in light prior to treatment renders it ineffective. Complete reversal of inhibition is not attained even with equimolar concentrations of its analog, aspartate (Mego, 1964).

8. Chloramphenicol [an inhibitor of mitochondrial and chloroplast ribosomal synthesis (Linnane and Stewart, 1967)] and cycloheximide and puromycin [inhibitors of cytoribosomal protein synthesis (Hutner *et al.*, 1968)] all inhibit chlorophyll synthesis. *Euglena* is very sensitive to cycloheximide. Puromycin does not differentially inhibit greening.

9. Nalidixic acid, an inhibitor of bacterial, mitochondrial, and chloroplast DNA synthesis, does not affect cell division, although bleaching occurs after 26 hr of treatment. Chloroplast DNA is virtually immeasurable by 16 hr. Inhibition of chloroplast DNA replication has been shown by absence of the formation of hybrid DNA in a Meselson–Stahl centrifugation experiment (Pienkos *et al.*, 1974). Nondividing, nonphotosynthesizing cells and bleached mutants are insensitive to this inhibitor. Light has been hypothesized to facilitate nalidixic acid uptake into plastids, modify the inhibitor, or modify an initiator of chloroplast division. No evidence is available in support of the last two suggestions (Lyman *et al.*, 1975). Myxin (1-hydroxy-6-methoxyphenazine 5,10-dioxide), an inhibitor of bacterial DNA synthesis, is also very effective in illuminated, dividing cells (McCalla and Baerg, 1969).

10. Nitrofuran derivatives permanently bleach illuminated cells. This role of light is unrelated to photosynthesis, for DCMU does not decrease the effect of nitrofurantoin when used in concentrations not inhibiting cell growth, and a mutant (P4-D) which lacks photosystem II is also bleached by nitrofurantoin. The reduction of nitrofurantoin, probably to nitrofuraldehyde, is a likely step in its action. This may occur by the preillumination of nitrofurantoin before inoculation; growth in the dark is then inhibited. Nitrofuran reductase, an oxygen-sensitive enzyme consisting of two components, has also been demonstrated in the cytosol of cells. It has not been demonstrated that enzyme activation is a prerequisite for bleaching. The reduction products may cause breaks in DNA (McCalla, 1965; McCalla and Reuvers, 1970; McCalla and Voutsinos, 1975).

Various bleached mutants of *Euglena* have been isolated using nalidixic acid, streptomycin, ethyl methylsulfonate, lincomycin, magnesium deficiency, uv radiation, and temperature. Most of these mutants have measurable amounts of 16 and 23 S rRNA, although less than or equal to that in dark-grown cells (Heizmann *et al.*, 1976).

Antihistamines are effective bleaching agents in *Chlamydomonas pseudococcum* (Gross *et al.*, 1955). Aflatotoxin has been used as a mutagenic agent for the isolation of streptomycin resistance in *C. reinhardii* (Schimmer and Werner, 1974). Canavanine inhibits cell division, RNA synthesis, and RNA accumulation in this organism as well (McMahon and Langstroth, 1972). Isonicotinic acid hydrazide inhibits growth in *C.*

paramecium. Adaptation to this compound has been described, and two resistant strains have been isolated. These two strains differ from each other with respect to sulfanilamide resistance and *p*-aminobenzoic acid utilization (Hall, 1967).

Hydroxyurea inhibits cell division in *P. agilis,* and prolonged exposure leads to cell death (Patterson, 1977). Chloramphenicol inhibits growth in batch cultures of *P. caeca* as well as decreasing the Q_{O_2}, concentrations of *a*- and *b*-type cytochromes, and the activity of succinoxidase, NADH oxidase, and cytochrome *c* oxidase (Lloyd *et al.,* 1970).

VI. LIFE CYCLES

For most acetate flagellates two daughter cells are produced as a result of mitotic division of the parent cell.

In *Chlamydomonas* sp. the life cycle involves both sexual and vegetative stages. Haploid vegetative cells of opposite mating types fuse to form a zygote, which following meiosis grows and divides into four cells. Alternatively, these cells may divide asexually by mitosis to establish clones. Before meiosis flagellar regression and formation of a wall around the meiotic cell occur. The demonstration of sexuality has been used in numerous genetic studies on nuclear and chloroplast genes. Mating has also been studied as an example of cell recognition involving flagellar surface components. Fusion involves the formation of aggregates resulting from adhesion between flagella of opposite mating types and apical fusion by papillar outgrowth (Mesland, 1976). Then a bridge between the cells forms and the papillae unite. Plasmogamy requires elongation and interaction between two papillae, not one. Cytoplasmic exchange involves plastid fusion. This is followed by secretion of the primary wall of the zygote and karyogamy. Gamete nuclei unite, and chromatin condensation occurs. In the mature zygote, microtubules are observed usually near the pyrenoid. Finally a secondary zygote wall is secreted (Brown *et al.,* 1968). During gametogenesis $^{14}CO_2$ incorporation into amino acids (mostly basic amino acids) occurs in the presence of exogenous NH_4^+. Gametes show a two- to fivefold increase in ornithine transcarbamylase activity compared to that in vegetative cells (Kates and Jones, 1966).

Polytomella sp. also has a complex life cycle. Asexual division is the mode of cell reproduction for most cells in the population. Cleavage begins posteriorly and proceeds to the cell anterior. The earliest descriptions of this species mentioned a cyst stage. Encystment begins as early as the midlogarithmic phase of growth and is complete by the beginning of the stationary phase. In *P. agilis*, ~10% of the motile cells encyst. No single cause for encystment has been defined (Sheeler *et al.*, 1970). Cysts are extremely hardy, surviving desiccation and storage at −150°C (Reed *et al.*, 1976), as well as sonication (Sheeler *et al.*, 1970). The earliest studies on the ultrastructural changes during encystment showed six stages, including precyst and cyst stages. In the former there were increases in the proplastidlike network and the amount of endoplasmic reticulum, and a thickening of the plasma membrane due to the deposition of an amorphous electron-dense material on its outer surface (Moore *et al.*, 1970). Starch accumulation occurs during the course of encystment. A thick cell wall is laid down during cyst maturation, having four to six layers (Brown *et al.*, 1976a; Moore *et al.*, 1970). The chemical composition of the wall has not been clearly defined (Gittleson *et al.*, 1969). The mature cyst contains a mitochondrion, although its structure is abnormal (Gittleson *et al.*, 1969; Moore *et al.*, 1970). During encystment the Golgi complex is initially longer and has more compact cisternae. Later the cisternae become distended and associated with vesicles containing fibrillar material. Finally they diminish in size and there is little vesicular association. Microtubular structures, including basal bodies, and plastids are not present in the mature cyst (Brown *et al.*, 1976a). Although there is the same number of electrophoretically distinguishable soluble proteins in motile cells and cysts, the latter contain proteins of greater mobility. Furthermore, motile cells contain three protein bands which show amylolytic activity, whereas cysts have only two such bands (Sheeler *et al.*, 1970). Excystment in *P. agilis* begins ~3 hr after cysts are placed in fresh medium at 25°C and is essentially complete after 18 hr. Individual excystment takes 30–40 min. In the earliest phase cellular polarity is noted with vacuoles at one end of the cell. Also seen are two of the four flagella, the differentiation of microtubules, and four wall layers. This is followed by differentiation of the Golgi complex, separation of the outer cell wall layer, corrugation and rupture of the cell wall, and extrusion of the protoplast (Brown *et al.*, 1976b).

Sexuality has been described in dense log-phase *P. caeca*. Cell fusion occurs anteriorly. The zygote is visible for 3–3.5 hr before dividing into four cells. The fate of the flagella during this process has not been clarified (Lewis *et al.*, 1974). Evidence for conjugation in *P. agilis* has been obtained, there being differences between this process in the two species (Moore and Cushing, 1979). If these observations could be confirmed, *Polytomella* would be useful for genetic studies, especially since its doubling time is only 4–5 hr. Sexuality in *P. uvella* has also been suggested (Dogiel, 1935).

VII. CELL CYCLES

A. Methods for Establishing Synchronous Cultures

Much of our information on the regulation of cell growth, metabolism, differentiation, and division has been from studies on synchronous populations of microorganisms. Some of the earliest studies on cell synchrony utilized *Astasia* and *Euglena*. Synchronization techniques have usually involved either of two approaches.

1. Induction Synchrony

Induction synchrony involves bringing a population of cells into register with respect to a particular phase of the cell cycle, e.g., cell division or DNA synthesis. Among the acetate flagellates the methods employed have been (a) repetitive temperature cycles—*A. longa* (Padilla and James, 1960), *E. gracilis* (Neal *et al.*, 1968), and *P. agilis* (Cantor and Klotz, 1971); (b) repetitive light–dark cycles—*E. gracilis* (Cook and James, 1960; Edmunds, 1964) and *C. reinhardii* (Bernstein, 1968); (c) inhibitors—*P. agilis* (Patterson, 1977); (d) starvation and refeeding—*A. longa* (Morimoto and James, 1969b) and *P. caeca* (Cantor and Lloyd, unpublished observations).

Induction synchrony usually involves the imposition of some metabolic or environmental stress, although attempts have been made to relate light or temperature cycles to natural conditions or to the normal behavior of cells under given conditions. In all cases it is difficult to determine whether observed events during a synchronous cycle are true cell cycle events or a consequence of the synchronizing procedure. With the exception of specific inhibitors, the exact mechanism of synchrony is unknown. Although inhibitors bring cells to specific points in the cell cycle, they frequently result in unbalanced growth and in cell death upon prolonged exposure. Starvation and refeeding usually result in an abnormally long lag period before the first division; unless the second cycle can be studied, it is difficult to conclude that the first cycle is truly reflective of normal cellular events. Thus, while *P. agilis* recovers from thiamine deprivation within one normal generation time, the first cycle necessarily involves the restoration of normal mitochondrial structure and function which has been disrupted during starvation (Cantor and Burton, 1975). These processes surely are not normal and usual events in the cell cycle.

2. Selection Synchrony

Selection synchrony involves selecting cells of a given age or size from a batch culture and using them to produce a synchronous population upon reinoculation into fresh medium. Since no overt manipulation of cellular metabolism or the environment is performed, it is assumed that these cells have not been subjected to any marked stress and that the synchronous cycle truly reflects the normal life cycle. Techniques used have included density gradient centrifugation, continuous-flow centrifugation, filtration, and mitotic selection. These techniques have been applied successfully to various species of yeast, bacteria, amebas, ciliates, and mammalian cells in culture, but not to acetate flagellates—a failure perhaps because of the particular swimming behavior of these organisms. For example, *P. agilis* and *P. caeca* cells swam through sucrose gradients when this technique was attempted.

Definition of the cell cycle has been attempted using *genetic approaches*. Temperature-sensitive yeast mutants, blocked at a particular point in the cell cycle, have been isolated. These mutants grow normally at low temperatures but not at elevated temperatures (Hartwell, 1974). This approach has been successfully applied to *C. reinhardii,* and at least nine cell cycle mutants have been isolated, each blocked at a different point in the cell cycle (Howell, 1974).

3. Inhibitors

Inhibitors, used to define and analyse the cell cycle in bath cultures, have been applied to acetate flagellates. This technique depends on the age distribution of a random cell population and considers the residual cell division occurring after introduction of the inhibitor. Cells before the inhibitor block do not divide, whereas those which have passed it divide once. Eventually all cells in the population accumulate at the block point. Because cells in an asynchronous culture are not uniformly distributed throughout the cell cycle, the age distribution is skewed so that there are more newly divided cells than older ones in the population. Thus, an inhibitor acting late in the cell cycle should arrest division soon after its addition. Inhibitors acting early in the cell cycle should stop cell division later (but within one generation). Cessation of cell division does not in itself indicate that the inhibitor is interfering with a cell cycle-specific event. Consider the inhibition of protein synthesis. If all protein synthesis were summarily inhibited in all cells, could one consider the delay prior to division arrest a reflection of the timing of the synthesis of a specific division protein? It could also act on DNA synthesis or some energy-yielding pathway, and the inhibition of division would be a summation of all these events. Specificity of action may be tested by considering the cell volume distribution in asynchronous populations. Usually this parameter is log-normally distributed about the mean. If a specific cell cycle event is blocked by an inhibitor, the cell volumes should become normally distributed about the mean concomitantly with the cessation of cell division. If the inhibitor acts uniformly throughout the cell cycle, the distribution should remain log-normally distributed.

This approach has been successfully employed in studying the *C. reinhardii* cell cycle. Inhibitors which block organellar DNA, RNA, and protein syntheses have block points in the second quarter of the cycle. Inhibitors of nuclear and cytoplasmic macromolecular syntheses have transition points in the fourth quarter. Thus organellar biogenesis is completed by midcycle, whereas components required for cell division are completed later in the cell cycle (Howell *et al.*, 1975). The same results have been obtained using this and genetic approaches.

B. The Cell Cycle in Acetate Flagellates

For actively dividing eukaryotic cells, the cell cycle is conventionally divided into several stages: a growth phase before the replication of bulk DNA (G_1), the period of DNA replication (S), a postreplicative growth phase before mitosis (G_2), mitosis (M), and cell division (D). In a variety of cell types, the G_1 and G_2 phases show considerable variability in duration. It must be reemphasized that the method of producing synchronous populations can lead to misinterpretations. Thus, *Tetrahymena pyriformis,* synchronized by repetitive heat shocks, appears to lack a G_1 phase. More recent studies, based on the feeding behavior of the organism and using centrifugation methods, indicate that there is a G_1 phase in this species (Wolfe, 1973).

With the use of a 24-hr cycle of 14 hr of light and 10 hr of dark growth, the *E. gracilis* cell cycle has been defined as follows: G_1 (0–7 hr)—during this period the large electron-dense chromosomes become compact: S (7–14 hr)—chromosome material doubles in this interval; G_2 + M (14–24 hr)—chromosomes become dispersed throughout the nucleus in G_2 and are arrayed perpendicularly or parallel to the division axis in M (Edmunds, 1964; Falchuk *et al.*, 1975b; Bertaux *et al.*, 1976). The fission time in *Euglena* varies with temperature, but is significantly less than 1 hr at 25°C (James, 1964). During the light period cell mass increases and the biosyntheses of DNA, RNA, protein, chlorophyll, and carotenoids occur. Cell division begins ~1 hr after the onset of the dark period (Padilla and Cook, 1964).

An alternative cycle has been postulated for *Euglena,* growing exponentially in the light, based on the microspectrophotometric analysis of DNA per nucleus. Since G_2 cells should contain twice as much DNA per nucleus as G_1 cells, at least three nuclear populations should be found, with *x, 2x,* and some other intermediate value(s) of DNA. The length of time spent in each portion of interphase should be reflected in the number of cells in each group. No cells could be found with a 2*x* amount of DNA, suggesting that the combined lengths of G$_2$ and M are very short and that cells begin cytokinesis at completion of replication. Thus, G$_1$ and S had durations of 4 and 6 hr, respectively (Christopher *et al.,* 1974).

The cell cycle of *C. reinhardii,* synchronized by a repetitive cycle of 12 hr of light and 12 hr of dark, has been defined as follows: G$_1$, 20 hr (previous dark) to 11 hr; S, 11–16.5 hr; G$_2$, too short to detect; M + D, 16.5–21 hr (Cattolico *et al.,* 1973). A cell wall-less mutant has also been synchronized and has a cell cycle very similar to that of the wild type (Lien and Knutsen, 1976). Other light–dark cycles have been used for synchrony of *Chlamydomonas* (12L:4D; 14L:10D). The timing of cellular events is clearly affected by the nature of the cycle. It should be kept in mind, therefore, that the events observed may be artefacts due to the synchronizing procedure.

Polytomella agilis may be synchronized by hydroxyurea treatment, which inhibits DNA synthesis. At 25°C untreated cells have a doubling time of 4–4.5 hr. The inhibitor arrested cell division after 3.5–4 hr. This suggests that the S period is at the beginning of the cell cycle. When cells are collected immediately after the cessation of residual cell division and reinoculated into fresh medium, a synchronous division occurs after 3 hr and the population doubles within 1–1.5 hr. Preliminary experiments on the uptake of [^{3}H]thymidine after release from hydroxyurea treatment show that it occurs immediately after cell division, supporting the notion that if there is a G$_1$ period it is very short. That the entire first cycle is of the same duration as a normal doubling time and that electron microscopy shows no alterations in cell structure as a result of the treatment suggest that hydroxyurea caused no major distortion of the cell cycle (Patterson, 1977; Cantor, unpublished observations). The known toxicity of hydroxyurea (Sinclair, 1967) must be kept in mind, especially since *P. agilis* did not undergo synchronous division when the inhibitor was not removed from the organisms shortly after the cessation of cell division.

C. Aspects of the Control of Cell Division

Sulfhydryl compounds have been implicated in regulation of the division process. In *A. longa* and *E. gracilis,* synchrony (i.e., the interval during which the entire population doubles) is significantly improved by the addition of sulfhydryl compounds to the medium (James, 1964; Padilla and Cook, 1964). Cysteine, methionine, thioglycolate, thiosulfate, and sodium sulfide effectively restored synchrony to a bleached mutant of

Euglena placed in continuous darkness after entrainment under a light–dark regime (Edmunds *et al.,* 1976). When *A. longa* was grown in continuous culture with sulfate as the limiting nutrient, the addition of a pulse of sulfate resulted in synchronous division. The sequence of events leading to division included an immediate increase in cellular DNA followed by a gradual decline, a marked increase in cellular protein followed by a sharp decline, and a step increase in cellular RNA. The increase in RNA was coincidental with cell division (Morimoto and James, 1969b). In temperature cycle-synchronized cultures, the rate of uptake of [^{35}S]thioglycolate increased markedly just before cell division, regardless of the temperature at which the uptake was measured. Label was incorporated at the anterior end of the cell, presumably into the resevoir or basal granule. The peak of incorporation did not coincide with nuclear migration, showing that label was not incorporated intranuclearly (James, 1964). If sulfhydryls are involved in the division process per se, it may be by way of regulating microtubule assembly or function. The basal granule may function in the regulation of mitotic microtubule biogenesis in *Astasia* in a manner analogous to that in *C. reinhardii* (Witman, 1975).

Nucleic acid metabolism regulates division. In temperature-synchronized *A. longa,* 8-azaguanine inhibits division. Inhibition was a function of the time of addition of inhibitor, being effective when added at the beginning or middle of the nondividing portion (cold period) of the cycle. When added at the end of the nondividing portion of the cycle or at the beginning of the warm period, immediate division was not blocked, but subsequent division was inhibited. How the inhibitor affects RNA or protein metabolism is unclear, since it was not incorporated into RNA, nor did gross changes in cellular RNA or protein occur (Padilla and Blum, 1963).

Phosphate may regulate the division of *C. reinhardii,* synchronized by cycles of 12 hr light and 4 hr dark. In normal cultures, the generation time was 12.6 hr, and nuclear and cell division began in the last 3 hr of the light period. In phosphate-free medium the generation time was 9.5 hr. There was decreased nuclear division synchrony, for karyokinesis occurred over a 4-hr interval. Sporulation began at 7 hr. To demonstrate a cell cycle transition point for inorganic phosphate, it was added to the culture at different times after the onset of zoospore formation; the number of cells present at 10 and 24 hr was noted. The addition of phosphate prior to the transition point should eliminate early division and sporulation. When added prior to 4 hr, there was complete recovery without early division but normal sporulation. From 4 to 8 hr fewer cells recovered, and more cells showed early sporulation. Since DNA accumulation in phosphate-free cultures began at 5 hr and continued until 10 hr, the transition point

could be shown to occur ~2 hr before the S phase. Interestingly, phosphatase derepression in these cells occurred maximally before the transition point (Lien and Knutsen, 1973a).

D. Aspects of Organelle Biogenesis

If one regards the cell cycle as a series of events leading to cell division,
increased production, if not doubling, of all components of cellular
metabolism must occur during the cell cycle for distribution to the progeny at cytokinesis. Here the following questions can be posed: (1) Is
biogenesis of a given organelle restricted to a specific portion of the cycle?
(2) What nucleocytoplasmic interactions, if any, are involved in
biogenesis? (3) Is organellar biogenesis required for cell division?

1. Ribosomes

Eukaryotic ribosomes necessarily depend on the presence of nucleolar
rRNA as well as the assembly of intact ribosomes either in the nucleus or
cytoplasm or both. Ribosome biogenesis must also consider mitochondrial
and chloroplast ribosomes.

In *Euglena,* cytoplasmic RNA is distinguishable from mitochondrial and
chloroplast RNA by molecular weight. Cytoplasmic RNAs have molecular weights of 1.35×10^6 and $0.64–0.85 \times 10^6$ (Brown and Haselkorn,
1971; Avadhani and Buetow, 1972).

Mitochondrial RNAs are of 0.93×10^6 and 0.43×10^6 MW (Avadhani
et al., 1975), and mature chloroplast RNAs are of 1.1 and 0.55×10^6 MW
(Brown and Haselkorn, 1971). Based on the time course of labelling, it has
been shown that cytoplasmic RNAs are derived from a common precursor
with a molecular weight of 3.5×10^6. An intermediate in the processing of
the 1.35×10^6 MW species has been demonstrated and has a molecular
weight of 2.2×10^6. Both the precursor and this intermediate hybridize
with DNA and can be displaced from hybrids by mature RNA (Brown and
Haselkorn, 1971). Thus, the processing of RNA seen here is not different
from that in most eukaryotic organisms.

In *C. reinhardii,* synchronized by repetitive light–dark cycles, there are
significant differences in the times of appearance of cytoplasmic and
chloroplast rRNA. Approximately 90% of cytoplasmic and chloroplast
RNA is synthesized in G_1. Cytoplasmic species increase during G_1 and
remain constant during S. The various studies do not agree on whether the
RNA is degraded after S phase. The ratio of chloroplast RNA to cytoplasmic RNA was maximal during the first two-thirds of G_1. These differences suggested independent synthesis of the two species (Cattolico *et al.,*
1973; Wilson and Chiang, 1977). ^{32}P incorporation into RNA and ribo-

somes were parallel, suggesting that there was no lag between RNA synthesis and utilization (Wilson and Chiang, 1977). Changes in the ribosomal DNA content during the cell cycle have been shown by determining the amount of DNA hybridizable with rRNA. Peaks of hybridization were seen at the beginning and end of G_1 (beginning and end of the dark period). When corrected for total cellular DNA, DNA per cell rose continually in G_1. The minimum estimates of 25 and 18 S RNA cistrons per cell are 90 and 140, respectively. Before cell division these values increased to 260 and 440, respectively. Pulse–chase experiments indicated that much of the RNA was degraded during the cell cycle (Howell, 1972).

2. Chloroplast

In addition to studies on chloroplast ribosome biogenesis, the synthesis of chloroplast membrane proteins has been followed in *C. reinhardii*. With the use of light–dark cycles of 12 hr of each, most proteins were synthesized between 4 and 10 hr of the light period. Photosystem I proteins appeared as follows: 0–1 hr—a peak with a 55,000 MW protein was observed; 5–6 hr—this peak continued to incorporate radioactive amino acids rapidly, falling off to a lower average rate of synthesis between 7 and 8 hr. No incorporation occurred after 10 hr. Three photosystem II proteins were distinguished: a 40,000 MW species was labelled rapidly beginning at 0–1 hr and showed higher peaks of synthesis at 7 and 9 hr; a 31,000 MW species was continuously labelled, but showed even higher rates at 5–6 hr; and a 27,000 MW protein attained a high rate of synthesis from 4 hr. Thus there is an asynchronous rate of synthesis of membrane proteins (Beck and Levine, 1974). Chloroplast and nuclear DNA were replicated at different times during the cell cycle: at 3–6 hr and 15–18 hr, respectively (Chiang and Sueoka, 1967). In both cases replication was semiconservative. Chlorophyll synthesis occurred between 6 and 12 hr. Other chloroplast markers were defined in the first half of the cycle: chloramphenicol sensitivity in the first quarter, and increased carboxylase and cytochrome b_{563} in the second quarter (Howell, 1974).

With a light–dark cycle of 12 and 4 hr, respectively, variations in the carotenoid content have been noted. All major carotenoids increase at a similar rate for most of the light period, although increases in lutein and violaxanthin precede those of β-carotene, neoxanthin, loroxanthin, and algal xanthophyll. All the carotenoids decline at 9 hr, corresponding to a decline in RNA accumulation and nucleolar disintegration (Francis *et al.*, 1975).

In *Euglena* synchronized by cycles of 14 and 10 hr of light and dark, respectively, incorporation of [³H]adenine into cytoplasmic DNA occurred at all cell ages, but increased fourfold during the light period and declined to minimal levels in the dark. There were two peaks of cytoplasmic incorporation, the second at the time of bulk DNA synthesis and the first completed before nuclear synthesis. Since mitochondrial DNA constitutes 0.5% of the total DNA and chloroplast DNA is 3% of the total DNA, it is suggested that the first peak represents chloroplast DNA synthesis (Cook, 1966a).

At the beginning of the light period, pyrenoids were not evident, although dense areas were seen in the stroma. After 3 hr of illumination, these dense areas had the appearance of pyrenoids. By 7.5 hr, pyrenoids were fully developed. During the light period, chloroplast lamellae were compactly arranged. In the dark, chloroplasts swelled and lamellae separated. This appearance persisted until 2 hr into the light period (Cook *et al.*, 1976).

In *Chlamydomonas* and *Euglena* the relationship between chloroplast-specific syntheses and cell division has been examined. Several cell cycle mutants of *C. reinhardii* have been defined in the second quarter of the cycle, suggesting that these events must occur in order to complete the cell cycle (Howell, 1974). In *Euglena,* there is evidence that chloroplast division is more tightly synchronized than cell division. This suggests that cell division does not regulate chloroplast division and that the latter is not the trigger for cell division. This is not to suggest a complete independence between chloroplast and cell division, for if the dark period is lengthened, both processes adapt within the same cycle. Furthermore, when streptomycin was added at the beginning of the light period, it had little effect on the first cell division, but both cell and chloroplast divisions were inhibited in the subsequent cell cycle (Boasson and Gibbs, 1973).

3. Mitochondria

The timing of increases in the respiratory rate has often been taken as a measure of increase in mitochondrial components. Mitochondrial biosynthesis and/or division have been proposed to explain the doubling of the Q_{O_2} in temperature-synchronized *A. longa* and *P. agilis* (James, 1965; Cantor and Klotz, 1971). Although division is not precluded, such doublings may be artefacts of the entrainment procedure. Changes in respiration or biochemical activity may reflect a regulation of enzyme activity, not biosynthesis. In *Polytomella* sp., synchronized by thiamine deprivation and replacement or hydroxyurea, oscillations in cell respiration occurred (Cantor, unpublished observations). Respiration maxima were observed in *Euglena* grown on lactic acid or glutamine plus malate and synchronized by transfer to fresh medium. These peaks occurred at the beginning and end of the nondividing portion of the cycle in lactate medium, and at the middle of the nondividing and division periods in glutamine plus malate medium. In the former case, an increase in respiration correlated with the electron microscopic demonstration of a "giant mitochondrion." No changes in mitochondrial structure were observed in the latter medium (Calvayrac, 1970). If changes in respiration reflected biosynthetic or morphogenetic activity, only one change would be expected per cycle and the form of change would be independent of the synchronizing procedure.

Mitochondrial DNA replication has been studied in synchronous populations of *E. gracilis* strains Z and B8 (an apoplastid strain). In strain Z, there was a major peak of [³H]adenine incorporation into nuclear DNA just before and at the beginning of cell division; this peak was always accompanied by a satellite peak. The satellite was also present in strain B8. When cycloheximide was added 10 min before the tritiated adenine, nuclear labelling was inhibited. However, a major peak of incorporation into mitochondrial DNA was noted in the first part of the cell division phase of the cycle, indicating that the two syntheses occurred at different times (Calvayrac *et al.*, 1972).

Electron microscopy during the cell cycle suggests division of existing mitochondria as the mode of organelle biogenesis. This division may also be accompanied by mitochondrial fusion. In *Chlamydomonas* (light–dark cycles of 12 hr of each), after 6 hr of light, mitochondria were seen in small groups. By 8 hr the mitochondria had a branched, ring, or irregular shape, and by 10 hr appeared as singular giant structures. At 12 hr mitochondrial profiles were scattered in the cytoplasm. Respiration increased during the light period and showed peaks at 6 and 9 hr; it remained constant during the dark period (Osafune *et al.*, 1972a). These studies were based on random sections through cells and were not accompanied by stereological analysis. From serial sections, highly branched mitochondria were demonstrated in gametes of *Chlamydomonas*, although one to four nonreticulated structures were also seen (Grobe and Arnold, 1975).

Giant mitochondria have been described in synchronous cultures of *Euglena* and probably result from fusion. The giant organelle appeared in the middle of the light period, and its presence correlated with decreased oxygen consumption. The Q_{O_2} then increased continuously into the dark period. Shortly after its formation the giant mitochondrion broke up into smaller bodies. This suggests a mitochondrial cycle of fusion and division (Osafune, 1973; Osafune *et al.*, 1975). Nondividing cells generally have reticulated mitochondria that divide at the time of cell division (Calvayrac and Lefort-Tran, 1976; Calvayrac *et al.*, 1974).

In both *P. papillatum* and *P. agilis,* serial reconstruction of single cells indicates one large, reticulated mitochondrion and possibly one or more smaller mitochondrion, suggesting the possibility of mitochondrial fusion and division (Gaffal and Kreutzer, 1977; Burton and Moore, 1974). In

Polytoma, newly divided cells have many (30 to 47) mitochondria of various shapes and sizes. Since neither of these studies involved synchronous populations, it is impossible to relate these observations to cell cycle events.

E. Enzyme Patterns during the Cell Cycle

The various types of enzyme patterns occurring during the cell cycle include the following: *continuous*—either linear with an abrupt doubling of the rate at some time during the cycle, or exponential; *step*—showing constant activity until a given time in the cycle, when it doubles and again becomes constant; *peak*—displaying a constant rate until given time(s) when a peak of activity is apparent. Step enzymes are usually stable. Peak enzymes are frequently unstable, and the decline in activity may be due to either instability or degradation. Another parameter considered is the enzyme potential or induction capacity, i.e., the ability of an enzyme to be induced or derepressed. This also shows continuous or periodic changes. The interpretations of the various patterns have been based on sequential reading of the genome, timing of gene replication, and variations in the concentrations of regulatory molecules. The last-mentioned has been applied to oscillatory (peak) patterns of activity.

Periodic—especially peak—patterns may also be interpreted in terms of the regulation of enzyme activity rather than concentration. The oscillations in NADH dehydrogenase in yeast may be due to changes in the concentrations of cellular NADH (Chance *et al.,* 1965). Respiration rates in *Crithidia fasciculata* may correlate with concentrations of the various adenine nucleotides (Edwards *et al.,* 1975). Regulation may be in terms of enzyme assembly from its constituent subunits. In the case of membrane-bound enzymes regulation may reflect integration of the enzymes into the membrane; this is especially true of mitochondrial or chloroplast enzymes.

Phosphatase synthesis has been studied in light-synchronized cultures of *Chlamydomonas* (12 hr light and 4 hr dark). Repressed acid phosphatase increased continuously during the light period and showed maximal activity between 10 and 12 hr (S phase). When phosphate was removed from the medium, there was a 2-hr lag before synthesis. Phosphatase synthesis was faster in derepressed cells, with maximum rates seen between 3 and 5 hr (G_1 phase). Phosphatase was derepressible at all times during the cell cycle (Lien and Knutsen, 1972). Since intracellular phosphate declined during the first 2.5 hr of derepression, it is suggested that phosphate is a corepressor. When phosphate was added to derepressed cells, synthesis was slower for 1 hr and remained constant before continuing its increase, suggesting mRNA stability. The enzyme itself was unstable for cycloheximide-decreased activity (Lien and Knutsen, 1973b).

Derepression of arylsulfatase in *Chlamydomonas* occurred at sulfate concentrations in the medium of $<5–10 \times 10^{-5} M$. It was derepressible at all times in the cell cycle regardless of whether light cycle induction synchrony or selection synchrony with a Rastgeldi centrifuge (Knutsen *et al.*, 1973) was employed. Unlike phosphatases, mRNA was unstable and the enzyme was stable. The addition of cycloheximide to derepressed cells halted the increase in activity, and a constant rate of catalysis ensued (Schreiner *et al.*, 1975).

The induction capacity of the enzymes of the glyoxylate bypass has been studied in synchronized *Euglena* (14 hr light and 10 hr dark). When acetate was added at 10 hr, there was a 1-hr lag, and then a linear increase in the activity of acetate thiokinase, malate synthase, and isocitrate lyase. They were inducible continuously over the entire cell cycle, but there was a step increase in the rate of induction from 10 to 14 hr corresponding to the timing of DNA accumulation. Cycloheximide completely inhibited induction, and *p*-fluorophenylalanine partly inhibited, when the inhibitors were added at 17 hr. Inhibitors of RNA synthesis (6-methylpurine, 5-fluorouracil, and actinomycin D) inhibited cell division but not induction unless cells were preincubated with the inhibitor (6-methylpurine) for 12 hr. Since induction was continuous throughout the cycle, sequential transcription could not explain the results. Thus, regulation seems explicable by gene duplication (Woodward and Merrett, 1975).

Astasia longa, synchronized by repetitive temperature cycles, shows a peak in oxygen consumption just before cell division (James, 1965). Pulse-labelling with radioactive sulfate and determination of its incorporation into mitochondrial proteins indicated that the rate of mitochondrial protein synthesis was also highest before cell division. There was also a correlation between mitochondrial and whole-cell protein synthesis. This suggests that mitochondrial synthesis occurs continuously through the cell cycle except for the peak of activity at the beginning of prophase (Kahn and Blum, 1967).

The activity of succinic dehydrogenase, NADH dehydrogenase, and cytochrome *c* oxidase has been examined in *P. agilis* synchronized by hydroxyurea. All these enzymes had peak patterns during the cell cycle. When chloramphenicol was added at the beginning of the first cycle, cell division was not inhibited, but only 50% of the cells divided at the end of the second cell cycle. This inhibitor of mitochondrial protein synthesis did not affect the timing of the peaks of succinate and NADH dehydrogenase activity, although their activity was reduced by approximately one-half.

In the first cell cycle, chloramphenicol had no effect on the magnitude of cytochrome oxidase activity, although the peaks were displaced to earlier times. There was a decrease in activity in the second cell cycle. Cycloheximide, an inhibitor of cytoribosomal protein synthesis, inhibited cell division completely. It had no effect on the magnitude or timing of the peaks of succinate or cytochrome activity. It resulted in the abolition of one peak of NADH dehydrogenase and a diminution of the second peak. In the first cell cycle, these data were attributed to activation of the enzymes or to their assembly and integration into mitochondrial membranes (Patterson, 1977).

REFERENCES

Aaronson, S., and Bensky, B. (1962). *J. Gen. Microbiol.* **27**, 75–88.
Aaronson, S., and Scher, S. (1960). *J. Protozool.* **7**, 156–158.
Anderson, E. (1962). *J. Protozool.* **9**, 380–395.
Aprille, J. R., and Buetow, D. E. (1973). *Arch. Microbiol.* **89**, 355–360.
Aprille, J. R., and Buetow, D. E. (1974). *Arch. Microbiol.* **97**, 195–201.
Arnold, C. G., Schimmer, O., Schötz, F., and Bathelt, H. (1972). *Arch. Microbiol.* **81**, 50–67.
Avadhani, N. G., and Buetow, D. E. (1972). *Biochem. J.* **128**, 353–365.
Avadhani, N. G., Lewis, F. S., and Rutman, R. J. (1975). *Sub-Cell. Biochem.* **4**, 93–145.
Baker, W. B., and Buetow, D. E. (1976). *J. Protozool.* **23**, 167–176.
Barea, J. L., and Cardenas, J. (1975). *Arch. Microbiol.* **105**, 21–25.
Barker, S. A., and Bourne, E. J. (1955). *In* "Biochemistry and Physiology of Protozoa" (S. H. Hutner and A. Lwoff, eds.), Vol. 2, pp. 45–56. Academic Press, New York.
Barras, D. R., and Stone, B. A. (1968). *In* "The Biology of *Euglena*" (D. E. Buetow, ed.), Vol. 2, pp. 149–191. Academic Press, New York.
Barras, D. R., and Stone, B. A. (1969a). *Biochim. Biophys. Acta* **191**, 329–341.
Barras, D. R., and Stone, B. A. (1969b). *Biochim. Biophys. Acta* **191**, 342–353.
Barry, S. C. (1962). *J. Protozool.* **9**, 395–400.
Beck, D. P., and Levine, R. P. (1974). *J. Cell Biol.* **63**, 759–772.
Beezley, B. B., Gruber, P. J., and Frederick, S. E. (1976). *Plant Physiol.* **58**, 315–319.
Bégin-Heick, N. (1973). *Biochem. J.* **134**, 607–616.
Bégin-Heick, N., and Blum, J. J. (1967). *Biochem. J.* **105**, 813–819.
Belsky, M. M., and Schultz, J. J. (1962). *J. Protozool.* **9**, 195–200.
Bernstein, E. (1968). *Methods Cell Physiol.* **3**, 119–145.
Bertaux, O., and Valencia, R. (1973). *C. R. Acad. Sci., Ser. D* **276**, 753–756.
Bertaux, O., Frayssinet, C., and Valencia, R. (1976). *C. R. Acad. Sci., Ser. D* **282**, 1293–1296.
Blum, J. J. (1966). *J. Gen. Physiol.* **49**, 1125–1137.
Blum, J. J., and Bégin-Heick, N. (1967). *Biochem. J.* **105**, 821–829.
Blum, J. J., and Buetow, D. E. (1963). *Exp. Cell Res.* **29**, 407–421.
Boasson, R., and Gibbs, S. P. (1973). *Planta* **115**, 125–134.
Brandes, D., Buetow, D. E., Bertini, F., and Malkoff, D. B. (1964). *Exp. Mol. Pathol.* **3**, 583–609.
Brawerman, G. (1968). *In* "The Biology of *Euglena*" (D. E. Buetow, ed.), Vol. 3, pp. 97–131. Academic Press, New York.

Bré, M. H., and Lefort-Tran, M. (1974). *C. R. Acad. Sci., Ser. D* **278**, 1349–1352.
Bré, M. H., Diamond, J., and Jacques, R. (1975). *J. Protozool.* **22**, 432–434.
Brody, M., and White, J. E. (1972). *FEBS Lett.* **23**, 149–152.
Brody, M., and White, J. E. (1973). *Dev. Biol.* **31**, 348–361.
Brody, M., and White, J. E. (1974). *FEBS Lett.* **40**, 325–330.
Brown, D. L., Leppard, G. G., and Massalski, A. (1976a). *Protoplasma* **90**, 139–154.
Brown, D. L., Massalski, A., and Leppard, G. G. (1976b). *Protoplasma* **90**, 155–171.
Brown, G. E., and Preston, J. F. (1975). *Arch. Microbiol.* **104**, 233–236.
Brown, R. D., and Haselkorn, R. (1971). *J. Mol. Biol.* **59**, 491–503.
Brown, R. M., Johnson, C., and Bold, H. C. (1968). *J. Phycol.* **4**, 100–120.
Buetow, D. E. (1965). *J. Cell. Comp. Physiol.* **66**, 235–242.
Buetow, D. E. (1968). *In* "The Biology of *Euglena*" (D. E. Buetow, ed.), Vol. 1, pp. 109–184. Academic Press, New York.
Buetow, D. E. (1970). *Methods Cell Physiol.* **4**, 83–115.
Buetow, D. E. (1974). *Methods Cell Physiol.* **13**, 283–311.
Buetow, D. E., and Buchanan, P. J. (1964). *Exp. Cell Res.* **36**, 204–207.
Buetow, D. E., and Buchanan, P. J. (1965). *Biochim. Biophys. Acta* **96**, 9–17.
Buetow, D. E., and Buchanan, P. J. (1969). *Life Sci.* **8**, 1099–1102.
Buetow, D. E., and Mego, J. L. (1967). *Biochim. Biophys. Acta* **134**, 395–401.
Buetow, D. E., and Padilla, G. M. (1963). *J. Protozool.* **10**, 121–123.
Buetow, D. E., and Schuitt, K. E. (1968). *J. Protozool.* **15**, 770–773.
Burton, M. D., and Moore, J. (1974). *J. Ultrastruct. Res.* **48**, 414–419.
Calvayrac, R. (1970). *Arch. Microbiol.* **73**, 308–314.
Calvayrac, R., and Butow, R. A. (1971). *Arch. Microbiol.* **80**, 62–69.
Calvayrac, R., and Lefort-Tran, M. (1976). *Protoplasma* **89**, 353–358.
Calvayrac, R., Van Lente, F., and Butow, R. A. (1971). *Science* **173**, 252–254.
Calvayrac, R., Butow, R. A., and Lefort-Tran, M. (1972). *Exp. Cell Res.* **71**, 422–432.
Calvayrac, R., Bertaux, O., Lefort-Tran, M., and Valencia R. (1974). *Protoplasma* **80**, 355–370.
Cantor, M. H., and Burton, M. D. (1974). *J. Protozool.* **21**, 420–421.
Cantor, M. H., and Burton, M. D. (1975). *J. Protozool.* **22**, 135–139.
Cantor, M. H., and James, T. W. (1965). *J. Cell. Comp. Physiol.* **65**, 285–292.
Cantor, M. H., and Klotz, J. (1971). *Experientia* **27**, 801–803.
Carell, E. F. (1969). *J. Cell Biol.* **41**, 431–440.
Cartledge, T. G., Cooper, R. A., and Lloyd, D. (1971). *In* "Separations with Zonal Rotors" (E. Reid, ed.), pp. V4.1–V4.16. Longman Group Ltd., London.
Cattolico, R. A., Senner, J. W., and Jones, R. F. (1973). *Arch. Biochem. Biophys.* **156**, 58–65.
Chance, B., and Sager, R. (1957). *Plant Physiol.* **32**, 548–561.
Chance, B., Schoener, B., and Elsaesser, S. (1965). *J. Biol. Chem.* **240**, 3170–3181.
Chapman, L. F., Cirillo, V. P., and Jahn, T. L. (1965). *J. Protozool.* **12**, 47–51.
Cheng, K. H., and Colman, B. (1974). *Planta* **115**, 207–211.
Chiang, K.-S., and Sueoka, N. (1967). *Proc. Natl. Acad. Sci. U.S.A.* **57**, 1506–1513.
Chiang, K.-S., Eves, E., and Swinton, D. (1975). *Dev. Biol.* **42**, 53–63.
Christopher, A. R., Dobrosielski-Vergona, K., Goetz, G., Johnson, P. L., and Carell, E. F. (1974). *Exp. Cell Res.* **89**, 71–78.
Cirillo, V. P. (1955). *Proc. Soc. Exp. Biol. Med.* **88**, 352–354.
Cirillo, V. P. (1956). *J. Protozool.* **3**, 69–74.
Cirillo, V. P. (1957). *J. Protozool.* **4**, 60–62.
Coakley, W. T., Bater, A. J., and Lloyd, D. (1977). *Adv. Microbial Physiol.* **16**, 279–341.
Codd, G. A., Lord, J. M., and Merrett, M. J. (1969). *FEBS Lett.* **5**, 341–342.

Cohen, D., and Parnas, H. (1976). *J. Theor. Biol.* **56**, 1–18.
Collins, N., and Merrett, M. J. (1975). *Biochem. J.* **148**, 321–328.
Collins, N., Brown, R. H., and Merrett, M. J. (1975). *Biochem. J.* **150**, 373–377.
Cook, J. R. (1966a). *J. Cell Biol.* **29**, 369–373.
Cook, J. R. (1966b). *Biol. Bull. (Woods Hole, Mass.)* **131**, 83–93.
Cook, J. R. (1968). *J. Cell. Physiol.* **71**, 177–184.
Cook, J. R. (1973). *J. Gen. Microbiol.* **75**, 51–60.
Cook, J. R., and Cook, B. (1962). *Exp. Cell Res.* **28**, 524–530.
Cook, J. R., and Heinrich, B. (1965). *J. Protozool.* **12**, 581–584.
Cook, J. R., and Hunt, W. (1965). *Photochem. Photobiol.* **4**, 877–880.
Cook, J. R., and James, T. W. (1960). *Exp. Cell Res.* **21**, 583–589.
Cook, J. R., Haggard, S. S., and Harris, P. (1976). *J. Protozool.* **23**, 368–373.
Cooper, R. A., and Lloyd, D. (1972). *J. Gen. Microbiol.* **72**, 59–70.
Cooper, R. A., Bowen, I. D., and Lloyd, D. (1974). *J. Cell Sci.* **15**, 605–618.
Cosgrove, W. B., and Swanson, B. K. (1952). *Physiol. Zool.* **25**, 287–292.
Cramer, M., and Myers, J. (1952). *Arch. Mikrobiol.* **17**, 384–402.
Danforth, W. F. (1953). *Arch. Biochem. Biophys.* **46**, 164–173.
Danforth, W. F. (1967). *In* "Research in Protozoology" (T. T. Chen, ed.), Vol. 1, pp. 201–306. Pergamon, London.
Danforth, W. F., and Wilson, B. W. (1957). *J. Protozool.* **4**, 52–55.
Datta, D. B., and Khan, J. S. (1977). *J. Protozool.* **24**, 187–192.
Davies, D. R., and Plaskitt, A. (1971). *Genet. Res.* **17**, 33–43.
deDuve, C. (1971). *J. Cell Biol.* **50**, 20D–55D.
Dinello, R. K., and Ernst-Fonberg, M. L. (1975). *In* "Lipids," Part B (J. M. Lowenstein, ed.), Methods in Enzymology, Vol. 35, pp. 110–114. Academic Press, New York.
Dogiel, V. (1935). *Arch. Zool. Exp. Gen.* **77**, 1–8.
Droop, M. R. (1974). *In* "Algal Physiology and Biochemistry" (W. D. P. Stewart, ed.), pp. 530–559. Blackwell, Oxford.
Ebringer, L. (1962). *J. Protozool.* **9**, 373–374.
Ebringer, L. (1964). *Folia Microbiol. (Prague)* **9**, 249–255.
Edelman, M., Epstein, H. T., and Schiff, J. A. (1966). *J. Mol. Biol.* **17**, 463.
Edmunds, L. N., Jr. (1964). *Science* **145**, 266–268.
Edmunds, L. N., Jr., Jay, M. E., Kohlmann, A., Liu, S. C., Merriam, V. H., and Sternberg, H. (1976). *Arch. Microbiol.* **108**, 1–8.
Edwards, C., Statham, M., and Lloyd, D. (1975). *J. Gen. Microbiol.* **88**, 141–152.
Egan, J. M., Jr., and Carell, E. F. (1972). *Plant Physiol.* **50**, 391–395.
Epstein, H. T., and Allaway, E. (1967). *Biochim. Biophys. Acta.* **142**, 195–207.
Epstein, S. S., Weiss, J. B., Causeley, D., and Bush, P. (1962). *J. Protozool.* **9**, 336–339.
Ernst-Fonberg, M. L., Dubinskas, F., and Jonak, Z. L. (1974). *Arch. Biochem. Biophys.* **165**, 646–655.
Eshelman, J. N., and Danforth, W. F. (1964). *J. Protozool.* **11**, 394–399.
Evers, A., and Ernst-Fonberg, M. L. (1974). *FEBS Lett.* **46**, 233–235.
Falchuk, K. H., Fawcett, D. W., and Vallee, B. L. (1975a). *J. Cell Sci.* **17**, 57–78.
Falchuk, K. H., Krishan, A., and Vallee, B. L. (1975b). *Biochemistry* **14**, 3439–3444.
Falchuk, K. H., Mazus, B., Ulpino, L., and Vallee, B. L. (1976). *Biochemistry* **15**, 4468–4475.
Ferrara, R., and Banchetti, R. (1976). *J. Exp. Zool.* **198**, 393–402.
Fonty, G., Grouse, E. J., Stutz, E., and Bernardi, G. (1975). *Eur. J. Biochem.* **54**, 367–372.
Francis, G. W., Strand, L. P., Lien, T., and Knutsen, G. (1975). *Arch. Microbiol.* **104**, 249–254.
Gaffal, K. P., and Kreutzer, D. (1977). *Protoplasma* **91**, 167–177.

Gerhardt, B. (1971). *Arch. Microbiol.* **80**, 205–218.
Gerhardt, B., and Berger, C. (1971). *Planta* **100**, 155–166.
Gibor, A. (1969). *J. Protozool.* **16**, 190–193.
Gibor, A., and Granick, S. (1962a). *J. Cell Biol.* **15**, 599–603.
Gibor, A., and Granick, S. (1962b). *J. Protozool.* **9**, 327–334.
Gittleson, S. M., Alper, R. E., and Conti, S. F. (1969). *Life Sci.* **8**, 591–599.
Goetz, G. H., Johnson, P. L., Dobrosielski-Vergona, K., and Carell, E. F. (1974). *J. Cell Biol.* **62**, 672–678.
Goodwin, T. W., and Gross, J. A. (1958). *J. Protozool.* **5**, 292–295.
Graham, D., and Smillie, R. M. (1971). *In* "Photosynthesis and Nitrogen Fixation," Part A (A. San Pietro, ed.), Methods in Enzymology, Vol. 23, pp. 228–241. Academic Press, New York.
Graves, L. B., Jr. (1971). *J. Protozool.* **18**, 543–546.
Graves, L. B., Jr., and Becker, W. M. (1974). *J. Protozool.* **21**, 771–774.
Graves, L. B., Jr., Trelease, R. N., and Becker, W. M. (1971). *Biochem. Biophys. Res. Commun.* **44**, 280–286.
Graves, L. B., Jr., Trelease, R. N., Grill, A., and Becker, W. (1972). *J. Protozool.* **19**, 527–532.
Greenblatt, C. L., and Sharpless, N. E. (1959). *J. Protozool.* **6**, 241–248.
Grobe, B., and Arnold, C. G. (1975). *Protoplasma* **68**, 291–294.
Gross, J. A. (1962). *J. Protozool.* **9**, 415–418.
Gross, J. A., Jahn, T. L., and Bernstein, E. (1955). *J. Protozool.* **2**, 71–75.
Gruel, D., Rawson, J. R. Y., and Haselkorn, R. (1975). *Biochim. Biophys. Acta* **414**, 20–29.
Gunnison, D., and Alexander, M. (1975). *Can. J. Microbiol.* **21**, 619–628.
Haapala, O. K., and Soyer, M. O. (1975). *Hereditas* **80**, 185–194.
Haigh, W. G., and Beevers, H. (1964a). *Arch. Biochem. Biophys.* **107**, 147–151.
Haigh, W. G., and Beevers, H. (1964b). *Arch. Biochem. Biophys.* **107**, 152–157.
Hall, R. P. (1967). *J. Protozool.* **14**, 164–167.
Hartwell, L. H. (1974). *Bacteriol. Rev.* **38**, 164–198.
Heinrich, B., and Cook, J. R. (1967). *J. Protozool.* **14**, 548–553.
Heizmann, P., Salvador, G. F., and Nigon, V. (1976). *Exp. Cell Res.* **99**, 253–260.
Hiyama, J., Nishimura, M., and Chance, B. (1969). *Plant Physiol.* **44**, 527–534.
Holowinsky, A. W., and Schiff, J. A. (1970). *Plant Physiol.* **45**, 339–347.
Holz, G. G. (1954). *J. Protozool.* **1**, 114–120.
Hommersand, H., and Thimann, K. V. (1965). *Plant Physiol.* **40**, 1220–1227.
Howell, S. H. (1972). *Nature (London), New Biol.* **240**, 264–267.
Howell, S. H. (1974). *In* "Cell Cycle Controls" (G. M. Padilla, I. L. Cameron, and A. Zimmerman, eds.), pp. 235–249. Academic Press, New York.
Howell, S. H., and Walker, L. L. (1976). *Biochim. Biophys. Acta* **418**, 249–256.
Howell, S. H., Blaschko, W. J., and Drew, C. M. (1975). *J. Cell Biol.* **67**, 126–135.
Hudock, G. A., Gring, D. M., and Bart, C. (1971). *J. Protozool.* **18**, 128–131.
Hughes, D. E., Wimpenny, J. W. T., and Lloyd, D. (1971). *In* "Methods in Microbiology" (J. R. Norris and D. W. Ribbons, ed.), Vol. 5B, pp. 1–54. Academic Press, New York.
Hunter, F. R., and Lees, J. W. (1962). *J. Protozool.* **9**, 74–78.
Hurlbert, R. E., and Rittenberg, S. C. (1962). *J. Protozool.* **9**, 170–182.
Hutner, S. H., and Provasoli, L. (1951). *In* "Biochemistry and Physiology of the Protozoa" (A. Lwoff, ed.), Vol. 1, pp. 27–128. Academic Press, New York.
Hutner, S. H., Zahalsky, A. C., and Aaronson, S. (1968). *In* "The Biology of *Euglena*" (D. E. Buetow, ed.), Vol. 2, pp. 193–214. Academic Press, New York.
Huth, W. (1967). *Flora (Jena)* **158**, 58–87.

Hyams, J., and Davies, D. R. (1972). *Mutat. Res.* **14**, 381–387.
James, T. W. (1964). *In* "Synchrony in Cell Division and Growth" (E. Zeuthen, ed.), pp. 323–349. Wiley (Interscience), New York.
James, T. W. (1965). *Exp. Cell Res.* **38**, 439–453.
James, T. W., and Padilla, G. M. (1959). *Proc. Natl. Biophys. Conf., 1st, Columbus, Ohio, 1957* pp. 694–700.
Jeener, R. (1953). *Arch. Biochem.* **43**, 381–388.
Johnson, B. F. (1962). *Exp. Cell Res.* **28**, 419–423.
Johnson, B. F., and James, T. W. (1960). *Exp. Cell Res.* **20**, 66–70.
Johnson, U. G., and Porter, K. R. (1968). *J. Cell Biol.* **38**, 403–425.
Kahn, J. S. (1973). *Arch. Biochem. Biophys.* **159**, 646–650.
Kahn, J. S. (1974). *Arch. Biochem. Biophys.* **164**, 266–274.
Kahn, V., and Blum, J. J. (1967). *Biochemistry* **6**, 817–826.
Karlan, A. W., and Russell, G. K. (1976). *J. Protozool.* **23**, 176–179.
Karn, R. C., and Hudock, G. A. (1973). *J. Protozool.* **20**, 316–320.
Kates, J. R., and Jones, R. F. (1966). *J. Cell. Physiol.* **67**, 101–105.
Keirns, J. J., Carritt, B., Freeman, J., Eisenstadt, J. M., and Bitensky, M. W. (1973). *Life Sci.* **13**, 287–302.
Kempner, E. S., and Miller, J. H. (1965a). *Biochim. Biophys. Acta* **104**, 11–17.
Kempner, E. S., and Miller, J. H. (1965b). *Biochim. Biophys. Acta* **104**, 18–24.
Kempner, E. S., and Miller, J. H. (1972a). *J. Protozool.* **19**, 343–346.
Kempner, E. S., and Miller, J. H. (1972b). *J. Protozool.* **19**, 678–681.
Kempner, E. S., and Miller, J. H. (1974). *J. Protozool.* **21**, 363–367.
Khan, A. A., and Kolattukudy, D. E. (1973a). *Biochemistry* **12**, 1939–1948.
Khan, A. A., and Kolattukudy, D. E. (1973b). *Arch. Biochem. Biophys.* **158**, 411–420.
Khan, A. A., and Kolattukudy, D. E. (1975). *Arch. Biochem. Biophys.* **170**, 400–408.
Kislev, N., and Eisenstadt, J. M. (1972). *Eur. J. Biochem.* **31**, 226.
Kivic, P. A., and Vesk, M. (1974). *Arch. Microbiol.* **96**, 155–159.
Kivic, P. A., Bart, C., and Hudock, G. A. (1969). *J. Protozool.* **16**, 743–744.
Klein, R. M., Morselli, M. F., and Wansor, J. (1963). *J. Protozool.* **10**, 223–225.
Klein, S., Schiff, J. A., and Holowinsky, A. W. (1972). *Dev. Biol.* **28**, 253–273.
Knutsen, G., Lien, T., Schreiner, O., and Vaage, R. (1973). *Exp. Cell Res.* **81**, 26–30.
Kramer, M. S., and Hutchens, J. O. (1969). *J. Protozool.* **16**, 295–297.
Krawiec, S., and Eisenstadt, J. M. (1970a). *Biochim. Biophys. Acta* **217**, 120–131.
Krawiec, S., and Eisenstadt, J. M. (1970b). *Biochim. Biophys. Acta* **217**, 132–140.
Lang, N. J. (1963). *J. Protozool.* **10**, 333–339.
Leedale, G. F. (1967). "Euglenoid Flagellates." Prentice-Hall, Englewood Cliffs, New Jersey.
Leedale, G. F., and Buetow, D. E. (1970). *Cytobiologie* **1**, 195–202.
Leedale, G. F., and Buetow, D. E. (1976). *J. Microsc. Biol. Cell* **25**, 149–154.
Levedahl, B. H. (1965). *Exp. Cell Res.* **39**, 233–241.
Levedahl, B. H. (1968). *In* "The Biology of *Euglena*" (D. E. Buetow, ed.), Vol. 1, pp. 85–96. Academic Press, New York.
Lewin, J. C. (1950). *Science* **122**, 652–653.
Lewis, E., Munger, G., Watson, R., and Wise, D. (1974). *J. Protozool.* **21**, 647–649.
Lien, T., and Knutsen, G. (1972). *Biochim. Biophys. Acta* **287**, 154–163.
Lien, T., and Knutsen, G. (1973a). *Exp. Cell Res.* **78**, 79–88.
Lien, T., and Knutsen, G. (1973b). *Physiol. Plant.* **28**, 291–298.
Lien, T., and Knutsen, G. (1976). *Arch. Microbiol.* **108**, 189–194.
Linnane, A. W., and Stewart, P. R. (1967). *Biochem. Biophys. Res. Commun.* **27**, 511–516.

Lloyd, D. (1974). "The Mitochondria of Microorganisms," pp. 54–81. Academic Press, New York.
Lloyd, D., and Chance, B. (1968). *Biochem. J.* **107**, 829–837.
Lloyd, D., Evans, D. A., and Venables, S. E. (1968). *Biochem. J.* **109**, 897–907.
Lloyd, D., Evans, D. A., and Venables, S. E. (1970). *J. Gen. Microbiol.* **61**, 33–41.
Loppes, R., and Deltour, R. (1975). *Arch. Microbiol.* **103**, 247–250.
Lucksch, I. (1933). *Beih. Bot. Zentralbl.* **A50**, 64–94.
Lwoff, A. (1932). "Récherches Biochimiques sur la Nutrition des Protozoaires." Masson, Paris.
Lwoff, A. (1944). "L'Évolution Physiologique." Hermann, Paris.
Lyman, H., Epstein, T., and Schiff, J. (1959). *J. Protozool.* **6**, 264–265.
Lyman, H., Jupp, A. S., and Larrinua, I. (1975). *Plant Physiol.* **55**, 390–392.
Lynch, M. J., and Buetow, D. E. (1975). *Exp. Cell Res.* **91**, 344–348.
Lynch, M. J., Leake, R. E., O'Connell, K. M., and Buetow, D. E. (1975). *Exp. Cell Res.* **91**, 349–357.
McCalla, D. R. (1965). *J. Protozool.* **12**, 34–41.
McCalla, D. R., and Baerg, W. (1969). *J. Protozool.* **16**, 425–428.
McCalla, D. R., and Reuvers, A. (1970). *J. Protozool.* **17**, 129–134.
McCalla, D. R., and Voutsinos, D. (1975). *J. Protozool.* **22**, 130–134.
McCracken, D. A., and Badenhuizen, N. P. (1970). *Staerke* **22**, 289–291.
McLennan, A. G., and Keir, H. M. (1975a). *Biochem. Soc. Trans.* **3**, 652.
McLennan, A. G., and Keir, H. M. (1975b). *Biochem. J.* **151**, 227–238.
McLennan, A. G., and Keir, H. M. (1975c). *Biochem. J.* **151**, 238–247.
McLennan, A. G., and Keir, H. M. (1975d). *Nucleic Acids Res.* **2**, 223–237.
McMahon, D., and Langstroth, P. (1972). *J. Gen. Microbiol.* **73**, 239–250.
Malkoff, D. B., and Buetow, D. E. (1964). *Exp. Cell Res.* **35**, 58–68.
Mangat, B. S., and Badenhuizen, N. P. (1970). *Staerke* **22**, 329–333.
Mangat, B. S., and Badenhuizen, N. P. (1971). *Can. J. Bot.* **49**, 1787–1792.
Manning, J. E., Wolstenholme, D. R., Ryan, R. S., Hunter, J. A., and Richards, O. C. (1971). *Proc. Natl. Acad. Sci. U.S.A.* **68**, 1169–1173.
Matagne, R. F., and Loppes, R. (1975). *Genetics* **80**, 239–250.
Matagne, R. F., Loppes, R., and Deltour, R. (1976). *J. Bacteriol.* **126**, 937–950.
Mego, J. L. (1964). *Biochim. Biophys. Acta* **79**, 221–225.
Mego, J. L., and Farb, R. M. (1974). *Biochim. Biophys. Acta* **350**, 237–239.
Mesland, A. M. (1976). *Arch. Microbiol.* **109**, 31–35.
Michaels, A., and Gibor, A. (1973). *J. Cell Sci.* **13**, 799–809.
Miller, D. H., Mellman, I. S., Lamport, D. T., and Miller, M. (1974). *J. Cell Biol.* **63**, 420–429.
Miller, J. H., and Kempner, E. S. (1976). *J. Protozool.* **23**, 444–446.
Mitchell, J. L. A. (1971). *Planta* **100**, 244–257.
Mo, Y., Harris, B. G., and Gracy, R. W. (1973). *Arch. Biochem. Biophys.* **157**, 580–587.
Mohanty, M. K., Hunter, F. R., and Myers, J. B. (1977). *J. Protozool.* **24**, 335–340.
Moore, J., and Cushing, S. D. (1979). *J. Protozool.* (in press).
Moore, J., Cantor, M. H., Sheeler, P., and Kahn, W. (1970). *J. Protozool.* **17**, 671–676.
Moriber, L. G., Hershenov, B., Aaronson, S., and Bensky, B. (1963). *J. Protozool.* **10**, 80–86.
Morimoto, H., and James, T. W. (1969a). *Exp. Cell Res.* **58**, 55–61.
Morimoto, H., and James, T. W. (1969b). *Exp. Cell Res.* **58**, 195–200.
Morosoli, R., and Bégin-Heick, N. (1974). *Biochem. J.* **141**, 469–475.
Morosoli, R., and LaFontaine, J. G. (1976). *Exp. Cell Res.* **99**, 88–94.

Müller, M. (1969). *Ann. N.Y. Acad. Sci.* **168**, 292–301.
Müller, M. (1975). *Annu. Rev. Microbiol.* **29**, 467–483.
Murray, D. R., Giovanelli, J., and Smillie, R. M. (1970). *J. Protozool.* **17**, 99–104.
Nass, M. M. K., Schori, L., Ben-Shaul, Y., and Edelman, M. (1974). *Biochim. Biophys. Acta* **374**, 283–291.
Neal, W. K., Funkhouser, E. A., and Price, C. A. (1968). *J. Protozool.* **15**, 761–763.
Nelson, P. E., and Surzycki, S. J. (1976). *Eur. J. Biochem.* **61**, 475–480.
Nicholas, P., and Nigon, V. (1974). *FEBS Lett.* **49**, 254–259.
Ohad, I., Siekevitz, P., and Palade, G. E. (1967). *J. Cell Biol.* **35**, 521–552.
Ohmann, E., Rindt, K. P., and Borriss, R. (1969). *Z. Allg. Mikrobiol.* **9**, 557–564.
Olson, C. B., and Buetow, D. E. (1977). *Int. Congr. Protozool., 5th, New York* Abstr. No. 109.
Ophir, I., Talmon, A., Polak-Charcon, S., and Ben-Shaul, Y. (1975). *Protoplasma* **84**, 283–295.
Osafune, T. (1973). *J. Electron Microsc. (Tokyo)* **22**, 51–61.
Osafune, T. (1975). *J. Electron Microsc. (Tokyo)* **22**, 51–55.
Osafune, T., Mihara, S., Hase, E., and Ohkuro, I. (1972a). *Plant Cell Physiol.* **13**, 211–227.
Osafune, T., Mihara, S., Hase, E., and Ohkuro, I. (1972b). *Plant Cell Physiol.* **13**, 981–989.
Osafune, T., Mihara, S., Hase, E., and Ohkuro, I. (1975). *J. Electron Microsc.* **24**, 33–39.
Padilla, G. M., and Blum, J. J. (1963). *Exp. Cell Res.* **32**, 289–304.
Padilla, G. M., and Cook, J. R. (1964). *In* "Synchrony in Cell Division and Growth" (E. Zeuthen, ed.), pp. 521–535. Wiley (Interscience), New York.
Padilla, G. M., and James, T. W. (1960). *Exp. Cell Res.* **20**, 401–415.
Parenti, F., DiPierro, S., and Perrone, C. (1972). *J. Protozool.* **19**, 524–527.
Parenti, J., Brawerman, G., Preston, J. F., and Eisenstadt, J. M. (1969). *Biochim. Biophys. Acta* **195**, 234–242.
Patterson, N. D. (1977). M.S. Thesis, California State Univ., Northridge.
Peak, J. G., and Peak, M. J. (1976). *J. Protozool.* **23**, 165–167.
Peak, J. G., and Peak, M. J. (1977). *J. Protozool.* **24**, 441–444.
Pellegrini, M., and Pellegrini, L. (1976). *C. R. Acad. Sci., Ser. D* **282**, 357–360.
Perini, F., Schiff, J. A., and Kamen, M. D. (1964). *Biochim. Biophys. Acta* **88**, 91–98.
Perl, M. (1974). *J. Biochem. (Tokyo)* **76**, 1095–1101.
Pienkos, P., Walfield, A., and Hershberger, C. L. (1974). *Arch. Biochem. Biophys.* **165**, 548–553.
Prask, J. A., and Plocke, D. J. (1971). *Plant Physiol.* **48**, 150–155.
Pringsheim, E. G. (1921). *Ber Dtsch. Bot. Ges.* **38**, 8–9.
Pringsheim, E. G. (1937). *Planta* **27**, 69–72.
Provasoli, L. (1938). *Boll. Zool. Agrar. Bachic.* **9**, 1–124.
Provasoli, L., Hutner, S. H., and Schatz, A. (1948). *Proc. Soc. Exp. Biol. Med.* **69**, 279–282.
Rahn, O. (1941). *Growth* **5**, 197–199.
Raison, J. K., and Smillie, R. M. (1969). *Biochim. Biophys. Acta* **180**, 500–508.
Rawson, J. R. Y. (1975). *Biochim. Biophys. Acta* **402**, 171–178.
Rawson, J. R. Y., and Boerma, C. (1976). *Proc. Natl. Acad. Sci. U.S.A.* **73**, 2401–2404.
Rawson, J. R. Y., and Haselkorn, R. (1973). *J. Mol. Biol.* **77**, 125–132.
Reed, R. B., Simone, F. P., and McGrath, M. S. (1976). *J. Gen. Microbiol.* **97**, 29–34.
Richards, O. C., and Ryan, R. S. (1974). *J. Mol. Biol.* **82**, 57–75.
Richards, O. C., Ryan, R. S., and Manning, J. E. (1971). *Biochim. Biophys. Acta* **238**, 190–201.
Ringo, D. L. (1963). *J. Protozool.* **10**, 167–173.
Robreau, G., and LeGal, Y. (1975). *Biochimie* **57**, 703–710.

Rosen, W. G., and Gawlik, S. R. (1961). *J. Protozool.* **8**, 90–96.
Rossetti, M. V., deGeralnick, A. A. J., and Batlle, A. M. C. (1977). *Int. J. Biochem.* **8**, 781–787.
Sager, R., and Granick, S. (1953). *Ann. N.Y. Acad. Sci.* **56**, 831–838.
Sato, C. (1976). *Exp. Cell Res.* **101**, 251–259.
Schimmer, O., and Werner, R. (1974). *Mutat. Res.* **26**, 423–425.
Schlosser, U. G., Sachs, H., and Robinson, D. G. (1976). *Protoplasma* **88**, 51–64.
Schötz, F., Bathelt, H., Arnold, C.-G., and Schimmer, O. (1972). *Protoplasma* **75**, 229–240.
Schreiner, O., Lien, T., and Knutsen, G. (1975). *Biochim. Biophys. Acta* **384**, 180–193.
Schuit, K. E., and Buetow, D. E. (1968). *J. Protozool.* **15**, 195–198.
Schuster, F. L., and Hershenov, B. (1974). *J. Protozool.* **21**, 33–39.
Senko, R. A., Moore, J., and Cantor, M. H. (1971). *J. Protozool.* **18**, Suppl., p. 20.
Sharpless, T. K., and Butow, R. A. (1970a). *J. Biol. Chem.* **245**, 50–57.
Sharpless, T. K., and Butow, R. A. (1970b). *J. Biol. Chem.* **245**, 58–70.
Sheeler, P., Cantor, M., and Moore, J. (1968a). *Life Sci.* **7**, 289–293.
Sheeler, P., Moore, J., Cantor, M. H., and Granik, R. (1968b). *Life Sci.* **7**, 1045–1051.
Sheeler, P., Cantor, M., and Moore, J. (1970). *Protoplasma* **69**, 171–185.
Shehata, T. E., and Kempner, E. S. (1977). *Appl. Environ. Microbiol.* **33**, 874–877.
Sinclair, W. K. (1967). *Cancer Res.* **27**, 297–308.
Siu, C.-H., Chiang, K.-S., and Swift, H. (1974). *Chromosoma* **48**, 19–40.
Siu, C.-H., Chiang, K.-S., and Swift, H. (1975). *J. Mol. Biol.* **98**, 369–891.
Siu, C.-H., Swift, H., and Chiang, K.-S. (1976a). *J. Cell Biol.* **69**, 352–370.
Siu, C.-H., Swift, H., and Chiang, K.-S. (1976b). *J. Cell Biol.* **69**, 371–382.
Siu, C.-H., Chiang, K.-S., and Swift, H. (1976c). *J. Cell Biol.* **69**, 383–392.
Smillie, R. M. (1968). *In* "The Biology of *Euglena*" (D. E. Buetow, ed.), Vol. 2, pp. 1–54. Academic Press, New York.
Smillie, R. M., and Rigopoulos, N. (1962). *J. Protozool.* **9**, 149–151.
Sommer, J. R., and Blum, J. J. (1965). *J. Cell Biol.* **24**, 235–251.
Stabenau, H. (1974). *Planta* **118**, 35–42.
Stabenau, H., and Beevers, H. (1974). *Plant Physiol.* **53**, 866–879.
Stageman, W. J., and Hoober, J. K. (1975). *Nature (London)* **25**, 244–246.
Tait, G. C. L., and Harris, W. J. (1977a). *Eur. J. Biochem.* **75**, 357–364.
Tait, G. C. L., and Harris, W. J. (1977b). *Eur. J. Biochem.* **75**, 364–372.
Talen, J. L., Sanders, J. P., and Flavell, R. A. (1974). *Biochim. Biophys. Acta* **374**, 129–135.
Thacker, A., and Syrett, P. J. (1972a). *New Phytol.* **71**, 423–433.
Thacker, A., and Syrett, P. J. (1972b). *New Phytol.* **71**, 435–441.
Tokunaga, M., Nakano, Y., and Kitaoka, S. (1976). *Biochim. Biophys. Acta* **429**, 55–62.
Tong, N. C. H. L., Gross, J. A., and Jahn, T. L. (1965). *J. Protozool.* **12**, 153–160.
Vogel, K., and Barber, A. A. (1968). *J. Protozool.* **15**, 657–662.
Votta, J. J., Jahn, T. L., and Levedahl, B. H. (1971). *J. Protozool.* **18**, 166–170.
Webster, D. A., and Hackett, D. P. (1965). *Plant Physiol.* **40**, 1091–1100.
Wells, R., and Sager, R. (1971). *J. Mol. Biol.* **58**, 611–622.
Wetherell, D. F. (1958). *Physiol. Plant.* **11**, 260–274.
White, J. E., and Brody, M. (1974). *FEBS Lett.* **40**, 325–330.
Wilson, B. W., and Danforth, W. F. (1958). *J. Gen. Microbiol.* **18**, 535–542.
Wilson, B. W., Buetow, D. E., Jahn, T. L., and Levedahl, B. H. (1959). *Exp. Cell Res.* **18**, 454–465.
Wilson, R., and Chiang, K. S. (1977). *J. Cell Biol.* **72**, 470–481.
Wise, D. L. (1955). *J. Protozool.* **2**, 156–158.
Wise, D. L. (1959). *J. Protozool.* **6**, 19–23.

Wise, D. L. (1968). *J. Protozool.* **15**, 528–531.
Wise, D. L. (1970). *J. Protozool.* **17**, 183.
Witman, G. B. (1975). *Ann. N.Y. Acad. Sci.* **253**, 178–191.
Wolfe, J. (1973). *Exp. Cell Res.* **77**, 232–238.
Wolken, J. J. (1961). "*Euglena*; an Experimental Organism for Biochemical and Biophysical Studies," p. 12. Inst. Biol., Rutgers Univ., New Brunswick, New Jersey.
Wolken, J. J. (1967). "*Euglena*," pp. 88–93. Appleton, New York.
Wolken, J. J., and Gross, J. A. (1963). *J. Protozool.* **10**, 189–195.
Woodward, J., and Merrett, M. J. (1975). *Eur. J. Biochem.* **55**, 555–559.
Zahalsky, A. C., Hutner, S. H., Keane, M., and Burger, R. M. (1962). *Arch. Mikrobiol.* **42**, 46–55.

Biological Rhythms in Protozoa

3

JOHN J. WILLE, JR.

BIOCHEMISTRY AND PHYSIOLOGY OF PROTOZOA
SECOND EDITION, VOL. 2

> The whole question of endogenous biological
> rhythms has become of increasing interest
> because it is apparent that organisms in
> general, even single-celled organisms, can
> keep time with a high degree of accuracy.
>
> *On Development*
> John Tyler Bonner

I. INTRODUCTION

In this chapter an attempt is made to give an up-to-date account of the progress in both experimental and theoretical work on circadian biological rhythms in protozoan cells.* The space allotted for this chapter precludes a comprehensive review of all contributions; we can only try to convey current and dominant themes and possible rewarding areas of future re-

---

* The terminology used in this chapter is defined in the Appendix.

search and hope that omissions and biases will be forgiven. With that caveat, our task is considerably aided by the appearance of past and recent comprehensive reviews dealing with important aspects of biological clocks in lower eukaryotic organisms (Hastings, 1959; Bruce, 1965; Sweeney, 1969a; Ehret and Wille, 1970; Ehret, 1974; Palmer *et al.*, 1976; Schweiger *et al.*, 1977). We recommend for general background on circadian biological rhythms one or more of the following: *Cold Spring Harbor Symposia on Quantitative Biology* (1960); Aschoff (1965); Menaker (1971); Sweeney (1972); Bünning (1973); Pavlidis (1973); Scheving *et al.* (1974); Palmer *et al.*, (1976); Goodwin (1976); Hastings and Schweiger (1976); *Proceedings International Society Study Chronobiol. Symp. 12th* (1977).

A. Generality of Circadian Regulation in Protozoan Cells

Over the past quarter of a century, a wealth of evidence has accumulated that protozoa, ciliates, dinoflagellates and other flagellates, and green algae display a spectrum of physiological, behavioral, and biochemical rhythms which under permissive conditions continue to exhibit oscillations with a period of about (*circa*) a day (*dies*). A wide variety of processes in protozoa have a circadian rhythm: mating behavior (*Paramecium bursaria*, Ehret, 1948, 1951, 1953, 1959; *P. aurelia*, Karakashian, 1965, 1968; *P. multimicronucleatum*, Barnett, 1965, 1966); phototactic response (*Euglena*, Pohl, 1948; Bruce, 1965; *Chlamydomonas*, Bruce, 1970); cell division (*Gonyaulax*, Sweeney and Hastings, 1958; *Euglena*, Edmunds, 1965, 1966; *P. bursaria*, Volm, 1964; *Tetrahymena pyriformis*, Wille and Ehret, 1968; Edmunds, 1974b; *P. multimicronucleatum*, Barnett, 1969; *Chlamydomonas reinhardi*, Bruce, 1970); photosynthetic capacity (*Gonyaulax*, Hastings *et al.*, 1961; *Acetabularia*, Vanden Driessche, 1966b; Schweiger *et al.*, 1977; *Phaeodactylum*, Palmer *et al.*, 1964); changes in chloroplast shape and ultrastructure (Vanden Driessche, 1966a; Schmitter, 1971; Herman and Sweeney, 1975); stimulated bioluminescence (Hastings and Sweeney, 1958); chronotypic transcriptotypes (Ehret, 1974); chronotypic enzymes (Sulzman and Edmunds, 1972); variations in concentrations of metabolic intermediates, energy charge carriers, and biogenic amines (Dobra and Ehret, 1977). Virtually all levels of cellular organization are subject to circadian regulation, and it is clear that the unicellular level of organization is sufficient for the expression of circadian rhythmicity. The question of whether eukaryotic cell organization is the minimal level of biological organization sufficient for manifesting a persistent circadian oscillation has not yet been resolved, but recent reports of circadian growth rates in *Escherichia coli* and *Klebsiella aerogenes* (Sturtevant, 1973a,b) under certain experimental conditions claim that even prokaryotes may be subject to circadian regulation.

B. Circadian Clock Properties and Basic Assumptions

Circadian rhythms have provided ample experimental systems in the search for the homeostatic mechanism(s) assumed to underlie the self-sustaining oscillations of a cellular pacemaker, i.e., an endogenous circadian clock. The elusiveness of this quest has not dimmed the ardor or inventiveness of researchers, who have spawned conflicting deterministic (Barnett et al., 1971a; Ehret, 1974; Ehret and Trucco, 1967) and stochastic (Cummings, 1975; Njus et al., 1974; Sweeney and Herz, 1977) models for the circadian oscillator, as well as provocative mathematical formulations of limit cycle dynamic systems of heuristic (Pavlidis, 1973) and experimental (Winfree, 1973, 1975) importance. More recently, several eclectic schemes comprising generally acceptable elements of both categories of models have been offered (Edmunds and Cirillo, 1974; Palmer et al., 1976; Ehret and Dobra, 1977; Schweiger and Schweiger, 1977).

In a recent report on the molecular basis of circadian rhythms, Hastings and Schweiger (1976) drew attention to the question of whether there was a single common molecular basis for all known circadian rhythms. They cautioned that "the formal and functional similarities in the observable behavior of circadian rhythms developed in adaptation to the *same* external cycle, and may represent the result of convergent evolution rather than the expression of a common molecular mechanism." If so, they suggest an approach which acknowledges that there may be more diversity of mechanism than function, and one which relies on arguments developed around a few well-studied organisms, rather than adopting the phylogenetically eclectic attitude of those who believe in one universal mechanism. As a further caveat, they point out that a common feature of biologically complex interacting, feedback-regulated systems is the rarity of nonoscillating, stable steady state operation. Such systems rather tend to instabilities and oscillations *unless* opposed by enormous selective pressure and often yield oscillations with longer periods than the characteristic time constants of the individual reactions involved. It can be argued that cellular systems, which also operate at far from thermodynamic equilibrium, are rarely capable of true steady state outputs, and that instabilities and oscillations of diverse biochemical reactions in cell types should be the rule rather than the exception. It may be that natural selection has acted to remove maladaptive oscillations, stabilize the remaining ones, and couple them to prevailing environmental periodicities. According to this view, no one metabolic, genetic, or epigenetic feedback regulatory pathway has escaped the pruning effects of natural selection. It remains, however, an open question whether redundant circadian escapement mechanisms lurk in complex biochemical oscillatory feedback loops of the cell, each capable of independently generating circadian outputs, or whether the economy of nature has settled on *one* basic circadian oscillator, conserved throughout the course of eukaryotic evolution.

As a working hypothesis, we assume that all eukaryotic cells possess an endogenous, light-entrainable biological clock characterized by the following set of properties: (1) the persistence of free-running rhythm with a circadian period τ of 24 hr; (2) entrainability and phase-responsiveness to single or multiple *Zeitgebers* (including light, temperature, nutrients, oxygen, and chronobiotic drugs, e.g., deuterium oxide, Li^+, K^+, ethanol, valinomycin, theobromine, theophylline, cycloheximide); and (3) a temperature-independent period. These and other properties have been well documented for the oscillatory mechanism of overt biological rhythms in higher organisms, and have been also shown for numerous protozoan cells. Furthermore, it is assumed that the clock is an entity distinct from the processes it causes to be rhythmic. Conceptually, it follows that the clock and the processes it drives are coupled through many diverse transducing mechanisms, which in virtually every case remain to be fully elucidated. It is precisely this question of distinguishing the "hands" of the clock from the oscillator, which transmits both phase and period information to the driven rhythms, that constitutes the core of the circadian biological clock problem. In Sections II–VII we explore the operation of the cellular circadian clock in depth in several protozoan types. Particular emphasis is placed on the *Tetrahymena* circadian clock and on aspects of circadian clock control in *Euglena, Gonyaulax, Acetabularia,* and *Chlamydomonas* systems. These observations are then considered in relation to parallel studies on the limit cycle biochemical oscillator model of the mitotic clock in *Physarum* and in relation to the formulation of a unifying clock dynamics for both mitotic and circadian oscillations in eucells.

II. CIRCADIAN CLOCK IN *TETRAHYMENA*

A. Cell Division and the Circadian–Infradian Rule

One of the most intensively investigated unicellular biological clocks is in the ciliated protozoan *T. pyriformis*. Earlier studies (Wille and Ehret, 1968; Ehret and Wille, 1970) established that a sudden increase in irradiance (i.e., a switch-up from continuous darkness to continuous light) or a sudden decrease in irradiance (i.e., a switch-down, from continuous light to continuous darkness) could synchronize a cell division rhythm with a free-running circadian period τ of 19–23 hr in axenic batch cultures of *T. pyriformis* strain W, provided the transitional signal occurs after a critical time in the late ultradian growth mode as the cells begin to enter the infradian growth mode. From these investigations, Wille and Ehret (1968) advanced the hypothesis that between the ultradian and infradian growth modes lies a special light-synchronizable state, i.e., when the average generation time $(\overline{GT})$ of the culture is equal to or somewhat greater than 24 hr, a cell in switching its state from the ultradian $(\overline{GT} \ll 24$ hr) to the infradian $(\overline{GT} \geq 24$ hr) growth mode is invariably capable of circadian outputs. This hypothesis has come to be known as the circadian–infradian rule (Ehret and Wille, 1970) and is seen in the ability of exponentially growing cells to enter into a synchronized circadian rhythm of cell division from asynchronous growth not only in *Tetrahymena* but also in *Euglena* (Edmunds, 1966) and *Gonyaulax* (Sweeney and Hastings, 1958); it is called the G-E-T effect, after the above organisms which show mitotic entrainment by a *Zeitgeber* at some particular phase of population growth (Figure 1).

Photoentrainment of an endogenous circadian cell division rhythm in *Tetrahymena* continuous cultures was also achieved by growing cells in an electronically controlled nephelostat which operates on the chemostat principle of controlled flow rate of nutrient medium (Wille and Ehret, 1968). The cell titer was kept constant by keeping the amount of scattered light at a fixed level through periodic monitoring of the culture and subsequent feeding events if the amount of scattered light changed upward with cell growth (Eisler and Webb, 1968). Following 4 days of photoentrainment on a $LD:10,14$ light–dark regimen, a cell division rhythm in which increases in cell titer were confined to the dark phase was shown to persist with a free-running period of 21 hr for 5 days. Again, photoentrainment of the circadian cell division rhythm was easily demonstrable when the washout rate of the cultures maintained in the nephelostat was infradian $(\overline{GT} = 40$ hr), but a similar photoentrainment failed to entrain at ultradian washout rates (Wille and Ehret, 1968). In subsequent experiments (Ehret and Wille, 1970), a continuous-culture device was employed which controls growth rate by a program of feeding events which simulated the demand feeding schedule obtained from circadian-synchronized infradian cultures grown in the nephelostat. These experiments provided additional evidence that light synchronization of an endogenous cell division rhythm is restricted to infradian cultures of *T. pyriformis* strain W. On a continuous-light regimen, synchronization of an endogenous circadian rhythm of cell division was achieved when batch cultures at the end of the ultradian exponential phase were converted to continuous cultures by the initiation of a symmetrical feeding program. If, however, symmetrical feeding was initiated in batch cultures in the early ultradian exponential phase of growth, synchronization of the cell division rhythm did not oc-

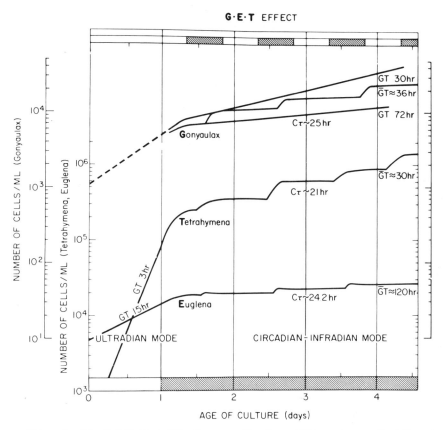

Figure 1. The G-E-T effect. Ultradian generation times (*GT*) are given on the left, and infradian generation times on the right. It is during the infradian mode that cells are light-synchronizable; although infradian generation times are long and variable, in each case the consequence is a circadian output; the circadian period (*Cτ*) is ≈1 day (by definition). *Gonyaulax* is shown in continuous light (top, open bar) as an asynchronous population with average infradian generation times in the range 30–70 hr and in light–dark cycles of 12 hr of each as a synchronized population with an integral generation time of ≈36 hr. The curves for *Euglena* and *Tetrahymena* might also have included asynchronous infradian slopes, but these have been omitted here for clarity; the synchronized populations represented in each of these cases show free-running endogenous circadian rhythms following photoinduction by a switch-down from light (bottom, open bar) to darkness (bottom, hatched bar) at a critical transition point between the two modes of growth. (After Ehret and Wille, 1970, reproduced by permission.)

cur, but could be induced by photoentrainment with a *LD:12*,12 light–dark entrainment cycle. Ehret and Wille (1970) concluded "that the critical condition for manifestation of the circadian rhythm of cell division in *Tetrahymena* is the transition from the ultradian to the infradian mode of

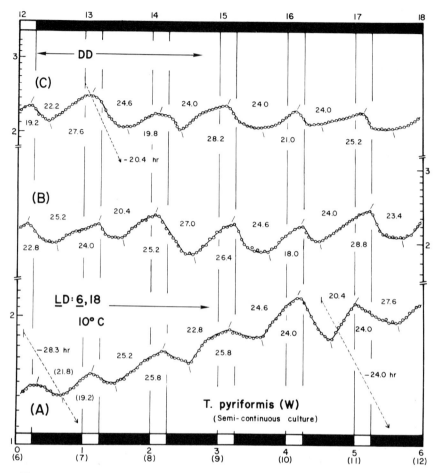

Figure 2. Entrainment and persistence of division rhythmicity in a semicontinuous culture of *T. pyriformis* strain W on PPL medium at 10° ± 0.05°C. The culture was continuously diluted with fresh medium supplied by a peristaltic pump at a predetermined rate designed to replenish exactly the aliquots withdrawn automatically at intervals of 72 min for determinations of cell concentration and to balance the rate of increase in cell titer. The three segments of the curve shown (A–C) represent a continuous monitoring of cell number over a time span of 18 days and should be read from left to right (starting with curve A) against the corresponding ordinate [number of cells per milliter ($\times 10^{-4}$)] and abscissa (time in days) scales. For the first 12 days the culture was maintained in *LD*:6, 18, with phase divisions as indicated. On Day 4 the effective "equivalent" dilution or washout rate (indicated by the dashed diagonal lines) was adjusted from −28.3 to −24.0 hr to balance exactly the calculated generation time of the culture (see curve B). On the twelfth day of entrainment, the culture was placed in continuous darkness (*DD*) for 6 days. This resulted in a decrease in the overall generation time to 20.4 hr and required a matching adjustment in the dilution rate as shown. Periods between inflection points (either troughs or peaks) of successive oscillations are given. (From Edmunds, 1974b, reproduced by permission.)

growth. Following that transition either light or food can act as *Zeitgeber* for induction of circadian rhythmicity in the entire population."

The validity of the circadian–infradian rule was tested and confirmed by investigating the phasing effect of light on cell division in exponentially increasing cultures of *Tetrahymena* grown at low temperatures in a semicontinuous-culture apparatus (Edmunds, 1974b). This work established that the growth rate of *T. pyriformis* (either strain W or GL) is markedly slower in constant illumination of moderate light intensity (850–1700 lx) than in constant darkness, at least at lower temperatures (8°–15°C).

Figure 2 demonstrates that photoentrainment of cell division by appropriately chosen repetitive light cycles (*LD:12*, 12 or *LD:6*, 18) was achieved when the cultures of *Tetrahymena* were grown at 10°C. A series of successive fission bursts is seen to alternate with periodic cessations in cell number increase of each 24-hr interval, which results in an approximate doubling of cell number in the 24-hr photoperiod. Division rhythmicity can also be seen to persist for at least six cycles with a circadian period in continuous darkness and at constant temperature following prior entrainment by a diurnal light–dark cycle. Last, light-induced phasing of cell division was shown not to be limited to stationary growth in the population but was routinely obtained in cultures in the early or midexponential growth mode provided that the overall generation time of the population approximately matched or was longer than the period T of the imposed light–dark cycle (i.e., infradian, $\overline{GT} > T$). As Wille and Ehret (1968) had found, entrainment by *LD:12*, 12 does not occur at higher temperatures associated with corresponding lower values of $\overline{GT}$ (e.g., at 15°C, $GT = 10$ hr; at 28°C, $GT = 2$–3 hr). These results largely substantiate the generality of the circadian-infradian rule and add the dimension of manipulation of the activation energies of chemical processes as a means of obtaining appropriate values of $\overline{GT}$ greater than 24 hr, as opposed to nutrient depletion of the culture, thereby forcing cells to enter the infradian mode of culture growth.

B. Circadian Parameters of the Infradian Growth Mode in Continuous Cultures

1. Circadian Rhythms of DNA and RNA Synthesis

Figure 3 shows the pattern of isotope incorporation ([³H]uridine into RNA, top curve; [³H]thymidine into DNA, lower curve) into a photoentrained continuous infradian mode culture of *T. pyriformis* strain W. The portion of the data shown (2 days of an *LD:12*, 12 light–dark cycle) repre-

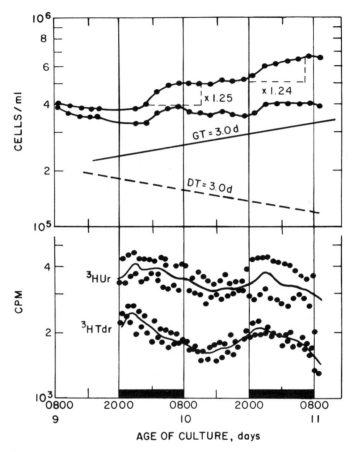

Figure 3. The isotopic uptake in counts per minute of [³H]thymidine (³HTdr) and [³H]uridine (³HUr) over a 36-hr period in a continuous culture of *T. pyriformis*. Top graph of cell titer (lower) during this period and cell titer have been corrected for dilution (upper). The long-term average generation time of this culture (*GT*) was 3 days; the dilution time (*DT*) of the culture is also given. (From Ehret *et al.*, 1974, reproduced by permission.)

sents an overall generation time of 72 hr in which about 25% of the cells divided synchronously each day (Ehret *et al.*, 1974). By making the assumption that the circadian S phase encompasses at least the period of rapidly accelerating rate of thymidine incorporation, they calculated the minimal duration of S to be about 12 hr. On this basis, there is virtually no G_2 phase, and mitosis (M) is thought to last about 1 hr. The curve for [³H]uridine incorporation shows that the rates of RNA synthesis are also circadian-synchronized in photoentrained continuous cultures, with suggestions of bimodality throughout the cycle, in which the peaks in RNA synthesis occur several hours later than those in DNA synthesis.

2. Oxygen Induction of the Ultradian Mode during Infradian Growth

Detailed studies on the approach of ultradian mode cells to the infradian growth mode in continuous cultures have been carried out (Ehret *et al.*, 1974). After inoculation, a lag of about 3 hr precedes the entry of cells into the ultradian mode. At a peak titer the cells then switch over to the infradian mode but, unlike batch cultures, early infradian cultures have an extended period of drastically diminished cell division during which the peak titer is reduced by washout, the duration of which is nearly proportional to the washout time of the culture. After this transit time, the culture either can resume asynchronous (but infradian) cell division or respond to photoentrainment via some external *Zeitgeber*. It had been noted earlier in the nephelostat studies (Wille and Ehret, 1968) that slight fluctuations in culture aeration were critical to the attainment of steady state culture titers. In the continuous-culture studies, Ehret and colleagues found that infradian cultures are characterized by a relatively anaerobic environment (about 5×10^{-3} atm O_2), which may represent a basic feature of the infradian mode, and by inference a possible prerequisite for circadian oscillation. When small-aliquot replicate cultures are removed from the parent continuous culture in the infradian mode and vigorously aerated, while at the same time the cell titer is monitored over a period of 8 hr, a drastic rise in cell number occurs after an initial lag of about 2.5–3 hr, leading eventually to an approximately 1.5- to 2-fold increase in the 7–8 hr after induction due to oxygen shift-up. Microscopic observation of aerated cells has confirmed that the bursts of cell division occur about 3.5 hr after the start of aeration, with about one-third of the cells showing a fission furrow stage. This rise in cell titer resulting from aeration of infradian cells has been called *oxygen induction* of the ultradian growth mode (Ehret *et al.*, 1974) and appears to represent the Pasteur effect in *Tetrahymena*. It was found that classical inhibitors of aerobic respiration, such as carbon dioxide, amytal, and rotenone were effective inhibitors of oxygen induction of the ultradian mode. Figure 4 clearly establishes that the time of induction of the transition from infradian growth is under circadian phase control (Ehret *et al.*, 1977). The potential for cell division occurred maximally only at permissive phases of the circadian cycle. At all circadian times when oxygen was tested for its capacity to increase cell titer, an increase was seen. However, the time of maximal increase occurred at midmorning (6 A.M.) and late evening (10 P.M.), coincident with circadian phases shortly after and just before nor-

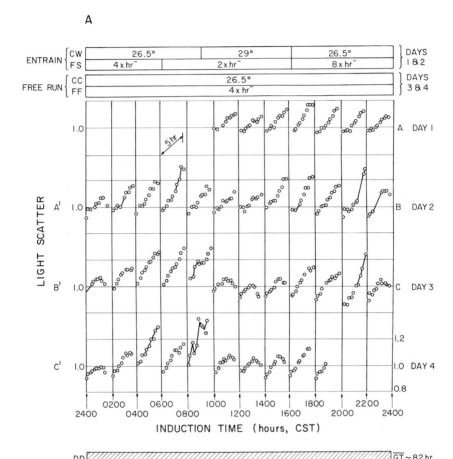

Figure 4. A 4-day measurement of the efficiency of oxygen inducibility as determined by light scattering in subcultures taken at 2-hr intervals from a master culture during entrainment (WC and FS, Days 1 and 2) and free-run (CC and FF, Days 3 and 4). The entrainment and free-run protocols are shown above each figure; in (A), $GT = 82$ hr, while in (B) $GT = 94$ hr. (From Ehret et al., 1977, reproduced by permission.)

mal expectations for circadian cell division. Hence, it appears that in *T. pyriformis* the Pasteur effect is under circadian phase control.

3. Thermal Cycle Entrainment of Circadian Rhythm of Cell Division

Thermal cycling has been shown to be as efficient a *Zeitgeber* as light–dark cycles in entraining a circadian rhythm of cell division in infradian cultures of *T. pyriformis* strain W grown under continuous-culture condi-

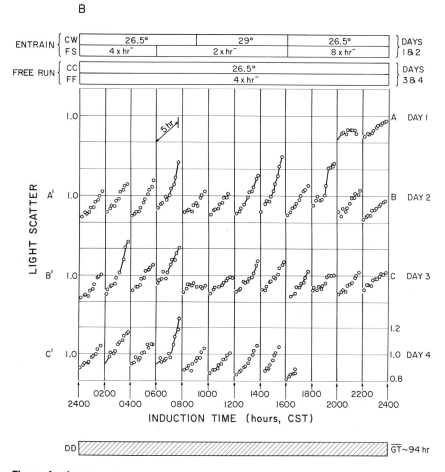

Figure 4. (continued)

tions (Meinert *et al.*, 1975) (Figure 5). Continuous cultures were subjected to warm (W)–cool (C) thermal cycles (WC:7,17; W = 31° or 29°C, C = 26.5°C). Cultures were grown in constant dim light and passed through the ultradian growth mode by Day 3.5. During the 6 days of thermal entrainment in the infradian mode, the switch-up to the W phase of the cycle was followed (within 3 hr) by a sharp transition in which cell division ceased and cell washout began. Cell division resumed about 3 hr after a switch-down to the C phase of the WC cycle. Under these conditions the average growth of the culture remained constant, i.e., there was little or no net change in cell titer per day. Therefore, the average generation time equaled the dilution rate (*DT*), and the periodicity of the oscilla-

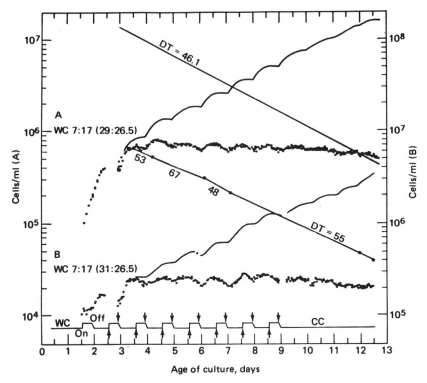

Figure 5. A plot of culture titer versus culture age for two continuous cultures of *T. pyriformis* strain W entrained by warm–cool (WC) cycles. The regimens for entrainment are given on the left and denote the durations of W and C in hours for both (A) and (B); temperatures in degrees Celsius for (A) and (B) are shown in parentheses. Closed circles represent the raw cell count data; dilution of the culture by the input medium is given in hours as the dilution time (*DT*). The lines ascending from left to right are the respective population densities calculated in the absence of dilution. The phase relationships of the WC cycle to the culture age are shown at the bottom. (From Meinert *et al.*, 1975, reproduced by permission.)

tion was equal to that of the entraining agent ($t = 24$ hr). The rhythm persisted as a barely detectable circadian rhythm ($\tau \sim 24$ hr) for 3–4 days of free run. The degree of cell synchrony in continuous cultures of *Tetrahymena* can be measured by estimating the percentage of the population which is in the same phase of the circadian cell cycle. The phasing index Φ_i was defined for circadian–infradian mode synchrony as equal to 1.0, when the time (t) taken for one-half of the daily step size (*ss*) was the smallest fraction of the cycle time (*ct*) for the circadian oscillation. For 24-hr rhythms this is given by the equation

$$\Phi_i t = 1 - t/12 \tag{1}$$

The phasing index Φ_i was calculated for cell synchrony obtained by thermal entrainment using the above WC cycle and was found to range from 0.38 to 0.49. However, the simultaneous application of both a feed–starve (FS) cycle and a WC cycle at various phase angles as dual entraining agents of infradian continuous cultures was found to improve significantly the phasing index. In particular, barrage feeding begun 4.2 hr after the switch-up in temperature (C to W), and which continued for 5 hr through the switch-down (W to C), increased the phasing index to 0.52 to 0.54, while barrage feeding (F phase) commencing 7 hr after the onset of the W phase led to increases in the phasing index of 0.57 to 0.66. Hence the use of multiple *Zeitgebers* cojoined at the appropriate phase relations can significantly improve the degree of cell synchrony of infradian mode-entrained cultures. Other phase angles of FS on WC gave lower phasing indexes. This effect of multiple entraining agents on phasing index highlights the authors' contention that in the natural environment more than one *Zeitgeber* is present, and that it is likely that phases of the circadian cell cycle depend on the phase angle relationships of two or more environmental periodicities.

4. Circadian Chronotypic Death

Transformation of growth curves to compensate for dilution in continuous cultures of *Tetrahymena,* so as to simulate the growth of batch cultures, involves the expression given by Meinert *et al.* (1975):

$$\overline{GT} = DT/\ln 2 \tag{2}$$

The use of this expression involves the hidden assumption that cells are lost only by washout and not by disappearance due to cell destruction. Negative slopes in the transferred data therefore imply cell death. Alternatively, if death is not observed, it can be attributed to inadequate stirring or circadian "settling," as suggested for similar findings in *Euglena* (Terry and Edmunds, 1970a,b). In *Tetrahymena* (Meinert *et al.,* 1975) cell death appears to occur each day of WC cycle entrainment at the time of the switch-up in temperature, as shown in Figure 6. In these cases, cell death was observed microscopically during the first few minutes after the removal of samples from the master culture to depression slides, as an explosive rupturing of a small percentage of the population. In circadian-synchronized cell cycles in free run, synchronous death of as much as 39% of the population results in a precipitous decline in the transformed cell titer. It occurs at the same phase of the circadian cell cycle as during entrainment. On the average, the entrainment of cultures at higher temperatures (31°C) results in losses in cell titer larger than those in cultures entrained at lower temperatures (29°C). The drop in cell titer can be largely accounted for by synchronous cell death, which is remarkably

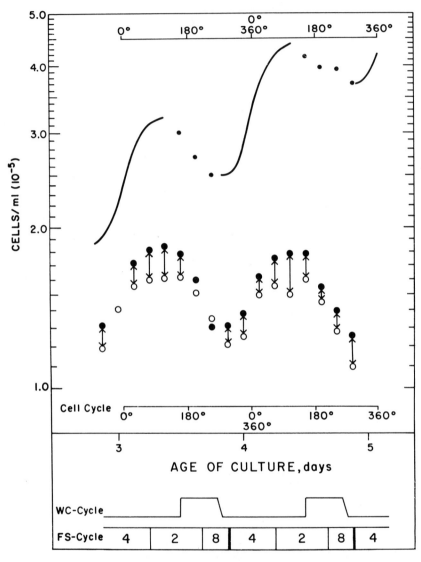

Figure 6. Chronotypic death in thermally entrained cultures of *Tetrahymena*. The circadian cell cycle is given at the top and bottom along with the entrainment regimens employed. Closed circles are cell counts from the third of a three-sample sequence and represent the true culture titer; open circles are the second count taken and indicate a lowered cell titer, reflecting the amount of hypoxic death that had occurred in the sample tubing. (From Meinert *et al.*, 1975, reproduced by permission.)

chronotypic in circadian-synchronized infradian continuous cultures. It is well known (Malecki *et al.*, 1971) that the death of *Tetrahymena* cultures is rapid once oxygen is completely removed from the growth medium, and that the attainment of a given culture cell titer is nearly proportional to the oxygen level. Meinert *et al.* (1975) argue that hypertitration (overproduction of cells for a given oxygen level) usually occurs at the ultradian-to-infradian transition; i.e., the cells exceed the oxygen support limit of the growth medium. When this happens, cell death occurs to reduce the number to an equilibrium condition sufficient for continued culture growth. This reduction is blocked by application of the WC cycle, which induces sensitive cells to die synchronously. This argument is strengthened by the fact that asynchronous populations of cells never exhibit death from hypoxia, as the cell number can never exceed the oxygen support level, while circadian-synchronized cells exhibit chronotypic death because periodic cell titer abruptly increases the oxygen requirement above the steady state oxygen support level. The putative oxygen-sensitive phase of the circadian-synchronized cell cycle coincides with the minimum probability for cell division, i.e., at roughly a 180° phase angle of the circadian cycle. In support of this interpretation, these authors report that application of an "oxygen cycle" prevents chronotypic death, if a higher concentration of oxygen (8% as compared with 4%) is given at the imputed phase angle (180°) for maximum oxygen requirement and, conversely, chronotypic death can be magnified for other phase angle relationships between the oxygen cycle and the circadian cycle.

C. Circadian Molecular Chronotypes

1. Phase-Specific RNA Transcriptotypes

Circadian-synchronized cultures of *Tetrahymena* in the infradian mode of growth typically display a rhythm in both DNA and RNA synthesis as discussed above. Initial tests of the chronon theory of circadian timekeeping (Ehret *et al.*, 1973) considered the question whether different RNAs were synthesized at different times of day (as predicted from the linear sequential transcription component of the chronon theory (Ehret and Trucco, 1967). For this purpose, cells of *P. multimicronucleatum* and *T. pyriformis* strain W were synchronized by an *LD:12,12* light–dark cycle in the infradian mode, and the cultures allowed to incorporate either [³H]uridine or ³²P to label RNA for short pulse periods at various times of the day. DNA–RNA molecular hybridization was employed to assess the binding capacity of purified time-of-day labeled RNAs with single-stranded homologous DNA immobilized on nitrocellulose membrane filters. The complex reaction kinetics of whole RNAs and sucrose gradient "cuts" of whole pulse-labeled and steady state labeled *Tetrahymena* RNAs has been reported elsewhere (Wille *et al.*, 1972; Barnett *et al.*, 1971b). Molecular hybridization experiments were performed with annealing reactions which compared the kinetics, saturation, and competition behavior of various time-of-day RNAs (Barnett *et al.*, 1971a). Figure 7 shows a family of curves from a competition experiment in which pulse-labeled RNA was prepared from *Tetrahymena* cells in circadian synchrony

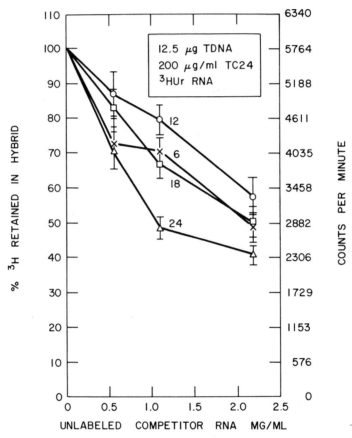

Figure 7.   Four chronotypically characteristic *Tetrahymena* circadian RNAs (from 6, 12, 18, and 24 hr circadian time) compete differentially with radioactive RNA (prepared at 24 hr circadian time) for sites on DNA. Annealing was for 24 hr in a formamide–citrate solution. Percentages (left) and mean counts plus or minus standard error for filters (right) are given. Counts on blank filters have been subtracted. (From Barnett *et al.*, 1971b, reproduced by permission.)

at t_{24} (i.e., from cells labeled at circadian hour 24 of a photoentrained culture) and allowed to compete with unlabeled circadian RNAs obtained from cells at t_6, t_{12}, t_{18}, and t_{24} (i.e., from circadian photoentrained cultures at hours 6, 12, 18, and 24). The data show clear evidence for the presence of chronotypically characteristic RNAs or circadian transcriptotypes. Inspection of Figure 7 reveals that unlabeled circadian t_{24} RNA competes best with its temporally homologous labeled RNA for the same sites on the DNA template. Additional studies (Barnett *et al.*, 1971a) have shown that the strong-competitor properties of independently prepared t_{24} RNAs cannot be due solely to the mere preponderance of rRNA or other kinetically fast-annealing components of whole t_{24} RNA, as preannealing the DNA with fast-component-enriched RNA does not abolish the strong-competitor properties of this stock of RNA. Most of the competition experiments (Barnett *et al.*, 1971a,b) confirm the prediction of the chronon theory that temporally characteristic RNAs are synthesized in circadian-synchronized cell cycles, although some unexpected rankings of competitor RNA in competition experiments occur (Barnett *et al.*, 1971a,b; Wille *et al.*, 1972). On balance, the molecular hybridization experiments strongly implicate the gene action system of the eukaryotic genome as the molecular mechanism underlying circadian regulation. Further critical tests of the chronon model for circadian timekeeping have been suggested (Barnett *et al.*, 1971a). The most straightforward test is direct visualization of DNA fibers hybridized to temporally distinct chronotypic transcriptotypes (e.g., t_6 and t_{24}, properly spaced on the same chronon replicon) by electron microscopic autoradiography. Another experimental design involves assessment of the efficiency of hybridization of temporally characteristic circadian RNAs for early- versus late-replicating DNA regions. Because of linear sequential transcription from long polycistronic DNA molecules, chronon theory predicts a rank ordering of hybridization efficiencies (e.g., $t_6 > t_{12} > t_{18} > t_{24}$ for early-replicating DNA); but for other regulatory mechanisms there is no obvious reason to expect such ordinality. At present, no one test is sufficient to prove that linear sequential transcription is the *only* molecular escapement mechanism underlying circadian oscillations. This would require evidence that an inversion of circadian transcriptotypes maps in a one-to-one correspondence with inversion of the phase ordinality of the controlled process to which it imparts phase information. The sensitivity of current methods of detecting select chronotypic transcriptotypes is not high enough to permit evaluation of this test. Other experiments on the apparent insensitivity of circadian rhythms to the inhibition of DNA-dependent RNA systems and the persistence of circadian regulation in apparently enucleate cells (see Section V) have raised doubts about the validity of the chronon model. However, the impressive tendency of clock mutations to map at the same locus in genetic experiments is consistent with its predictions.

2. Oxygen Consumption and Glyconeogenesis

When 2.5-liter batch cultures of *T. pyriformis* strain W are assayed for dissolved oxygen content and oxygen consumption per culture per minute from the time of inoculation through nearly 2 days of ultradian growth, and then through a subsequent $2\frac{1}{2}$ days of infradian growth, it has been found (Ehret *et al.*, 1977) that the respiratory rate of the cells is greatest in the early ultradian mode (approximately 40 μl $O_2/10^6$ cells/min). This rate declines exponentially later and stabilized at from 0.25 to 1 μl $O_2/10^6$ cells/min in the infradian mode. The total respiration of the culture ceased to increase about midway through the decline, while the dissolved oxygen dropped below 10% and the cells reached a titer of approximately 2×10^5 cells/ml. Interestingly, at this point in batch culture growth transition from ultradian to infradian growth there also occurred a dramatic decrease in the rate of increase in both RNA and protein (Ehret *et al.*, 1977). Nevertheless, cell titer continued to increase at the ultradian rate of $GT = 4$ hr for nearly two more divisions. Glycogen, however, continued to increase at the ultradian rate for a full 12 hr after cell division had ceased. Electron microscopic studies of infradian mode *Tetrahymena* cells (Sutherland *et al.*, 1973) reveal that cells literally fill up with glycogen at this phase of population culture growth. Other studies (Levy, 1973; Levy and Scherbaum, 1965) have also drawn attention to the high glyconeogenic capacity of *Tetrahymena* cells entering this portion of the batch culture growth phase. Cells in the infradian mode of growth thus become richly endowed with a store of glycogen which they use as an energy reserve for many days prior to any decline in cell titer (Ehret *et al.*, 1977). Figure 8 shows that, if the cells are in circadian synchrony, the depletion of glycogen is also circadian and is accompanied by a pattern of glycogen synthesis which is likewise circadian and has a free-running period of about 19 hr. This circadian variation in glycogen stores persists for at least 5 days of batch culture growth under conditions where the average generation time is greater than 10 days. Apparently, as cells enter the infradian mode of growth, which is generally accompanied by a precipitous decline in dissolved oxygen since the respiratory rate of the cells exceeds the oxygen capacity of the growth medium and the diffusion rate of oxygen into the medium, the ability to synthesize energy storage compounds (e.g., glycogen) reaches a level 20-fold higher than the level achieved by well-aerated cells in the ultradian mode of growth. This glyconeogenic capacity appears to prepare the cells for the relatively anaerobic infradian mode in

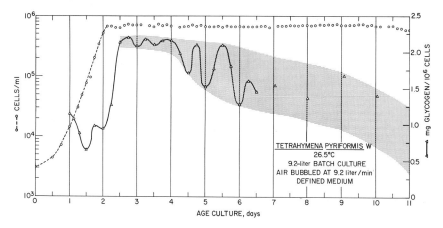

Figure 8. Growth curve of a batch culture of *T. pyriformis* strain W. The upper data points are the population densities, whereas the lower line and the shaded area represent the range of glycogen content during the infradian mode of growth. (From Antipa *et al.*, 1972, reproduced by permission.)

which the energy available through aerobic respiration is only 5% of that available during the ultradian mode.

3. *Circadian Regulation of Glycogen Metabolism*

Recently, Dobra and Ehret (1977) devised a method for culturing *T. pyriformis* strain W as monolayers on the surface of solid agar containing enriched protease–peptose medium, which permits a more convenient assay of the rate of exchange of respiratory gases and controlled uniform illumination by visible light as a *Zeitgeber*. Under these conditions it was possible to photoentrain infradian cultures of *Tetrahymena* (LD:8, 16 at 20°C) and control the flow rate of oxygen gas (21%) into the culture at 250 ml/min. A circadian rhythm for both oxygen consumption and carbon dioxide evolution was demonstrated.

Previous investigators (Levy and Scherbaum, 1965) had shown that glycogen synthesis and storage was partly determined by the level of oxygen and the availability of glucose and acetate in the medium. In studies (Dobra and Ehret, 1977; Ehret *et al.*, 1977) on highly oxygenated (21% O_2) plate cultures of *Tetrahymena*, glycogen storage approached a maximum of only 40–50 μg/ml/10^5 cells and underwent a stepwise decrease during entrainment and free run with a period of 24 hr during the infradian mode while under more restrictive oxygen conditions (i.e., 1%) glycogen depletion still occurred just preceding the onset of light, but there was also a second phase of net glycogen synthesis during the late dark phase of the light–dark cycle. Moreover, the decrease in glycogen appeared to be in phase with the increase in carbon dioxide evolution.

Stimulated by the report of Janakidevi *et al.* (1966a,b) that *Tetrahymena* has measurable amounts of the neurotransmitter substances epinephrine and serotonin and that glycogen levels might be altered by reserpine, dichloroisoproterenol, and aminophylline (Blum, 1967), Dobra and Ehret (1977) investigated the possibility that circadian regulation of glycogen metabolism in *Tetrahymena* may be similar to that found in the vertebrate hepatic system. A number of formal similarities were sought, including a circadian oscillation in the amount of tyrosine aminotransferase (TAT), which in the rat oscillates with a period of 24 hr and is subject to regulation by light–dark and feed-starve cycles as well as changes in dietary tyrosine and tryptophan and the administration of such drugs as norepinephrine, theophylline, and quinolinic acid. To make the picture complete, they examined the control of the associated pathways for glycogen storage and release in *Tetrahymena*, which in the liver involves increases in intracellular levels of cyclic AMP (cAMP) through norepinephrine-activated membrane-bound adenylate cyclase. Norepinephrine-stimulated adenylate cyclase activity had been previously demonstrated in cultures of *Tetrahymena* (Rosensweig and Kindler, 1972). It therefore devolved upon Dobra and Ehret (1977) to observe whether the levels of glycogen in circadian-synchronized infradian plate-grown cultures of *Tetrahymena* correlated with circadian changes in TAT activity and the adenylate system, along with determination of circadian changes in respiration observed during light–dark entrainment.

Figure 9 shows that the circadian rhythm of cAMP levels peaks just prior to peak increases in glycogen depletion and carbon dioxide evolution. Moreover, the level of ATP undergoes a twofold increase toward the end of the light phase during the time of greatest respiration and glycogen utilization. Maximal TAT activity occurs in the early dark phase and precedes the increases in both cAMP levels and glycogen depletion. Since TAT can influence the rate of synthesis of biogenic amines when the substrate is limiting, its phase relationship to the adenylate system was taken as evidence for a regulatory role of biogenic amines in glycogen metabolism. This idea was tested by Ehret *et al.* (1977) in an experiment in which 0.1 m*M* norepinephrine was applied topically to entrained solid agar plate cultures of *Tetrahymena*. They found that the time of greatest suppression of TAT activity by norepinephrine was in phase with the time of greatest synthesis in the controls.

These findings have led Ehret and his collaborators to formulate a so-called *energy reserve escapement* mechanism for the circadian clock (Ehret and Dobra, 1977), which intimately links glycogen metabolism with the circadian–infradian rule (see Section II,E).

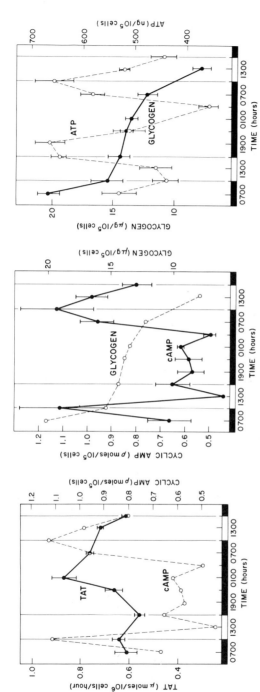

Figure 9. The measurements of four parameters: ATP, cAMP, glycogen, and TAT are represented for circadian-synchronized cultures of *T. pyriformis*. Each abscissa refers to the scale on the bottom of the figure which is the time of day. The vertical bars represent the dark phase of the light–dark cycle. Left panel: Circadian rhythms of TAT activity and cAMP concentrations. Ordinate is in picomoles cAMP per 10^5 cells, and micromoles of TAT per 10^6 cells. Middle panel: Circadian rhythms in stepwise decrease of total cellular glycogen and cAMP concentration. Ordinate is in micrograms of glycogen per 10^5 cells and picomoles cAMP per 10^5 cells. Right panel: Circadian rhythms of stepwise glycogen decrease and ATP concentration. Ordinate is in micrograms of glycogen per 10^5 cells, and nanograms of ATP per 10^5 cells. (From Dobra and Ehret, 1977, reproduced by permission.)

D. Circadian Cell Cycle: A Limit Case of the Cell Developmental Cycle

Deeper reflection on the particular significance of the circadian–infradian phenomenon has led Ehret *et al.* (1977) to propose that the cell developmental cycle is so fundamental to the life of the cell that it persists, albeit sometimes in a truncated form, regardless of the $\overline{GT}$ value of the cell population. The capacity of a given cell to divide is viewed as a finite probability for any $\overline{GT}$ value of the population. During the ultradian mode, growth conditions do not impose any growth-contingent limitations on the capacity of all the cells of the population to divide. This is expressed in the equivalence of the average generation time with the length of the interdivisional period ($\overline{GT} = ct$). While in the infradian mode, under limiting growth-related contingencies, the cycle time can dissociate from the average generation time ($\overline{GT} \gg ct$) while remaining constant at approximately 24 hr ($ct \approx 24$ hr), i.e., "ct arrives at a nearly temperature-independent and apparently genetically determined limit value, which is never longer than 'about a day' and is thus by definition *circadian*" (Ehret *et al.*, 1977b). In the infradian mode, the interdivisional period is always a simple harmonic of the circadian period; i.e., the cell cycle is always an integral multiple of the cycle time and only accidentally equal to the average generation time. In essence this proposal is one form of the central clock hypothesis for circadian oscillation. The underlying oscillator is supposed to be a "genomically ordained" transcription algorithm consisting of a basic transcription cycle (pretranscriptional, transcriptional, and post-transcriptional phase: $P_1 \rightarrow T \rightarrow P_2$). Evidence in support of this contention was discussed in Section II,C, as phase-specific circadian chronotypic transcriptotypes, and the circadian pattern of RNA synthesis within the basic cell developmental cycle, regardless of whether or not cell division occurs. Debate centers around the question whether a yet more fundamental circadian oscillator exists which can operate in the absence of the transcriptional cycle (e.g., in cells *devoid* of transcribing DNA and in cells in which RNA transcription has been *completely* blocked).

E. Circadian Energy Reserve Escapement Hypothesis

As noted above, *Tetrahymena* cells can pass into the infradian growth mode under conditions where one or more growth-restricting environmental factors become limiting, either by temperature ($T = 10°C$, Edmunds, 1974b), tryptophan (10^{-5}–10^{-6} M, Groh and Ehret, 1974), dissolved oxygen (10–30 ppm, Antipa *et al.*, 1972), and dramatic decreases in DNA and protein synthesis occur while glycogen increases exponentially at the

same rate for at least two more ultradian cell division steps. These findings have led to the discovery that circadian glycogen metabolism is regulated in infradian mode cells in much the same manner both in a protistan
(*Tetrahymena*) and in mammalian cells (Ehret and Potter, 1974; Haus
and Halberg, 1966), lending support to the unifying proposal that chronotypic gene action resulting in phase-specific molecular transcripts
and chronotypic enzymes (Edmunds, 1974a; Ehret, 1974; Hardeland *et al.*,
1973; Peraino and Morris, 1975) are coupled to overt driven circadian
rhythms through causally interconnected oscillations in glycogen storage
and depletion and in catecholamine and indoleamine metabolism. These
interrelationships are shown in Figure 10. By analogy with the period-

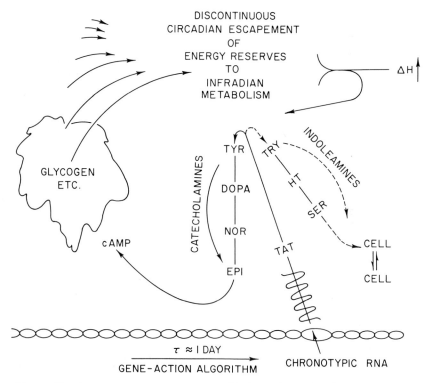

Figure 10. An energy reserve escapement with alternative path options (the one choice
being "*if* catecholamines, *then* release"; versus "*if* indoleamines, *then* hold") is coupled to a
gene action circadian oscillator that generates chronotypic enzymes (such as TAT) that
control the path of choice. This scheme results in long-term conservation of energy reserves
through circadian parceling during infradian growth. (From Ehret and Dobra, 1977, reproduced by permission.)

generating and phase-specifying components of a grandfather clock, in circadian–infradian mode cells a reliable and continuous energy source is provided by a store of energy reserve polymers (glycogen), which is metered out periodically by an energy reserve escapement component (*either* the catecholamine pathway, tyrosine to epinephrine, *or* the indoleamine pathway, tryptophan to serotonin). This store serves the needs of infradian intermediary metabolism; it is the power source for the clock mechanism and is presumably coupled to various transducing mechanisms (gears) which act as servomechanisms for expressing overt rhythmicity (hands). The period and phase of the circadian oscillator are guaranteed by the pendulum component which has its molecular embodiment in the linear sequential transcription component of the chronon.

Ehret *et al.* (1977a) present this proposal with additional speculation based on evolutionary perspectives. He argues that the circadian–infradian option is exclusively a eukaryotic invention evolved by cells to solve the perennial problem of how to endure the famine that follows a feast. The circadian–infradian mode of growth has been achieved by the fittest and allows the eucell to uncouple the hands of the cell division rhythm and to retain the more basic cell cycle oscillator. This dissociation is signaled by declining "cheap" energy sources (aerobic) at time of nutrient depletion and is accompanied by a switch from metabolic circuits regulating steady state production of cells to oscillatory metabolic circuits regulating *daily* rationing of the infradian stores of energy reserve compounds. It remains to be proven whether the chronotypic appearance of enzymes regulating the circadian energy reserve escapement system is driven by a sequential tape-reading gene action clock as proposed (Ehret *et al.*, 1977a), or by feedback inhibition of enzyme activity via a membrane-bound adenylate system which generates limit cycle oscillations in levels of cAMP (Cummings, 1975), or via direct coupling of chronotypic enzyme activity levels through an ion concentration-gated membrane clock (Sweeney, 1974a; Njus *et al.*, 1974). In this regard, Ehret *et al.* (1977a) have pointed out the inevitable involvement of membranes in each of the component processes of the energy reserve escapement clock.

III. CIRCADIAN CLOCK IN *EUGLENA*

A. Cell Division Rhythm in Wild-Type *Euglena* Cultures

Cell division of photoautotrophically grown cultures of the flagellate *E. gracilis* Klebs strain Z can be synchronized by appropriate light–dark cycles with various photofractions (*LD:16,8*; *LD:12,12*; and *LD:14,10*), and under the proper conditions of photoautotrophic growth (dim light, 800 lx) cell division rhythm with a free-running period of about 24 hr (τ = 24 hr) is routinely obtained in the infradian mode of growth ($\overline{GT}$ = 5 days, Edmunds, 1966). The circadian rhythm of cell division persists for at least 10 days under constant dim illumination, and small fission bursts (average $\overline{ss}$ = 14%) occur at integer multiples of about 24 hr (Figure 11; Edmunds, 1966).

Photoentrainment by light–dark cycles having $t \neq 24$ hr (e.g., *LD:10,10*) may also occur within certain limits (Edmunds and Funch, 1969b), and "skeleton" photophases comprising the framework of a normal full-photophase cycle (e.g., *LD:4, 4,4:12* or *LD:3,6,3,12*) also entrain the rhythm to a precise 24-hr period (Edmunds and Funch, 1969b). Interestingly, high-frequency (e.g., *LD:1,2*) and even "random" illumination regimens are like continuous dim light and permit circadian division periodicities (Edmunds and Funch, 1969a).

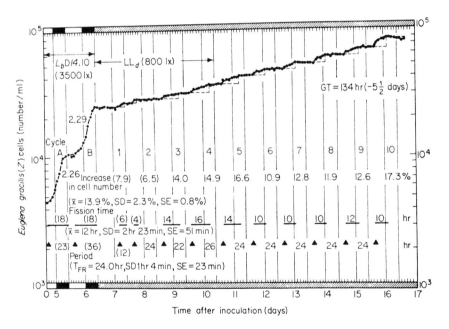

Figure 11. The persistence of rhythmic cell division in a population of *E. gracilis* under constant conditions of dim light (800 lx) and 25°C, following 6 days in a *LD:14,10* (3500 lx) light–dark cycle. For each division burst, the increase in cell number, fission time, and period are given. Step size is shown for cycles A and B. The generation time (*g*) of the population in dim light is indicated. (From Edmunds, 1966, reproduced by permission.)

B. Thermal Cycle Entrainment of Circadian Rhythm of Cell Division and Motility

Photoautotrophically grown cultures of *Euglena gracilis* strain Z cultured under continuous illumination (7500 lx) on a diurnal thermal cycle (WC:12,12; W = 25°C, C = 18°C) induce cell division synchrony having a period of rhythm equal to that of the entrainer (t = 24 hr), as well as a circadian rhythm of cell settling (Figure 12; Terry and Edmunds, 1970a,b). The motility rhythm or rhythm of cell settling, which may occur either concurrently with or in the absence of cell division, has the same phase relationship to the entraining thermal cycle at the two temperature ranges tested (18°–25°C and 28°–35°C), whereas the cell division rhythm has a 180° phase difference between the two thermal regimens. The cell settling rhythm lasts for as long as 9 days in culture at a constant temperature of 25°C after thermal cycle entrainment and can persist in continuous bright light; the cell division rhythm does not persist in free run following thermal cycling. Persistence of both rhythms may occur in dim light but was not explored.

C. Persistence of Circadian Rhythm of Cell Division in Photosynthetic Mutants

A mutant strain of *Euglena gracilis* var. *bacillaris* strain Z designated P_4ZUL, which is unable to carry out the Hill reaction owing to a block at or near photosystem II in the photosynthetic electron transport chain, was grown heterotrophically and subjected to photoentrainment on a *LD:10,* 14 light–dark cycle at two different temperatures (19° and 25°C) (Jarrett and Edmunds, 1970). Figure 13 shows that photoentrainment occurred only at the lower temperature (19°C), where the average generation time was nearly 24 hr as compared to $\overline{GT}$ = 10 hr at 25°C. This is clearly the expectation of the circadian–infradian rule, as outlined above. When a culture of the mutant that has been growing exponentially ($\overline{GT}$ = 26 hr) in continuous darkness at 19°C is subjected to a sudden switch-up in irradiance by a single transition from dark to light, a synchronous circadian rhythm of cell division ensues with a free-running period of 23 hr and an overall generation time of 26 hr.

Recently (Edmunds *et al.*, 1976; Edmunds, 1977) a circadian rhythm of cell division was shown in a completely bleached mutant (W_6ZHL) obtained from wild-type *E. gracilis* strain Z grown photoorganotrophically at 34°C for 3 weeks. This mutant entirely lacked chloroplasts, which were lost by growth dilution at the high growth temperature. When exposed to an imposed *LD:10,* 14 cycle, the bleached population of the W_6ZHL mu-

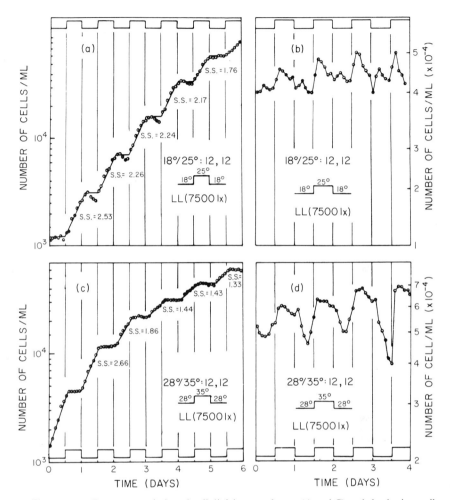

Figure 12. Temperature-induced cell division synchrony (A and C) and rhythmic motility (B and D) in photoautotrophically grown *Euglena* cultured under continuous illumination
with a 24-hr, 18°/25°C temperature cycle. Populations in the growth phase (a and c) and in
the stationary phase (b and d) are shown. (From Terry and Edmunds, 1970a,b, reproduced
by permission.)

tant was photoentrainable at 19°C ($GT \geqq 24$ hr) but not at 25°C ($\overline{GT} < 24$
hr). Following several days of light–dark entrainment, a free-running cell
division rhythm ($\tau = 21$ hr) persisted for at least 4–5 days in constant
illumination (7500 lx). These data are consistent with those of Mitchell
(1971) for the apochloristic strain WZUL, obtained by uv treatment, and
the pale green, nitrosoguanidine-mutagenized strain P_7ZNL.

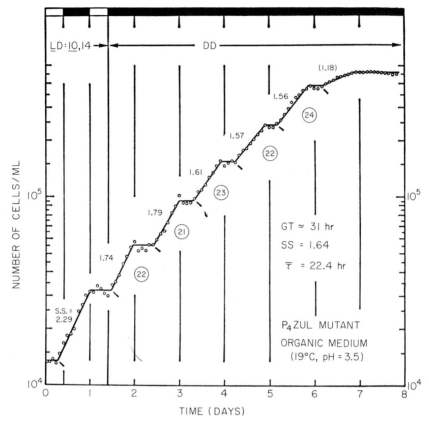

Figure 13.   Persisting, free-running, circadian rhythmicity in the cell division of cultures of the P_4ZUL mutant of *Euglena* grown heterotrophically at 19°C in continuous darkness. The culture had been previously entrained to a 24-hr period by a $LD:10,14$ regimen. The average free-running period ($\bar{\tau}$) of the population rhythm is 22.4 hr. The average step size (*ss*) of the successive fission bursts is substantially less than 2.00, indicating that not all the cells divide during any one circadian cycle. (From R. M. Jarrett and L. N. Edmunds, Jr., 1970, *Science* **167**, 1730–1733. Copyright 1970 by the American Association for the Advancement of Science.)

D. Other Overt Circadian Rhythms in *Euglena*

A persistent rhythm of motility which free-runs with a circadian period in infradian (stationary-phase) cultures under constant dim light but not in constant illumination has been reported by Brinkmann (1966). Bruce and Pittendrigh (1956) reported that photoautotrophic cultures of *Euglena* possess a circadian rhythm of phototactic sensitivity which is relatively temperature-independent and whose period can be drastically lengthened from 23.5 to 28 hr by growth of infradian cultures in heavy water (Bruce and Pittendrigh, 1960).

Nondividing cultures of *Euglena* maintained autotrophically in *LD:12,* 12 at 25°C display a daily rhythm of amino acid ([¹⁴C]phenylalanine) incorporation which persists for at least two cycles in constant darkness (Feldman, 1968). A circadian rhythm of photosynthetic capacity has been discovered by Walther and Edmunds (1973). It is demonstrable in photoautotrophically grown cultures of *Euglena* synchronized by a *LD:10,* 14 cycle at 25°C, and the rhythm of carbon dioxide fixation and peak activity of glyceraldehyde-3-phosphate dehydrogenase occurs in the same phase relationship to the entraining light–dark cycle. The approximate timing of the external acrophases for these and many other variables of the circadian system of *Euglena* is shown in Figure 14.

E. Nutritional Conditions Affecting Expression of Rhythm

Brinkmann (1966) has reported that the type of nutrition, mixotrophic or autotrophic, has an influence on the degree to which temperature can modulate the length of the free-running period of the dark motility rhythm in infradian cultures of *Euglena*. Under autotrophic conditions the circadian period is relatively independent of temperature (15°–35°C), while under mixotrophic conditions it decreases with increasing temperature. Brinkmann (1974, 1976a) also observed that long-duration applications of ethanol (10–100 m*M* for several weeks) increase the free-running period of the motility rhythm of autotrophic cultures of *E. gracilis* by several hours, depending on the temperature. Correlated metabolic studies revealed induction of the glyoxalate pathway for the breakdown of ethanol, an increase in the cellular ATP/ADP ratio, and an increase in respiratory activity for 2 days following the addition of ethanol. Brinkmann (1974, 1976a) suggests that these effects indicate primarily a nutritional effect of ethanol rather than evidence in favor of an essential role of membranes in the regulation of circadian period. In a later study (Brinkmann, 1976b), evidence was obtained for a circadian rhythm in the energy of activation required to produce acid denaturation of the outer cellular membrane in free-running cultures of *E. gracilis* previously photoentrained on an *L,D:12,* 12 regimen. By examining the dependence of the acid denaturation energy of activation on temperature, it was shown that the sensitive component of the membrane is probably a protein(s) which undergoes circadian turnover in the membrane. This provides direct support for circadian regulation of the outer plasma membrane and perhaps identifies the cell membrane as the primary target for mediating environmental signals to the circadian clock.

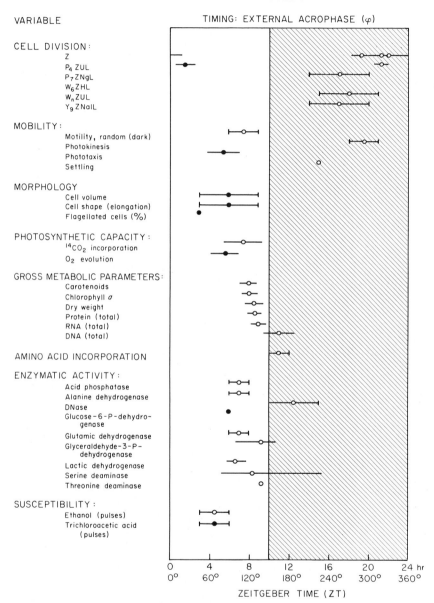

Figure 14. The circadian system of *Euglena*. Location of the time in the circadian cycle (*Zeitgeber* time) when a given variable displays its acrophase. (From Edmunds and Halberg, 1979, reproduced by permission.)

Feldman and Bruce (1972) have reported that acetate added to auto-trophic *E. gracilis* cultures caused an increase in the period length of the circadian rhythm of phototaxis to 27 hr at a 100 mM concentration. In addition, acetate pulses (10 mM) administered at different phases of the circadian rhythm of phototaxis induced phase shifts. In fact, a pulse of 36-hr duration resulted in a nearly arrhythmic culture. Other carbon sources added as a pulse (succinate, lactate, pyruvate) caused a temporary cessation of the rhythm and variable phase shifts upon resumption of the rhythm. Feldman and Bruce (1972) argue that, since each carbon source is effective by itself, the changes are probably caused by a general metabolic switch rather than by any specific effect of the carbon source. These results are also compatible with the suggestion by Brinkmann (1966) that phase shift may occur when cells are induced to make a transition from one metabolic state to another, particularly if the component reactions of each ensemble are characterized by slightly different Q_{10} responses. These results, along with other reports (Edmunds, 1964, 1965) concerning the breakdown of division synchrony in autotrophic cultures of wild-type *Euglena* maintained on a minimal salt medium with a light–dark cycle at 25°C ($\overline{GT}$ = 24 hr) by exposure either to continuous illumination (3500 lx) or by the addition of acetate (0.025 M), ethanol (0.006 M), or glutamic acid (0.34 M) to the medium, are consistent with the circadian–infradian rule as pointed out earlier by Ehret and Wille (1970). In all such cases growth perturbation caused a transient return of the culture to the ultradian mode of growth ($\overline{GT}$ = 13–15 hr).

F. Circadian Oscillation of Enzyme Activities

Periodic changes in enzyme activity during synchronous growth of wild-type *Euglena* batch cultures on a photoentrainment regimen (*LD:10*, 14) have been sought and found for a variety of enzymes (Walther and Edmunds, 1970, 1973; DNase; ribulose-1,5-diphosphate carboxylate, both NADH- and NADPH-dependent glyceraldehyde-3-phosphate dehydrogenase). Other chronotypic enzymes include alanine, glutamic, and lactic dehydrogenase, L-serine and L-threonine deaminase, glucose-6-phosphate dehydrogenase, and acid and alkaline phosphatase, all observed in photoorganotrophically batch-cultured *Euglena* synchronized by a *LD:10*, 14 cycle (Sulzman and Edmunds, 1972). Presumably all activities would continue to oscillate under continuous light or continuous darkness and constant temperature inasmuch as other rhythms persist. Figure 15 shows that 24-hr oscillations in the activities of several enzymes occur in nondividing *Euglena* cultures, i.e., under conditions where no net change in cell number occurs (Sulzman and Edmunds, 1972). It is inferred from

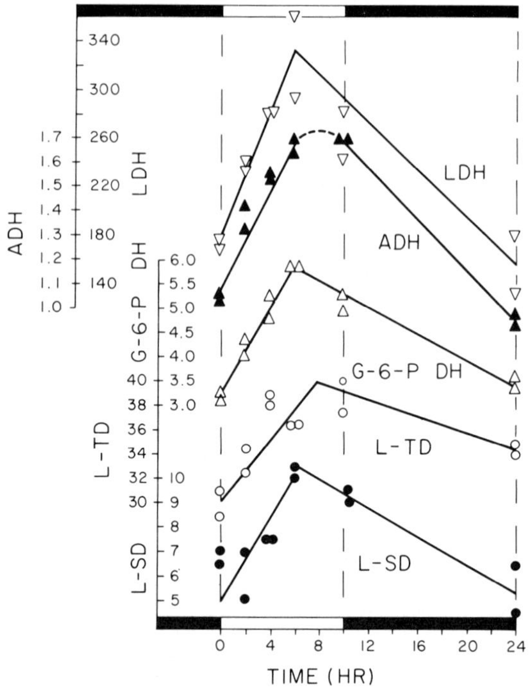

Figure 15. Twenty-four-hour oscillations in the activities of several enzymes in nondividing, stationary cultures of *Euglena* in *LD:10,14* at 19°C on a low pH glutamate–malate medium. Enzyme activities are given on the ordinate. L-SD, L-Serine deaminase activity (micromoles pyruvate produced per minute); L-TD, L-threonine deaminase activity (micromoles α-ketobutyrate produced per minute); △, G-6-PDH, glucose-6-phosphate dehydrogenase activity (micromoles × 10^2 NADH oxidized per minute); ▲, ADH, alanine dehydrogenase activity (micromoles × 10^2 NADH oxidized per minute); ▽, LDH, lactic dehydrogenase activity (micromoles × 10^2 NADH oxidized per minute). (From Edmunds, 1974c, reproduced by permission.)

these results that, since these rhythmic changes in enzyme activities occur in the absence of division, autogenous enzyme oscillation (Goodwin, 1966) entrained by the sequential replication of structural genes coding for the enzymes does not account for the time-ordered circadian appearance of the enzyme activities. Of course these results do not exclude the central transcription clock of the chronon model (Ehret and Trucco, 1967).

Sulzman and Edmunds (1973) report on a persistent circadian rhythm of alanine dehydrogenase (ADH) activity in nondividing populations of *Euglena gracilis* strain Z. In an earlier report (Edmunds *et al.*, 1974) a circadian rhythm of ADH activity was described for cultures synchronized in the infradian (nondividing) mode by two light–dark cycles (*LD:10,* 14), which continued to oscillate for at least 7 days in continuous darkness (Figure 16). Some of the possible factors that could generate the circadian oscillations in ADH activity were examined. No differences in K_m, pH optimum, or electrophoretic mobility could be demonstrated between enzyme extracted from either the minimum or the maximum phase of the oscillation. Mixing experiments involving high- and low-activity enzyme extracts gave no evidence for the presence of activators or inhibitors. The application of low doses of cycloheximide suppressed the oscillation in enzyme activity but, following removal of the inhibitor after a 12-hr treatment, the rhythm resumed with no apparent change in phase. However, greater amounts of enzyme were present at the maximal point of the oscillation than at the minimal point. Sulzman and Edmunds (1973)

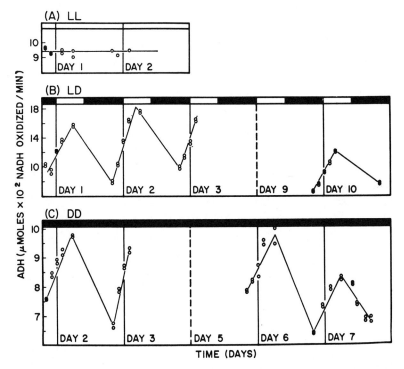

Figure 16. Activity ADH in nondividing, heterotrophically cultured *Euglena* under various lighting regimens. (A) Under constant bright illumination (LL). (B) In *LD:10,* 14 cycles. (C) In constant darkness (DD) after many cycles of *LD:10,* 14. Data are shown from Days 2, 3, 6, and 7 after the transition from LD to DD. Vertical lines are 24 hr apart. Double points are duplicate determinations. (From Sulzman and Edmunds, 1972, reproduced by permission.)

speculate that the simplest hypothesis consistent with these results is that new ADH molecules are synthesized during a specific phase of the oscillation (i.e., a chronotypic enzyme is synthesized but does not rule out the possibility that circadian oscillations in ADH activity are generated by cyclic turnover of a portion of the preexisting pool of enzyme. Such an interpretation is consistent with the observation that prolonged exposure to cycloheximide can alter the phase and prolong the period of the circadian oscillator controlling the phototactic response in *Euglena* (Feldman, 1967) and is also compatible with the expectations of the chronon model (Ehret and Trucco, 1967). These authors conclude that a biological clock based on coupled oscillations of existing proteins as proposed by Pavlidis (1971), in which a small number of enzymes are coupled to one another through allosteric interactions and through feedback interactions of small-molecule effectors and substrates, and the chronon model are not mutually exclusive. The long-term operation of the clock might depend on the synthesis of these proteins through genetically controlled mechanisms such as induction and repression, as postulated in the chronon model, while the minute-to-minute operation of the clock might occur in the absence of protein synthesis for short time spans via the coupled oscillation of existing proteins, as in the Pavlidis model.

G. Loss and Restoration of Overt Rhythmicity in Photosynthetic Mutants

Several years after the observation of persisting circadian rhythm of cell division in photosynthetic mutant strains of *Euglena* (P_4ZUL, Y_9Z-Na1L, and W_6ZHL, Edmunds, 1974), Edmunds *et al.* (1976; Edmunds, 1977) found that a gradual loss occurred in these cultures: first, a loss of the ability to display persisting overt rhythm of cell division, and then a complete loss of photoentrainment of a synchronous rhythm of cell division. The loss of rhythmic properties became progressively more pronounced the longer the cultures were maintained over a 7-year period of culture, but at the same time no diminution in overall generation time of the population growth was noted. A dramatic restoration of the capacity for both entrainment and free run of cell division rhythm was attained by simply adding cysteine and methionine ($1 \times 10^{-5} M$) or thioglycolic acid ($5 \times 10^{-5} M$) to the medium for the uv-induced mutant strain P_4ZUL, as shown in Figure 17. Similar results were obtained for both the apochloristic strain W_6ZHL and the naladixic acid-induced mutant Y_9Na1L by adding exogenous cysteine and methionine to the initial medium, or by adding these amino acids to the arrhythmic culture after several days of exponential growth in *LD:10,14*. Other sulfur-containing compounds

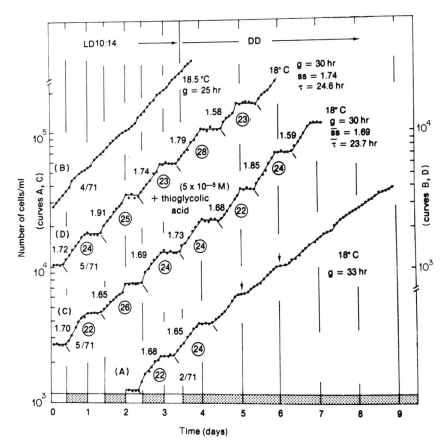

Figure 17. Gradual loss and subsequent restoration of circadian division rhythmicity in organotrophically batch-cultured populations of the ultraviolet light-induced P_4ZUL photosynthetic mutant of *Euglena*. (A) Initial loss of persistence of the rhythm in continuous darkness (DD). (B) Loss of capacity for entrainment of the rhythm by a LD:10, 14 cycle a few months later. (C) Restoration of the capacity for both entrainment and a DD free run by the addition of cysteine and methionine (1×10^{-5} M) to the medium. (D) Restoration of synchrony by the addition of thioglycolic acid (5×10^{-5} M) to the medium. The generation time (g), the mean step size ($\overline{ss}$) or fractional population increase for the successive division bursts, and the mean free-running period (τ) in DD for the growth curves are indicated (using the onsets of division bursts as the phase reference points). (From Edmunds *et al.*, 1976, reproduced by permission.)

which were tested and found effective in restoring rhythmicity to ar-rhythmic cultures of the P_4ZUL mutant were dithiothreitol, sodium thiosulfate, sodium sulfite, and even sodium monosulfide.

Edmunds (1977) has addressed himself to the basic question of whether the loss of rhythmicity in these mutants is a result of stoppage of the circadian oscillator, or of a more subtle uncoupling of the *hands* of the clock from the underlying circadian pacemaker. The latter interpretation is favored, based on two lines of evidence. First, it was found not only that cysteine or methionine addition to arrhythmic cultures restored the capacity for entrainment and subsequent free run of the cell division rhythm, but also that addition of the amino acids to arrhythmic cultures under constant illumination following prior exposure to *LD:10,*14 ultimately restored division synchrony to the culture with the phase angle relationship expected when the underlying (but unexpressed) clock had been running undisturbed throughout the course of the experiment. These results are similar to the finding that a low-dose pulse (Sulzman and Edmunds, 1973) of cycloheximide uncoupled the free-running rhythm of ADH activity from the circadian oscillator, which became coupled back into phase after return of the treated culture to inhibitor-free culture conditions. Moreover, it was found that in some cases a rhythm of cell settling was present even in exponentially dividing arrhythmic cultures of the P_4ZUL mutant under constant illumination following diurnal temperature cycling. The simplest hypothesis consistent with these results is that cell division rhythm and cell settling are two different hands of the same underlying clock, and that pleiotrophic effects of the mutant genome lead to dissociation of the overt rhythm of cell division, leaving the underlying clock unaffected and able to drive other overt rhythms. The restoration of cell division synchrony through supplementation of the growth medium by sulfur-containing compounds merely permits coupling of the circadian oscillator to one of the overt driven processes (cell division). These findings may, however, shed some light on the particular biochemical lesions which prevent gating of the cell division rhythm in *Euglena*. At present there is little evidence to suggest that the sulfur-containing compounds overcome a primary lesion in the sulfate-utilizing pathways of the mutants. Other possibilities envisioned involve a less direct effect operating via changes in the redox potential or distribution of energy charge in different cell compartments (Edmunds, 1977), or, perhaps, lesions in membrane biosynthesis resulting in arrhythmicity in membrane-mediated rhythms as suggested by Brinkmann (1976b).

IV. CIRCADIAN CLOCK IN *GONYAULAX*

Various physiological and behavioral phenomena have been shown to exhibit a free-running circadian oscillation in infradian mode photoautotrophic cultures of the dinoflagellate *Gonyaulax polyedra*. There is a persistent diurnal rhythm of cell division (Sweeney and Hastings, 1958), photosynthetic capacity (Hastings *et al.*, 1961), stimulated bioluminescence (Hastings and Sweeney, 1958), and luminescence or glow intensity (Hastings and Bode, 1962).

Recently, the presence of a circadian rhythm in photosynthesis was extended to two other genera of marine dinoflagellates, *Glenodinium* sp. and *Ceratium furca* (Prezelin *et al.*, 1977). The similarities between these rhythms and the well-studied photosynthetic rhythm in *Gonyaulax* may indicate that the mechanism of circadian regulation is the same in each case.

A. Lack of Desynchronization of Phase among Multiple Overt Rhythms

The simultaneous expression of four different circadian outputs from a photoentrained culture of *Gonyaulax* has permitted McMurry and Hastings (1972a) to test whether the characteristic phase relationships among them persist under constant conditions. Figure 18 shows that photosynthetic capacity, stimulated bioluminescence, and the glow rhythm maintained their phase relationships unchanged for several weeks under constant illumination following prior LD:*12,*12 photoentrainment. Since no drift in phase occurred, this result implies that the periods of the underlying oscillators controlling each rhythm are immensely precise, or that the several processes are tightly coupled through some as yet unknown mechanism, *or* that there is only one underlying clock which drives the period and phase of the four separate overt rhythms (hands). Direct support for the last-mentioned interpretation is shown in Figure 19. A 6-hr exposure to darkness was initiated under circadian free-run conditions in constant light in cultures of *Gonyaulax* previously synchronized by a LD:*12,*12 light–dark cycle. All the overt rhythms were phase-advanced to the same extent whether the cultures were subjected to resetting of the phase by total darkness (0 foot-candles) or dim light (3 foot-candles) during the free-running portion of the experiment. These authors conclude that the data are consistent with the hypothesis that a single "master" oscillator drives all the overt rhythms. Further support adduced by McMurry and Hasting (1972a) for this contention are the facts that the Q_{10} values for the effect of temperature on the frequency are approximately the same for three of the overt rhythms tested (Sweeney and Hastings, 1960) and that the rhythms of stimulated bioluminescence and cell division respond similarly to phase shifting by ultraviolet light (Sweeney, 1963).

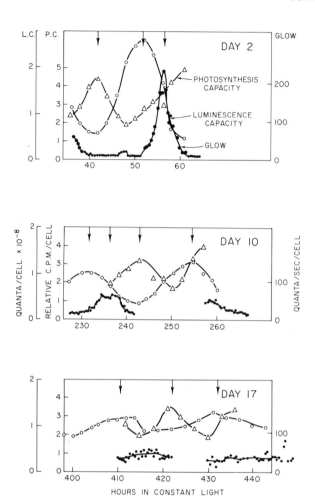

Figure 18. Luminescence capacity (LC), glow, and photosynthesis capacity (PC) rhythms after 2, 10, and 17 days in constant light (200 foot-candles, cool white fluorescent). Arrows indicate rhythm maxima, estimated visually. A flask containing 1200 ml of culture was transferred from a LD:12,12 cycle at the end of a light period (Time 0) into constant light. At the times shown, samples for the assay of all three rhythms were withdrawn. Glow samples were placed upon the scintillation counter turntable 32 min apart (three replicates) or 24 min apart (four replicates) and the glow intensity was thus measured every 32 or 24 min in between times of sampling the flask (every 2–3 hr). Three replicates for each rhythm were assayed for the Day 2 and Day 17 measurements; four were done for the Day 10 measurements. Some glow points are missing because of turntable malfunction. The culture increased linearly in cell number from 4500 cells/ml on Day 2 to 16,000/ml on Day 17. The cell division rhythm was not measured but in a similar experiment was not detectable 17 days after Time 0. (After L. McMurray and J. W. Hastings, 1972, *Science* **175**, 1137–1139. Copyright 1972 by the American Association for the Advancement of Science.)

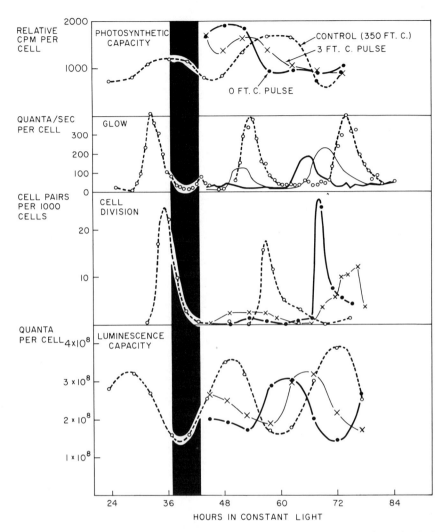

Figure 19. Phase shift of four rhythms caused by a 6-hr dark interruption. The dark interruption is represented by the wide vertical bar. Ten-milliliter samples from a *LD:12,12* culture were pipetted into vials during the light phase. At the end of the next light phase the vials were placed on the turntable in a constant light of 350 foot-candles (FT.C.), and the turntable was started. Sampling began as shown after 23 hr. The appropriate vials were exposed to a lower light intensity by putting them in transparent containers wrapped with the appropriate number of white towels (3 FT.C.) or by putting them in a light-tight box (0 FT.C.). In assaying for the rhythms, samples were removed from the vials on a rotating schedule; only the vials used for measurement of glow intensity were undisturbed. All determinations except those of the number of cell pairs were made in triplicate from vials spaced around the turntable and, except for the glow intensity values, were averaged. (From L. McMurry and J. W. Hastings, 1972, *Science* **175**, 1137–1139. Copyright 1972 by the American Association for the Advancement of Science.)

B. Chronotypic Luciferase Enzyme Activity

As noted earlier, interest in the mechanism of circadian oscillations has led to a search for molecular correlates which might shed light on the biochemical mechanism of the circadian clock. One approach is to see if oscillations in enzyme activity are due to the *de novo* synthesis and/or daily destruction of the primary structure of the enzyme. The mechanism of luciferase activity changes was investigated in the *Gonyaulax* system by McMurry and Hastings (1972b) under conditions where luciferase manifests circadian oscillations in activity levels. They found that the polypeptide backbone of a 40,000 MW proteolytic fragment of the luciferase molecule is conserved over several circadian cycles, while at the same time no diffusible small-molecule activators or inhibitors of luciferase activity were detected. They also extracted night-phase and day-phase cells in $5\,M$ guanidine hydrochloride which fully denatures and unfolds the protein and then allowed renaturation. Under identical conditions the renatured materials still showed day and night activity differences.

C. Effects of Heavy Water on the Period of an Overcompensated Clock

The effect of heavy water in biological systems has often been described as comparable to the effect of lowering the temperature. The lengthening effect on the free-running period of the circadian clock (Bruce and Pittendrigh, 1960) has been likened to a diminishing of the apparent temperature in the low-temperature equivalence hypothesis. The latter was supported by earlier observations of Pittendrigh *et al.* (1973) that two different features of the *Drosophila* eclosion rhythm are similarly affected by heavy water and temperature (i.e., the period of the rhythm under constant conditions is almost temperature-independent) and similarly unaffected by heavy water but that the phase angle of the rhythm is delayed by lower temperatures and by heavy water. Recently, a test of the low-temperature equivalence hypothesis was made in the *Gonyaulax* system (McDaniel *et al.*, 1974) which exhibits overcompensation; i.e., the frequency of the free-running circadian rhythm, τ, is greater at lower temperatures. As previously noted, the circadian rhythm of bioluminescence in *Gonyaulax* has a shorter free-running period at 18.5°C constant temperature (Sweeney and Hastings, 1958). This result was again confirmed (McDaniel *et al.*, 1974) for cultures of *Gonyaulax* assayed for glow intensity rhythm at either 22° or 16°C in constant dim light (1180 lx) following a prior photo-entrainment on a *LD:12,12* cycle, while the addition of heavy water to the medium at two different concentrations (6 and 12%) correspondingly increased the period of the rhythm in a dose-dependent fashion at either constant temperature (Figure 20). Moreover, in contrast to the low temperature–heavy water-induced phase angle delay in the *Drosophila* rhythm with respect to the light–dark cycle, the glow rhythm in *Gonyaulax* displayed a phase angle advance at the lower temperature in the presence of heavy water and a phase angle delay at the higher temperature in the presence of heavy water. It is clear that these initial results fail to support the low-temperature equivalence hypothesis.

D. Transducing Mechanisms between the Circadian Oscillator and Overt Rhythms

The possible mechanisms which transmit phase and period information from the circadian oscillator to the overt rhythms of luminescence and photosynthetic capacity in the *Gonyaulax* system have been explored (Sweeney, 1969b). Conditions were chosen in which the rhythmicity per se of the photosynthetic capacity was uncoupled (growth at low light intensities) from expression of the rhythmic changes in photosynthetic capacity (exposure of cells to high light intensities), so that the actual rate of photosynthesis remained constant. Hence any phase-dependent changes in the rates of the partial processes or amounts of intermediates during the rhythmic cycle could be attributed to the underlying circadian oscillator rather than to differences in photosynthetic rates. In this way it was shown that transduction between the oscillator and photosynthesis did not involve reactions of photosystem II, as the Hill reaction proceeds at the same rate in cells 180° out of phase both in their responses to quinone as a Hill reagent and in their sensitivity to inhibition by dichlorophenol indolphenol (DCMU) treatment. Likewise, the absence of a phase-dependent sensitivity of cells to the specific phosphorylation inhibitor carbonyl cyanide *m*-chlorophenyl hydrazone (CCmP) suggests no involvement of photosystem I reactions in the transduction of phase information. However, circadian variation in the activity level of ribulosediphosphate carboxylase, the first enzyme of the Calvin cycle, implicate it as the probable site of control. The variation in activity during the circadian cycle was of the right magnitude to account for an *in vivo* photosynthetic capacity. Mixtures of enzyme extracts in various concentrations from the night phase and the day phase gave the intermediate enzyme activity level expected from the levels of either enzyme preparation alone, ruling out diurnal fluctuations of enzyme activators or inhibitors as the cause of activity differences. Interestingly, the low nighttime enzyme level could be raised to the higher daytime level by increasing the

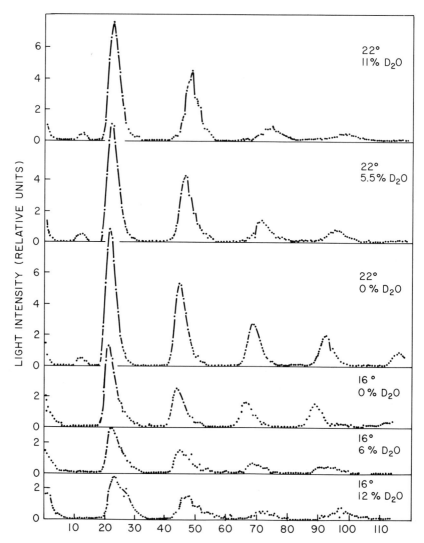

Figure 20. Free-running period of the *Gonyaulax* glow rhythm. The top three curves illustrate the glow rhythm for cells grown in *LD:12,*12 at 22°C and assayed at 22°C. For the three bottom curves the cells were grown and assayed at 16°C. Heavy water was added at the end of the dark period, at which time the cells were transferred to constant dim light, 110 foot-candles (1180 lx). Each of these curves represents an average of the glow of four separate vials. The percentage deuteration for each culture is given above each curve. (From McDaniel *et al.*, 1974, reproduced by permission.)

bicarbonate ion concentration of the buffer used in *in vitro* assays, and the *in vivo* amplitude of the photosynthetic capacity rhythm was markedly diminished by growth in medium exogenously supplemented with higher concentrations of bicarbonate ions. These results imply that the changing activity levels of ribulosediphosphate carboxylase may not be due simply to larger amounts of the enzyme in day-phase cells, but to changes in the enzyme activity at constant enzyme concentration at only low bicarbonate ion concentration. Neither the K_m or thermal inactivation properties of crude enzyme prepared from day-phase or night-phase populations of cells showed any differences. Hence the manner in which the circadian oscillator exerts control over the rate of photosynthesis by changing the activity of ribulosediphosphate carboxylase remains unknown.

Other studies on *G. polyedra* (Prezelin and Sweeney, 1977) examined the light absorption spectra of midnight ($ct = 1800$ hr) and midday ($ct = 0600$ hr) cells. No significant differences were found. Examination of the photosynthesis–irradiation ($P-I$) curves for cells entrained to a 12:12 hr light–dark cycle showed that the initial slopes of the $P-I$ curves for cell cultures examined at $ct = 0300$, $ct = 0600$, $ct = 0900$, and $ct = 1200$ hr were different and were directly proportional to the maximum percentage of photosynthesis at saturation for a given time of day. The results of the $P-I$ curves were explained in terms of the proportion of total photosynthetic units functional in the thylakoid membrane at any one time in the circadian cycle, each photosynthetic unit being defined as the smallest unit of membrane which can carry out the light reactions of photosynthesis, and comprising both photosystem I and II, plus assemblies of the light-harvesting pigments. They also argue that, since the absorption capabilities and half-saturation constants do not change significantly, changes in photosynthetic potential reflect changes in the activity of the existing photosynthetic apparatus. Further considerations based on the demonstration of a temporal change in the relative quantum yield of photosynthesis led Prezelin and Sweeney (1977) to suggest that the site of uncoupling of the inactivated photosynthetic units in dark-phase cells occurred between the reaction center and the light-harvesting pigments. These authors offer a partial explanation of the photosynthetic rhythm in *Gonyaulax,* based on a membrane feedback model for circadian rhythms (Sweeney, 1974a), in which it is proposed that circadian ion fluxes across the thylakoid membrane generate reversible conformational changes which couple or uncouple entire photosynthetic units in the membrane, and thereby the circadian ion fluxes control the energetic states of the membrane and transduce the rhythm of photosynthesis.

Previous studies had demonstrated (Hastings and Sweeney, 1957) a rhythm of luciferase enzyme activity and luciferin substrate concentration in cell-free extracts of *Gonyaulax* cells obtained from circadian-synchronized populations. *In vitro* enzyme kinetics studies of the luciferin–luciferase system show that both day-phase and night-phase luciferase enzyme preparations saturate the bioluminescence output at virtually the same luciferase concentration, indicating that the difference in activity levels is probably not due to different amounts of inhibitor in the luciferin preparation. It appears that concentration differences in luciferin and luciferase alone do not account for the rhythm of bioluminescence observed *in vivo*. Earlier observations (Bode *et al.*, 1963) led to the discovery that higher yields of cell-free luminescence could be obtained by prior exposure of night-phase cells to bright light before extraction; however, yields from day-phase extracts, which gave high yields without prior illumination, were not increased. These results indicate that the rhythm of luminescence may be due in part to a rhythm of sensitivity to stimulation. This suggestion appears to have been confirmed by Sweeney (1969b). When cell suspensions of *Gonyaulax* are treated with 0.0005 M acetic acid (pH = 4.2), maximal light production is elicited in both night-phase and day-phase cells, and the difference in response is slightly greater for the night-phase cells and on the same order of magnitude as that obtained from soluble luciferin and luciferase. But at higher concentrations of acetic acid the day-phase and night-phase cells yield the same efficiency of light emission, and prior illumination of night-phase cells does not inhibit the luminescence achieved by acid addition. These results led Sweeney to the conclusion that transduction of the luminescent rhythm occurs via changes in the mechanism by which luminescence is stimulated *in vivo*, perhaps by an underlying rhythm of proton fluxes, which along with permeability changes in the particulate "scintillon" fraction allow conversion of luciferin to an active anionic form capable of interacting with the membrane-bound form of luciferase. Direct support for this contention is not available. However, these data encourage speculation that transduction of the rhythms of ribulosediphosphate carboxylase and luciferase activities converge at membrane-mediated events, resulting in periodic permeability changes in chloroplast or other subcellular membrane envelopes (Sweeney, 1974a).

E. Circadian Rhythm of Thylakoid Membrane Spacing in Chloroplasts

A circadian rhythm of chloroplast ultrastructure has been documented in circadian-synchronized populations of the marine dinoflagellate *G. polyedra* (Herman and Sweeney, 1975). Cultures were maintained at 21°C on a $LD:12,12$ light–dark cycle, transferred to constant illumination (600 lx), fixed, sectioned, and examined in the electron microscope. Longitudinal sections of whole *Gonyaulax* cells fixed at different circadian times revealed that the nucleus and Golgi bodies are concentrically arranged with respect to a point in the center of the cell where a large aggregation of cytoplasmic ribosomes are seen. No chloroplasts were found within the radius of this Golgi sphere, but beyond it they project radially toward the cell periphery. The apical portions of chloroplasts which project toward the cell center undergo a circadian variation in structure in a constant-light environment, and at a time corresponding to midday (ct = 0600 hr) the stacks of thylakoids in the inner parts of the chloroplasts are further apart than those at the distal part of the chloroplast. This configuration was termed "expanded" thylakoid spacing and is not seen in cells fixed at midnight (ct = 1800 hr), which remain in the "contracted" phase. Expansion is initiated first at the innermost portion of the chloroplast and proceeds distally from 0200 to 0600 hr circadian time. The expanded portion of the thylakoid stacks is characterized by relative depletion of chloroplast ribosomes which are seen in greater abundance in the stroma between thylakoid stacks in completely unexpanded chloroplasts, or in particularly great abundance between thylakoids at the transition zone between widely separated and closely spaced thylakoids. Herman and Sweeney (1975) have discussed the significance of these ultrastructural circadian correlates in terms of a possible protein synthesis gradient originating at the dense ribosome cluster located in the cell center and have suggested experimental tests of the dependence of the chloroplast form rhythm on protein and RNA biosyntheses as well as photosynthesis. Studies on the possible relationship of this rhythm of chloroplast form to the well-known rhythm of chloroplast swelling in *Acetabularia mediterranea* (Sweeney and Haxo, 1961; Vanden Driessche and Hars, 1972a,b) would be of great interest, as they might reveal a common basis for circadian oscillator control of similar processes in phylogenetically different species.

F. Circadian Changes in Particle Size and Number of Peripheral Vesicle Membranes

The possible importance of membranes to the control of circadian oscillations has given rise to several hypotheses by which membranes may generate circadian periodicity (Njus *et al.*, 1974, 1976; Sweeney, 1974a; Sweeney and Herz, 1977). Sweeney (1976b) investigated the ultrastructure of the thecal membranes of *G. polyedra* by freeze–fracture techniques and observed significant changes in membrane particle distribution and

sizes with time in the circadian cycle. Freeze–fracture techniques applied
to cellular material create two half-membranes by splitting the lipid
bilayer. The fracture face of the half-membrane closest to the extracellular space is designated the extracellular (E) face, while the fracture face of
the half-membrane closest to the cytoplasm is designated the protoplasmic (P) face (Branton *et al.*, 1975). In particular, the number of particles
on the E face of the peripheral vesicle doubled at 1800 hr circadian time
compared with 0600 hr circadian time, and the 120 Å class particles were
more numerous at 1200 and 1800 hr circadian time than at 0000 and 0600
hr. Because of the relative scarcity of P faces of these thecal membranes,
no measure could be reliably made that would rule out interconversion of
particles between the two faces of the membranes. Further speculations
(Sweeney, 1976b) regarding the functional significance of circadian particle
frequency and size changes in the peripheral membrane involve correlations of time of maximal luminescence increase and time of maximal
increase with the increase in particle number and size ($ct = 1800$ hr); this
may correspond to the appearance of membrane-bound luciferase during
the time of peak luminescence capacity. Equally probable is her suggestion that changes in particle number and size distribution reflect the incorporation of new protein molecules into the membrane as well as changes
in the orientation of particles between the different faces. These new
findings are consistent with certain expectations of the membrane model
(Njus *et al.*, 1974), namely, that circadian oscillations are the result of
feedback mechanisms, one component of which is the arrangement of
transport proteins within one or more unspecified cellular membranes,
and the other component, the distribution of an ion gradient across the
membrane, which is in turn regulated by the transport proteins of the
membrane. However, the changes in membrane-associated particles may
be only another example of a process driven by the circadian oscillator,
rather than part of the clock mechanism itself. Further information on the
generality of membrane-associated particle size and number changes in
other circadian synchronized cells is needed before circadian-associated
membrane states can be attributed to the operation of a membrane clock.
This is particularly so, as no circadian changes in the P face of the cyclo-
plasmic membrane were observed in *Gonyaulax* (Sweeney, 1976), and
while the possibility remains that additional freeze–fracture studies will
reveal circadian variations in other membranes it is unlikely that all overt
rhythms are demonstrably linked to ultrastructural changes in membranes. A more fruitful approach suggested by Sweeney and Herz (1977)
is to try to correlate particle number and size changes with phase-
resetting perturbations under permissive and nonpermissive conditions for
expression of a given overt rhythm.

G. Circadian Rhythm of Intracellular Potassium Content: Phase Resetting by Ethanol and Valinomycin Pulses

A circadian rhythm in the intracellular content of K^+ has been reported in circadian synchronized cultures of *G. polyedra* (Sweeney, 1974b). Interest in this discovery centers on the possible significance of specific ion fluxes and membrane-bound transport proteins in the generation of circadian rhythmicity as described above. For this reason, Sweeney (1974b) examined the effect of short exposures to ethanol on free-running circadian-synchronized *Gonyaulax* and showed that it causes phase shifting in the rhythm of stimulated bioluminescence. This agrees with previous findings which showed that ethanol causes phase shift in rhythms in *Phaseolus* (Bünning and Baltes, 1963), as well as increases in the length of the period of the free-running rhythm in *Phaseolus* (Keller, 1960). Similarly, the antibiotic valinomycin, which is known to enhance specifically the transport of K^+ through membranes (Grell *et al.*, 1973; Pressman, 1968), and which has been shown to phase-shift the rhythm of circadian leaf movements in *Phaseolus* (Bünning and Moser, 1972), caused phase shifts in *Gonyaulax* (Sweeney, 1974b). The phase response curves for the effects of ethanol (0.1%) and valinomycin (0.1 μg/ml) on the phase of the circadian rhythm of stimulated bioluminescence are shown in Figure 21. The phase response curve for ethanol closely resembles that for light in cultures grown under similar conditions. In these experiments, when valinomycin was administered along with ethanol, the action of the former was found to negate directly the effect of the latter on phase; i.e., the phase of the bioluminescence rhythm returned to that of the untreated cell suspension. Measurements on the intracellular K^+ content immediately after a 4-hr pulse of ethanol indicated a 50% reduction as compared to untreated controls, while in the presence of valinomycin (0.1 μg/ml) and ethanol (0.1%) only a slight reduction was observed in treated cells compared to untreated controls. However, exogenously added increased concentrations of K^+ or Na^+ for 4 hr had little or no apparent effect on the rhythm of stimulated bioluminescence. The greatest phase delay after ethanol treatment occurred at about 12 hr circadian time, at the peak of the rhythm in intracellular K^+ in untreated cells.

The resemblance of the ethanol phase response curve to that obtained for light pulses given at different times during the free-running circadian oscillation suggests that phase-shifting effects of light on the underlying circadian oscillator may be membrane-mediated, in agreement with the hypothesis that the mechanism by which circadian rhythms are generated involves oscillation in the physical properties of biological membranes.

In other experiments with *Gonyaulax* (Sweeney and Herz, 1977;

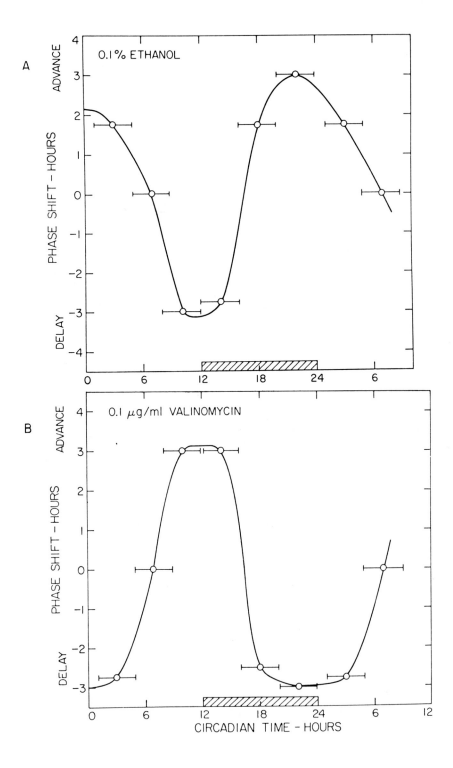

Sweeney, 1976a) no phase-shifting effects were observed for the ionophore gramicidin ($2 \times 10^{-7} M$) when it was administered as a pulse between 1100 and 1500 hr circadian time, i.e., the time in the circadian cycle when valinomycin causes the large positive phase shift. This result appears to rule out a combined increase in fluxes of both K^+ and Na^+ through membranes in generating phase shifts. Also, when the ionophore A23187, which enhances transport for divalent cations such as Ca^{2+} and Mg^{2+}, was assayed at two concentrations and several circadian times, no significant phase shifts were observed. These observations point to a specific involvement of K^+ transport in circadian timekeeping rather than a general ion involvement. Indeed, when valinomycin and K^+ were administered for a 4-hr pulse (at 1800 to 2200 hr circadian time), the phase delay was found to be twice that compared to valinomycin alone. The effect appears to be limited to the time in the cycle when phase delays were expected; advances were not obtained by additional potassium. The application of acetoazolamide (5 mM), an inhibitor of carbonic anhydrase, the enzyme which controls the rate of carbon dioxide take into cells and organelles, caused a 3-hr advance when given between 0600 and 1000 hr circadian time but caused only inconclusive shifts at other times. The photosynthesis inhibitor CCmP delayed the phase when given as a pulse between 0000 and 0600 hr circadian time but was ineffective at other times of the circadian cycle. The latter observation might indicate that active transport in membranes, mediated by ATP availability, is required for normal functioning of the circadian oscillator at the beginning of the circadian oscillation, as CCmP is known to inhibit active transport through the reduction of available ATP. These various inhibitors have in common the membrane physiology as a target, and their specific effects on the circadian phase are accounted for, according the membrane model, by assuming that one portion of the phase response curve is associated with active ion pumping and a second portion with passive ion diffusion. In the *Gonyaulax* system, Sweeney and Herz (1977) have tentatively identified these two segments of the phase response curve as segment C (0000–0600 hr circadian time) and segment AB (0600–24 hr circadian time).

Figure 21.   Phase response curves for the effects on the circadian rhythm phase in stimulated bioluminescence in *G. polyedra* strain 60P brought about by 4-hr exposures to 0.1% ethanol (A) and to 0.1 μg/ml valinomycin plus 0.1% ethanol calculated with reference to ethanol only (B). Note that phase advances are considered positive in sign and are plotted upward from zero on the ordinate. The abscissa is circadian time, using the convention that dawn equals 0 hr. The time of the dark period of the previous entraining light–dark cycle is shown as a hatched bar on the abscissa. Cells in all experiments were in 60–75 foot-candles of continuous light at 22°C. (From Sweeney, 1974b, reproduced by permission.)

H. Loss and Restoration of Circadian Rhythmicity in Either Constant Bright Light or at a Critical Temperature

When cultures of *G. polyedra* are placed at 11° and 4°C, the circadian rhythm of bioluminescence is lost, even though the cells continue to luminesce at a low temperature. Rhythmicity is restored when the cultures are returned to the higher temperature (20°C), with the phase apparently determined by the time of transfer to the higher temperature (Njus *et al.*, 1977). Previous reports of loss of circadian rhythmicity in *Gonyaulax* involved exposure of circadian-synchronized cells to constant bright light (Bruce, 1960; Hastings, 1964), whereupon transfer back to dim light or constant darkness resulted in resumption of the rhythm, the phase of the rhythm again being determined by the time of return (Sweeney and Hastings, 1957; Hastings and Sweeney, 1958). New observations on the combined exposure or circadian-synchronized populations to a subcritical temperature and subcritical light intensities indicate that effects of two *Zeitgebers* are additive; i.e., rhythmicity is lost under combined low temperature and light intensity treatments that would be ineffective if individually applied. In these experiments, following the return of the culture from either bright light or low temperature, the new phase entirely depends on the time of return to pretreatment conditions.

The lack of persistence of circadian rhythm at a critical temperature (T_c = critical temperature = 12.5°C) or in constant bright light is referred to by Njus *et al.*, (1977) as conditional arrhythmicity; i.e., the overt rhythm is inhibited. They contrast this with historical arrhythmicity which does not involve inhibition of the rhythm, such as is observed when an organism is raised under constant conditions from seed, and where the rhythm can be established by a single sudden change in light or temperature (*Zeitgeber*). The former is associated with nonpermissive conditions and the latter with permissive condition for expression of circadian rhythmicity. In the above studies, transferring the *Gonyaulax* culture at or below the critical temperature appears to stop the circadian oscillation and to hold it at a unique phase (*ct* = 12 hr); this is also the circadian phase at which bright light stops the clock in *Gonyaulax* (Hastings and Sweeney, 1958). Conditional arrhythmicity thus appears to act under diverse nonpermissive conditions during the same phase of the circadian oscillation. The explanation offered for this phenomenon is that the oscillation is stopped at some unique phase point under nonpermissive conditions, and that a return to permissive conditions simply allows it to start up again. This satisfactorily explains the phase dependence of the oscillation on the return time. However, loss of rhythmicity under nonpermissive conditions may result from failure of expression of the rhythm, while leaving the oscillation free to continue unaltered. Failure of expression might occur through uncoupling of the transducing mechanism or could result from subthreshold amplitude oscillations, as predicted from limit cycle models (Njus et al., 1974). If the latter is correct, return to permissive conditions will result in resumption of the rhythm with the phase related to the underlying oscillation and will depend on the isochron structure of the oscillator. Njus et al. (1977) raise the interesting possibility that failure of the expression could be also due to desynchronization of a population of oscillators which continue to oscillate independently under the nonpermisssive conditions. A return to the permissive conditions results in phase shifts which resynchronize the population of oscillators. They argue that the latter interpretation is not supported by the data, which show that short-duration exposure of an already synchronized culture yields only moderate phase shifts from any phase of the cycle, whereas a long-duration stimulus of critical intensity (12°C for 12 hr) is required for loss of rhythmicity. A final possibility raised by the authors is that the oscillation continues under nonpermissive conditions but that the transfer to permissive conditions resets the clock to $ct = 12$ hr from any phase of the oscillation. They point out that this is not likely, since a temperature step between 13° and 20°C within the permissive temperature range does not result in such resets. If the clock is unaltered, then the fixed time of return from a unique phase point of the oscillation must occur by a resetting effect of the permissive condition, while if the clock is altered by the perturbation of nonpermissive conditions but continues to oscillate, it is necessary for the return to permissive conditions to reset the phase of the oscillation as well as to recouple it to the mechanism of expression.

V. CIRCADIAN CLOCK IN *ACETABULARIA*

The unicellular alga *A. mediterranea* displays several circadian rhythms related to chloroplast structure and function: photosynthesis, RNA synthesis, polysaccharide content, chloroplast shape, number of plastids per cell, and ATP content (Vanden Driessche and Hars, 1974). The rhythms of photosynthesis and chloroplast shape are correlated, as their expression depends on the same external light conditions (Vanden Driessche, 1966a,b). The substructural organization of chloroplast was also shown to vary in a circadian rhythm (Vanden Driessche and Hars, 1972a,b). At 0900 hr circadian time the thylakoids are located peripherally and resemble chloroplasts of algae kept in constant darkness. At 1500 hr circadian time (the acrophase of the photosynthetic rhythm) and at 1800 hr circadian time (the acrophase of the polysaccharide content), the thylakoids are dispersed throughout the plastid, and increases in the number of polysaccharide granules are seen. At 0300 hr circadian time, the thylakoids again return to the outer peripheral arrangement, and a reduction in polysaccharide granules also occurs. A division rhythm of chloroplasts has also been reported, which has a generation time of about 7 days (Vanden Driessche and Hars, 1972a,b). The correlations between the rhythm of chloroplast shape and functional capacity of the chloroplast indicate that the changes in shape are related to its photosynthetic state.

A. Persistence of Circadian Rhythm in Enucleate Cells

Sweeney and Haxo (1961) demonstrated that a persistent rhythm of the maximal rate of photosynthesis occurs in enucleate fragments of *Acetabularia*. These results were later confirmed (Richter, 1963), and it was also shown that algae grown under continuous illumination prior to enucleation could be photoentrained by *LD:12,* 12 light–dark cycle. Grafting experiments involving nucleate and enucleate portions or circadian-synchronized *Acetabularia* in different phases of the oscillation (e.g., having a 180° phase difference) demonstrated that the phase of the nucleus determines the ultimate phase of the free-running rhythm (Schweiger *et al.*, 1964). Following actinomycin D treatment, the rhythm of photosynthesis is abolished in intact *A. mediterranea* but not in enucleate fragments (Vanden Driessche, 1966b, 1971). This observation was confirmed by Sweeney, *et al.* (1967) for enucleate fragments of the alga *Acetabularia crenulata*. This apparently paradoxical result requires further explanation, but has been attributed to differences in the control system for mRNA degradation; i.e., short-lived mRNA decays in nucleate algae but is preserved in enucleate fragments.

B. Rhythm of Oxygen Evolution in Individual Cells and Cell Fragments

Further evidence for the persistence of a circadian rhythm in enucleate *A. mediterranea* has been obtained by measuring the oxygen evolution of whole individual cells and cell fragments with polarographic electrode techniques (Mergenhagen and Schweiger, 1973, 1975). Figure 22 shows that a rhythm of oxygen evolution continues under constant illumination in both a capless apical enucleate fragment (B) and in a nucleated basal fragment (C) with nearly identical amplitudes and periods. In fact, pieces of basal fragments as small as 10 mm long without a nucleus display a circadian rhythm of photosynthesis with correspondingly smaller ampli-

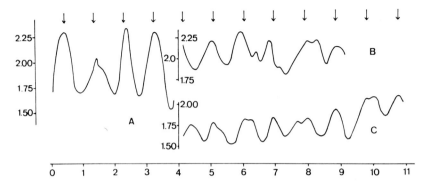

Figure 22. Photosynthesis rhythm in a whole cell, in an apical fragment, and in a basal nucleate fragment. A whole cell was fragmented on Day 4. The oxygen evolution was recorded in the whole cell (A), in the anucleate capless apical fragment (B), and in the nucleate basal fragment (C). Abscissa: time (days); ordinate: oxygen evolution (relative units). The arrows indicate the circadian time. (From Mergenhagen and Schweiger, 1975, reproduced by permission.)

tudes but similar periods. Apical fragments consisting primarily of the cap portion of the algae display a circadian rhythm of photosynthetic activity which can persist for at least 6 days. In later stages of development, when the cap is undergoing cellularization and meiosis, the rhythm of oxygen production is lost and a new steady state level of oxygen evolution is achieved. Since cyst formation initiates multicellular compartmentalization in the cap, it may be inferred that diffusion-dependent processes between nuclei are limited by compartmentalization and that the rhythm is lost through the desynchronization of oscillators, each compartment generating its own circadian oscillation independently of the others (Mergenhagen and Schweiger, 1975).

C. Period and Phase Control in Individual Nucleate and Enucleate Cell Fragments

The method of measuring oxygen evolution by polarographic recording from individual cells and cell fragments permitted Karakashian and Schweiger (1976a) to examine the phase-shifting effect of a single dark pulse interrupting the free-running circadian rhythm of photosynthesis in both nucleate and enucleate cells of *A. mediterranea*. They showed that the average period of the free-running rhythm in continuous light (2500 lx and a constant temperature of 20°C) was not significantly different in enucleate versus nucleate cells. Considerable variability was observed in the period of an individual cell from cycle to cycle, most of which could be attributed to inherent intracellular differences in the activity of the circadian oscillator. Individual cells also exhibited pronounced differences in average period, which appeared to be uncorrelated with size, growth rate, or prior history of individual algae. As predicted from the inherent differences in average periods of individual cells in the free-running rhythm of photosynthetic activity, the synchrony of rhythmicity in a population of five cells damps out after several cycles of free run in constant light.

The effect of two different external temperatures, 18° and 25°C, on the period of the free-running rhythm was determined for individual nucleated and enucleated *Acetabularia* cells (Karakashian and Schweiger, 1976). Overcompensation was observed; i.e., the period at the higher temperature was longer on average than at the lower temperature, with an apparent Q_{10} of about 0.8, regardless of whether a nucleus was present or not.

Individual algal cells, demonstrating a well-defined persisting rhythm of oxygen evolution under conditions of constant illumination, were subjected to a dark pulse of 8 hr and returned to constant illumination for an additional week to establish the posttreatment phase. Figure 23 shows the phase response curve for both nucleated (+) and enucleated (•) cells. Both were similarly sensitive to darkening during the first 12 hr of the circadian cycle and gave roughly equal magnitude phase delays in the earlier interval, while a few hours later dark pulses caused large phase advances. The maximal advances coincided with time of maximal oxygen production during the circadian cycle. These results provide undeniable evidence that some of the basic features of the circadian control mechanism found in higher multicellular organisms, including temperature compensation of the period and phase control of the free-running rhythm by single brief light changes, can be observed in individual *Acetabularia* cells whether they possess a nucleus or not.

D. Coupled Translation-Membrane Model

Schweiger *et al.* (1977) have formulated a coupled translation-membrane model for the molecular basis of circadian rhythms. This model is based on recent experiments testing the effect of the protein synthesis inhibitor cycloheximide on the phase-resetting of the oxygen evolution rhythm in *Acetabularia* (Karakashian and Schweiger, 1976b). They administered pulses of cycloheximide (0.1 μg/ml) for various durations at many different phases of the rhythm, and found a cycloheximide-sensitive phase (between $ct = 0000$ hr and $ct = 1200$ hr), and a cycloheximide-insensitive phase (between $ct = 1200$ hr and $ct = 2400$ hr). The phase of the rhythm is shifted only if the presence of the drug coincides with the cycloheximide-sensitive portion of the cycle. The extent of

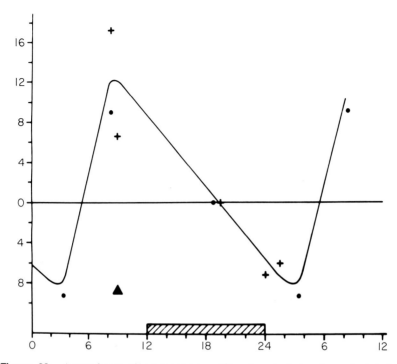

Figure 23. Approximate phase response profile of nucleated and enucleated *Acetabularia* cells subjected to single dark pulses 8 solar hours in length at various times in the circadian cycle. Phase changes are plotted with reference to the midpoint of the dark pulse given. Both nucleated (plus signs) and enucleated cells were phase-sensitive to darkening during the first 12 hr of the circadian cycle; dark pulses given early in this interval induced phase delays, while pulses a few hours later caused phase advances. Similar pulses given late during the second half of the circadian cycle (hatched bar) effected no change in phase. For orientation in the circadian cycle, the maximum oxygen production detected by the electrode is shown at hour 9. The results for nucleated cells were determined with cells maintained at a constant temperature of 25°C, while those for enucleated cells were derived from cells monitored at 20°C. Two of the points obtained for enucleated cells are reproduced on both sides of the profile for clarity of presentation. Abscissa: time in the cycle (circadian hours); ordinate: phase differences relative to original phase (solar hours). Advances are plotted above the horizontal line (0 change) and delays below. (From Karakashian and Schweiger, 1976a, reproduced by permission.)

phase shift is correlated significantly with the length of exposure to the drug. These results establish definitely a dependence of the rhythm on the synthesis of protein. In a later report, Schweiger *et al.* (1977) provided evidence for a periodically appearing membrane protein which has a molecular weight of 30,000 and appears at different times in the cycle, depending on temperature. The protein is synthesized in the presence of cycloheximide, and does not satisfy the criterion of an "essential" membrane protein component of the clock mechanism, but may be an acceptor of an essential membrane protein.

In the coupled translation-membrane model proposed by Schweiger (Schweiger *et al.*, 1977), there are two central components of the oscillating system, membranes and "essential" membrane proteins. Synthesis of the latter is autoregulated by feedback control. Circadian oscillations depend on quantitative and qualitative changes in the pattern of proteins composing the membrane, and those proteins which direct the rhythm are termed essential. Changes in the pattern of proteins composing the membrane are brought out by two processes, assembly (involving synthesis of essential proteins and loading into the membrane), and disassembly (the degradation of essential proteins and unloading of the membrane). The oscillation results from circadian variation in either the rate of loading or unloading. If the rate of assembly exhibits circadian oscillations, then the rate of unloading is assumed to be constant, or vice versa. It is proposed that the degree of loading and the resulting change in the functional state of the membrane are coupled to assembly through a feedback mechanism. When a critical threshold in loading is reached, this reduces the rate of assembly. Regulation operates in a switch-like fashion: the upper threshold leads to a switch-off, whereas a lower threshold leads to a turn-on of the supply of essential proteins. The basis of the circadian period of 24 hr is assumed to result from time lags generated in the loading process (membrane fluidity, etc.), and to be rate limited by gene transcription and translation of essential proteins. An unconsidered possibility is the limit-cycle oscillation in the dynamics of essential protein complexes in the assembly step. A possible candidate for such an assembly-step complex is the tubulin–microtubule anchorage system that is involved in the cytokinetic system of regulating membrane receptor surface distribution. (Wille, unpublished observations).

VI. CIRCADIAN CLOCK IN *CHLAMYDOMONAS*

A circadian rhythm of phototactic response has been described by Bruce (1970) in wild-type strains of *C. reinhardi*. It has the following characteristics. The rhythm can be initiated by a sudden single shift of cultures from continuous light to continuous darkness, a switch-down in the level of irradiance. The rhythm can be entrained to a light–dark cycle, and the phase of the rhythm shifted by exposure to a short interval of dark during certain phases of the free-running circadian cycle. Temperature compensation of the period occurs and it is slightly longer at 18° and 28°C than the characteristic 24-hr period seen at 22°C.

Genetic Dissection of Period Control of the Phototactic
Response Rhythm

Several wild-type strains of *Chlamydomonas,* when characterized for their free-running circadian period lengths (τ) and the phase angle (Φ) relationship of the phototactic response maxima to the onset of the entraining light–dark photoperiod, had a shorter than normal average period (strain 90^-, $\tau = 21$ hr; strain w^-c_1, $\tau = 21$–22 hr). There was one long-period mutant (strain Lo^- 104, $\tau = 26.5$–28 hr) and one phase mutant (strain Lo^-39, $\Phi = 13$ hr, compared to $\Phi = 4$ hr for wild-type strains) (Bruce, 1972). The Lo^- strains are characterized by an inability to utilize nitrate; all other strains can be grown on nitrate. The two short-period strains and the long-period strain were examined for the effect of temperature on the length of the free-running period, and temperature compensation of the period was observed between 15° and 29°C. Interestingly, both the short-period and long-period strains displayed compensation; the periods were slightly longer at the higher temperatures than at the lower temperatures. The result of genetic crosses between wild-type strains and the long-period strain yielded progeny with either a normal period or a long period, and a backcross of long-period progeny with a long-period parent produced only long-period progeny, suggesting Mendelian segregation of the period character at single locus. A cross between the short-period (wild-type strain 90^-) and normal-period (strain 89^+) strains yielded a short-period progeny strain ($Z12^+$) which produced predominantly (11 out of 18 tested progeny) short-period progeny. In addition, a cross between the short-period strain $Z12^-$ and the long-period strain Lo^-104 yielded all three possible categories of segregants—normal-period, short-period, and long-period. However, two of the long-period segregants reverted to a 24-hr period after subculturing, and several progeny displayed atypical behavior, including arrhythmicity, and beat phenomena. A cross between the spontaneous short-period mutant (strain w^-c_1) and the wild-type Lo^+ normal-period strain produced either normal-period or short-period progeny. Experiments with the phase mutant Lo^-39 indicate that a stably expressed characteristic is not involved; sometimes it displays the normal phase instead of the mutant phase.

As the result of further genetic analysis involving a number of independently obtained long-period mutants, Bruce (1974) has established that in four mutants the long-period characteristic is controlled by single genes at separate loci. This contrasts sharply with earlier genetic studies on clock period in *Drosophila* (Konopke and Benzer, 1971) and in *Neurospora* (Feldman and Hoyle, 1973; Feldman *et al.,* 1973), where mutations all mapped on a single gene or on genes so closely linked that no recombinants were found. This is not the case in *Chlamydomonas,* and it has been possible to determine the phenotype of double, triple, and other multiple mutants. Four long-period mutants (per-1, per-2, per-3, and per-4) were characterized for their average periods at different temperatures under free-running conditions. All four displayed overcompensation of the period, in contrast to the previously analyzed period variants (Bruce, 1972).

Three of the four mutants (per-1, per-3, and per-4), when crossed with a normal-period strain (Lo⁻ or Lo⁺), yielded both normal- and long-period progeny, whereas a cross between Lo⁻ and the fourth mutant (per-2) produced short-period (22.5–23 hr) as well as normal- and long-period progeny, and a cross between normal-period progeny from a per-1 × Lo⁺ cross and per-2 produced only long- and normal-period phenotypes. In pairwise crosses between all six possible combinations of the four long-period mutants, both wild-type and double mutant recombinant types were scored, which had either a normal period (23–25 hr) or an extra-long period (30–33 hr). In general, it seems that the period-lengthening effect of the genes is additive and, since both normal and extra-long period recombinants were recovered, no close linkages are indicated between any of the four loci. In crosses between double mutants involving three or four genes no normal progeny were recovered, as expected, while some progeny with periods longer than that of the double mutant were found, i.e., the expected triple mutants.

These results indirectly imply that the circadian clock in *Chlamydomonas* involves many genes, but since none of these independently segregating loci has been mapped, the genetic organization of clock mutants may or may not reveal an underlying genetic mechanism for the circadian clock, e.g., a clustering of clock mutations at a single loci is expected based on the simplest expectations of the chronon model, but chronon redundancy can also account for nonlinked clock mutants as well (Ehret, 1974). The additive effect of long-period-producing genes fits a single master oscillator control system rather than a series of self-oscillatory systems which are mutually synchronizing, since mutations blocking only one of the component clocks should desynchronize the coupled clocks from one another. This situation was not found for any of the mutants analyzed.

VII. THE MITOTIC CLOCK IN CELLS

A. Deterministic and Stochastic Models

The cell division cycle and the mitotic cycle are characterized by a recurrent sequence of discrete cellular events which under certain conditions continue to occur at regular interdivisional periods. In this sense, the cell cycle has the properties of a cellular clock. For example, the interphase of a cell cycle is classically subdivided into three distinct phases on the basis of time of occurrence of nuclear DNA synthesis: a discontinuous synthetic phase, S, which is preceded and followed by gaps, nonsynthetic G_1 and G_2 phases. Indeed, the DNA duplication cycle may itself be decomposed into stages during which only a special fraction of the genome is uniquely replicated in a fixed temporal sequence (Plaut, 1969; Mueller and Kajiwara, 1966; Braun et al., 1965; Adegoke and Taylor, 1977). Likewise, specific inhibitors of RNA and protein synthesis have demonstrated the need for de novo RNA and protein synthesis at a critical transition point in the cell cycle (Mitchison, 1971). Other examples of timed enzyme protein synthesis abound in the cell cycle literature. Control of the timing of cell division in prokaryotes (Cooper and Helmstetter, 1968; Masters and Pardee, 1965) and in eukaryotes (Mitchison, 1971; Hartwell et al., 1974; Thormar, 1959; Zeuthen, 1971; Frankel, 1970; Rasmussen, 1967; Rusch et al., 1966; Rao and Johnson, 1970) has been extensively studied. An excellent review of the literature and a discussion of various models are found in Mitchison (1971). Two broad classes of models have emerged over the past decade. The first proposes that the events of the cell cycle are themselves part of the clock timing mitosis, i.e., that each prior event in the cycle is a necessary condition for the subsequent occurrence of all later events of the cycle. Events of the cell cycle may be grouped into one or more causal sequences which join finally to a closed loop of states. According to this model the periodic occurrence of any event in the cycle is guaranteed by the closed loop, while the regular phase relationship of events of the cycle is accounted for by the causal sequence. Hartwell's (Hartwell et al., 1974) elegant genetic analysis of a yeast division cycle with temperature-sensitive mutations arresting cells at specific phases of the division cycle has led him to propose this "dependent-pathways" type of model. According to this deterministic model the cell cycle is an example of a simple clock (Winfree, 1975); i.e., to go forward or back in time it must pass through the same fixed sequence of events. The second broad class of models proposes the continuous accumulation of a division protein or mitogen, during the cell cycle, which reaches a critical concentration, perhaps converts to a new division structure, initiates mitosis, is used up in the process, and reaccumulates during the subsequent cycle. Extensive heat-shock experiments on the ciliated protozoan Tetrahymena have led Zeuthen and his colleagues (Zeuthen, 1971) to such a division protein model, and other variants of this model have been suggested (Donachie and Masters, 1969). The mitogen model is one form of a central clock hypothesis which accounts for the periodicity of mitosis by the gated appearance of a signal from a central timer, and the phase relations of other events in the cycle are somehow driven by the clock, or by mitosis, but the events of the cycle are not themselves an integral part of the clock or how the clock manages to exhibit periodicity. This differs strikingly from the dependent-pathways model in which the events of the cycle are parts of the clock and its periodicity, not merely phenomena driven by it. A third class of hypothesis derived from mammalian tissue culture studies has recently been proposed by Smith and Martin (1973). According to this, hypothesis, the cell cycle is composed of both an indeterminate part (A state) and a determinate part (B state); the A state is characterized by the random entry of G_1 cells into the S phase, and the B state by programmed DNA duplication and cell division sequences. It should be recalled that this model was designed to account for the distribution of cell generation times of various mammalian tissue culture cells, which predominately have circadian–infradian average generation times. According to this transition probability model the transition from A state to B state could depend on a critical amount of a single substance, regulated by a number of coupled feedback loops. The instantaneous amount of the initiator substance could vary cyclically, but if the number of initiator molecules were rare, the variation would be subject to random fluctuations, and only at irregular intervals would the critical threshold for triggering entry into S be exceeded. Based on these data (Smith and Martin, 1973; Shields, 1977) it has been argued that regulation of the transition probability is a key function that controls cell proliferation rates. Recently, an attempt to apply the transition probability model to the G_1–S transition in yeast (Shilo et al., 1976) by identifying the "start" event as the rate-limiting step of the cycle has come under serious criticism (Nurse and Fantes, 1977). Shilo et al. (1977) argue that the kinetics of bud emergence after release from a block at the start event are the same as in steady state culture, while Nurse and Fantes (1977) argue that this interpretation is only possible if the kinetics of bud emergence after release from the block is due solely to traverse of the start event and not to physiological recovery from the block. In fact, release from the block at the start event, imposed by the alpha factor (a mating substance) or temperature, do not follow the same first-order kinetics. Since the nature of the block at the start event affects the kinetics of bud emergence, this negates comparison with a steady state culture. In addition, the fraction of unbudded cells plotted against time was not continued below 2% (Wheals, 1977), leaving open a choice between several different distributions including the first-order kinetics, log-normal, and reciprocal normal distributions, while only the former is consistent with the transition probability model. Both log-normal and reciprocal normal distributions are consistent with the critical cell size hypothesis (Fantes *et al.*, 1975). According to this hypothesis the rate of entry of cells into the budded stage could be due to the heterogeneity in the size of the unbudded cells and the time needed to grow to a critical size. However, Shilo *et al.* (1977) examined the proportion of unbudded yeast cells growing exponentially on solid agar plates and found that the frequency of unbudded cells declined exponentially from 18 to 1%, according to first-order kinetics. Shilo *et al.* (1977) also contend that the transition probability model and the critical cell size hypothesis are not mutually exclusive. They admit that it is possible that the physiological condition of the cell could affect the start event by imposing a threshold requirement such as minimal size. They further argue that heterogeneity in the size of unbudded cells cannot be the sole reason for an asynchronous rate of cell cycle initiation, as the second cycle of budding following release of the start event from the temperature block is asynchronous and the first budding is synchronous, which is not predictable from the critical cell size model. Moreover, Shilo *et al.* (1976) found that the kinetics of cycle initiation was the same for cells arrested without growth and those which were arrested at the start event and continued to grow.

Perhaps the simplest interpretation of asynchronous transit of yeast cells, from the start event to the initiation of the budding cycle, is that the accumulation of a threshold level of initiation factors for entry into S is normally coupled to the growth cycle, but under nonphysiological conditions the two processes are uncoupled. It would be interesting to know whether the period of additional cell growth during the block of the start event led to a shortened second budding cycle in the fraction of unbudded cells that were arrested.

B. Control of Mitotic Timing in the Slime Mold *Physarum*

The hypothesis of a cytoplasmic mitogen, or division protein, is strongly supported by the fact that most binucleate cells resulting from somatic fusion show mitotic synchrony (Rao and Johnson, 1970), and by the striking spontaneous synchrony of nuclear mitosis in syncytial organisms such as *Physarum* (Rusch, 1970).

The plasmodial stage of the acellular slime mold *Physarum polycephalum* typically displays a natural synchrony of nuclear mitosis among up to 10^8 nuclei in a common cytoplasm (Rusch, 1970). The rhythm of mitosis is endogenous and has a relatively temperature-dependent ultradian period of 8–12 hr, which persists for several cycles of plasmodial growth. The phase of the rhythm in a given plasmodium is determined by the time of fusion of numerous microplasmodial fragments, which coalesce within 40–60 min to form the growing macroplasmodium that then undergoes its first postfusion synchronous mitosis 6–7 hr later. Thereafter, a recurrent rhythm of synchronous nuclear mitoses occurs at 8- to 12-hr intervals. Massive nuclear and cytoplasmic mixing between any two plasmodia at all possible phases of the mitotic cycle can be accomplished by grafting together pieces of different plasmodia. Two methods are currently practiced. In the first, one piece is placed directly on top of the other, and in the second the cut edge of one piece is placed close to the other. The first method ensures rapid and effective mixing of the contents of both pieces, while the second allows slow mixing and preserves the original parental geometries. With the use of the first method it has been shown that, if a plasmodium scheduled to mitose at 3 P.M. is fused to a plasmodium due to undergo mitosis at 1 P.M., in approximately equal mass ratios, the nuclei in the mixture synchronously divide at 2 P.M. In similar fusions which employed 2 volumes of the 1 P.M. plasmodium to 1 volume of the 3 P.M. plasmodium, the nuclei in the mixture synchronized closer to the 1 P.M. parent's time of mitosis, and vice versa (Rusch et al., 1966). These results suggest that the phase of the nuclear rhythm is internally reset in fusions by the various concentrations of a finely graded diffusible cytoplasmic substance, a mitogen or division protein. Further fusion experiments involving plasmodia a half-cycle apart at many different phases effected by the rapid-mixing method (Sachsenmaier et al., 1972) showed that, when synchronization was achieved in the mixture, nuclei early in the cycle at time of fusion were phase-advanced by about a half-cycle, while those late in the cycle at time of fusion were phase-retarded (delayed). The compromise phase arrived at through the synchronization of nuclei in the mixture is approximately the arithmetic average of the parental phases (Figure 24). A discontinuity in phase resetting is seen when fusions occur between one plasmodium which is within 1 hr of its expected mitosis and another plasmodium a half-cycle earlier in the cycle. The locus of this phase discontinuity is coincident with events associated with entry of nuclei into early prophase. These observations were taken as evidence for the gradual accumulation throughout the mitotic cycle of a mitogen protein which reaches a critical threshold concentration just prior to mitosis. According to this "mitogen accumulation" model, mitogen decline occurs when a threshold level of mitogen titrates a fixed number of nuclear sites (Sachsenmaier et al., 1972).

The familiar mitogen model accounts for the periodic behavior of the central clock by assuming that the triggering of mitosis uses up all mitogen which must accumulate to a threshold during the next cycle. This relationship is depicted in Figure 25. Phase varies smoothly throughout the cycle

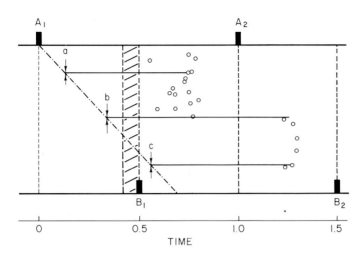

Figure 24. Fusion of plasmodial pieces derived from two differently phased macro-plasmodia (A and B) at various time intervals. A summarized plot for four independent experiments is shown. Two macroplasmodia (parent cultures A and B) were cultivated with a mitotic phase difference of approximately a half-cycle. At various intervals after mitosis (A_1) small pieces (~ 1.5 cm^2) were excised from A and B and placed on top of each other. The horizontal projection of each point onto the oblique line indicates the time when fusion was started. Three examples are marked with double arrows (a–c). The time unit is the duration of one division cycle (~ 10 hr). Variation among different experiments was observed for the actual length of the division cycle (9–12 hr) and for the phase difference (2.8–5 hr) of the parent cultures A and B. The time of mitosis (telophase) in the parent cultures is shown by the bars, and mitosis in mixed plasmodia occurs at zero. The shaded area indicates the entire period of mitosis (B_1) (prophase to telophase). (After Sachsenmaier *et al.*, 1972, reproduced by permission.)

as a function of mitogen concentration, with an abrupt change at mitosis. Mitogen accumulation has a sawtooth waveform and is an example of an extreme relaxation oscillation. Past studies on the effect of the protein synthesis inhibitor cycloheximide on mitotic timing of *Physarum* (Cummins *et al.*, 1966; Murakami and Ohta, 1971) suggest that mitogen accumulation requires continuous protein synthesis throughout most of the G_2 phase. Recently, Scheffey and Wille (1978) found that cycloheximide pulses applied in G_2 cause a delay in mitosis which is linearly dependent on the phase in the mitotic cycle at which the pulse is applied (Figure 26). A 30-min pulse of 10 μg/ml cycloheximide starting in G_2 at a time after mitosis induces an excess delay (the delay minus the duration of the pulse) of the next mitosis of $(0.55)t - 1.3$ hr, while pulses applied less than 30 min before mitosis induce only small delays. The linear dependence of delay upon phase can be interpreted as evidence for the synthesis of an

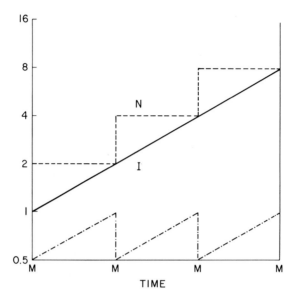

Figure 25. Schematic diagram of the kinetics of the total mass (solid line) and the number of nuclei (dashed line) of exponentially growing synchronous plasmodia of *P. polycephalum*. The ratio of the total mass to the number of nuclei is shown by the dotted-dashed line. N, Nuclear receptor sites; I, initiator; M, synchronous nuclear mitoses. (From Sachsenmaier *et al.,* 1972, reproduced by permission.)

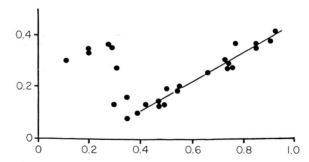

Figure 26. Excess mitotic delays induced by 30-min pulses of 10 μg/ml cycloheximide. Ordinate: excess delay (delay in excess of pulse duration) plotted as proportion of a control cell cycle; abscissa: proportion of cell cycle completed when pulses begin. Mitosis occurs at 0 and 1.0, and S ends at about 0.3. (From Scheffey and Wille, 1978, reproduced by permission.)

unstable mitogen protein which accumulates gradually during the cycle. This is in accord with earlier heat-shock experiments (Brewer and Rusch, 1968) which support the hypothesis of an unstable mitogen substance which gradually accumulates to a peak about 2 hr prior to mitosis. One possible candidate for the mitogen protein(s), which has been offered by Bradbury *et al.* (1974a,b) is the nuclear histone H1, which undergoes peak G_2 levels of phosphorylation at the same time as peak sensitivity to heat shocks. Further evidence in support of phosphorylated H1 histone as a putative mitogen protein is phase advance of mitosis by application of calf thymus H1 phosphorylating enzyme to the surface of late G_2, plasmodia (Inglis *et al.*, 1976). Phase advance of mitosis in late G_2 plasmodia by water-soluble extracts of G_2 plasma plasmodia has also been reported (Oppenheim and Katzir, 1971). More recently, Matsumoto (1977) found that extracts of late G_2 plasmodia can offset some of the delay induced by ultraviolet treatment of G_2 plasmodia. Unfortunately, unpublished observations in several independent laboratories (Wille, Sachsenmaier, Mohberg) have been unable to repeat the mitotic advance effect of *Physarum* extracts prepared according to the procedures of Oppenheim and Katzir (1971). Further investigations are required to resolve these discrepancies. Indeed, such experiments suffer from the technical disadvantage that the maximal amount of observable phase advance is often less than 1 hr. Given the variance of mitotic timing among replicates of treated and untreated plasmodia, the variance of the average amount of phase advance is generally within the limits of experimental error.

In summary, at present there is no direct evidence linking the timing of the mitotic cycle with a particular mitogen protein.

C. Evidence for a Limit Cycle Oscillator Controlling Mitotic Timing in *Physarum*

Kauffman and Wille (1975) have proposed a limit cycle biochemical oscillator model for the control timing of mitosis in plasmodia of the slime mold *Physarum*. The model postulates the synthesis of an inactive mitogen precursor X at a constant rate throughout the cycle, which is converted enzymatically to an active form Y at a constant rate. The active form of mitogen autocatalyzes more of itself from the inactive form at another rate and decays with first-order kinetics. This system of biochemical reactions leads to a sustained oscillation in the concentrations of the precursor and active forms of the mitogen, and mitosis is triggered when the level of the active form exceeds a critical concentration Y_c (Figure 27). The model predicts that the periodicity of mitosis is independent of the

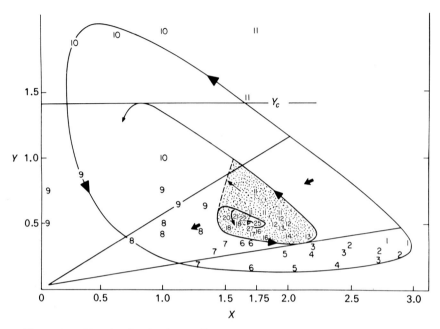

Figure 27. Plot showing the time until the next crossing of Y_c from any state inside the XY limit cycle oscillation. The shaded region shows the concentric ring of states centered around the steady state, S, of the oscillations which are more than one or two cycles away from their next crossing of Y_c. (From Wille *et al.,* 1977, reproduced by permission.)

occurrence of mitosis itself. The model has been analyzed in detail elsewhere (Tyson and Kauffman, 1975). A number of predicted consequences of the oscillator model have been tested and confirmed by both fusion and heat-shock experiments on syncytially synchronous *Physarum* plasmodia (Kauffman and Wille, 1975; Wille *et al.,* 1977). In the heat-shock experiments, it was assumed that the effect of heat shock was to destroy both X and Y equally, and at a rate proportion to their respective concentrations. Then, the effect of destroying various percentages of X and Y for each phase of the oscillation should be to displace the biochemical system toward the state $X = Y = 0$, as shown in Figure 28. The model predicts (1) a phase advance of the next mitosis if shocks are applied shortly after mitosis. (2) From 2 hr after mitosis on, short duration shocks should produce a delay which increases gradually to a maximum in late G_2 and then declines to no delay just before mitosis. (3) Heat shock shortly before mitosis displaces the oscillation ''inside'' the limit cycle toward the nonoscillating phaseless steady state point, S; a short-duration shock should leave the system in a state which is followed by a trajectory crossing Y_c inside the limit cycle, rather than in which the limit cycle trajectory crosses Y_c; since the distance from this crossing of Y_c to the next crossing of Y_c is shorter than the full distance around the limit cycle, such plasmodia should exhibit a shortened cycle, the first full cycle after the heat shock. (4) From the immediate vicinity of S, a trajectory takes three full cycles to spiral out with sufficient amplitude to cross Y_c; thus the inside of the limit cycle can be divided into regions from which a trajectory crosses Y_c less than one full cycle later, crosses Y_c between one and two full cycles later, and between two or three full cycles later. These regions are roughly concentric around S (see shaded zone in Figure 27). (5) Heat shocks late in G_2, but not several hours after mitosis, drive the oscillating system inside the limit cycle toward the singularity, S. If longer heat shocks drive the system further toward $X = Y = 0$ than shorter-

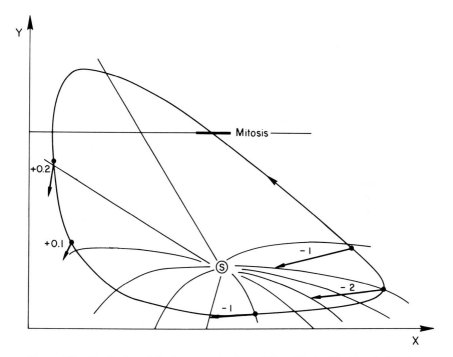

Figure 28. The limit cycle in the concentrations of X and Y. A critical level of Y triggers mitosis. The nearly radial lines emanating from a point inside the cycle are "isochrons" separating equal intervals of time along trajectories, and along the limit cycle path, and indicates hours before mitosis. The effect of 20% destruction of X and Y from five phases is shown as destruction vectors (arrows) which indicate the direction and magnitude in delay $(-)$ or advance $(+)$ of the next mitosis. All isochrons meet at the steady state singularity, S, inside the limit cycle.

duration shocks, then only in late G_2, as heat shock duration increases, mitotic delay should abruptly jump from a slight delay to slightly more than one full cycle of delay, or even two full cycles of delay; i.e., shocks applied in late G_2 should yield plasmodia which skip mitosis for two full cycles, while similar duration shocks applied to plasmodia in early G_2 whose oscillating system is on the side of the limit cycle closest to the state $X = Y = 0$ should never cause skipping of a full cycle or even two full cycles, as the oscillating system should be driven outside the limit cycle toward $X = Y = 0$. (6) As heat shock duration increases for shocks applied in late G_2, it should be possible to drive the biochemical oscillation past the singularity, S, in the direction going toward $X = Y = 0$. As a point of the limit cycle is perturbed successively further toward $X = Y = 0$, its delay until the next crossing of Y_c should suddenly jump from 2–3 hr to about 13 hr, and then to about 27 hr and, as it is driven well past S, its delay should suddenly drop to about 9 hr. The results of a large number of short-duration heat-shock experiments (Kauffman and Wille, 1975) and long-duration heat-shock experiments (Wille *et al.*, 1977) have largely confirmed these predictions, as well as retrodicting the results of other past heat-shock experiments on *Physarum* plasmodia (Brewer and Rusch, 1968). In addition to accounting for the various effect of heat shocks, the biochemical oscillator model has successfully accounted for the results of a large number of fusion experiments involving effects on mitosis, as well as predicting the ultimate phase of the fused pair as a function of the parental phases fused. The oscillator model has been highly successful in describing a number of important features of the mitotic clock in *Physarum*. As a caveat it must be emphasized that the experimental data published so far on the mitotic control system in *Physarum* (Kauffman and Wille, 1975; Wille *et al.*, 1977) do not unequivocally reveal the existence of a nonoscillating or phaseless steady state as predicted by the limit cycle model. This is *the* central distinction between the limit cycle biochemical oscillator model and any other alternative model. It may well be that the initiation of mitosis in *Physarum* is a genuine phase discontinuity occurring once each mitosis when nuclei enter prophase. This has been previously suggested as an extreme relaxation oscillator form of the mitogen accumulation model (Sachsenmaier, 1976). Moreover, there is some question whether the slow-mixing fusion technique employed by Kauffman and Wille (1975) adequately ensures uniform phase homogeneity throughout the fused plasmodium. For this reason, Sachsenmaier (1976), Tyson and Sachsenmaier (1978), and Winfree (1978, in preparation) have raised the possibility that the phase compromise data published by Kauffman and Wille (1975) do not represent permanent phase shifts but transients. An additional criticism has been voiced (Tyson and Sachsenmaier, 1978) against the interpretation of mitotic delays produced by prolonged heat shocks at elevated temperatures. They have shown that the normal growth rate of plasmodia as measured by the protein/DNA ratio is abnormally high immediately after a 1-hr heat shock and does not decline to normal levels until well after the first full postshock synchronous mitosis. They argue, rightly, that one can only draw the conclusion that mitosis and DNA synthesis are downstream events triggered by an central timer like the biochemical oscillator model of Kauffman and Wille, when heat shocks only act directly and instantaneously on the control variables [the assumption made but not verified by Kauffman and Wille (1975)]. Since this is not the case, a phenomenon such as "catch-up" shortened second cycle lengths following the first postshock mitosis could as easily be explained by the experimentally demonstrated high protein/DNA ratio induced by heat shock. Given this current state of uncertainty, it would be advisable to refrain from readily accepting the limit cycle interpretation of mitotic timing control in *Physarum*.

VIII. GENERAL CONSIDERATIONS OF MITOTIC AND CIRCADIAN CLOCK DYNAMICS: HOW MANY CLOCKS?

A. Interplay of Several Loosely Coupled Clocks

It is apparent from the foregoing discussion that cells have at least one or more independent clocks which may be only loosely coupled to the respective processes they drive. The integrative behavior of cells as expressed through these homeostatic mechanisms may result from an interplay of partly independent, partly interconnected dynamic systems, each of which has its own characteristic time constants, yet is coupled through common variables. This type of interaction has been suggested by Mitchison (1971) for the yeast cell cycle which appears to consist of a causal sequence driving DNA duplication and cytokinesis, loosely coupled to a growth cycle. This is borne out by the genetic analysis of yeast cell division cycle made by Hartwell *et al.* (1974). Their temperature-sensitive mutants show two apparent causal sequences: (1) initiation of DNA synthesis, continuation of DNA, medial nuclear division, and late nuclear division; and (2) bud initiation, bud elongation, and nuclear migration. Hartwell *et al.* propose that the two sequences must join to cause cytokinesis, but that the first sequence alone suffices to close a causal loop to a start event which is itself a necessary event for the initiation of both sequence 1 and sequence 2. Interestingly, Hartwell has a mutant unable to initiate DNA synthesis, which undergoes repeated rhythmic budding, and a second mutant blocking bud initiation which undergoes at least two rounds of S. Hence neither sequence 1 nor sequence 2 is required for the temporary recurrence of the other. These results suggest rather that periodicity is independent of the two causal sequences and seems to support a central clock model. Actually, Hartwell favors a combination of a dependent-pathways clock for the DNA duplication–cell division sequence and a central clock for the bud growth sequence.

A possible interplay of the cell developmental cycle, circadian clock, and membrane-associated oscillatory phenomena has been proposed and discussed at length by Edmunds (Palmer *et al.*, 1976; Edmunds and Cirillo, 1974). In the coupling between any two of these oscillatory systems, there exists a formal relationship in which either one system is passively driven by the other or itself serves as the driving oscillation for the other system. For the cell cycle–circadian oscillator combination it is clear that the cell cycle can be uncoupled from the circadian clock as evidenced by ultradian cell cycle periods. In the reciprocal relationship it appears that the cell cycle can be gated by a circadian oscillator, and as the frequency of cell division declines to zero, the cirdadian oscillator continues to drive other overt thythms uncoupled from the rhythmic recurrence of cell division. Yet it cannot be concluded that the circadian oscillator is entirely independent of the cell division clock, as the capacity to manifest circadian regulation is restricted to cells in the circadian–infradian mode of growth. One notable exception to the circadian–infradian rule was found in the clock regulating circadian rhythm of sex reversals in *P. multimicronucleatum* which may be capable of surviving ultradian cell division without losing phase information (Barnett, 1966). Alternative explanations for such exceptions have been offered (Ehret and Wille, 1970).

The existence of circadian oscillations in nondividing cells (Edmunds, 1977) is the only evidence that observable events of mitosis and cell division are not necessary for persistence of the circadian oscillation. We have already seen that the mitotic oscillation controlling the timing of mitosis in *Physarum* continues in the absence of mitosis for at least two cycles (Wille *et al.*, 1977). It is entirely conceivable that the mitotic oscillation continues in nondividing cells which have circadian–infradian average generation times and are circadian synchronized. This is in fact the limit case of the cell cycle proposed by Ehret (Ehret and Dobra, 1977) and discussed above. In this situation, it is proposed that the events which trigger cell division are uncoupled from the central transcription clock which continues to generate transcripts according to a genetically determined program.

B. Cellular Oscillator Hypothesis

Another hypothesis is that there is just one central timer, a cellular oscillator which controls both the timing of mitosis and circadian timing in cells. It is a limit cycle dynamic system which has at least two different periodic modes. In the ultradian mode, the period of the limit cycle is always equal to the cell cycle time; i.e., there is a mitotic oscillator. In the circadian–infradian mode the period of the cellular oscillator is independent of the average generation time and is equivalent to the circadian cell cycle period; i.e., there is a circadian oscillator. If we further posit that the cellular oscillator is a biochemical oscillator of the sort proposed for regulation of mitotic timing in *Physarum* (Kauffman and Wille, 1975; Wille *et al.*, 1977), we might anticipate that transition from one periodic mode to another is expected in a limit cycle oscillator control system, when the constant kinetics characteristic of one set of parameter values is changed to another set. Multiple nonequilibrium stationary states and sustained oscillations have been found in a variety of theoretical and experimental systems (Higgins, 1967; Othmer and Scriven, 1974; Turing, 1952; Zaiken and Zhabotinskii, 1970; Othmer, 1975). Othmer (1975) has presented a detailed mathematical treatment of the dynamic system underlying the Zhabotinskii–Belousev reaction. The number and stability of stationary and periodic solutions of the semiphenomenological reaction mechanism that qualitatively models the Zhabotinskii–Belousev reaction were studied as functions of the parameters in the model. This procedure is known as phase plane analysis of the model's kinetics. Othmer showed that both multiple stationary solutions and multiple periodic solutions can exist simultaneously, and that periodic solutions bifurcate in at least several ways, depending on the amplitude. In each case the period and amplitude of the periodic solutions were dependent on the parameters. The finding of the existence of bifurcations of multiple limit cycles at finite amplitude in theoretical models of isothermal systems is especially relevant to the contention raised above that one dynamic system can admit of a variety of temporal behaviors. The transitions between critical points and periodic orbits may require large perturbations in parameter values ("hard" bifurcations). The system can display hysteresis in that whether or not it flips between two distinct limit cycles at certain parameter values depends on whether the value of the controlling parameter is increasing or decreasing. These hysteresis effects are more pronounced the smaller the amplitude of the disturbance. Interestingly, Othmer's (1975) analysis led him to conclude that the applicability of deterministic equations is questionable near parameter values that correspond to bifurcations of multiple periodic or stationary solutions, and that stochastic analysis may be more appropri-

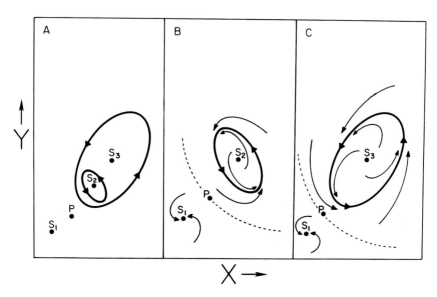

Figure 29. Phase portrait of a hypothetical *XY* biochemical oscillator showing multiple critical points and limit cycles which result from parametric control of a hypothetical two-variable biochemical cellular oscillator. (A) The possible stable behaviors of the dynamic system as a value of the parameter, σ, are varied over a wide range. A single stationary point, S_1, exists for values of σ in the neighborhood of σ_1. For very small changes in the value of σ near the bifurcation point, P, the system can flip from asymptotic stability at S_1 or to the low-amplitude limit cycle which orbits around the unstable focus, S_2. This possibility is shown in (B). For increasing values of the parameter σ greater than σ_2, the dynamic system has a periodic solution with a high-amplitude limit cycle oscillation around another unstable focus, S_3. The latter situation is depicted in (C). Heavy lines show the trajectories of the limit cycle path. Light arrows show the path of trajectories from the state of the system not on the limit cycle as they approach the limit cycle in time. The dotted lines extending away from P show the separatrix which divides the regions of attraction of the stable focus, S_1, from the basins of attraction of the limit cycles.

ate. The latter finding may bear on the transition probability model proposed by Smith and Martin (1973). Whether or not there is a stochastically determined part of the cell cycle appears to be intimately associated with the coupling of growth with cytokinesis. Growth-contingent limitations also are intimately associated with the transition of ultradian mode cells to the circadian–infradian mode prior to the cells' acquisition of the potential to express circadian rhythms (Ehret and Dobra, 1977). In both instances the entry of cells into either the mitotic cycle or circadian cell cycle is preceded by a growth-sensitive stage. These similarities may appear to be merely fortuitous, but for the fact that the transition between the circadian cell cycle and the mitotic cycle is gated by the circadian oscillator to those phases of the oscillation which correspond to the G_1–S boundary of the ultradian cell cycle (Ehret *et al.,* 1974). On the basis of these considerations, I suggest that the mitotic control system and the circadian control system are two periodic modes of the same underlying dynamic system. The appearance of a growth-sensitive phase in the mitotic cycle of cells coincident with a G_1 period and from which cells exit randomly into the deterministic part of the cycle is consistent with the parameter-dependent bifurcation of the dynamic control system from a critical point (a G_0-like state) into either the mitotic cycle or the circadian cycle, the choice being controlled by the strength of the perturbation leading to an increase or decrease in parameter values. This situation is diagrammatically represented in Figure 29 for our hypothetical two-variable (XY) cellular oscillator. Large perturbations leading to increasing parameter values from initially low-amplitude values, yield hard bifurcations resulting in a large-amplitude oscillation characteristic of the mitotic oscillation (Figure 29C). However, perturbations which cause decreasing parameter values starting initially from a large-amplitude limit cycle of the mitotic oscillation undergo hysteresis as the parameter value approaches the bifurcation point, P, and yield a "soft" bifurcation resulting in the longer-period limit cycle characteristic of the circadian oscillation (Figure 29B). The lower amplitude oscillation of the circadian oscillator would account for the different spectra of coupled or driven processes characteristic of the circadian versus the mitotic oscillator. Hysteresis in the system at a low amplitude of the oscillation would account for the different growth-related contingencies which determine entry from the mitotic cycle into the circadian cycle, and vice versa.

EPILOGUE

The era of accumulation of examples of circadian biological rhythms in protozoan cells is drawing to a close. A new phase of interest in possible physiological mechanisms regulating biological timekeeping in cells has arisen.

This chapter has focused on a few well-studied protozoans, and from these it is possible to outline some empirical generalizations for a circadian biological clock. General agreement on such semiphenomenological rules would go a long way toward realization of the appropriate class of explanations for circadian regulation.

The following generalizations seem justified: (1) Phase and period control of multiple circadian rhythms in intact cells is consistent with the operation of a single master clock—the *one cell–one clock rule.* (2) The expression of circadian rhythmicity is restricted to cells whose average generation time is greater than 24 hr—the *circadian–infradian* rule. (3) There is a common metabolic cycle in cells expressing circadian rhythmicity, involving alternate phases of storage and synthesis of energy reserve polymers—the *one clock–one circadian metabolic cycle* rule. (4) Circadian regulation can and often does involve more than one cell compartment beside the nucleus, but the latter compartment ultimately provides the specific substrates for phase and period control of the rhythm. (5) Regulation of mitotic timing resembles in many features the regulation of circadian rhythms in cells. These include phase-independent loss and restoration in overt rhythms coupled to the oscillator, and similar responses in their phase-resetting behavior after exposure to stimuli of varying intensities and durations at different phases of the oscillation.

On the basis of these considerations, and in view of the fact that there is no compelling evidence that cells need more than one underlying dynamic control system, we suggest that the eukaryotic cell has but one *cellular* oscillator for cell division and circadian rhythms. This is best modeled as a nonlinear dynamic system which can regulate the frequency of mitosis in the ultradian mode of growth, through one high-amplitude limit cycle oscillation, and regulate in the circadian–infradian mode the phase and period of circadian rhythms, through a second stable periodic mode, a low-amplitude limit cycle nested within the mitotic oscillator. The grounds for proposing this unifying cellular control system stem from observations relating the kinetics of the growth-dependent transitions between the two modes.

APPENDIX: CIRCADIAN GLOSSARY

The following glossary is derived largely from extant glossaries prepared either from the organismic point of view (Aschoff *et al.*, 1965) or with a cellular and molecular emphasis (Ehret, 1974; Ehret *et al.*, 1977).

Acrophase: The time in the circadian cycle when the maximum amplitude of a circadian parameter occurs at a fixed phase (ϕ) or the circadian cycle.

Chronobiotic: Any agent or drug which can permanently change the phase or period of a circadian rhythm.

Chronon: A polycistronic DNA replicon common to all eukaryotes and naturally selected to be of such length as to rate-limit the eukaryotic infradian cell cycle of RNA transcription to circadian periods; a component of the circadian escapement in the chronon theory of Ehret and Trucco (1967).

Chronon theory: The circadian clock escapement includes two component functions: (1) the sequential transcription component, or chronon, which is relatively temperature-independent during infradian growth, and (2) chronon recycling and initiation (Ehret *et al.*, 1973).

Chronotype: The temporal phenotype of an organism.

ct: Circadian cycle time; i.e., a full circadian cycle has a duration τ hours.

Circadian: About a day; the cell cycle of chronotypic events, including cell division, may be less than a day (ultradian) but (excluding cell division) is never longer than circadian in active and viable cell populations, even though the probability of cell division is very low and the average generation time $(\overline{GT})$ is extremely long (infradian).

Circadian–Infradian Rule: A eukaryotic cell is incapable of endogenous circadian oscillations during the ultradian (fast exponential) mode of growth; a cell, in switching its state from the ultradian to the infradian mode, is capable of circadian outputs in the latter mode. The circadian component of an infradian culture is revealed through the daily application of an appropriate *Zeitgeber* such as light, heat, or food.

Circadian transcriptotype: The circadian chronotype of temporally characteristic gene products.

Circa diem: About a day. The adjectival forms *ultradian, circadian,* and *infradian* are etymological derivatives used to connote periods of cyclical processes or durations of biological time constants without bias as to underlying mechanisms (e.g., whether the oscillation is of endogenous or exogenous origin does not influence the metrical application of the term).

DD: Continuous darkness.

Free-running rhythms: Self-sustained undamped oscillations under constant conditions.

FS: Programmed feeding cycle (feed–starve).

Infradian: More than a day, e.g., the average generation time of undernourished cells, developmental stages requiring more than a day, ovulatory and menstrual cycles.

Infradian growth mode: The mode of growth during which cells divide with an *average* generation times $(\overline{GT})$ longer than (and often *considerably* longer than) 1 day. Preferred to archaic terms such as "stationary phase," "quiescent phase," and "G_0," because "phase" should be reserved to index the cycle, and further (1) it is inappropriate to say that cells do not divide simply because the *probability* that they will divide is very low, (2) it is better to measure and define *that* probability, however low, and (3) as pointed out in the circadian-infradian rule, the cell cycle continues even without cell division and cells are neither quiescent or, as implied by G_0, noncyclical during infradian growth. Failure to reckon with this has led to unrealistically long estimates of G_1 and G_2 phases in infradian cells whose cell cycle time is circadian.

Isochron: In periodic processes described by limit-cycle dynamics, the set of all points (states) whose solution paths (trajectories) are at least one or more integral periods away from a given representative point (or phase) on the limit cycle. As time goes forward, they all eventually reach a limiting trajectory at the same phase and continue to rotate in synchrony with it along the limiting trajectory (Winfree, 1970). It is an attempt to define the phase for all points in the state space of the variables which are not members of the set of states on the limit cycle.

LD: Programmed light–dark cycle.

LL: Continuous light.

Period, T: Period of the entraining or forcing oscillation.

Period, τ: Time after which a definite phase (ϕ) of an oscillation recurs. One should avoid confusing τ with ϕ in cell cycle terminology (e.g., in the $G_1 \rightarrow S \rightarrow G_2 \rightarrow M$ cycle, the connotation "S phase" is the correct form).

Phase angle: Let the circadian cycle be represented as a circle of 360°, and let the phase vary continuously from 0° to 360°. If dawn is arbitrarily chosen as 0° phase of the circadian cycle, and dusk 12 hr later at 180° phase, then the phase angle measures the phase at which any other circadian parameter reaches its maximum or minimum relative to dawn or dusk.

Phase response curve: The relationship between the amount and direction of a phase shift and the phase within which a stimulus (*Zeitgeber*) is applied.

Phase shift: A single displacement ($\Delta\phi$) of an oscillation along the time axis.

Ultradian: Less than a day, e.g., heartbeat, pulse rate, generation time of a well-nourished prokaryote such as *E. coli* or the eukaryote *T. pyriformis*.

Ultradian growth mode: The mode of growth during which cells divide with generation times shorter than a day (fast exponential). Preferred to the archaic phrase "log phase for growth" because (1) "phase" is a term in cycles research reserved for defining explicitly a point in a cycle, and (2) cell population growth can proceed just as "logarithmically" during the infradian growth mode (q.v.) and therefore "log" does not uniquely characterize cells whose generation times are ultradian.

WC: Programmed thermal cycle (warm–cool).

Zeitgeber: A forcing oscillation, entraining agent, or synchronizer that can entrain a biological rhythm.

ACKNOWLEDGMENTS

I thank Dr. Charles F. Ehret for his invaluable assistance in the many phases of the preparation of this chapter and acknowledge the helpful criticism of portions of this article generously offered by Dr. Leland Edmunds. I am grateful to Drs. Beatrice Sweeney and J. Woodland Hastings for the use of unpublished material and also thank Dr. Stuart Kauffman for the many stimulating discussions relating to our common interest in cellular clocks.

REFERENCES

Adegoke, J., and Taylor, J. (1977). *Exp. Cell Res.* **104,** 47–54.

Antipa, G., Ehret, C., Eisler, W., and Blomquist, J. (1972). *Annu. Rep., Argonne Natl. Lab.* **ANL-7970,** 142–144.

Aschoff, J. (1965). "Circadian Clocks." North-Holland Publ., Amsterdam.

Aschoff, J., Klotter, K., and Wever, R. (1965). *In* "Circadian Clocks (J. Aschoff, ed.), pp. x–xix. North-Holland Publ., Amsterdam.

Barnett, A. (1965). *In* "Circadian Clocks" (J. Aschoff, ed.), pp. 305–308. North-Holland Publ., Amsterdam.

Barnett, A. (1966). *J. Cell. Physiol.* **67,** 239–270.

Barnett, A. (1969). *Science* **164,** 1417–1419.

Barnett, A., Ehret, C. F., and Wille, J. (1971a). *In* "Biochronometry" (M. Menaker, ed.), pp. 637–650. Natl. Acad. Sci., Washington, D.C.

Barnett, A., Wille, J. J., and Ehret, C. F. (1971b). *Biochim. Biophys. Acta* **247,** 243–261.

Blum, J. J. (1967). *Proc. Natl. Acad. Sci. U.S.A.* **58,** 81–88.

Bode, V. C., De Sa, R., and Hastings, J. W. (1963). *Science* **141,** 913–915.

Bradbury, E. M., Inglis, R. J., and Matthews, H. R. (1974a). *Nature (London)* **247,** 257–261.

Bradbury, E. M., Inglis, R. J., Matthews, H. R., and Langan, T. (1974b). *Nature (London)* **249,** 553–556.

Branton, D., Bullivant, S., Gilula, N., Karnovsky, M., Moor, H., Muhlethaler, M., Northcote, D., Packer, L., Satir, B., Satir, P., Speth, V., Staehlin, L., Steere, R., and Weinstein, R. (1975). *Science* **190**, 54–56.

Braun, R., Mittermayer, C., and Rusch, H. (1965). *Proc. Natl. Acad. Sci. U.S.A.* **53**, 924–931.

Brewer, E. N., and Rusch, H. P. (1968). *Exp. Cell Res.* **49**, 79–86.

Brinkmann, K. (1966). *Planta* **70**, 344–389.

Brinkmann, K. (1974). *J. Interdiscipl. Cycle Res.* **5**, 186.

Brinkmann, K. (1976a). *J. Interdiscipl. Cycle Res.* **7**, 149.

Brinkmann, K. (1976b). *Planta* **129**, 221–227.

Bruce, V. G. (1960). *Cold Spring Harbor Symp. Quant. Biol.* **25**, 29–48.

Bruce, V. G. (1965). *In* "Circadian Clocks" (J. Aschoff ed.), pp. 125–138. North-Holland Publ., Amsterdam.

Bruce, V. C. (1970). *J. Protozool.* **17**, 328–334.

Bruce, V. G. (1972). *Genetics* **70**, 537–548.

Bruce, V. G. (1974). *Genetics* **77**, 221–230.

Bruce, V. G., and Pittendrigh, C. S. (1956). *Proc. Natl. Acad. Sci. U.S.A.* **42**, 676–682.

Bruce, V. G., and Pittendrigh, C. S. (1960). *J. Cell. Comp. Physiol.* **56**, 25–31.

Bünning, E. (1973). "The Physiological Clock." Springer-Verlag, Berlin and New York.

Bünning, E., and Baltes, J. (1963). *Naturwissenschaften* **50**, 622–623.

Bünning, E., and Moser, I. (1972). *Proc. Natl. Acad. Sci. U.S.A.* **69**, 2732–2733.

Cold Spring Harbor Symposia on Quantitative Biology (1960). "Biological Clocks," Vol. 25. Cold Spring Harbor Laboratory, Cold Spring Harbor, New York.

Cooper, S., and Helmstetter, C. (1968). *J. Mol. Biol.* **31**, 519–540.

Cummings, F. W. (1975). *J. Theor. Biol.* **55**, 455–470.

Cummings, J., Blomquist, J., and Rusch, H. (1966). *Science* **154**, 1343–1344.

Dobra, K. W., and Ehret, C. F. (1977). *Proc. Int. Soc. Study Chronobiol., Int. Symp., 12th, Washington, D.C.* pp. 589–594.

Donachie, W. D., and Masters, M. (1969). *In* "The Cell Cycle" (G. M. Padilla, G. Whitson, and I. L. Cameron, eds.), pp. 37–49. Academic Press, New York.

Edmunds, L. N., Jr. (1964). *Science* **145**, 266–268.

Edmunds, L. N., Jr. (1965). *J. Cell. Comp. Physiol.* **66**, 147–158.

Edmunds, L. N., Jr. (1966). *J. Cell. Comp. Physiol.* **67**, 35–44.

Edmunds, L. N., Jr. (1974a). "Les Cycles Cellulaires et Leurs Blocages Chez Plusieurs Protistes," pp. 53–67. CNRS, Paris.

Edmunds, L. N., Jr. (1974b). *Exp. Cell Res.* **83**, 367–379.

Edmunds, L. N., Jr. (1974c). *In* "Mechanisms of Regulation of Plant Growth" (R. Bieleski, A. Ferguson, and M. Cresswell, eds.), Bull. No. 12, pp. 287–297. R. Soc. N.Z., Wellington.

Edmunds, L. N., Jr. (1977). *Proc. Int. Soc. Study Chronobiol., Int. Symp., 12th, Washington, D.C.* pp. 571–577.

Edmunds, L. N., Jr., and Cirillo, V. P. (1974). *Int. J. Chronobiol.* **2**, 233–246.

Edmunds, L. N., Jr., and Funch, R. (1969a). *Science* **165**, 500–503.

Edmunds, L. N., Jr., and Funch, R. (1969b). *Planta* **87**, 134–163.

Edmunds, L. N., Jr., and Halberg, F. (1979). In preparation.

Edmunds, L. N., Jr., Sulzman, F. M., and Walther, W. G. (1974). *In* "Chronobiology" (L. E. Scheving, F. Halberg, and J. Pauly, eds.), pp. 61–66. Igaku Shoin, Tokyo.

Edmunds, L. N., Jr., Jay, M. E., Kohlmann, A., Liu, S. C., Merriam, V. H., and Sternberg, H. (1976). *Arch. Microbiol.* **108**, 1–8.

Ehret, C. F. (1948). *Anat. Rec.* **101**, 654.

Ehret, C. F. (1951). *Anat. Rec.* **111**, 528.

Ehret, C. F. (1953). *Physiol. Zool.* **26**, 274–300.

Ehret, C. F. (1959). *Fed. Proc., Fed. Am. Soc. Exp. Biol.* **18,** 1232–1240.
Ehret, C. F. (1974). *Adv. Biol. Med. Physics.* **15,** 47–77.
Ehret, C. F., and Dobra, K. W. (1977). *Proc. Int. Soc. Study Chronobiol., Int. Symp., 12th, Washington, D.C.* pp. 563–570.
Ehret, C. F., and Potter, V. R. (1974). *Int. J. Chronobiol.* **2,** 321–326.
Ehret, C. F., and Trucco, E. (1967). *J. Theor. Biol.* **15,** 240–262.
Ehret, C. F., and Wille, J. J. (1970). *In* "Photobiology of Microorganisms" (P. Halldal, ed.), pp. 369–416. Wiley (Interscience), New York.
Ehret, C. F., Wille, J. J., and Trucco, E. (1973). *In* "Biological and Biochemical Oscillators" (B. Chance, E. K. Pye, A. K. Ghosh, and B. Hess, eds.), pp. 503–512. Academic Press, New York.
Ehret, C. F., Barnes, J. H., and Zichal, K. E. (1974). *In* "Chronobiology" (L. Scheving, F. Halberg, and J. Pauly, eds.), pp. 44–50. Igaku Shoin, Tokyo.
Ehret, C. F., Meinert, J. C., Groh, K. R., Dobra, K. W., and Antipa, G. (1977). *In* "Growth Kinetics and Biochemical Regulation of Normal and Malignant Cells" (B. Drewinko and R. Humphrey, eds.), pp. 49–76. Williams & Wilkins, Baltimore, Maryland.
Eisler, W. J., Jr., and Webb, R. B. (1968). *Appl. Microbiol.* **16**(9), 1375–1380.
Fantes, P. A., Grant, W. D., Prichard, R. H., Sudbery, P. E., and Wheals, A. E. (1975). *J. Theor. Biol.* **50,** 213–244.
Feldman, J. F. (1967). *Proc. Natl. Acad. Sci. U.S.A.* **57,** 1080–1087.
Feldman, J. F. (1968). *Science* **160,** 1454–1456.
Feldman, J. F., and Bruce, V. G. (1972). *J. Protozool.* **19,** 370–373.
Feldman, J. F., and Hoyle, M. N. (1973). *Genetics* **75,** 605–613.
Feldman, J. F., Hoyle, M. N., and Shelgren, J. (1973). *Genetics* **74,** 577–578.
Frankel, J. (1970). *J. Exp. Zool.* **173,** 79–100.
Goodwin, B. C. (1966). *Nature (London)* **209,** 479–481.
Goodwin, B. C. (1976). "Analytical Physiology of Cells and Developing Organisms." Academic Press, New York.
Grell, E., Funck, T., and Sauter, H. (1973). *Eur. J. Biochem.* **34,** 415–424.
Groh, K., and Ehret, C. (1974). *Annual Rep., Argonne Natl. Lab.* **ANL-7530,** 193–195.
Hardeland, R., Hohmann, D., and Rensing, L. (1973). *J. Interdiscip. Cycle Res.* **4,** 89–118.
Hartwell, L., Culotti, J., Pringle, J. R., and Reid, B. J. (1974). *Science* **183,** 46–51.
Hastings, J. W. (1959). *Annu. Rev. Microbiol.* **13,** 297–312.
Hastings, J. W. (1964). *In* "Photophysiology: Action of Light on Living Materials" (A. C. Giese, ed.), Vol. 1, pp. 333–361. Academic Press, New York.
Hastings, J. W., and Bode, V. C. (1962). *Ann. N.Y. Acad. Sci.* **98,** 876–889.
Hastings, J. W., and Schweiger, H. (1976). *Dahlem Conf. Circadian Rhythms, Berlin, 1975, Mol. Basis Circadian Rhythms,* pp. 49–62. Dahlem.
Hastings, J. W., and Sweeney, B. M. (1957). *J. Cell. Comp. Physiol.* **49,** 209–226.
Hastings, J. W., and Sweeney, B. M. (1958). *Biol. Bull. (Woods Hole, Mass.)* **115,** 440–458.
Hastings, J. W., Astrachan, L., and Sweeney, B. M. (1961). *J. Gen. Physiol.* **45,** 69–76.
Haus, E., and Halberg, F. (1966). *Experientia* **33,** 113–114.
Herman, E., and Sweeney, B. M. (1975). *J. Ultrastruct. Res.* **50,** 347–354.
Higgins, J. (1967). *Ind. Eng. Chem.* **59,** 19.
Inglis, R. J., Langan, T. A., Matthews, H. R., Hardie, D. G., and Bradbury, E. M. (1976). *Exp. Cell Res.* **97,** 418–425.
Janakidevi, K., Dewey, V. C., and Kidder, G. W. (1966a). *J. Biol. Chem.* **241,** 2576–2578.
Janakidevi, K., Dewey, V. C., and Kidder, G. W. (1966b). *Arch. Biochem. Biophys.* **113,** 758–759.
Jarrett, R. M., and Edmunds, L. N., Jr. (1970). *Science* **167,** 1730–1733.

Karakashian, M. W. (1965). *In* "Circadian Clocks" (J. Aschoff, ed.), pp. 301–304. North-Holland Publ., Amsterdam.
Karakashian, M. W. (1968). *J. Cell. Physiol.* **71**, 197–210.
Karakashian, M. W., and Schweiger, H. G. (1976a). *Exp. Cell Res.* **97**, 366–377.
Karakashian, M. W., and Schweiger, H. G. (1976b). *Exp. Cell Res.* **98**, 303.
Kauffman, S., and Wille, J. J. (1975). *J. Theor. Biol.* **55**, 47–93.
Keller, S. (1960). *Z. Bot.* **48**, 32–57.
Konopke, R. J., and Benzer, S. (1971). *Proc. Natl. Acad. Sci. U.S.A.* **68**, 2112–2116.
Levy, M. (1973). *In* "Biology of Tetrahymena" (A. M. Eliot, ed.), pp. 227–257. Dowden, Hutchinson & Ross, Stroudsburg, Pennsylvania.
Levy, M., and Scherbaum, O. (1965). *J. Gen. Microbiol.* **38**, 221–230.
McDaniel, M., Sulzman, F. M., and Hastings, J. W. (1974). *Proc. Natl. Acad. Sci. U.S.A.* **71**, 4389–4391.
McMurry, L., and Hastings, J. W. (1972a). *Science* **175**, 1137–1139.
McMurry, L., and Hastings, J. W. (1972b). *Biol. Bull. (Woods Hole, Mass.)* **143**, 196–206.
Malecki, M. V., Ličko, V., and Eiler, J. J. (1971). *Curr. Mod. Biol.* **3**, 291–298.
Masters, M., and Pardee, A. B. (1965). *Proc. Natl. Acad. Sci. U.S.A.* **54**, 64–70.
Matsumoto, S. (1977). *J. Cell Struct. Funct.* **2**, 101–109.
Meinert, J. C., Ehret, C. F., and Antipa, G. A. (1975). *Microbiol. Ecol.* **2**, 201–214.
Menaker, M. (1971). "Biochronometry." Natl. Acad. Sci., Washington, D.C.
Mergenhagen, D., and Schweiger, H. G. (1973). *Exp. Cell Res.* **81**, 360–364.
Mergenhagen, D., and Schweiger, H. G. (1975). *Exp. Cell Res.* **92**, 127–130.
Mitchell, J. L. A. (1971). *Planta* **100**, 244–257.
Mitchison, J. (1971). "The Biology of the Cell Cycle." Cambridge Univ. Press, London and New York.
Mueller, G., and Kajiwara, K. (1966). *Biochim. Biophys. Acta* **114**, 108–115.
Murakami, K., and Ohta, J. (1971). *Plant Cell Physiol.* **12**, 797–801.
Njus, D., Sulzman, F., and Hastings, J. W. (1974). *Nature (London)* **248**, 116–120.
Njus, D., Gooch, V. D., Mergenhagen, D., Sulzman, F., and Hastings, J. W. (1976). *Fed. Proc., Fed. Am. Soc. Exp. Biol.* **35**, 2353–2357.
Njus, D., McMurry, L., and Hastings, J. W. (1977). *J. Comp. Physiol.* **117**, 335–344.
Nurse, P., and Fantes, P. (1977). *Nature (London)* **267**, 647.
Oppenheim, A., and Katzir, N. (1971). *Exp. Cell Res.* **68**, 224–226.
Othmer, H. G. (1975). *Math. Biosci.* **24**, 205–238.
Othmer, H. G., and Scriven, L. E. (1974). *J. Theor. Biol.* **43**, 83–112.
Palmer, J. D., Livingston, L., and Zusy, D. (1964). *Nature (London)* **230**, 1087–1088.
Palmer, J. D., Brown, F. A., Jr., and Edmunds, L. N., Jr. (1976). "Introduction to Biological Rhythms." Academic Press, New York.
Pavlidis, T. (1971). *J. Theor. Biol.* **33**, 319–338.
Pavlidis, T. (1973). "Biological Oscillators: Their Mathematical Analysis." Academic Press, New York.
Peraino, C., and Morris, J. E. (1975). *Chronobiologia* **2**, Suppl., p. 54.
Pittendrigh, C. S., Caldarola, P. C., and Cosbey, E. S. (1973). *Proc. Natl. Acad. Sci. U.S.A.* **70**, 2037–2041.
Plaut, W. (1969). *Genetics* **61**, Suppl. 1, Part 2, 239–244.
Pohl, R. (1948). *Z. Naturforsch. Teil* **3b**, 367.
Pressman, B. C. (1968). *Fed. Proc. Fed. Am. Soc. Exp. Biol.* **27**, 1283.
Prezelin, B. B., and Sweeney, B. M. (1977). *Plant Physiol.* **60**, 388–392.
Prezelin, B. B., Meeson, B. W., and Sweeney, B. M. (1977). *Plant Physiol.* **60**, 384–387.
Rao, P. T., and Johnson, R. T. (1970). *Nature (London)* **225**, 159–164.

Rasmussen, L. (1967). *Exp. Cell Res.* **48**, 132–139.
Richter, G. (1963). *Z. Naturforsch., Teil* **18b**, 1085–1089.
Rosensweig, Z., and Kindler, S. H. (1972). *FEBS Lett.* **25**, 221–223.
Rusch, H. P. (1970). *Adv. Cell Biol.* **1**, 297–327.
Rusch, H. P., Sachsenmaier, W., Behrens, K., and Gruter, V. (1966). *J. Cell Biol.* **31**, 204–209.
Sachsenmaier, W. (1976). *Dahlem Conf. Circadian Rhythms, Berlin, 1975, Mol. Basis Circadian Rhythms,* pp. 410–420. Dahlem.
Sachsenmaier, W., Remy, U., and Plattner-Schobel, R. (1972). *Exp. Cell Res.* **73**, 41–48.
Scheffey, C., and Wille, J. (1978). *Exp. Cell Res.* **113**, 259.
Scheving, L., Halberg, F., and Pauly, J., eds. (1974). "Chronobiology." Igaku Shoin, Tokyo.
Schmitter, R. E. (1971). *J. Cell Sci.* **9**, 147–173.
Schweiger, E., Wallraff, H. G., and Schweiger, H. G. (1964). *Science* **146**, 658–659.
Schweiger, H. G., and Schweiger, M. (1977). *Int. Rev. Cytol.* **51**, 315–342.
Shields, R. (1977). *Nature (London)* **267**, 704–707.
Shilo, B., Shilo, V., and Simchen, G. (1976). *Nature (London)* **264**, 767–770.
Shilo, B., Shilo, V., and Simchen, G. (1977). *Nature (London)* **267**, 648–649.
Smith, J. A., and Martin, L. (1973). *Proc. Natl. Acad. Sci. U.S.A.* **70**, 1263–1267.
Sturtevant, R. P. (1973a). *Int. J. Chronobiol.* **1**, 141–146.
Sturtevant, R. P. (1973b). *Anat. Rec.* **175**, 453.
Sulzman, F. M., and Edmunds, L. N. Jr. (1972). *Biochem. Biophys. Res. Commun.* **47**, 1338–1344.
Sulzman, F. M., and Edmunds, L. N., Jr. (1973). *Biochim. Biophys. Acta* **320**, 594–609.
Sutherland, A., Antipa, G., and Ehret, C. F. (1973). *J. Cell Biol.* **58**, 240–244.
Sweeney, B. M. (1963). *Plant Physiol.* **38**, 704–708.
Sweeney, B. M. (1969a). "Rhythmic Phenomena in Plants." Academic Press, New York.
Sweeney, B. M. (1969b). *Can. J. Bot.* **47**, 299–308.
Sweeney, B. M. (1972). *Proc. Int. Symp. Circadian Rhythmicity, Wageningen, 1971* pp. 137–156.
Sweeney, B. M. (1974a). *Int. J. Chronobiol.* **2**, 25–33.
Sweeney, B. M. (1974b). *Plant Physiol.* **53**, 337–342.
Sweeney, B. M. (1976a). *Dahlem Conf. Circadian Rhythms, Berlin, 1975, Mol. Basis Circadian Rhythms,* pp. 267–281. Dahlem.
Sweeney, B. M. (1976b). *J. Cell Biol.* **68**, 451–461.
Sweeney, B. M., and Hastings, J. W. (1957). *J. Cell Comp. Physiol.* **49**, 115–128.
Sweeney, B. M., and Hastings, J. W. (1958). *J. Protozool.* **5**, 217–224.
Sweeney, B. M., and Hastings, J. W. (1960). *Cold Spring Harbor Symp. Quant. Biol.* **25**, 87.
Sweeney, B. M., and Haxo, F. T. (1961). *Science* **134**, 1361–1363.
Sweeney, B. M., and Herz, J. N. (1977). *Proc. Int. Soc. Study Chronobiol., Int. Symp., 12th, Washington, D.C.* pp. 751–761.
Sweeney, B. M., Tuffli, C. F., and Rubin, R. H. (1967). *J. Gen. Physiol.* **50**, 647–659.
Terry, O. W., and Edmunds, L. N., Jr. (1970a). *Planta* **93**, 106–127.
Terry, O. W., and Edmunds, L. N., Jr. (1970b). *Planta* **93**, 128–142.
Thormar, H. (1959). *C. R. Trav. Lab. Carlsberg* **31**, 207–226.
Turing, A. M. (1952). *Philos. Trans. R. Soc. London, Ser. B* **237b**, 37–72.
Tyson, J., and Kauffman, S. (1975). *J. Math. Biol.* **1**, 289–310.
Tyson, J., and Sachsenmaier, W. (1978). *J. Theor. Biol.* **73**, 723–738.
Vanden Driessche, T. (1966a). *Exp. Cell Res.* **42**, 18–30.
Vanden Driessche, T. (1966b). *Biochim. Biophys. Acta* **126**, 456–470.

Vanden Driessche, T. (1971). *In* "Biochronometry" (M. Menaker, ed.), pp. 612–622. Natl. Acad. Sci., Washington, D.C.

Vanden Driessche, T., and Hars, R. (1972a). *J. Microsc. (Paris)* **15**, 85.

Vanden Driessche, T., and Hars, R. (1972b). *J. Microsc. (Paris)* **15**, 91.

Vanden Driessche, T., and Hars, R. (1974). *In* "Chronobiology" (L. E. Scheving, F. Halberg, and J. E. Pauly, eds.), pp. 55–60. Igaku Shoin, Tokyo.

Vanden Driessche, T., Bonotto, S., and Brachet, J. (1970). *Biochim. Biophys. Acta* **224**, 631–634.

Volm, M. (1964). *Z. Vgl. Physiol.* **48**, 157.

Walther, W. G., and Edmunds, L. N., Jr. (1970). *J. Cell Biol.* **46**, 613–617.

Walther, W. G., and Edmunds, L. N., Jr. (1973). *Plant Physiol.* **51**, 250–258.

Wheals, A. E. (1977). *Nature (London)* **267**, 647.

Wille, J. J., and Ehret, C. F. (1968). *J. Protozool.* **15**, 785–789.

Wille, J. J., Barnett, A., and Ehret, C. F. (1972). *Biochem. Biophys. Res. Commun.* **46**, 685–691.

Wille, J. J., Scheffey, C., and Kauffman, S. A. (1977). *J. Cell Sci.* **27**, 91–104.

Winfree, A. (1970). *In* "Lectures on Mathematics in Life Sciences" (M. Gerstenhaber, ed.), Vol. 2, pp. 111–150. Providence, Rhode Island.

Winfree, A. (1973). *In* "Biological and Biochemical Oscillators" (B. Chance, E. K. Pye, A. K. Gosh, and B. Hess, eds.), pp. 461–502. Academic Press, New York.

Winfree, A. (1975). *Nature (London)* **253**, 315–319.

Zaikin, A. N., and Zhabotinskii, A. M. (1970). *Nature (London)* **225**, 535–537.

Zeuthen, E. (1971). *Exp. Cell Res.* **68**, 49–60.

Microtubules

4

JOANNE K. KELLEHER AND ROBERT A. BLOODGOOD

I. INTRODUCTION

The evolution of the eukaryotic cell was accompanied by major changes in cellular organization. In comparison to their prokaryotic ancestors, eukaryotic cells possess a more complex arrangement of cellular membranes, a greater variety of asymmetrical cell shapes, many innovative secretory, motility, and intracellular transport systems, and a novel method for the segregation of genetic material. Many of these developments may have been critically dependent upon the evolution of the mi-

BIOCHEMISTRY AND PHYSIOLOGY OF PROTOZOA
SECOND EDITION, VOL. 2

Table I Review Articles and Edited Volumes on Microtubules

Reference	Structure	Biochemistry	Distribution	Assembly	Resorption	Drug binding	Purification	Motility	Mitosis	Cilia and flagella	Cytoplasmic microtubules	Cytoskeletal microtubules	Animal	Plant	Nervous tissue	Protozoa	Fungi	Algae
Bajer and Mole-Bajer (1972)	X		X	X				X	X	X	X	X						
Bardele (1973)		X							X	X	X					X	X	X
Bloodgood (1974)					X													
Bloodgood (1976)							X											
Bouck and Brown (1976)				X									X		X			
Fuller (1976)									X								X	
Gaskin and Shelanski (1976)	X	X	X	X											X			
Hepler (1976)		X	X	X				X	X			X		X				
Hepler and Palevitz (1974)	X	X	X	X		X		X	X		X	X		X				
Margulis (1973)		X						X			X							X
Mohri (1976)	X	X		X				X		X								

Newcomb (1969)	X	X	X						X			X		X				X
Olmsted and Borisy (1973b)	X	X	X	X		X		X	X	X	X	X						
Pickett-Heaps (1974)	X								X		X	X		X			X	X
Porter (1966)	X	X						X	X		X	X						
Roberts (1974)	X	X		X		X	X	X	X	X	X	X						
Samson *et al*. (1973)		X				X	X								X			
Shelanski (1973)		X		X		X									X			
Snyder and McIntosh (1976)	X	X		X		X												
Stephens and Edds (1976)	X	X	X	X				X	X	X	X	X						
Summers (1975)		X						X		X								
Tilney (1971)	X		X	X				X			X	X						
Warner (1972)	X	X						X		X								
Williams (1975)			X	X	X					X						X		
Wilson and Bryan (1974)		X		X		X												
Wuerker and Kirkpatrick (1972)		X	X					X			X	X			X			
Borgers and DeBrabander (1975)[a]	X	X	X	X		X			X		X	X						
Goldman *et al*. (1976)[a]	X	X	X	X		X	X	X	X	X	X	X						
Inoué and Stephens (1975)[a]	X	X		X				X	X	X	X	X						
Soifer (1975b)[a]	X	X	X	X		X	X	X	X	X	X	X						
Sleigh (1974)[a]	X	X	X	X			X	X		X								

[a] Edited volumes containing numerous articles on microtubules.

crotubule, a remarkably versatile protein polymer found within virtually all eukaryotic cells sometime during their life cycle or developmental history. The fact that microtubules are a unique and ubiquitous property of eukaryotes has resulted in interesting speculations as to their origin (Margulis, 1970, 1975). It is perhaps not surprising that the greatest variety of microtubule form and function is displayed among protozoa, a very heterogeneous group representing the extant remains of an intriguing Precambrian evolutionary experiment.

Although microtubules were first recognized in thin-section electron micrographs of cilia in 1954 (Fawcett and Porter, 1954), most were not preserved by electron microscopic techniques until the advent of aldehyde fixatives (Sabatini et al., 1963). During the past 15 years, the literature in this field has mushroomed, and with it the number of reviews of this topic. For this reason, another comprehensive review on microtubule structure, biochemistry, function, and distribution is clearly unnecessary. Instead, Table I is provided as a key to the review articles on microtubules and should allow an investigator access to any of the available literature. The major thrust of this chapter is to consider the important and timely question of what controls the assembly of microtubules in time and space. This challenging problem of regulation, first outlined by Porter (1966), still remains the focal point of most research on microtubules. In addressing this topic, we do not restrict ourselves to information obtained from protozoan systems. Prior to coming to grips with the problem of the regulation of microtubule assembly, some preliminary information about microtubules is necessary.

II. ESSENTIAL CHARACTERISTICS OF MICROTUBULES

A. Structure

A microtubule is a hollow cylinder 24–26 nm in diameter and of widely variable length, which is assembled from a native protein dimer of 110–120,000 MW called tubulin. A cross-sectional view of this cylinder reveals a ring of 13 distinct subunits 5 nm in diameter surrounding a 15-nm hollow central core. Each of these 13 subunits is part of a protofilament that runs the length of the microtubule. The protofilaments of the microtubule wall are most likely arranged in a three-start left-handed helix (Erickson, 1974; Linck and Amos, 1974). Microtubules assemble as single tubules, as doublets (in cilia and flagella), or as triplets (in basal bodies and centrioles). A large number of bridge or armlike structures are observed in association with microtubules from many sources (Tilney, 1971; Bardele, 1973; McIntosh, 1974). These accessory structures function in different cases to (1) stabilize the microtubule, (2) cross-link microtubules into definite arrays, (3) cross-link microtubules to other structures, principally membranes, and (4) generate force, as in the case of dynein arms associated with ciliary and flagellar doublet microtubules. A variety of these armlike structures are observed in the protozoa (see Figures 1–5 for examples of protozoan microtubules).

B. Properties of Tubulin

Tubulin exists as two monomeric species, α- and β-tubulin, each with an apparent molecular weight of approximately 55,000. The native subunit from which microtubules are assembled is probably a heterodimer, 110,000 to 120,000 MW, containing one molecule each of α- and β-tubulin (Bryan and Wilson, 1971; Luduena et al., 1975). Partial amino acid sequences of α- and β-tubulins from echinoderm and vertebrate sources indicate that each of these proteins has been highly conserved during evolution (Luduena and Woodward, 1973), although major differences exist between the α and β chains. There may be multiple α and β chains within a single species (Stephens, 1975). There is a major question whether separate tubulin pools exist for each microtubule organelle (mitotic apparatus, cilia, cytoskeletal microtubules) within a particular cell. If there are distinguishable pools of tubulin subunits for different organelles, does this reflect separate gene products (as discussed in Fulton and Kowit, 1975; Bibring et al., 1976) or simply posttranslational modifications?

C. Molecules Associated with Tubulin

The tubulin dimer has the capacity to interact with a bewildering array of molecules, including ions, nucleotides, a variety of drugs, other tubulin molecules, and a variety of nontubulin proteins. Some of these interactions figure prominently in our discussion of the ways in which microtubule assembly may be regulated in vivo and in vitro.

1. Guanine Nucleotides and Divalent Cations

The tubulin dimer contains two binding sites for guanine nucleotides (Berry and Shelanski, 1972; Shelanski and Taylor, 1968). One of the nucleotide-binding sites, the e site, exchanges with GTP in solution much more readily than the other, the n site (Weisenberg et al., 1968; Berry and Shelanski, 1972). As a rule, the in vitro polymerization of tubulin requires GTP in the medium (Olmsted and Borisy, 1975), although it has been

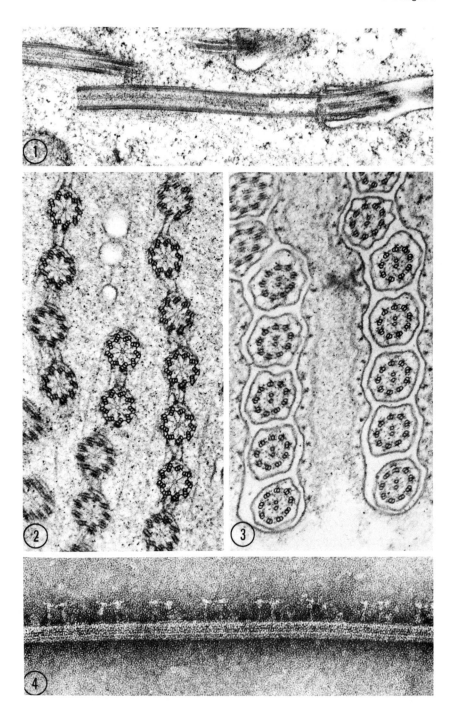

reported that polymerization occurs without added GTP in high concentrations of glycerol or sucrose (Shelanski *et al.*, 1973). ATP has been reported by a number of workers to support microtubule assembly *in vitro*, but this appears to result from the activity of a nucleoside diphosphokinase that converts bound GDP to GTP by transphosphorylation from ATP in solution (Berry and Shelanski, 1972; Weisenberg *et al.*, 1976). When GTP is associated with the *e* site of tubulin, it is hydrolyzed during assembly (Maccioni and Seeds, 1977). However, there are conflicting reports about whether or not hydrolysis of bound nucleotide is essential for *in vitro* polymerization. It has been reported that brain tubulin does (Arai and Kaziro, 1976; Penningroth and Kirschner, 1977; Weisenberg *et al.*, 1976) and does not (Olmsted and Borisy, 1975; Maccioni and Seeds, 1977) assemble when a nonhydrolyzable analog of GTP is substituted for GTP. Interestingly, microtubules formed in the absence of nucleotide hydrolysis have dramatically increased stability as compared to microtubules assembled in the presence of GTP (Arai and Kaziro, 1976; Weisenberg *et al.*, 1976).

Divalent cations also bind to the tubulin molecule. Magnesium is bound to tubulin, 1 mole/mole of dimer, possibly in the form of a magnesium–GTP complex (Olmsted and Borisy, 1975). Purified brain tubulin also binds 1 mole of calcium per mole of dimer, and this binding is inhibited by magnesium (Solomon, 1977).

2. Binding of Drugs to Tubulin

Observation of the binding of drugs previously characterized as spindle poisons, e.g., colchicine, was an important step in the identification and early purification of tubulin and its presumptive identification as a component of neurotubules and the mitotic apparatus (Borisy and Taylor,

---

Figure 1. Longitudinal section through a basal body of the flagellate protozoan *Trichonympha*, which is found as a symbiote in the hindgut of the wood-eating roach *Cryptocercus*. Magnification: ×32,000.

Figure 2. Cross section through basal bodies in the flagellate protozoan *Urinympha* from the hindgut of *Cryptocercus*. Note the extensive interconnections between the basal bodies. These delicate structures may play a role in determining the arrangement of the basal bodies or in coordinating the motility of the flagella arising from the basal bodies. Magnification: ×50,000.

Figure 3. Cross section through the basal portions of the flagella of *Trichonympha*. These portions of the flagella lie within grooves in the cell surface. Note also the attachments between the doublet microtubules in the flagella and the flagellar membrane. Magnification: ×50,000.

Figure 4. Negative-stain electron micrograph of one outer doublet microtubule from the flagellum of *Chlamydomonas*. Note the radial links arranged in pairs along one side of the microtubule. Magnification: ×156,000. Micrograph courtesy of Dr. L. I. Binder.

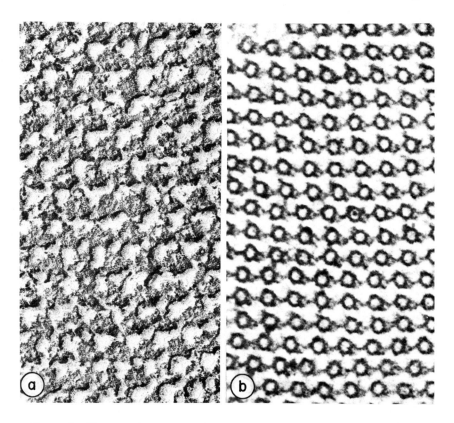

Figure 5. Electron micrographs of the axostyle from the protozoan *Saccinobaculus* which lives as a symbiont within the hindgut of *Cryptocercus*. The axostyle is composed solely of microtubules and two types of cross-bridges. (a) Freeze–fracture image of the axostyle. (b) Thin-section image of an axostyle in the same orientation as in (a). Magnification: (a) ×150,000; (b) ×170,000.

1967a,b; Shelanski and Taylor, 1968; Weisenberg *et al.*, 1968). Several drug-binding sites have now been recognized on the tubulin molecule. The colchicine-binding site (1 mole/mole of dimer) is probably located in a hydrophobic pocket (Bryan, 1972) and may be altered or blocked when tubulin polymerizes (Wilson and Meza, 1973). Podophyllotoxin, a plant lignan, is a competitive inhibitor of colchicine binding (Wilson, 1970); recent data indicate that colchicine and podophyllotoxin occupy overlapping binding sites (Cortese *et al.*, 1977). *Vinca* alkaloids (e.g., vinblastine, vincristine) bind at two high-affinity sites per tubulin dimer (Bhattacharyya and Wolff, 1976). The colchicine- and vinblastine-binding sites are independent; in fact, vinblastine increases colchicine binding somewhat by stabilizing the tubulin molecule. It has also been reported that maytansine (Mandelbaum-Shavit *et al.*, 1976), chlorpromazine (Cann and Hinman, 1975), oncadazole (Hoebeke *et al.*, 1976), and 1-anilino-8-naphthalene sulfonate (Bhattacharyya and Wolff, 1975) all bind to tubulin. Maytansine competes for the vinblastine-binding site, while oncadazole competes for the colchicine-binding site. Since these drugs inhibit microtubule polymerization *in vitro* and *in vivo*, it is generally assumed that, in binding to tubulin, they alter the subunit structure of the soluble dimers so that polymerization is prevented (see Wilson and Bryan, 1974, for review). Microtubule polymerization *in vitro* is inhibited at concentrations of colchicine (Olmsted and Borisy, 1973a), podophyllotoxin, and vinblastine (Wilson *et al.*, 1976) which tie up only a small fraction of the subunits. Margolis and Wilson (1977) have demonstrated that this substoichiometric poisoning by colchicine occurs because a colchicine–dimer complex is added to the end of a growing microtubule, effectively blocking further dimer addition.

Tubulin-binding drugs have been widely used as a tool for the identification of microtubule-dependent processes *in vivo*. However, caution must be exercised when interpreting *in vivo* drug effects, for certain antimitotic drugs clearly have other effects independent of microtubule depolymerization. Colchicine and podophyllotoxin inhibit nucleoside transport *in vivo* (Mizel and Wilson, 1972), and colchicine affects phosphatidylinositol turnover (Schellenberg and Gillespie, 1977). This caution regarding the use of antimitotic drugs is especially relevant when experimenting with unicellular eukaryotic organisms because, as a group, these cells are less sensitive to these drugs than mammalian cells, and tubulin isolated from protists does not readily bind colchicine (Haber *et al.*, 1972; Burns, 1973).

III. MICROTUBULE POLYMERIZATION

A. *In Vitro* Assembly and the Role of Accessory Proteins

Despite considerable previous effort by a number of investigators, the conditions necessary for the *in vitro* assembly of bona fide microtubules (based on structural criteria and sensitivity to cold and calcium) were not discovered until 1972 when Weisenberg (1972) reported that GTP, magnesium, EGTA, and one of the Good organic buffers (Good *et al.*, 1966) were required. This observation has now been repeated by numerous investigators using brain tissue from various mammals, from chickens, and even from dogfish. Although brain continues to be the easiest system in which to demonstrate microtubule assembly *in vitro,* microtubules have been assembled from extracts of sperm tail outer doublets (Kuriyama, 1976), platelets (Castle and Crawford, 1975), sea urchin eggs (Kuriyama, 1977), *Drosophila* embryos (Green *et al.*, 1975), Ehrlich ascites tumor cells

(Doenges *et al.*, 1977), and renal medulla (Barnes *et al.*, 1975). The effects of solution variables on the *in vitro* microtubule assembly reaction using brain have been studied by Olmsted and Borisy (1973a, 1975). Microtubule assembly *in vitro* occurs by a condensation polymerization mechanism with distinct nucleation and elongation steps (Johnson and Borisy, 1975). Tubulin dimers are in equilibrium with assembled microtubules. The rate of assembly of microtubules is a function of tubulin dimer concentration and of the number of microtubules (Johnson and Borisy, 1977). Recent evidence suggests that microtubule assembly *in vitro* occurs at one end of the elongating microtubule and disassembly occurs at the other end (Margolis and Wilson, 1978).

The ability to assemble microtubules reversibly *in vitro* has been used as a basis for purifying microtubule proteins (Shelanski *et al.*, 1973; Borisy *et al.*, 1974). When microtubules obtained by multiple cycles of assembly and disassembly *in vitro* are examined by acrylamide gel electrophoresis, the prominent protein bands are α- and β-tubulin and two high molecular weight proteins (approximately 300,000 and 350,000 MW) referred to as microtubule-associated proteins 1 and 2 (MAPs 1 and 2) or as high molecular weight proteins 1 and 2 (HMWs 1 and 2) (Sloboda *et al.*, 1975; Borisy *et al.*, 1975b). MAPs can be separated from tubulin by ion-exchange chromatography using DEAE (Murphy and Borisy, 1975) or phosphocellulose (Sloboda *et al.*, 1976). In the absence of accessory proteins tubulin assembles poorly under the conditions used, but the addition of MAPs to the tubulin dramatically increases both the initial rate of assembly and the total amount of microtubule polymer formed (Murphy and Borisy, 1975; Sloboda *et al.*, 1976). Stimulation of the initial rate of assembly results from an increase in the number of initiation events occurring (Sloboda *et al.*, 1976). The increase in the equilibrium amount of polymer results from the ability of the MAPs to stabilize the formed polymer and thereby reduce the rate of disassembly (Murphy *et al.*, 1977a).

Microtubules purified by cycles of assembly and disassembly *in vitro* contain both tubulins and both MAPs. When examined by thin-section electron microscopy, the entire surface of these microtubules is decorated with filamentous or fuzzy material (Dentler *et al.*, 1975; Murphy and Borisy, 1975). When microtubules assembled in the absence of MAPs are examined by the same techniques, the surface of the microtubules is smooth (Dentler *et al.*, 1975; Murphy and Borisy, 1975), suggesting that the surface fuzz is composed, at least in part, of MAPs. The surface material can be removed from fuzzy microtubules by trypsin treatment (Vallee and Borisy, 1977), and isolated MAPs can be reassociated with smooth-walled microtubules (Sloboda and Rosenbaum, 1979). Fluorescence-labeled antibody to brain MAPs stains microtubules *in vivo* (Sherline and Schiavone, 1977; Connolly *et al.*, 1978).

Other workers have found that the primary nontubulin protein copurifying with microtubules cycled *in vitro* is not the high molecular weight MAP but rather a protein or group of proteins they call tau, which have a molecular weight of 55,000–62,000 (Weingarten *et al.*, 1975; Kirschner *et al.*, 1975; Cleveland *et al.*, 1977a,b). Tau can be separated from tubulin using phosphocellulose chromatography, and when added back to the tubulin stimulates both the initial rate of assembly and the equilibrium amount of polymer (Cleveland *et al.*, 1977a). Fluorescence-labeled antibody to tau labels microtubules *in vivo* (Connolly *et al.*, 1977).

A serious discrepancy exists among reports from different laboratories in terms of the protein present after several cycles of microtubule assembly *in vitro* by rather similar techniques. Kirschner's laboratory observed much tau protein and little or no high molecular weight MAP, whereas Borisy and Rosenbaum's laboratories observed considerable amounts of MAP and little tau. The possibility exists that there may well be two different molecular weight classes of microtubule accessory proteins that are both capable of affecting tubulin assembly *in vitro*, and that slight differences in technique in different laboratories may select for one class versus the other. Murphy *et al.* (1977b) reported that, after fractionation of the accessory proteins, they found 60% of the tubulin assembly stimulatory activity in the MAP fraction and the remaining 40% in the other accessory proteins. Cleveland *et al.* (1977a) found one-third of the tubulin assembly stimulatory activity associated with the tau fraction and the remaining associated with the other accessory proteins. Sloboda *et al.* (1976) provide data suggesting that tau may be a proteolytic cleavage product of the high molecular weight MAPs. Cleveland *et al.* (1977b) argue from one-dimensional peptide maps that tau is unrelated to MAPs. These same authors report that antibodies to tau do not react with MAPs. Recently, Herzog and Weber (1978) have effected an elegant purification of MAP_2 and tau from the same starting preparation. Their results clearly demonstrate that a similar, if not identical, capacity for stimulating brain tubulin assembly *in vitro* is associated with each of their protein preparations. However, only microtubules stimulated to assemble by the presence of MAP_2 are observed to have their surface decorated with filamentous material. It is likely, therefore, that both tau and one of the MAPs are microtubule accessory proteins capable of regulating brain microtubule assembly *in vitro*.

Nontubulin proteins that stimulate brain tubulin assembly *in vitro* have also been obtained from *Chlamydomonas* flagella (Bloodgood and Rosenbaum, 1976), *Polytomella* microtubule rootlets (Stearns and Brown, 1976), and neuroblastomas (Seeds and Maccioni, 1978). In the case of neuroblastomas, the stimulating activity was only found in extracts of differentiated cells which were assembling microtubule-filled processes.

It has become increasingly apparent that tubulin itself has the capacity, in the absence of accessory proteins, to assemble into microtubules *in vitro*. This can be accomplished in the presence of high concentrations of magnesium and glycerol (Lee and Timasheff, 1975, 1977), a high concentration of magnesium alone (Herzog and Weber, 1977), dimethyl sulfoxide (Himes *et al.*, 1976), polycations (Erikson, 1976; Erikson and Voter, 1976), or simply high concentrations of tubulin alone (Sloboda and Rosenbaum, 1979).

When preparations of brain microtubules assembled *in vitro* are disassembled in the cold and examined by gel filtration chromatography or analytical ultracentrifugation, tubulin is found in two rather different sized species, one having a sedimentation coefficient of 6 S (tubulin dimers) and the other with one of 30–36 S. When these preparations are examined by negative-stain electron microscopy, large numbers of ring-shaped structures (340–380 Å in diameter are observed [first by Borisy and Olmsted (1972) and subsequently by numerous workers]. These structures disappear soon after microtubule assembly is initiated *in vitro* by raising the temperature to 37°C and are gone before most of the polymer formation occurs (Olmsted *et al.*, 1974; Sloboda *et al.*, 1976). Microtubule accessory proteins stimulate the conversion of 6 S tubulin into ring-shaped aggregates, the number of rings formed being proportional to the concentration of MAPs (Sloboda *et al.*, 1976). With increasing numbers of rings in the initial preparation, a faster initial rate of assembly of microtubules is observed after warming to 37°C (Sloboda *et al.*, 1976). These and other observations have led to the suggestion that the ring-shaped aggregates of tubulin are involved in the initiation of microtubule assembly, although some investigators believe the rings function only as a storage form of tubulin (Weisenberg, 1974).

B. *In Vivo* Assembly

It is clear the microtubule assembly and disassembly *in vivo* are events that are highly regulated temporally and spatially. The factors that may be responsible for the ability of a cell to exercise this high degree of control are discussed in Section IV. Although much information is available concerning the mechanism of brain microtubule assembly *in vitro* (Section III,A), it has been difficult to obtain comparable information concerning microtubule assembly *in vivo*.

Mitotic apparatus formation in the giant ameba *Chaos carolinesis* has been used by Goode (1973) to examine the mechanism of microtubule polymerization *in vivo*. Virtually all the microtubules in this spindle are destroyed by cooling the cell to 2°C. Upon rewarming, the mitotic spindle completely reforms within 10 min. Goode (1973) determined that the rate of spindle microtubule elongation was 1.5 μm/sec. Since the mitotic spindle of this ameba contained the equivalent of 6000 microtubules 5 μm long and the average number of spindles per cell was known, the number of 4-nm tubulin subunits released into solution upon cooling was estimated to be 7.5×10^7. Based on a cell volume of 7.25×10^{-6} cm^3, the concentration of subunits is 1×10^{15} 4-nm monomeric subunits per cubic centimeter. With a monomer molecular weight of 55,000, this gives a protein concentration of 0.09 mg/ml. This represents the minimum concentration of tubulin that must be present in the ameba after cold-induced disassembly of the spindle. Employing a diffusion kinetics equation such as is used for virus assembly (Setlow and Pollard, 1962) Goode calculated the concentration of tubulin molecules necessary to allow the observed rate of spindle microtubule growth; this turned out to be 1×10^{15} molecules/cm^3, in close agreement with the calculated value for the minimum concentration of tubulin monomers provided by cold-induced dissassembly, assuming the tubulin was spread uniformly over the entire volume of the cell. Goode concluded that the observed data were consistent with the diffusion kinetics of Setlow and Pollard (1963). This means that the kinetics of microtubule growth *in vivo* may be accounted for by the simple diffusion of subunits to a single growth site at the microtubule end and that more complex polymerization models are not required to fit the data. Models in which the number of growth points is proportional to the polymer length are incompatible both with the data observed for *Chaos* and with *in vitro* polymerization kinetics (Bryan, 1976; Johnson and Borisy, 1975). There are reasons to believe that the actual tubulin concentration in the nuclei of *Chaos* may actually be considerably higher than the minimum estimate provided by Goode's calculation. Goode neglected any tubulin that may have been in a soluble form prior to cold-induced depolymerization of the spindles. Further, the assumption that the tubulin released from the spindles was distributed over the entire cell cytoplasm may well have been wrong. In *Chaos,* the nuclear envelope remains intact during mitosis, and the tubulin released by cold depolymerization may well remain within the nuclei, which would result in a much higher local concentration of tubulin than that predicted by Goode. In fact, his calculated concentration, 0.09 mg/ml, is beneath the critical concentration for microtubule assembly *in vitro*.

Some of the most informative studies on microtubule assembly *in vivo* have involved work on the mitotic apparatus of marine eggs, in which the state of assembly of the microtubules can be manipulated *in vivo* by temperature, pressure, and antimitotic drugs while the extent of polymerization is being quantitated by polarization microscopy (Inoué and Sato,

1967; Inoué *et al.*, 1975). These studies suggest that the polymerization of tubulin is entropy-driven and mediated by hydrophobic interactions; the various thermodynamic parameters have been calculated.

Although most cytoplasmic and mitotic microtubules are in equilibrium with a pool of disassembled subunits, there are certain classes of very stable microtubules that are not. These include the microtubules of cilia, flagella, axostyles, kinetodesmal (km) fibers, and similar structures. The disassembly of these structures assumes particular interest because the process appears to be rather different from the assembly process [see Bloodgood (1974) for a review]. In particular, it is interesting that certain of these organelles (hence their constituent microtubules) can be assembling while others of the same organelle in the same cell are disassembling. In *Chlamydomonas,* one flagellum can be shortening while the other one is elongating (Coyne and Rosenbaum, 1970); a similar situation has been observed for ciliate cytopharyngeal baskets (Tucker, 1970) and protozoan axostyles (Cleveland, 1956). This kind of observation imposes constraints on our thinking about models for the assembly and disassembly of these organelles. Even in more labile microtubule systems, such as the mitotic spindle, certain microtubules (the continuous spindle microtubules) are elongating at the same time that other microtubules (the kinetochore microtubules) are shortening. Recent work on the turnover of tubulin subunits in *in vitro* assembled microtubules suggests that a single microtubule can simultaneously be assembling at one end and disassembling at the other end (Margolis and Wilson, 1978).

IV. POSSIBLE MECHANISMS REGULATING MICROTUBULE ASSEMBLY *IN VIVO*

A. Tubulin Concentration

Tubulin concentration directly affects microtubule assembly *in vitro;* both the initial rate of assembly and the extent of polymerization are functions of the initial tubulin concentration. Below the critical concentration (approximately 0.2 mg/ml for a physiological ratio of tubulin and MAPs), no assembly occurs (Johnson and Borisy, 1975).

However, most of the available information suggests that microtubule assembly *in vivo* is not initiated by increasing the cytoplasmic concentration of polymerizable tubulin subunits. When neuroblastoma cells are induced to differentiate, extensive polymerization of microtubules occurs, but the total concentration of tubulin is the same in both undifferentiated and differentiated cells (Morgan and Seeds, 1975). Spindle assembly at mitosis is clearly not initiated by the synthesis of tubulin. Tubulin is synthesized well before mitosis (Forrest and Klevecz, 1972), and the inhibition of protein synthesis for at least 1 hr before mitosis occurs does not impede formation of the spindle. Nunez *et al.* (1975) showed that the levels of tubulin in fetal and in 30-day-old rat brain were similar, but that the characteristics of assembly of microtubules from brain extracts were very different. Tubulin from the fetal brain assembled more poorly than tubulin from the 30-day-old brain, but this could be overcome by the addition of nucleating structures. Stephens (1972) showed that the bulk of ciliary proteins, including tubulin, was made well before the onset of ciliogenesis.

There is one striking example in which microtubule assembly appears to be strictly coupled to new tubulin synthesis and in addition suggests the existence of separate recognizable tubulin pools for different microtubular organelles. Kowit and Fulton (1974a,b), using a combination of radioactive and immunological labeling techniques, studied the source of tubulin for the new flagella assembled when *Naegleria* undergoes its ameboflagellate transformation. The cytoplasmic and flagellar tubulins of *Naegleria* appeared to be immunologically distinguishable, a result different from that obtained with some other systems. The outer doublet tubulin constituted a small fraction of the total cell tubulin, and nearly all the flagellar outer doublet tubulin was synthesized *de novo* at the time of transformation. However, *Naegleria* may represent an exceptional system. In the other cases studied, cilia and flagella could be partially (Rosenbaum and Child, 1967; Rosenbaum *et al.*, 1969) or completely (Auclair and Siegel, 1966) assembled in the absence of new protein synthesis.

It is reasonable to conclude from most of the available studies that microtubule assembly *in vivo* is probably not regulated by tubulin concentration.

B. Nontubulin Microtubule-Associated Proteins

As discussed in Section III,A, brain microtubule accessory proteins clearly can regulate the initiation and the total amount of assembly of brain tubulin *in vitro* in the absence of structured microtubule-organizing centers (MTOCs). However, accessory proteins are not obligatory for microtubule initiation. Under *in vitro* conditions where brain tubulin does not self-assemble, the addition of seeds allows the assembly of soluble brain tubulin. Various materials may serve as seeds, including brain microtubule pieces (Olmsted *et al.*, 1974; Dentler *et al.*, 1974), flagellar axonemes (Allen and Borisy, 1974; Binder *et al.*, 1975), centrioles (Gould and Borisy, 1977), kinetochores (Telzer *et al.*, 1975), and basal bodies

(Snell *et al.*, 1974). These results argue that the accessory proteins need not play a role in microtubule initiation *in vivo;* nevertheless, there is reason to believe that they do play a significant role. We know that tau protein and MAPs are present in brain tissue, that they cycle stoichiometrically with *in vitro* cycles of microtubule assembly and disassembly, and that antibodies to both uniformly label the entire length of microtubules in cultured neuroblastoma and fibroblast cells (Sherline and Schiavone, 1977; Connolly *et al.*, 1977, 1978). At present, the best hypothesis is that the microtubule accessory proteins isolated from brain may act *in vivo* to stabilize formed microtubules or to modify the activity of structured MTOCs.

In cilia and flagella, the doublet microtubules have many nontubulin proteins associated with them (Linck, 1976); it is conceivable that some of these proteins may play a role in regulating flagellar microtubule assembly. In support of this view, Stephens (1972) reports that the synthesis of nexin, a poorly characterized protein linking adjacent doublet microtubules, is synthesized in an amount sufficient for one round of ciliary growth immediately prior to ciliogenesis in sea urchin embryos. He suggests that the appearance of this protein may be a trigger for ciliogenesis. There is evidence, from the study of genetic mutants, that at least some of the microtubule-associated components of the flagellum, including the radial links and dynein arms, are not necessary for the assembly of flagellar microtubules (Afzelius, 1976; Witman *et al.*, 1978).

C. Guanine Nucleotide Levels

Tubulin isolated from any source examined contains bound GDP and/or GTP, and the presence of bound GTP is generally necessary for *in vitro* microtubule assembly (see Section II,C, for a more detailed discussion of this issue). Hence the level of guanine nucleotide in the cell could possibly serve as a means of regulating the assembly of microtubules *in vivo,* but there is little if any evidence supporting this hypothesis. At most 2 moles of nucleotide are hydrolyzed per mole of dimer assembled. With the example of microtubule assembly in *Chaos* (Goode, 1973) discussed in Section III,B, it can be calculated that the concentration of nucleotide hydrolyzed would be at the micromolar level. This amount of hydrolysis is probably small compared to the typical cytoplasmic concentration of GTP, reported to be on the order of 0.1 mM (Lovtrup-Rein *et al.*, 1974). This argument of course does not take into account local variations in GTP concentration that may exist in different compartments of the cell. In general, it seems unlikely that microtubule assembly *in vivo* is regulated by temporal changes in the concentration of guanine nucleotide.

D. Divalent Cations: Calcium and Magnesium

The first successful polymerization of brain tubulin *in vitro* indicated that the process required Mg^{2+} ions and was inhibited by physiological levels of calcium, approximately $1 \times 10^{-5} M$ (Weisenberg, 1972). Later studies with purified brain tubulin reported a much lower sensitivity to divalent cations. Polymerization was maximal in the presence of Mg^{2+} as high as 1.0 mM, and Ca^{2+} inhibited polymerization only at concentrations greater than about 1.0 mM (Olmsted and Borisy, 1975). However, Rosenfeld *et al.* (1976) reported that the inhibition of polymerization by calcium was a function of the magnesium concentration. At physiological magnesium concentrations (millimolar), significant inhibition of polymerization occurred at micromolar concentrations of calcium. Nishida and Sakai (1977) reported on an uncharacterized endogenous factor that increases the sensitivity of tubulin to inhibition by calcium over several orders of magnitude. This factor is apparently lost on purification of tubulin and this may be the reason that the polymerization of purified tubulin is rather insensitive to calcium at physiological concentrations ($1–10 \mu M$). Marcum *et al.* (1978) observed that $10 \mu M$ calcium both inhibited and reversed the polymerization of purified brain tubulin, but only in the presence of stoichiometric concentrations of calcium-dependent regulator protein (CDR, calmodulin), a calcium-binding protein of 17,000 MW found in many cell types. It is likely, then, that cells contain mechanisms for amplifying the effect of low concentrations of calcium. Therefore, calcium remains a viable candidate for *in vivo* regulation of microtubule assembly.

In vivo effects of divalent cations have been assessed with the aid of ionophores which render cells permeable to ions in solution. Using this technique, Schliwa (1976) showed that the axopods of the heliozoan *Actinosphaerium eichhorni* were retracted and microtubules were dissolved by $1 \times 10^{-5} M$ Ca^{2+} ions in the presence of the ionophore A23187. This effect was easily reversed by moving the cells to a calcium-free solution containing EGTA. Mg^{2+} ions, in contrast, are much less effective in shortening axopods, and only slight shortening was observed at 1 mM Mg^{2+}. On the assumption that this ionophore rendered the cell totally permeable to divalent cations, these results suggest that microtubule formation *in vivo* is influenced by physiological levels of calcium. If this is true, then any mechanism which causes local changes in the calcium concentration might regulate microtubule assembly. In this context it is interesting to note that Petzelt and von Ledebur-Villeger (1973) have proposed that the activity of a calcium-dependent ATPase in or near the mitotic apparatus may sequester enough calcium to lower its concentration and to favor polymerization. Kiehart and Inoue (1976) have demonstrated the local depolymerization of spindle microtubules *in vivo* by microinjection of calcium.

E. Cyclic Nucleotides

It has long been thought that cyclic nucleotides directly or indirectly regulate certain microtubule-associated cell processes such as mitosis (Abell and Monahan, 1973), cell shape changes (Porter *et al.*, 1974), and lysosomal enzyme release (Weissman *et al.*, 1975). It has recently become clear that cGMP and cAMP usually have antagonistic effects on the same system. It is very unlikely that cyclic nucleotides exert their effects on the cell cycle via direct regulation of the assembly of mitotic apparatus microtubules (Abell and Monahan, 1973). The effect of cAMP on the state of assembly of cytoplasmic microtubules is very dependent upon the cell type being studied. In Chinese hamster ovary (CHO) cells, dibutyrl cAMP results in an increase in cell asymmetry, an increase in the number of cytoplasmic microtubules (Porter *et al.*, 1974), and an increase in the percentage of tubulin in the polymerized form (Rubin and Weiss, 1975). In the case of polymorphonuclear leukocytes (PMNs), cAMP acts like colchicine and inhibits lysosomal enzyme release, while cGMP promotes enzyme release. There are many more microtubules in the cells in the presence of cGMP than in the presence of cAMP (Weissman *et al.*, 1975). These latter authors have presented evidence suggesting that the cyclic nucleotide effects in their system may be mediated through protein kinase activity.

The Chediak–Higashi syndrome in man and in several animal species is characterized by spontaneous redistribution (capping) of concanavalin A (Con A) receptors on PMNs, whereas capping of Con A must be induced in PMNs from normal individuals by the disassembly of microtubules (by treatment with colchicine or vinblastine). Chediak–Higashi PMNs are characterized by a reduced number of microtubules. Raising the cellular cGMP level prevents spontaneous capping and results in a large increase in the number of microtubules, but the actual cyclic nucleotide defect in Chediak–Higashi cells appears to be an excess of cAMP and not a deficiency of cGMP (Oliver, 1976).

It has been reported that cAMP inhibits ciliary regeneration in *Tetrahymena* (Wolfe, 1973) and flagellar regeneration in *Chlamydomonas* (Rubin and Filner, 1973).

Although microtubule-associated processes are undoubtedly affected by cytoplasmic levels of cyclic nucleotides and the amount of assembled microtubules in cells can be manipulated by varying cyclic nucleotide levels, there is little reason to believe that cyclic nucleotides directly interact with microtubules or the proteins composing them or that they directly regulate microtubule assembly *in vivo*. No effect of cyclic nucleotide levels on microtubule assembly *in vitro* has been observed, although cAMP-dependent protein kinases have been reported to phosphorylate tubulin (Goodman *et al.*, 1970) and MAPs (Sloboda *et al.*, 1975) *in vivo* and *in vitro*. In conclusion, the action of cyclic nucleotides on cellular processes involving microtubules is probably indirect and may involve alterations in calcium levels, protein kinase activities, and the oxidation–reduction state of sulfhydryl compounds (for reviews, see Rebhun *et al.*, 1976; Rebhun, 1977).

F. Sulfhydryl State of Tubulin

The relative oxidation or reduction state of sulfhydryl groups on tubulin influences microtubule polymerization *in vitro*. Kuriyama and Sakai (1974) found that polymerization was completely inhibited when 2 moles of free sulfhydryl groups per tubulin monomer were blocked with mercaptide-forming, alkylating, or oxidizing agents. Mellon and Rebhun (1976) showed that the sulfhydryl oxidizing agent diamide reversibly inhibits microtubule assembly *in vitro* by oxidizing tubulin sulfhydryl groups. Oxidized glutathione inhibits *in vitro* microtubule assembly, presumably by forming mixed disulfides with tubulin (tubulin–SSG), whereas reduced glutathione (GSH) protects the system from the effects of diamide.

Nath and Rebhun (1976) showed that diamide inhibits sea urchin mitotic spindle assembly and disassembles spindle microtubules *in vivo*, presumably through glutathione oxidation, since glutathione is much preferred over tubulin as a substrate for oxidation and the cells contain high concentrations of glutathione.

Oliver *et al.* (1976) showed that the treatment of human PMNs with the glutathione-oxidizing agents *tert*-butyl hydroperoxide (BHP) and diamide (1) promotes capping of Con A receptors (like an antimitotic agent), (2) decreases the number of microtubules in the cells, and (3) decreases the level of GSH.

Burchill *et al.* (1978) showed that phagocytosis in human PMNs stimulates microtubule assembly. After phagocytosis is complete, microtubule disassembly occurs; this depolymerization is preceded by a rise in oxidized glutathione (GSSG) and is coincident with the appearance of mixed disulfides of glutathione and protein (protein–SSG).

The above examples strongly suggest that sulfhydryl oxidation–reduction plays an important role in the cellular regulation of microtubule assembly (see review in Rebhun *et al.*, 1976). Although there is no direct evidence for normal *in vivo* alterations in the level of GSH being a regulator of microtubule assembly *in vivo*, the activity of glutathione reductase fluctuates during periods associated with microtubule polymerization in PMNs (Strauss *et al.*, 1969) and sea urchin eggs (Ii and Sakai, 1974).

G. Posttranslational Modifications

1. Tyrosylation of Tubulin

Brain extracts incorporate tyrosine posttranscriptionally onto the C-terminal glutamate residue of α-tubulin (Arce *et al.*, 1975; Barra *et al.*, 1974; Raybin and Flavin, 1975a). The enzyme tubulin–tyrosine ligase, which performs this modification, has been isolated from brain tissue and characterized (Raybin and Flavin, 1977). Ligase activity was found in every rat tissue examined, in several sources of brain (mammalian and avian), and in neuroblastoma cells, but was not detected in sea urchin eggs, *Tetrahymena* cells or cilia, or yeast (Raybin and Flavin, 1975b). Argarana *et al.* (1977) showed that α-tubulin could be tyrosylated *in vivo*. The possible significance of tubulin tyrosylation as a regulator of cytoplasmic microtubule assembly is not clear, because tubulin polymerization *in vitro* does not appear to be influenced by the extent of tyrosylation (Raybin and Flavin, 1957b). With the use of neuroblastoma cells, it was observed that *in vivo* tyrosylation was confined primarily to insoluble (presumably membrane-bound) tubulin (Raybin and Flavin, 1975b). It is conceivable that tyrosylation serves to regulate the partition of tubulin among different compartments within the cell. Rodriguez and Borisy (1977) observed that the extent to which chick brain tubulin could be tyrosylated *in vitro* varied with the stage of development.

2. Phosphorylation of Tubulin and Accessory Proteins

Mammalian brain tubulin isolated by DEAE chromatography contains 0.8 mole of covalently bound phosphate per mole of tubulin dimer, attached to the serine residues of β-tubulin (Goodman *et al.*, 1970; Eipper, 1972). It has been demonstrated that phosphate incorporation occurs *in vivo* by incubating brain slices with [^{32}P]orthophosphate (Eipper, 1972), and *in vitro* by incubating purified brain tubulin with [^{32}P]ATP (Eipper, 1974; Goodman *et al.*, 1970). Phosphorylation was accomplished by the activity of a cAMP-dependent protein kinase (Goodman *et al.*, 1970; Eipper, 1974; Lagnado *et al.*, 1975; Soifer, 1975a; Sloboda *et al.*, 1975). Although it was suggested that tubulin itself was the protein kinase (Soifer, 1975a), the enzyme activity has now been definitely separated from tubulin (Eipper, 1974; Piras and Piras, 1974; Sloboda *et al.*, 1975; Shigekawa and Olsen, 1975; Sandoval and Cuatrecasas, 1976). However, since the protein kinase activity cycles stoichiometrically with *in vitro* assembled microtubules (Sloboda *et al.*, 1975), it must be considered a MAP. Protein kinase activity has also been purified from *Tetrahymena* cilia and used to phosphorylate solubilized ciliary doublet tubulin (Murofuschi, 1973).

More significantly, Sloboda *et al.* (1975) showed, using *in vitro* assembled brain tubulin preparations containing endogenous protein kinase activity, that one of the two high molecular weight MAPs (MAP_2) incorporated 650 times more phosphate (specific activity) than the tubulin. The rate of phosphorylation of MAP_2 was stimulated four- to sixfold by cAMP. A total of 1.0 mole phosphate/mole MAP_2 was incorporated in the absence of cAMP, and 1.9 moles phosphate/mole MAP_2 was incorporated in the presence of cAMP. They also demonstrated *in vivo* incorporation of phosphate into both MAP_1 and MAP_2 following the injection of $[^{32}P]$orthophosphate into chick brain. Sheterline and Schofield (1975) obtained preparations of tubulin and accessory proteins from bovine anterior pituitary by cycles of *in vitro* assembly and disassembly. These preparations incorporated phosphate into 70,000 and 280,000 MW components in addition to tubulin. These same authors showed that their preparations contained a phosphatase activity capable of dephosphorylating these components, something which is absent from the brain tubulin preparations.

Although the phosphorylation of tubulin, and particularly the high molecular weight MAPs, immediately suggests a convenient regulatory system, it has not been possible to implicate *in vitro* phosphorylation in the regulation of microtubule assembly. The phosphorylation of microtubule proteins does not affect the rate or extent of microtubule assembly attained *in vitro* (Sheterline, 1976; Sloboda, unpublished observations). The possibility remains that phosphorylation may be related to the functioning of the intact microtubules (a binding or motility function of the MAP-containing surface projections).

3. Glycosylation of Tubulin

Routine preparations of tubulin from various sources, when run on acrylamide gel electrophoresis, show no detectable staining with the periodic acid–Schiff reagent used for the demonstration of carbohydrate. Feit and Shelanski (1975) reported that at least a portion of the tubulin obtained from brain is a glycoprotein. When they injected $[^{14}C]$glucosamine into mouse brain, isolated the tubulin, subjected it to routine acrylamide gel electrophoresis or to isoelectric focusing, and autoradiographed the gels, they found radioactivity associated with the tubulin. This radioactivity could be recovered as a mixture of glucosamine and galactosamine. Although this glycosylation may serve as a mechanism for regulating the assembly of tubulin into microtubules, it is much more reasonable to suggest that it regulates the distribution of tubulin among different compartments within the cell. Stephens (1977) reports that ciliary membrane tubulin is glycosylated while ciliary axonemal tubulin is not, suggesting that glycosylation may occur coincident with and perhaps only at the time of tubulin insertion into a membrane.

H. Polyanions: Cellular RNA as an Inhibitor of Assembly

Polyanions in the form of RNA, carboxymethylcellulose, and phosphocellulose inhibit microtubule polymerization *in vitro* (Bryan *et al.*, 1975). Since tubulin is an anionic protein and MAPs are cationic, it has been proposed that cellular polyanions inhibit microtubule assembly by competing with tubulin for the binding of accessory proteins (MAPs) that promote microtubule assembly, and that RNA of an unspecified type could serve as a natural inhibitor of microtubule assembly (Bryan *et al.*, 1975; Bryan and Nagle, 1975). This proposal could explain why microtubule self-assembly occurs poorly from extracts of tissue culture cells and sea urchin eggs (which have a high RNA/tubulin ratio) but self-assemble well from extracts of neural tissue (which have much lower RNA/tubulin ratios). Although RNA is an unlikely candidate for the regulation of microtubule assembly *in vivo* (since RNA concentrations probably do not fluctuate fast enough to be useful), it may serve the cell (perhaps coincidentally) as a mechanism totally preventing random self-nucleation of microtubules from ever occurring *in vivo*. It is essential that cells carefully control the spatial arrangement of microtubules, and this is done by requiring that all microtubule initiation occur from a structured MTOC (see Section IV,I).

I. Microtubule-Organizing Centers

Much of the work on the assembly of purified microtubule proteins into microtubules *in vitro* has involved self-nucleation of microtubules utilizing only components derived from the disassembly of previously formed microtubules. There is serious doubt about the applicability of much of the information derived from these totally *in vitro* systems to an understanding of how microtubule assembly is regulated within cells. In almost all cases where they have been carefully examined, microtubules within cells are associated at one or both ends with another structural element of the cell. These structures fall into a small number of classes: (1) basal bodies, (2) centrioles (and other structures associated with spindle poles such as yeast spindle plaques), (3) kinetochores on chromosomes, and (4) membranes. In some cases, microtubules terminate (or begin, depending on the point of view) at rather unstructural but discrete areas of electron-dense material. That these structural elements to which microtubules are attached *in vivo* are actually capable of nucleating microtubule assembly has been confirmed in studies in which membranes (Becker *et al.*, 1975), basal bodies (Snell *et al.*, 1974), centrioles (Gould and Borisy, 1977), chromosome kinetochores (Telzer *et al.*, 1975), and yeast spindle plaques (Borisy *et al.*, 1975a) have been isolated, incubated with purified brain tubulin under conditions that minimize self-nucleation, and observed to nucleate the assembly of bona fide microtubules. These semi-*in vitro* systems do not necessarily operate with complete fidelity, since basal bodies nucleate singlet, not doublet, microtubules. However, this is undoubtedly due to the absence of certain proteins, not found in brain, that are essential for ciliary or flagellar doublet microtubule formation. Another way to demonstrate that these structures are capable of nucleating microtubule assembly was used by Heidemann and Kirschner (1975), who injected basal bodies isolated from *Tetrahymena* or *Chlamydomonas* into mature eggs of *Xenopus* and observed the subsequent induction of microtubule-containing asters in the cytoplasm of the eggs. Weisenberg and Rosenfeld (1975) reported the induction of asters in *Spisula* egg extracts, but only when a centriole-containing fraction was present. They demonstrated by thin-section electron microscopy that a centriole was present at the center of each aster.

The use of structured MTOCs for the nucleation of most microtubules *in vivo* allows the cell to exercise careful control over the positioning and orientation of microtubules, which is essential if the microtubules are to perform their proper functions. In fact, it appears that many cells may possess mechanisms for actively preventing self-nucleation of microtubules from ever occurring (Bryan *et al.*, 1975; Bryan and Nagle, 1975). In addition to spatial control, the cell must be able to exert careful temporal control over microtubule assembly. If most or all microtubule assembly occurs from structured MTOCs, the cell may possess mechanisms for turning the nucleation capacity of MTOCs on and off. Little information is available concerning this level of control, but it is conceivable that some of the microtubule accessory proteins observed *in vitro* may function *in vivo* to regulate the activity of structured MTOCs. Guttman and Gorovsky (1975) showed that the regeneration of cilia in starved *Tetrahymena* is accompanied by the rapid synthesis of an 80,000 MW protein fraction, although this newly synthesized protein fraction is not found in any appreciable amount within the regenerated cilia. One interpretation of their data is that a protein must be synthesized which modifies the basal body before the initiation of assembly of new cilia can occur. Heidemann *et al.* (1977) reported that RNase treatment of isolated basal bodies abolished their capacity to induce aster formation after injection into *Xenopus* eggs. At least one antimitotic drug, isopropyl-*N*-phenyl carbamate, may have an effect specifically at the level of the MTOC (Coss and Pickett-Heaps, 1974). It is conceivable that genetics can be brought to bear on the problem of the regulation of microtubule assembly through MTOCs. Raff *et al.* (1976) reported on a maternal effect mutation in the axolotl *Ambystoma* with a defect in microtubule assembly in the egg. This defect could be overcome by injecting into the egg pieces of microtubules or isolated *Chlamydomonas* basal bodies, suggesting that if involved the absence of a structured MTOC competent to initiate microtubule assembly. Goodenough and St. Clair (1975) obtained a mutant of *Chlamydomonas* unable to assemble the flagellum; this mutant has a structural defect in the basal body.

It seems likely that structured MTOCs play an important role in the cellular regulation of microtubule assembly. More research effort should be expended in addressing questions concerning the chemical composition, assembly, and modification of MTOCs.

V. CONCLUSIONS AND PROSPECTS

This chapter has placed strong emphasis on the assembly of microtubules *in vivo* and *in vitro* and the mechanisms that may regulate this assembly and allow it to be carefully controlled in time and space. The ability to assemble purified tubulin molecules into microtubules *in vitro* has provided an enormous impetus to the field of microtubule assembly and has resulted in the acquisition of a large amount of information concerning *in vitro* assembly and disassembly and their regulation.

Most of the available evidence, both from *in vivo* and *in vitro* studies, suggests that the regulation of assembly is twofold: (1) Control through initiation, where it appears that structured MTOCs play the dominant role, although it is still a mystery how the initiating ability of these MTOCs is itself turned on and off; and (2) control through the assembly competence of the microtubule subunits, where it appears likely that the sulfhydryl oxidation state and/or association with calcium-binding proteins will turn out to be important steps in controlling tubulin assembly competence *in vivo*.

A serious question exists about the validity of extending *in vitro*-derived information to descriptions of how microtubule assembly occurs within cells. It is essential that emphasis be turned away from *in vitro* studies and toward studies on microtubule assembly within cells. The regeneration of cilia and flagella has become one of the most useful systems for these types of studies, although these organelles contain relatively stable microtubules whose assembly and disassembly may be very different from that in systems of labile microtubules. In addition, it is essential that emphasis be placed on using genetics as a tool to overcome some of the problems involved in studying microtubule assembly and its regulation *in vivo*. Considerable progress is beginning to be made in several laboratories using genetics to study the assembly and stability of the *Chlamydomonas* flagellum (see Bloodgood, 1978).

The mitotic apparatus deserves continued attention because it is one of the most carefully controlled and highly generalized examples of microtubule assembly and disassembly. There exists here very careful regulation of microtubule assembly in time, in space, and even in orientation. The mitotic apparatus contains very different populations of microtubules possessing different stabilities, different MTOCs, and different temporal programs of assembly and disassembly.

Interest in microtubule assembly continues to gain momentum, and this area will remain a source of exciting research problems. However, other areas of microtubule research should not be ignored, in particular those concerned with how microtubule-associated structures (arms and bridges) transduce force for biological movements. Here, as in the past, emphasis will be on ciliary and flagellar systems.

REFERENCES

Abell, C. W., and Monahan, T. M. (1973). *J. Cell Biol.* **59**, 549–558.
Afzelius, B. A. (1976). *Science* **193**, 317–319.
Allen, C., and Borisy, G. G. (1974). *J. Mol. Biol.* **90**, 381–402.
Arai, T., and Kaziro, Y. (1976). *Biochem. Biophys. Res. Commun.* **69**, 369–376.
Arce, C. A., Barra, H. S., Rodriguez, J. A., and Caputto, R. (1975). *FEBS Lett.* **50**, 5–7.
Argarana, C. E., Arce, C. A., Barra, H. S., and Caputto, R. (1977). *Arch. Biochem. Biophys.* **180**, 264–268.
Auclair, W., and Siegel, B. W. (1966). *Science* **154**, 913–915.
Bajer, A. S., and Mole-Bajer, J. (1972). *Int. Rev. Cytol.* **34**, Suppl. 3, 1–271.
Bardele, C. F. (1973). *Cytobiologie* **7**, 442–488.
Barnes, L. D., Engel, A. G., and Dousa, T. P. (1975). *Biochim. Biophys. Acta* **405**, 422–433.
Barra, H. S., Arce, C. A., Rodriquez, J. A., and Caputto, R. (1974). *Biochem. Biophys. Res. Commun.* **60**, 1384–1390.
Becker, J. S., Oliver, J. M., and Berlin, R. D. (1975). *Nature (London)* **254**, 152–154.
Berry, R. W., and Shelanski, M. L. (1972). *J. Mol. Biol.* **71**, 71–80.
Bhattacharyya, B., and Wolff, J. (1975). *Arch. Biochem. Biophys.* **167**, 264–269.
Bhattacharyya, B., and Wolff, J. (1976). *Proc. Natl. Acad. Sci. U.S.A.* **73**, 2375–2378.
Bibring, T., Baxandall, J., Denslow, S., and Walker, B. (1976). *J. Cell Biol.* **69**, 301–312.
Binder, L. I., Dentler, W. L., and Rosenbaum, J. L. (1975). *Proc. Natl. Acad. Sci. U.S.A.* **72**, 1122–1126.
Bloodgood, R. A. (1974). *Cytobios* **9**, 143–161.

Bloodgood, R. A. (1976). *In* "Cell Biology Data Book" (P. L. Altman and D. D. Dittmer, eds.), pp. 356–358. Fed. Am. Soc. Exp. Biol., Bethesda, Maryland.

Bloodgood, R. A. (1978). *Cell Biol. Int. Rep.* **2**, 299–302.

Bloodgood, R. A., and Rosenbaum, J. L. (1976). *J. Cell Biol.* **71**, 322–331.

Borgers, M., and DeBrabander, M., eds. (1975). "Microtubules and Microtubule Inhibitors." North-Holland Publ., Amsterdam.

Borisy, G. G., and Olmsted, J. B. (1972). *Science* **177**, 1196–1197.

Borisy, G. G., and Taylor, E. W. (1967a). *J. Cell Biol.* **34**, 525–533.

Borisy, G. G., and Taylor, E. W. (1967b). *J. Cell Biol.* **34**, 533–548.

Borisy, G. G., Olmsted, J. B., Marcum, J. M., and Allen, C. (1974). *Fed. Proc., Fed. Am. Soc. Exp. Biol.* **33**, 167–174.

Borisy, G. G., Peterson, J. B., Hyams, J. S., and Ris, H. (1975a). *J. Cell Biol.* **67**, 38a.

Borisy, G. G., Marcum, J. M., Olmsted, J. B., Murphy, D. B., and Johnson, K. A. (1975b). *Ann. N.Y. Acad. Sci.* **253**, 107–132.

Bouck, G. B., and Brown, D. L. (1976). *Annu. Rev. Plant Physiol.* **27**, 71–94.

Bryan, J. (1972). *Biochemistry* **11**, 2611–2616.

Bryan, J. (1976). *J. Cell Biol.* **71**, 749–767.

Bryan, J., and Nagle, B. W. (1975). *Mosbach Colloq., 26th* pp. 161–174.

Bryan, J., and Wilson, L. (1971). *Proc. Natl. Acad. Sci. U.S.A.* **68**, 1762–1766.

Bryan, J., Nagle, B. W., and Doenges, K. H. (1975). *Proc. Natl. Acad. Sci. U.S.A.* **72**, 3570–3574.

Burchill, B. R., Oliver, J. M., Pearson, C. B., Leinbach, E. D., and Berlin, R. D. (1978). *J. Cell Biol.* **76**, 439–447.

Burns, R. G. (1973). *Exp. Cell Res.* **81**, 285–292.

Cann, J. R., and Hinman, N. D. (1975). *Mol. Pharmacol.* **11**, 256–267.

Castle, A. G., and Crawford, N. (1975). *FEBS Lett.* **51**, 195–200.

Cleveland, D. W., Hwo, S.-Y., and Kirschner, M. W. (1977a). *J. Mol. Biol.* **116**, 207–225.

Cleveland, D. W., Hwo, S.-Y., and Kirschner, M. W. (1977b). *J. Mol. Biol.* **116**, 227–247.

Cleveland, L. R. (1956). *J. Protozool.* **3**, 161–180.

Connolly, J. A., Kalnins, V. I., Cleveland, D. W., and Kirschner, M. W. (1977). *Proc. Natl. Acad. Sci. U.S.A.* **74**, 2437–2440.

Connolly, J. A., Kalnins, V. I., Cleveland, D. W., and Kirschner, M. W. (1978). *J. Cell Biol.* **76**, 781–786.

Cortese, F., Bhattacharyya, B., and Wolff, J. (1977). *J. Biol. Chem.* **252**, 1134–1140.

Coss, R. A., and Pickett-Heaps, J. D. (1974). *J. Cell Biol.* **63**, 84–98.

Coyne, B., and Rosenbaum, J. L. (1970). *J. Cell Biol.* **47**, 777–781.

Dentler, W. L., Granett, S., Witman, G. B., and Rosenbaum, J. L. (1974). *Proc. Natl. Acad. Sci. U.S.A.* **71**, 1710–1714.

Dentler, W. L., Granett, S., and Rosenbaum, J. L. (1975). *J. Cell Biol.* **65**, 237–241.

Doenges, K. H., Nagle, B. W., Uhlmann, A., and Bryan, J. (1977). *Biochemistry* **16**, 3455–3459.

Eipper, B. A. (1972). *Proc. Natl. Acad. Sci. U.S.A.* **69**, 2283–2287.

Eipper, B. A. (1974). *J. Biol. Chem.* **249**, 1398–1406.

Erickson, H. P. (1974). *J. Cell Biol.* **60**, 153–167.

Erickson, H. P. (1976). *In* "Cell Motility," Book C (R. Goldman, T. Pollard, and J. Rosenbaum, eds.), Cold Spring Harbor Conferences on Cell Proliferation, Vol. 3, pp. 1069–1080. Cold Spring Harbor Lab., Cold Spring Harbor, New York.

Erickson, H. P., and Voter, W. A. (1976). *Proc. Natl. Acad. Sci. U.S.A.* **73**, 2813–2817.

Fawcett, D. W., and Porter, K. R. (1954). *J. Morphol.* **94**, 221–281.

Feit, H., and Shelanski, M. L. (1975). *Biochem. Biophys. Res. Commun.* **66**, 920–927.

Forrest, G. L., and Klevecz, R. R. (1972). *J. Biol. Chem.* **247**, 3147–3152.
Fuller, M. S. (1976). *Int. Rev. Cytol.* **45**, 113–153.
Fulton, C., and Kowit, J. D. (1975). *Ann. N.Y. Acad. Sci.* **253**, 318–332.
Gaskin, F., and Shelanski, M. L. (1976). *Essays Biochem.* **12**, 115–146.
Goldman, R., Pollard, T., and Rosenbaum, J., eds. (1976). "Cell Motility," Books A,B,C. (R. Goldman, T. Pollard, and J. Rosenbaum, eds.), Cold Spring Harbor Conferences on Cell Proliferation, Vol. 3. Cold Spring Harbor Lab., Cold Spring Harbor, New York.
Good, N. E., Winget, G. D., Winter, W., Connolly, T. N., Izawa, S., and Singh, R. M. M. (1966). *Biochemistry* **5**, 467–477.
Goode, D. (1973). *J. Mol. Biol.* **80**, 531–538.
Goodenough, U. W., and St. Clair, H. S. (1975). *J. Cell Biol.* **66**, 480–491.
Goodman, D. B. P., Rasmussen, H., DiBella, F., and Buthrow, C. E., Jr. (1970). *Proc. Natl. Acad. Sci. U.S.A.* **67**, 652–659.
Gould, R. R., and Borisy, G. G. (1977). *J. Cell Biol.* **73**, 601–615.
Green, L. H., Brandis, J. W., Turner, F. R., and Raff, R. A. (1975). *Biochemistry* **14**, 4487–4491.
Guttman, S. D., and Gorovsky, M. A. (1975). *J. Cell Biol.* **67**, 149a.
Haber, J. E., Peloquin, J. G., Halvorson, H. O., and Borisy, G. G. (1972). *J. Cell Biol.* **53**, 355–367.
Heidemann, S. R., and Kirschner, M. W. (1975). *J. Cell Biol.* **67**, 105–117.
Heidemann, S. R., Sander, G., and Kirschner, M. W. (1977). *Cell* **10**, 337–350.
Hepler, P. K. (1976). *In* "Plant Biochemistry" (J. Bonner and J. E. Varner, eds.), 3rd ed., pp. 147–187. Academic Press, New York.
Hepler, P. K., and Palevitz, B. A. (1974). *Annu. Rev. Plant Physiol.* **25**, 309–362.
Herzog, W., and Weber, K. (1977). *Proc. Natl. Acad. Sci. U.S.A.* **74**, 1860–1864.
Herzog, W., and Weber, K. (1978). *Eur. J. Biochem.* **92**, 1–8.
Himes, R. H., Burton, P. R., Kersey, R. N., and Pierson, G. B. (1976). *Proc. Natl. Acad. Sci. U.S.A.* **73**, 4397–4399.
Hoebeke, J., Van Nijen, G., and DeBrabander, M. (1976). *Biochim. Biophys. Res. Commun.* **69**, 319–324.
Ii, I., and Sakai, H. (1974). *Biochim. Biophys. Acta* **350**, 151–161.
Inoué, S., and Sato, H. (1967). *J. Gen. Physiol.* **50**, 259–288.
Inoué, S., and Stephens, R. E., eds. (1975). "Molecules and Cell Movement." Raven, New York.
Inoué, S., Fuseler, J., Salmon, E. D., and Ellis, G. W. (1975). *Biophys. J.* **15**, 725–744.
Johnson, K. A., and Borisy, G. G. (1975). *In* "Molecules and Cell Movement" (S. Inoué and R. E. Stephens, eds.), pp. 119–141. Raven, New York.
Johnson, K. A., and Borisy, G. G. (1977). *J. Mol. Biol.* **117**, 1–31.
Kiehart, D. P. and Inoué, S. (1976). *J. Cell Biol.* **70**, 230a.
Kirschner, M. W., Honig, L. S., and Williams, R. C. (1975). *J. Mol. Biol.* **99**, 263–276.
Kowit, J. D., and Fulton, C. (1974a). *Proc. Natl. Acad. Sci. U.S.A.* **71**, 2877–2881.
Kowit, J. D., and Fulton, C. (1974b). *J. Biol. Chem.* **249**, 3638–3646.
Kuriyama, R. (1976). *J. Biochem. (Tokyo)* **80**, 153–165.
Kuriyama, R. (1977). *J. Biochem. (Tokyo)* **81**, 1115–1125.
Kuriyama, R., and Sakai, H. (1974). *J. Biochem. (Tokyo)* **76**, 651–654.
Lagnado, J., Tan, L. P., and Reddington, J. (1975). *Ann. N.Y. Acad. Sci.* **253**, 577–597.
Lee, J. C., and Timasheff, S. N. (1975). *Biochemistry* **14**, 5183–5187.
Lee, J. C., and Timasheff, S. N. (1977). *Biochemistry* **16**, 1754–1764.
Linck, R. W. (1976). *J. Cell Sci.* **20**, 405–439.
Linck, R. W., and Amos, L. A. (1974). *J. Cell Sci.* **15**, 551–559.

Lovtrup-Rein, H., Nelson, L., and Lovtrup, S. (1974). *Exp. Cell Res.* **86,** 206–209.
Luduena, R. F., and Woodward, D. O. (1973). *Proc. Natl. Acad. Sci. U.S.A.* **70,** 3594–3598.
Luduena, R. F., Wilson, L., and Shooter, E. M. (1975). *In* "Microtubules and Microtubule Inhibitors" (M. Borgers and M. DeBrabander, eds.), pp. 47–58. North-Holland Publ., Amsterdam.
Maccioni, R., and Seeds, N. W. (1977). *Proc. Natl. Acad. Sci. U.S.A.* **74,** 462–466.
McIntosh, J. R. (1974). *J. Cell Biol.* **61,** 166–187.
Mandelbaum-Shavit, F., Wolpert-DeFilippes, M. K., and Johns, D. G. (1976) *Biochem. Biophys. Res. Commun.* **72,** 47–54.
Marcum, J. M., Dedman, J. R., Brinkley, B. R., and Means, A. R. (1978). *Proc. Natl. Acad. Sci. U.S.A.* **75,** 3771–3775.
Margolis, R. L., and Wilson, L. (1977). *Proc. Natl. Acad. Sci. U.S.A.* **74,** 3466–3470.
Margolis, R. L., and Wilson, L. (1978). *Cell* **13,** 1–8.
Margulis, L. (1970). "Origin of Eukaryotic Cells." Yale Univ. Press, New Haven, Connecticut.
Margulis, L. (1973). *Int. Rev. Cytol.* **34,** 333–361.
Margulis, L. (1975). *In* "Microtubules and Microtubule Inhibitors" (M. Borgers and M. DeBrabander, eds.), pp. 3–18. North-Holland Publ., Amsterdam.
Mellon, M. G., and Rebhun, L. I. (1976). *J. Cell Biol.* **70,** 226–238.
Mizel, S. B., and Wilson, L. (1972). *Biochemistry* **11,** 2573–2578.
Mohri, H. (1976). *Biochim. Biophys. Acta* **456,** 85–127.
Morgan, J. L., and Seeds, N. W. (1975). *J. Cell Biol.* **67,** 136–145.
Murofuschi, H. (1973). *Biochim. Biophys. Acta* **327,** 354–364.
Murphy, D. B., and Borisy, G. G. (1975). *Proc. Natl. Acad. Sci. U.S.A.* **72,** 2696–2700.
Murphy, D. B., Johnson, K. A., and Borisy, G. G. (1977a). *J. Mol. Biol.* **117,** 33–52.
Murphy, D. B., Vallee, R. B., and Borisy, G. G. (1977b). *Biochemistry* **16,** 2598–2605.
Nath, J., and Rebhun, L. I. (1976). *J. Cell Biol.* **68,** 440–450.
Newcomb, E. (1969). *Annu. Rev. Plant Physiol.* **20,** 253–288.
Nishida, E., and Sakai, H. (1977). *J. Biochem. (Tokyo)* **82,** 303–306.
Nunez, J., Fellous, A., Francon, J., and Lennon, A. M. (1975). *In* "Microtubules and Microtubule Inhibitors" (M. Borgers and M. DeBrabander, eds.), pp. 269–279. North-Holland Publ., Amsterdam.
Oliver, J. M. (1976). *Am. J. Pathol.* **85,** 395–412.
Oliver, J. M., Albertini, D. F., and Berlin, R. D. (1976). *J. Cell Biol.* **71,** 921–932.
Olmsted, J. B., and Borisy, G. G. (1973a). *Biochemistry* **12,** 4282–4289.
Olmsted, J. B., and Borisy, G. G. (1973b). *Annu. Rev. Biochem.* **42,** 507–540.
Olmsted, J. B., and Borisy, G. G. (1975). *Biochemistry* **14,** 2996–3005.
Olmsted, J. B., Marcum, J. M., Johnson, K. A., Allen, C., and Borisy, G. G. (1974). *J. Supramol. Struct.* **2,** 429–450.
Penningroth, S. M., and Kirschner, M. W. (1977). *J. Mol. Biol.* **115,** 643–673.
Petzelt, C., and von Ledebur-Villeger, M. (1973). *Exp. Cell Res.* **81,** 87–94.
Pickett-Heaps, J. D. (1974). *In* "Dynamic Aspects of Plant Ultrastructure" (A. W. Robards, ed.), pp. 219–255. McGraw-Hill, New York.
Piras, M. M., and Piras, R. (1974). *Eur. J. Biochem.* **47,** 443–452.
Porter, K. R. (1966). *In* "Principles of Biomolecular Organization" (G. E. W. Wolstenholme and M. O'Connor, eds.), pp. 308–345. Churchill, London.
Porter, K. R., Puck, T. T., Hsie, A. W., and Kelley, D. (1974). *Cell* **2,** 145–162.
Raff, E. C., Brothers, A. J., and Rass, R. A. (1976). *Nature (London)* **260,** 615–617.
Raybin, D., and Flavin, M. (1975a). *Biochim. Biophys. Res. Commun.* **65,** 1088–1095.
Raybin, D., and Flavin, M. (1975b). *J. Cell Biol.* **73,** 492–504.

Rabyin, D., and Flavin, M. (1977). *Biochemistry* **16**, 2189–2194.

Rebhun, L. I. (1977). *Int. Rev. Cytol.* **49**, 1–54.

Rebhun, L. I., Miller, M., Schnaitman, T. C., Nath, J., and Mellon, M. (1976). *J. Supramol. Struct.* **5**, 199–219.

Roberts, K. (1974). *Prog. Biophys. Mol. Biol.* **28**, 371–420.

Rodriquez, J. A., and Borisy, G. G. (1977). *J. Cell Biol.* **75**, 296a.

Rosenbaum, J. L., and Child, F. M. (1967). *J. Cell Biol.* **34**, 345–364.

Rosenbaum, J. L., Moulder, J. E., and Ringo, D. L. (1969). *J. Cell Biol.* **41**, 600–619.

Rosenfeld, A. C., Zackroff, R. V., and Weisenberg, R. C. (1976). *FEBS Lett.* **65**, 144–147.

Rubin, R. W., and Filner, P. (1973). *J. Cell Biol.* **56**, 628–635.

Rubin, R. W., and Weiss, G. D. (1975). *J. Cell Biol.* **64**, 42–53.

Sabatini, D. D., Bensch, K., and Barrnett, R. J. (1963). *J. Cell Biol.* **17**, 19–58.

Samson, F. E., Redburn, D. A., and Himes, R. H. (1973). *Methods Neurochem.* **5**, 115–146.

Sandoval, I. V., and Cuatrecasas, P. (1976). *Biochemistry* **15**, 3424–3432.

Schellenberg, R. R., and Gillespie, E. (1977). *Nature (London)* **265**, 741–742.

Schliwa, M. (1976). *J. Cell Biol.* **70**, 527–540.

Seeds, N. W., and Maccioni, R. B. (1978). *J. Cell Biol.* **76**, 547–555.

Setlow, R. B., and Pollard, E. C. (1962). "Molecular Biophysics." Addison-Wesley, Reading, Massachusetts.

Shelanski, M. L. (1973). *In* "Proteins of the Nervous System" (D. J. Schneider, R. H. Angeletti, R. A. Bradshaw, A. Grass, and B. W. Moore, eds.), pp. 227–241. Raven, New York.

Shelanski, M. L., and Taylor, E. W. (1968). *J. Cell Biol.* **38**, 304–315.

Shelanski, M. L., Gaskin, F., and Cantor, C. R. (1973). *Proc. Natl. Acad. Sci. U.S.A.* **70**, 765–768.

Sherline, P., and Schiavone, K. (1977). *Science* **198**, 1038–1040.

Sheterline, P. (1976). *Biochem. Soc. Trans.* **4**, 789–791.

Sheterline, P., and Schofield, J. G. (1975). *FEBS Lett.* **56**, 297–302.

Shigekawa, B. L., and Olsen, R. W. (1975). *Biochem. Biophys. Res. Commun.* **63**, 455–462.

Sleigh, M. A., ed. (1974). "Cilia and Flagella." Academic Press, New York.

Sloboda, R. D., and Rosenbaum, J. L. (1979). *Biochemistry* **18**, 48–55.

Sloboda, R. D., Rudolph, S. A., Rosenbaum, J. L., and Greengard, J. L. (1975). *Proc. Natl. Acad. Sci. U.S.A.* **72**, 177–181.

Sloboda, R. D., Dentler, W. L., and Rosenbaum, J. L. (1976). *Biochemistry* **15**, 4497–4505.

Snell, J. J., Dentler, W. L., Haimo, L. T., Binder, L. I., and Rosenbaum, J. L. (1974). *Science* **185**, 357–360.

Snyder, J. A., and McIntosh, J. R. (1976). *Annu. Rev. Biochem.* **45**, 699–720.

Soifer, D. (1975a). *J. Neurochem.* **24**, 21–33.

Soifer, D., ed. (1975b). "The Biology of Cytoplasmic Microtubules." *Ann. N.Y. Acad. Sci.* **253**.

Solomon, F. (1977). *Biochemistry* **16**, 358–363.

Stearns, M. E., and Brown, D. L. (1976). *J. Cell Biol.* **70**, 242a.

Stephens, R. E. (1972). *Biol. Bull. (Woods Hole, Mass.)* **142**, 489–504.

Stephens, R. E. (1975). *In* "Molecules and Cell Movement" (S. Inoué and R. E. Stephens, eds.), pp. 181–206. Raven, New York.

Stephens, R. E. (1977). *Biochemistry* **16**, 2047–2058.

Stephens, R. E., and Edds, K. T. (1976). *Physiol. Rev.* **56**, 709–777.

Strauss, R. R., Paul, B. B., Jacobs, A. A., and Sbarra, A. J. (1969). *Arch. Biochem. Biophys.* **135**, 265–271.

Summers, K. (1975). *Biochim. Biophys. Acta* **416**, 153–168.

Telzer, B. R., Moses, M. J., and Rosenbaum, J. L. (1975). *Proc. Natl. Acad. Sci. U.S.A.* **72**, 4023–4027.
Tilney, L. G. (1971). *In* "Origin and Continuity of Cell Organelles" (J. Reinert and H. Ursprung, eds.), pp. 220–260. Springer-Verlag, Berlin and New York.
Tucker, J. B. (1970). *J. Cell Sci.* **6**, 385–429.
Vallee, R. B., and Borisy, G. C. (1977). *J. Biol. Chem.* **252**, 377–382.
Warner, F. D. (1972). *Adv. Cell Mol. Biol.* **2**, 193–235.
Weingarten, M. D., Lockwood, A. H., Hwo, S.-Y., and Kirschner, M. W. (1975). *Proc. Natl. Acad. Sci. U.S.A.* **72**, 1858–1862.
Weisenberg, R. C. (1972). *Science* **177**, 1104–1105.
Weisenberg, R. C. (1974). *J. Supramol. Struct.* **2**, 451–456.
Weisenberg, R. C., and Rosenfeld, A. C. (1975). *J. Cell Biol.* **64**, 146–158.
Weisenberg, R. C., Borisy, G. G., and Taylor, E. W. (1968). *Biochemistry* **7**, 4466–4479.
Weisenberg, R. C., Deery, W. J., and Dickinson, P. J. (1976). *Biochemistry* **15**, 4248–4254.
Weissman, G., Goldstein, I. Hoffstein, S., and Tsung, P.-K. (1975). *Ann. N.Y. Acad. Sci.* **253**, 750–762.
Williams, N. E. (1975). *Int. Rev. Cytol.* **41**, 59–86.
Wilson, L. (1970). *Biochemistry* **9**, 4990–5007.
Wilson, L., and Bryan, J. (1974). *Adv. Cell Mol. Biol.* **3**, 22–72.
Wilson, L., and Meza, I. (1973). *J. Cell Biol.* **58**, 709–719.
Wilson, L. Anderson, K., and Chin, D. (1976). *In* "Cell Motility," Book C (R. Goldman, T. Pollard, and J. Rosenbaum, eds.), Cold Spring Harbor Conferences on Cell Proliferation, Vol. 3, pp. 1051–1064. Cold Spring Harbor Lab., Cold Spring Harbor, New York.
Witman, G. B., Plummer, J., and Sander, G. (1978). *J. Cell Biol.* **76**, 729–747.
Wolfe, J. (1973). *J. Cell Physiol.* **82**, 39–48.
Wuerker, R. B., and Kirkpartick, J. B. (1972). *Int. Rev. Cytol.* **33**, 45–75.

Chemosensory Transduction in *Dictyostelium discoideum*

5

JOSÉ M. MATO AND THEO M. KONIJN

I. INTRODUCTION

For all interested in chemotaxis there are three fundamental questions: (1) What are chemotactic signals and how are they detected? (2) What regulatory molecules change their levels in response to an attractant? (3) How do these molecules modulate the motile response or microfilament and microtubule functions? This chapter is limited to recent work on chemotaxis in cellular slime molds as related to these questions.

By far the best-studied species of cellular slime mold is *Dictyostelium discoideum*. Interest in this species as a model for the study of a variety of

BIOCHEMISTRY AND PHYSIOLOGY OF PROTOZOA
SECOND EDITION, VOL. 2

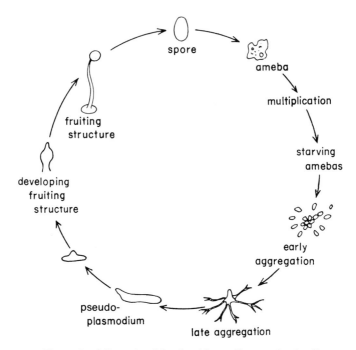

Figure 1. Life cycle of *D. discoideum*. See text for details.

processes in eukaryotic cells has been clear since its discovery by Raper (1935). The life cycle of *D. discoideum* is shown in Figure 1.

In a suitable environment *D. discoideum* spores germinate, yielding small amebas (about 10 μm in diameter) which feed on bacteria and divide as long as food is available. Triggered by starvation, amebas aggregate to form a slug from which cells differentiate into spores and stalk cells. The mechanism of cell aggregation is chemotaxis (Runyon, 1942; Bonner, 1947), and the chemotactic molecules are called acrasins. In the case of *D. discoideum* the attractant is adenosine 3′,5′-cyclic monophosphate (cAMP), which is released periodically. cAMP not only mediates cell aggregation by acting as a chemoattractant (Konijn *et al.*, 1968) but also propagates the signal outward from cell to cell (Robertson *et al.*, 1972; Roos *et al.*, 1975; Shaffer, 1975).

Different aspects of the life cycle of cellular slime molds have been reviewed by Bonner (1967), Raper (1973), Olive (1974), and Loomis (1975). Recent reviews on chemotaxis and cell aggregation in cellular slime molds by Konijn (1975), Gerisch and Malchow (1976), and Newell (1977) are also available.

II. DEMONSTRATION OF CHEMOTAXIS IN CELLULAR SLIME MOLDS

Evidence for chemotaxis in cellular slime molds was presented more than 30 years ago by Runyon (1942) and by Bonner (1947). Since then two different assays have been developed to show chemotaxis unequivocally in these organisms: the small-population assay (Konijn and Raper, 1961; Konijn, 1965) and the cellophane square test (Bonner et al., 1966). Both techniques have been successfully used to show chemotaxis in many different species of cellular slime molds. A detailed discussion of these different techniques can be found elsewhere (Konijn, 1975); both techniques are described here briefly.

In the small-population assay a small drop containing a suspension of amebas is placed on a hydrophobic agar surface. The amebas remain homogeneously spread within the boundaries of the drop. Chemotaxis is assayed by placing the test solution close to the ameba drop and observing microscopically the accumulation of cells at the edge of the drop closest to the attractant (Figure 2). This test becomes quantitative when the threshold concentration at which cells still react positively is determined. Conditions have been described (Konijn, 1970) for *D. discoideum* which permit positive chemotaxis at $10^{-9} M$ cAMP (Figure 2). In the cellophane square test a small square of cellophane containing amebas is placed in a petri dish containing agar into which the attractant has been incorporated (Bonner et al., 1966). By inactivating the attractant, amebas build a gradient which they follow, moving away from the cellophane square. Quantitation can be achieved by measuring the threshold concentration of attractant at which cells move away from the cellophane square over larger distances than in control plates.

III. CHARACTERIZATION OF ATTRACTANTS

Konijn et al. (1967) identified cAMP as an attractant of *D. discoideum*. Aggregative cells of *Dictyostelium rosarium*, *D. purpureum*, and *D. mucoroides* are also attracted by this molecule (Konijn, 1972). The threshold concentration of cAMP is at least 100-fold lower in aggregative cells than in vegetative cells of *D. discoideum* (Bonner et al., 1969). Pan et al. (1972) identified folic acid as a second attractant of *D. discoideum*. Folic acid, unlike cAMP, is more active in the vegetative phase than in the aggregative phase. In addition, folic acid attracts vegetative cells of all the species so far tested. Therefore, it has been proposed that folic acid is a food-searching cue for cellular slime molds (Pan et al., 1972).

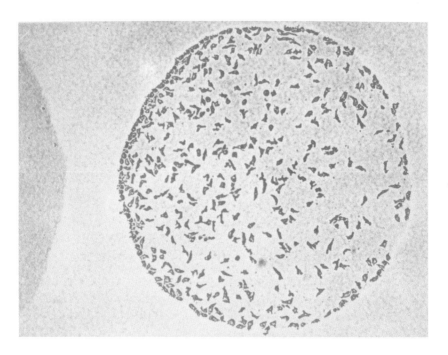

Figure 2. Chemotactic response of *D. discoideum* cells to cAMP. cAMP, 0.1 μl ($10^{-9}M$), was applied three times at 5-min intervals at the left side of the amebal population. Photograph was taken 5 min after the last application. Magnification: ×95. (From Konijn, 1970, reproduced by permission.)

No other attractants have been identified so far. Yet the acrasins of *Polysphondylium violaceum, Dictyostelium lacteum,* and *D. minutum* have been partially purified and characterized (Wurster *et al.,* 1976; Mato *et al.,* 1977a; Kakebeeke *et al.,* 1978). The attractant of *P. violaceum* seems to be a small peptide. Whether it is circular or blocked at both ends is not yet known, nor has its exact amino acid composition been determined. This compound attracts aggregative amebas of *Polysphondylium violaceum* and *P. pallidum* but not their vegetative cells (Wurster *et al.,* 1976). Furthermore, no attraction is observed with various species of *Dictyostelium*. Mato *et al.* (1977a) have purified an attractant of *D. lacteum* from yeast extract. This compound only attracts aggregative amebas of *D. lacteum* and is inactive with various other species of *Dictyostelium* and *Polysphondylium*. It has an aromatic heterocycle and is negatively charged above pH 7.0 (Mato *et al.,* 1977a; van Haastert, unpublished observations). Recently, Kakebeeke *et al.* (1978) purified an attractant of *D. minutum* which attracts only aggregative *D. minutum* cells and is inactive with other species of *Dictyostelium* and *Polysphondylium*. The attractant of *D. minutum* is negatively charged above pH 3.0. Most of the attractants seem to have an aromatic moiety; the possibility that the attractant of *P. violaceum* also has such a moiety has not been excluded. We proposed (Mato *et al.*, 1978a) that interaction between the aromatic ring and the receptor generates the chemotactic signal and that the other moieties (phosphate, ribose, etc.) are responsible for the species specificity for the attractant (see Section V). As long as the attractants are not identified, it cannot be excluded that constituents assumed to belong to acrasins are actually contaminants.

IV. THE AMEBAL CHEMOTACTIC SYSTEM

Sensitive *D. discoideum* amebas react chemotactically within 5–10 sec after application of a cAMP signal (Gerisch *et al.*, 1975a). After a rapid contraction a pseudopod is extended in the direction of the attractant. During aggregation, cells move toward the aggregation center in periods of 100 sec (Cohen and Robertson, 1971; Alcantara and Monk, 1974). If the attractant source is not pulsatile, cells move toward it continuously (Alcantara and Monk, 1974). In the cellophane square test, chemotaxis is not accompanied by changes in cell motility (Bonner *et al.*, 1970). On agar, however, cells also respond to cAMP with increased speed of movement (Alcantara and Monk, 1974; Mato, unpublished observations). The reason for these differences is not clear, though these results indicate that chemotaxis can result from a change from random movement to directed movement with or without a concomitant change in speed.

One of the most intriguing questions about chemotaxis concerns how a cell detects a chemical signal and consequently sends a pseudopod in the direction of the attractant source. In 1947 Bonner proposed that the input signal for chemotaxis was a spatial concentration gradient. Evidence indicating that this may be the case has been obtained by Mato *et al.* (1975). In *D. discoideum* the threshold cAMP signal has been calculated to be about $10^{-9}\,M$/mm. *Dictyostelium discoideum* cells are about 10 μm in diameter, resulting in a difference of only $10^{-11}\,M$ cAMP between the two ends of a cell triggered by a threshold chemotactic signal. This difference must be measured against a background concentration of about $10^{-9}\,M$ cAMP. This indicates that to respond chemotactically a cell has to measure differences of 1 part in 100 parts (Mato *et al.*, 1975). Two models have been proposed involving (1) comparison of the concentration of attractant along the cell surface (Bonner, 1947; Mato *et al.*, 1975), and (2) measurement of spatial gradients by extending pilot pseudopods in different directions

(Gerisch *et al.*, 1975b). Cells possess small pseudodigits 0.5 μm in length (Eckert *et al.*, 1977). If such fingerlike projections are used as sensors, the analysis is extended to measure differences of 1 part in 2000 parts. To permit the measurement of cAMP without interference from diffusion, Mato *et al.* (1975) have proposed a very short equilibration time (1 msec) for the cAMP–receptor complex. Gerisch *et al.* (1977) have interpreted this short equilibration time as a limitation to their model of spatial gradient recognition based on the extension of pilot pseudopods. Whatever the model, the measurement of a gradient by a cell has to be faster than the diffusion time for cAMP over its length or that of a pilot pseudopod. Therefore a short equilibration time helps both models, rather than ruling out either one of them. Obviously it is not yet clear which one of these two models operates *in vivo*. Yet the sensory mechanism may not be unique and, as Bonner (1977) has proposed, both models may work together and cooperate in the analysis of a chemotactic signal.

In the small-population assay, at high cAMP concentrations, cells do not move preferentially toward the attractant source but rather radially, away from their original confinement. It was first proposed that by inactivating the cAMP the amebas built a spatial gradient which they followed by moving radially up the gradient (Bonner *et al.*, 1969; Konijn, 1969). However, a mutant lacking a normal reaction toward the attractant source was able to show a normal radial response (Mato and Konijn, 1977a). These results indicate that both responses, radial and toward the attractant source, differ and are the result of a different analysis by the cell. The radial response might be rather an "alarm" response by which cells escape from a high level of attractant. The radial response in the small-population assay should not be confused with the radial movement outward of cells reacting chemotactically in the cellophane square test.

Evidence has been presented which suggests the existence of negative chemotaxis in *D. discoideum* (Samuel, 1961; Keating and Bonner, 1977). When two small populations containing vegetative *D. minutum* cells are placed close to each other, one with high cell density and the other with low cell density, cells in the latter move away from the drop containing high cell concentration (Figure 3). A similar response is observed when a concentrated amebal extract is deposited close to vegetative *D. minutum* cells. Vegetative *D. discoideum* and *P. violaceum* cells behave similarly (Kakebeeke, unpublished observations). These experiments clearly show the existence of negative chemotaxis in the cellular slime molds. Identification of the repellents should prove helpful to elucidate the pathways of sensory transduction during chemotaxis. Whether the radial response triggered by high cAMP concentrations is induced by the release of a repellent by the cells is not known.

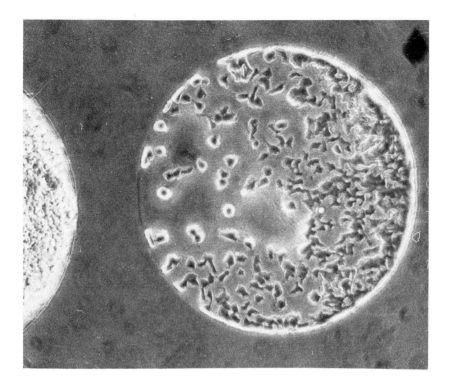

Figure 3. Negative chemotaxis in *D. minutum*. Two drops, one containing a high cell density and the other a low cell density of vegetative *D. minutum*, were placed close to each other. The photograph was taken 30 min after deposition of the dense amebal population and 1 hr after the deposition of the low-density population. Magnification: ×130 (P. I. J. Kakabeeke, unpublished observations).

V. A MODEL OF THE cAMP–CHEMORECEPTOR INTERACTION

Few biological molecules have received as much attention from organic chemists as cAMP. Since its first modification by Posternak *et al.* (1962), about 20 years ago, more than 400 different cAMP derivatives have been synthesized. This large variety of derivatives has permitted a detailed study of the relationship between structure and chemotactic activity in *D. discoideum* (Konijn, 1972, 1973; Konijn and Jastorff, 1973; Mato and Konijn, 1977b). Based on the chemotactic activity of about 50 different cAMP derivatives, a model of the cAMP–chemoreceptor interaction has been proposed (Mato *et al.*, 1978a). To define the sites at which cAMP binds to the active site of the receptor the following assumption has been made. When the chemotactic activity of cAMP is reduced at least 10^3-fold by the substitution of one of its atoms, this atom is considered to bind to the receptor. With the use of this criterion the hypothesis has been proposed that three hydrogen bonds and one ionic bond bind the cAMP molecule to the receptor. Two of these hydrogen bonds are localized on the base (6-amino and 7-nitrogen) and one on the phosphate ring (3'-oxygen). The ionic bond is localized on the negatively charged phosphate (Figure 4). In addition, evidence has been presented indicating that the receptor binds to the anti conformation of the base (Mato and Konijn, 1977a; Mato *et al.*, 1978a). That several noncyclic AMP derivatives, despite their noncyclic structure and consequent phosphodiesterase resistance, are chemotactically active clearly indicates the receptor is not a phosphodiesterase (Mato and Konijn, 1977b). cAMP phosphorothioate is chemotactically nearly as active as cAMP (Gerisch and Malchow, 1976; Mato and Konijn, 1977b), whereas as a phosphodiesterase inhibitor it is about 100-fold less active (Gerisch and Malchow, 1976). Also, these results indicate

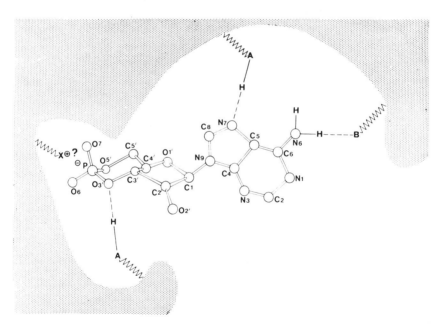

Figure 4. A model of the cAMP–chemoreceptor interaction in *D. discoideum*. The cAMP molecule binds first to the receptor by three hydrogen bonds at, respectively, the 6-amino, 7-nitrogen, and 3'-oxygen positions, and one ionic interaction at the negative charge of the phosphorus atom; then the base moiety binds by interaction between its π electrons and a corresponding acceptor (Phe, Tyr, Trp, His) at the active site (not shown). A, Amino acid site chain donor of the hydrogen bond (Ser, Thr, Lys, Tyr); B, amino acid site chain acceptor of the hydrogen bond (Asp, His, or an acceptor at the protein backbone, –CO · NH–); X, positively charged amino acid site chain (Lys, Arg). (From Mato *et al.*, 1978, reproduced by permission.)

that the chemoreceptor is not a phosphodiesterase. To explain the various chemotactic activities of derivatives with modifications in the base moiety we have proposed that the base also binds to the receptor by interactions between its π electrons and a corresponding acceptor at the active site (Mato *et al.*, 1978a). These could be hydrophobic interactions or stacking of the base with an aromatic amino acid side chain. Furthermore, the cAMP–receptor interaction might be biphasic; in the first phase the cAMP molecule binds to the receptor by three hydrogen bonds and one saltlike bond, and in the second phase the purine moiety interacts with an acceptor at the active site and generates the chemotactic signal. It is interesting to note here that the attractants seem to have an aromatic nucleus. In addition, in the case of folic acid the chemotactic specificity resides in the pterin ring (Pan *et al.*, 1975). Thus, all attractants may generate a chemotactic signal by interaction between their aromatic moiety and an acceptor at the active site of the chemoreceptor.

This model of the cAMP–receptor interaction in *D. discoideum*, although different, shows striking similarities to the model proposed for the cAMP receptor in higher organisms (Jastorff *et al.*, 1978); the hydrogen bond at the 3'-oxygen, the ionic bond at the phosphorus, and the hydrophobic interaction between the aromatic base and an acceptor at the active site seem to have been preserved during evolution.

VI. CHARACTERIZATION OF cAMP CHEMORECEPTORS

Two types of cAMP-binding sites exist at the cell surface of aggregative *D. discoideum* amebas: a cyclic nucleotide phosphodiesterase (Pannbacker and Bravard, 1972; Malchow *et al.*, 1972) and a nonenzymatic receptor (Malchow and Gerisch, 1974; Mato and Konijn, 1975a; Green and Newell, 1975; Henderson, 1975). Gerisch and Malchow (1976) and Newell (1977) have reviewed this subject in detail; the two sites are discussed briefly here.

Both cAMP-binding sites are developmentally regulated, reaching maximal activity at the onset of aggregation. The chemotactic activity observed with various cyclic and noncyclic AMP derivatives rules out the possibility that phosphodiesterase is a chemoreceptor (Gerisch and Malchow, 1976; Mato and Konijn, 1977b). The membrane-bound phosphodiesterase shows negative cooperativity-like kinetics (Malchow *et al.*, 1975). Such kinetics has theoretically been shown to be able to steepen a spatial gradient locally along a cell (Nanjundiah and Malchow, 1976). If so, the function of this membrane-bound phosphodiesterase may not only be to clear the background cAMP concentration but also to amplify the input signal, thus facilitating its analysis.

The binding of cAMP by aggregative *D. discoideum* cells is reversible and saturable. About 10^5 to 10^6 receptors per cell with a dissociation constant of from 10^{-8} to 10^{-7} M have been calculated (Malchow and Gerisch, 1974; Mato and Konijn, 1975a; Green and Newell, 1975; Henderson, 1975). A cell with such cAMP-binding characteristics can detect an input signal of a magnitude of $10^{-9} M$/mm against a background concentration of $10^{-9} M$ (Mato *et al.*, 1975). The specificity of binding is the same as that observed for chemotaxis (Malchow and Gerisch, 1974). Because of all these lines of evidence, this cAMP receptor is an obvious candidate for mediating chemotaxis in *D. discoideum*.

The appearance of cAMP receptors can be accelerated by pulsating cells with cAMP at about 5- to 8-min intervals (Mato *et al.*, 1977b; Roos *et al.*, 1977a; Klein and Darmon, 1977). Pulses of cAMP also stimulate the appearance of aggregation-specific cell-to-cell contact sites (Gerisch et al., 1975c), and cell surface phosphodiesterase (Roos *et al.*, 1977a), and in several aggregateless mutants, normal aggregates are observed when pulsed with cAMP (Darmon *et al.*, 1975). These results indicate that cAMP not only functions as a chemoattractant but also controls the differentiation program. The other chemoattractant of *D. discoideum*, folic acid, when applied in pulses, also accelerates the onset of aggregation (Wurster and Schubiger, 1977). ATP also speeds up cell aggregation, probably by plasma membrane phosphorylation (Mato and Konijn, 1975b, 1976; Perekalin, 1977). The possible role of plasma membrane phosphorylation during chemosensory transduction is discussed in Section IX.

cAMP binding is also increased by the presence of 1 mM Ca^{2+} and by a short incubation (5–10 min) with 100 μg/ml of concanavalin A (Con A) (Juliani and Klein, 1977). Con A also induces the appearance of phosphodiesterase at the cell surface of *D. discoideum* amebas (Gillette and Filosa, 1973; Darmon and Klein, 1976). The incubation of cells with high concentrations of cAMP ($10^{-6} M$) reduces the binding of cAMP (Klein and Juliani, 1977). This contrasts with the enhancing effect on cAMP binding observed when cells are pulsed with 10^{-7} M cAMP (see above). These differences indicate that a complicated mechanism regulates the number and/or the affinity of the cAMP receptors at each moment during cell aggregation. Induction, repression, and desensitization of the cAMP receptors are among the mechanisms used by a cell to adapt itself to different environmental conditions. The existence of a nonlinear Scatchard plot for cAMP binding (Green and Newell, 1975) suggests the possibility of there being more than one different cAMP receptor or the regulation of binding by negative cooperativity.

Recent data from Mullens and Newell (1978) favor the hypothesis of negative cooperativity during cAMP binding to cell-surface receptors.

Further investigation of the cAMP receptors of *D. discoideum* awaits its purification, which may lead to a better understanding of the first step of chemosensory transduction.

VII. cGMP CHANGES DURING CHEMOTACTIC TRANSDUCTION

The identification of cAMP as an attractant or first messenger during chemotaxis in *D. discoideum* leads to the question of what signal(s) the cAMP–receptor interaction generates that induces a cell to move toward the attractant source. Formation of a pseudopod may be controlled by the level of a regulatory molecule(s). Recently, evidence has accumulated which points to guanosine 3′,5′-cyclic monophosphate (cGMP) as one of the molecules that may be involved in the control of pseudopod formation. The addition of cAMP to a cell suspension of aggregative *D. discoideum* cells induces a fast, brief increase in the cellular cGMP content (Figure 5), which is monophasic in the wild-type *D. discoideum* NC-4 (Mato *et al.*, 1977c) and biphasic in the axenic strain Ax-2 (Wurster *et al.*, 1977). At 22°C cGMP levels reach a peak 10 sec after the addition of cAMP and recover basal levels within 30 sec (Mato *et al.*, 1977c). Vegetative cells are at least 100-fold less sensitive than aggregative cells to cAMP as an attractant (Bonner *et al.*, 1969). The addition of cAMP to vegetative *D. discoideum* cells does not change the cGMP content of the

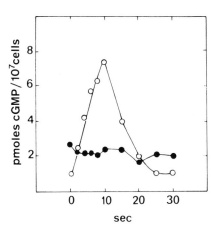

Figure 5. Time course of cGMP formation after *D. discoideum* cells were triggered with $5 \times 10^{-8} M$ cAMP. cAMP was added at Time 0, and the formation of cGMP was followed. ○, Aggregative cells chemotactically sensitive to cAMP; ●, vegetative cells insensitive to cAMP. (From Mato *et al.*, 1977c, reproduced by permission.)

cells (Mato *et al.*, 1977c) (Figure 5). Vegetative cells are attracted by folic acid (Pan *et al.*, 1972) and respond to this attractant with a fast increase in the cGMP content, which is monophasic in both *D. discoideum* NC-4 and Ax-2 (Wurster *et al.*, 1977; Mato *et al.*, 1977d). Folic acid-mediated cGMP accumulation shows kinetics similar to that observed with cAMP. Folic acid induces chemotaxis in vegetative cells of *D. minutum* (Pan *et al.*, 1972) and *D. vinaceo-fuscum* (Konijn, 1975) and also induces a similar cGMP elevation in these species (Kakebeeke, unpublished observations). The purified attractants of *D. lacteum* and *P. violaceum* also induce, at physiological concentrations, a fast increase in the cellular cGMP content (Mato and Konijn, 1977a; Wurster *et al.*, 1978). These results indicate that an increase in the cGMP content is a common response to an attractant independent of the species or of the developmental stage used. *Dictyostelium discoideum* cells respond chemotactically to cAMP within 5–10 sec after addition of the nucleotide (Gerisch *et al.*, 1975a). Attractant-mediated cGMP accumulation precedes pseudopod formation (Figure 5), which suggests a function for this nucleotide in pseudopod formation.

cAMP and folic acid-mediated cGMP accumulation are concentration-dependent (Mato *et al.*, 1977c,d). The cAMP concentration necessary for 50% maximal cGMP accumulation closely agrees with the dissociation constant of the cAMP receptor (Mato *et al.*, 1977c). Although no folic acid receptors have been characterized so far, the concentration of this attractant required for half-maximal cGMP accumulation is similar to the threshold concentration for chemotaxis (Mato *et al.*, 1977d). These results show that at physiological concentrations of the attractants for chemotaxis cGMP accumulation also takes place. The next question is whether the same cAMP receptor mediates chemotaxis and cGMP accumulation. A close correlation exists between the potency of several cAMP and AMP derivatives as attractants and as stimulators of cGMP accumulation (Mato *et al.*, 1977d). Further evidence has been obtained with a mutant of *D. discoideum*, Agip 55, which is defective in its chemotactic response and its cAMP-binding activity (Darmon *et al.*, 1975; Mato *et al.*, 1977b). This chemotactic defect is accompanied by a low cGMP output in response to cAMP (Mato *et al.*, 1977b). After a cell suspension of this mutant is pulsed with cAMP, a normal chemotactic response is observed (Darmon *et al.*, 1975; Mato *et al.*, 1977b), which is accompanied by normal cAMP binding and cGMP accumulation in response to cAMP. These results show unequivocally that the same cAMP receptor is involved in chemotaxis and cGMP accumulation.

In various systems ligand-induced cGMP accumulation depends on the presence of extracellular Ca^{2+} (Schultz and Hardman, 1975). In *D. dis-*

coideum a normal cGMP elevation occurs in the presence of 1 mM EGTA when cells are stimulated by either cAMP (Mato *et al.*, 1977c) or folic acid (Krens, van Haastert, and Mato, unpublished observations). These results indicate that an influx of Ca^{2+} is not a prerequisite for obtaining cGMP accumulation. Several smooth muscle relaxants also induce cGMP accumulation in the absence of extracellular Ca^{2+} (Schultz *et al.*, 1977). Yet a chemoattractant may induce Ca^{2+} translocation from intracellular deposits, and the increase in the concentration of free Ca^{2+} may activate guanylate cyclase, as proposed in other systems (Goldberg and Haddox, 1977). The Ca^{2+} ionophore A23187, in the presence or absence of extracellular Ca^{2+}, has no effect on the cGMP content of *D. discoideum* cells (Mato and Konijn, 1977a), whereas the same concentration of ionophore accelerates cell aggregation (Brachet and Klein, 1977). Moreover, in the presence of the ionophore (7 μM) a normal cGMP accumulation in response to cAMP or folic acid is observed (van Haastert, Krens, and Mato, unpublished observations). These results argue against a major role for Ca^{2+} as a regulator of cGMP levels. Results with guanylate cyclase also support this hypothesis (see Section VIII).

VIII. GUANYLATE CYCLASE

Guanylate cyclase activity can be assayed in crude homogenates of *D. discoideum* cells and its activity increases two- to fivefold during differentiation to aggregation competence (Ward and Brenner, 1977; Mato *et al.*, 1978a). According to Ward and Brenner (1977), the enzyme from *D. discoideum* V-12 requires Mn^{2+} to be active and Mg^{2+} cannot replace this cation. Mato and Malchow (unpublished observations) have found that in *D. discoideum* Ax-2 Mg^{2+} alone could support about 50–80% of the enzyme activity measured in the presence of Mn^{2+}. Basal guanylate cyclase activity shows Michaelian kinetics with a K_m of 500 μM (Ward and Brenner, 1977) or 1000 μM (Figure 6). The mean intracellular GTP concentration in *D. discoideum* V-12 is 0.2 mM (Ward and Brenner, 1977), which is two- to fivefold lower than the K_m of the enzyme. Guanylate cyclase is stimulated two- to threefold by 0.3 mM ATP (Mato and Malchow, 1978). ATP activates the enzyme by lowering the K_m from 1 mM GTP to the physiological range of 0.2 mM GTP (Figure 6). The ATP derivatives adenylyl-iminodiphosphate and adenylyl(β,γ-methylene)diphosphonate cannot replace ATP in activating guanylate cyclase (Mato and Malchow, 1978; and Mato, unpublished observations). Although these results suggest a phosphorylation mechanism, they need further confirmation. Be-

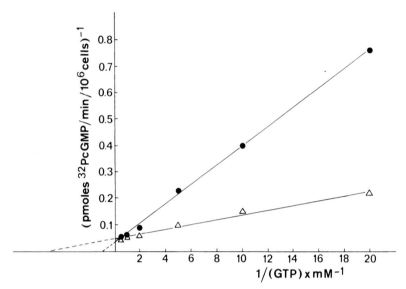

Figure 6. Double reciprocal plots of guanylate cyclase activity in the presence or absence of 0.3 mM ATP. Guanylate cyclase activity was measured as described by Mato and Malchow (1978) at different concentrations of GTP in the absence (●) or presence (△) of 0.3 mM ATP. Aggregation-competent *D. discoideum* Ax-2 cells were used in these experiments.

cause ATP stimulation of guanylate cyclase occurs at physiological concentrations, its effect might be important in regulating this enzyme activity *in vivo*.

Attractant-mediated cGMP accumulation can occur as a result of increased synthesis and/or inhibited degradation. Recently, Mato and Malchow (1978) showed that cAMP stimulated guanylate cyclase. cAMP was added to a suspension of sensitive *D. discoideum* cells, which was immediately sonicated, and guanylate cyclase was assayed. Since oscillations in the cGMP level occur (Wurster *et al.,* 1977), it is important to know the dynamic state of the cells prior to chemotactic stimulation. Autonomous oscillations and chemotactic stimulation are both accompanied by a change in the extracellular pH (Malchow *et al.,* 1978a,b). Therefore, extracellular pH recording was used to monitor cellular activities. Figure 7 shows the pH change in response to cAMP and the guanylate cyclase activity of samples taken before and immediately after stimulation with cAMP. cAMP induces a fast, up to sevenfold, stimulation of guanylate cyclase that decays to almost basal activity within 1–2 min. This stimulation of guanylate cyclase activity by cAMP is sufficient to explain the elevation of cGMP induced by this nucleotide. Similar stimulation by cAMP of guanylate cyclase activity was observed when cells were homogenized and assayed in the presence of EGTA (Mato and Malchow, 1978). These results indicate that Ca^{2+} does not support the activated stage of guanylate cyclase and agree with the previously discussed results on cGMP accumulation (see Section VII). No stimulation could be elicited by pulsing a homogenate of cells with $1 \times 10^{-7} M$ cAMP. Whether ATP is necessary for attractant-mediated cGMP accumulation is not yet known. The cAMP and ATP effects on guanylate cyclase are additive (Mato and

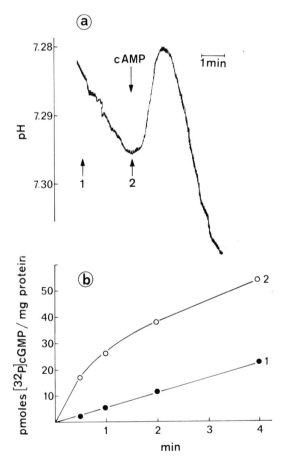

Figure 7. (a) pH recording of a cell suspension before and after stimulation with $3 \times 10^{-8} M$ cAMP. As indicated at the arrows, samples were taken for the guanylate cyclase assay. (b) Guanylate cyclase activity of the samples taken from the cell suspension shown in (a) 2 min before (●, 1) and immediately after (○, 2) cAMP addition. (From Mato and Malchow, 1978, reproduced by permission.)

Malchow, 1978). cAMP induces desensitization of the cGMP response (see Section IX); it is therefore possible that ATP plays a role during the sensitization–desensitization cycle of the cGMP response.

After homogenization by freezing and thawing, about 50% of the enzyme activity is recovered in the pellet after centrifugation at 9000 g for 1 min; 20% remains soluble, and 30% is lost (Mato and Malchow, 1978). Different protocols of homogenization—e.g., Douncer homogenization—can lower the amount of enzyme present in the pellet (Ward and Brenner, 1977). Both fractions are activated by ATP and inhibited by Triton X-100 (Mato and Malchow, 1978), indicating structural similarities. Therefore, it is quite possible that all the enzyme originates from membrane structures and that the soluble fraction appears during homogenization. Which fraction, membrane and/or soluble, is activated by cAMP is not known. No stimulation by cAMP of guanylate cyclase has been observed after cell homogenization (Ward and Brenner, 1977; Mato et al., 1978b; Mato and Malchow, 1978). Furthermore, it is also not known whether guanylate cyclase is a plasma membrane-bound enzyme in D. discoideum. Yet the results with cAMP and guanylate cyclase are of broad interest, because they showed for the first time that a substance acting at the cell surface elevates cGMP via stimulation of guanylate cyclase. Although not yet confirmed, the evidence accumulated with D. discoideum strongly supports the hypothesis of a second-messenger function for cGMP during sensory transduction in Dictyostelium.

All the detergents so far tested inhibit the enzyme activity (Triton X-100, Lubrol PX, NP-14, deoxycholate, and cholate acid), suggesting that a lipid–protein interaction could have the same effect in vivo. The polyene antibiotic filipin, which is known to bind cholesterol, inhibits basal guanylate cyclase activity twofold, further suggesting a role for lipids in the regulation of this enzyme activity (Mato, unpublished observations). After stimulation of guanylate cyclase activity, a fast inhibition of the enzyme is necessary to obtain the type of transient cGMP accumulation shown in Figure 5. Whether lipids are involved in the mechanism that switches off guanylate cyclase activity is not known.

IX. ATTRACTANT-INDUCED DESENSITIZATION OF CHEMOTACTIC TRANSDUCTION

The binding of [^{3}H]cAMP to aggregating D. discoideum cells is an oscillatory process with a periodicity of about 2 min at 25°C (Klein et al., 1977; King and Frazier, 1977). These oscillations can be due to changes in binding or to changes in dilution of the isotope by the periodic release of cAMP from the cells. Recently, King and Frazier (1977) reported evidence supporting the first alternative—that is, desensitization of cAMP receptors following chemotactic stimulation.

Using the cGMP response, Mato *et al.* (1977c) have obtained evidence of the existence of desensitization of cAMP-induced cGMP synthesis. When *D. discoideum* cells are repeatedly stimulated by $5 \times 10^{-8} M$ cAMP for 1 min, cGMP elevation only occurs in response to the first pulse. Under these experimental conditions cAMP hydrolysis is fast, 90% in 30 sec, indicating that the reason cells do not "see" a second cAMP pulse is not because of a high background concentration of attractant. Furthermore, when cells are triggered under conditions where the concentration of attractant is kept constant for several minutes (by using high concentrations of weak cAMP agonists or phosphodiesterase-resistant derivatives), the cGMP response is nevertheless fast and transient (Mato and Konijn, 1977a; Mato *et al.*, 1977d). These results clearly indicate a fast adaptation of the chemotactic response to changes in the background concentration of attractant. All the attractants so far tested induce in a cell suspension of sensitive cells a fast decrease in light scattering (Gerisch and Hess, 1974; Wurster and Schubiger, 1977; Wurster *et al.*, 1978). Changes in light scattering might be the expression of attractant-mediated cell contraction and one of the components of the chemotactic response (Wurster *et al.*, 1978). In contrast to the results obtained with cGMP accumulation, a second pulse of cAMP or folic acid—given 30 sec after the first—induces a normal decrease in light scattering (Mato *et al.*, 1978c) (Figure 8). In their vegetative stage *D. discoideum* cells show a low sensitivity to cAMP as an attractant (Bonner *et al.*, 1969), which is accompanied by a low cAMP-binding capacity (Malchow and Gerisch, 1974). Yet vegetative cells respond to cAMP with a monophasic change in light scattering (Gerisch and Hess, 1974), which is not accompanied by changes in the level of cGMP (Mato *et al.*, 1977c). There are various possible explanations of these observations. Changes in light scattering and cGMP accumulation may be triggered by the same receptor. Although relatively few receptors may be sufficient to trigger a decrease in light scattering—as shown by the results with vegetative cells—cGMP elevation might depend on the presence of a larger number of receptors. If so, desensitization of 90% or more of the receptors will still leave enough receptors active to trigger changes in light scattering but not enough for a significant change in the cGMP content. However, changes in light scattering and cGMP may be mediated by two different receptors. As we have seen, there is no conclusive evidence for the existence of either one or more types of receptors (see Section VI). What seems to be clear is that most of the receptors that appear during differentiation to aggregation competence are related to

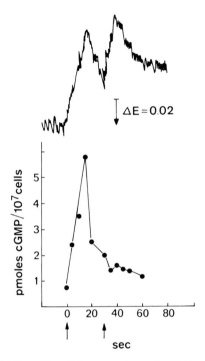

Figure 8. Effect of folic acid on light scattering and cGMP content when cells of *D. discoideum* are stimulated repeatedly. Vegetative cells were stimulated at Time 0 with 3×10^{-6} *M* folic acid and, for a second time, with a pulse of the same magnitude, 30 sec later. cGMP was measured by radioimmunoassay as described by Mato *et al.* (1977c) and changes in light scattering according to Gerisch and Hess (1974). (Unpublished data of J. M. Mato, F. A. Krens, and P. J. M. van Haastert.)

the generation of cGMP. Thus, incubation of cAMP-sensitive cells for 10 min with Con A almost doubles cAMP binding (Juliani and Klein, 1977) and increases the cGMP output by the same magnitude in response to a given cAMP signal (Mato *et al.*, 1978c).

Desensitization may not only occur at the cAMP receptor level but also at any other step during chemosensory transduction (guanylate cyclase, adenylate cyclase, etc.). In higher systems, hormone-induced desensitization is known to occur at the hormone receptor and at the adenylate cyclase level (Johnson *et al.*, 1978).

There is some evidence suggesting that plasma membrane phosphorylation may regulate the transduction of a chemotactic signal. A cAMP-independent protein kinase is present at the cell surface of *D. discoideum* (Weinstein and Koritz, 1973; Mato and Konijn, 1975b, 1976), which catalyzes the phosphorylation of several proteins (Mato and Konijn, 1975b, 1976; Parish *et al.*, 1977a; King and Frazier, 1977). Phosphorylation of living *D. discoideum* cells is an oscillatory process, its maximum coinciding with the minimum of cAMP binding (King and Frazier, 1977). In the presence of 1 m*M* ATP the interval between starvation and the onset of aggregation in *D. discoideum* can be shortened up to 50% (Mato and Konijn, 1975b, 1976). Similar results have been observed with cAMP-sensitive and -insensitive species of cellular slime molds (Perekalin, 1977). A loss of cell-to-cell contact sites during aggregation and decreased sensitivity to cAMP as an attractant are also observed in the presence of 1 m*M* ATP (Mato and Konijn, 1976). Furthermore, 1 m*M* ATP increases up to 50% the output of cGMP in response to either cAMP or folic acid (Mato, 1978). All these results indicate that ATP may work at different levels during chemosensory transduction (receptor, coupler, guanylate cyclase). More work is needed before a model on the mechanism of action of ATP during chemosensory transduction can be proposed. It is interesting to note that plasma membrane phosphorylation may regulate the transduction of signals triggered by neurotransmitters (Gordon *et al.*, 1977; Teichberg *et al.*, 1977).

What is the function of desensitization? Local application of cAMP in the proximity of an ameba with a microcapillary induces the formation of a pseudopod, and within 10 sec a new pseudopod can be induced at a different cell site (Gerisch *et al.*, 1975a). These results, like those of Alcantara and Monk (1974), argue against desensitization, yet the number of occupied receptors under these conditions is small—a number of 1000 has been calculated to be sufficient to trigger chemotaxis (Mato *et al.*, 1975). However, *D. discoideum* cells not only react to cAMP by responding chemotactically, but also release this nucleotide into the medium after being signaled (Roos *et al.*, 1975; Shaffer, 1975). Furthermore, cells are able to relay up to 100-fold more cAMP than they have received (Gerisch and Malchow, 1976). This means that, when a cell relays a cAMP signal, many of its receptors might be temporarily occupied, inducing thereby a significant refractory period. When a neighboring cell reacts to this cAMP and releases its own cAMP, the former cell is still in a refractory period and does not react chemotactically to the new signal. This induces a unidirectional propagation of the signal, which is indeed what one sees, and it is also a requirement for any of the existing models for cell aggregation (Cohen and Robertson, 1971; Parnas and Segel, 1977; Cohen, 1977). An obvious conclusion is that desensitization is not necessary for species such as *D. minutum* and *D. lacteum* that do not show a relay (Gerisch, 1968; Mato *et al.*, 1977a). These species seem to respond to a continuous source of attractant generated at the aggregation center (Gerisch, 1968).

X. INTRACELLULAR TARGETS OF CYCLIC NUCLEOTIDES

Binding of cAMP to intracellular soluble proteins of *D. discoideum* homogenates was first reported by Malkinson *et al.* (1973). During differentiation there is up to a sevenfold increase in the specific activity of cAMP binding (Malkinson *et al.,* 1973; Sampson, 1977; Rahmsdorff and Gerisch, 1978). Two cAMP binding proteins have been partially purified by chromatography on DEAE-cellulose (Sampson, 1977; Veron and Patte, 1978; Rahmsdorff and Gerisch, 1978). When chromatography on Sephadex G-200 was used, the first fraction from the DEAE-cellulose chromatogram eluted as a single peak with molecular weight of about 160,000 and the second DEAE-cellulose fraction was separated into two peaks with molecular weights of, respectively, 180,000 and 40,000 (Rahmsdorff and Gerisch, 1978). Early vegetative cells have been reported to bind cAMP either specifically (Sampson, 1977; Rahmsdorff and Gerisch, 1978) or to be inhibited by 5'-AMP (Veron and Patte, 1978). According to the latter authors, binding in vegetative cells is affected by a dialyzable factor. Rahmsdorff and Gerisch (1978) reported no increase of binding activity after extensive dialysis or by the addition of a heated cell extract. The reason for these differences remains unknown. Another point of controversy involves protein kinases. Whereas Sampson (1977) reported the existence of two soluble cAMP-dependent developmentally regulated protein kinases, Veron and Patte (1978) and Rahmsdorff and Pai (1979) did not observe protein kinase activity stimulated by cAMP. Four major soluble protein kinases and one membrane-bound protein kinase have been identified in homogenates of *D. discoideum* cells strain Ax-2 (Rahmsdorff and Pai, 1979). Rahmsdorff *et al.* (1978) showed elegantly that, after the addition of cAMP to a suspension of *D. discoideum* cells, a membrane-bound protein kinase is stimulated which catalyzes the phosphorylation of a membrane-bound protein with a molecular weight similar to that of myosin. No phosphorylation of this myosinlike protein has been observed after the addition of cGMP, cAMP, or Ca^{2+} to a homogenate of cells. Thus, while cAMP acting as an extracellular factor stimulates phosphorylation of a specific membrane-bound protein, no clear evidence has been obtained for a relationship of the soluble cAMP binding proteins to protein kinase.

In *D. discoideum* cGMP binding activity of sonicated cells is found in the cytosol (Mato *et al.,* 1978d; Rahmsdorff and Gerisch, 1978). This binding activity is not only present in homogenates of *D. discoideum* cells but is also found in homogenates of *D. rosarium* and *P. violaceum* cells (Mato *et al.,* 1979). cGMP binding activity in *D. discoideum* is 10^3-fold more specific for cGMP than for cAMP; millimolar concentrations of ATP,

GTP, and 5'-GMP do not inhibit cGMP binding (Mato *et al.*, 1978d; Rahmsdorff and Gerisch, 1978). Rahmsdorff and Gerisch (1978) have reported a value of 2×10^4 cGMP binding sites per cell with a dissociation constant of $2 \times 10^{-7} M$ for *D. discoideum* Ax-2. A value of 3×10^3 cGMP binding sites per cell with a dissociation constant of approximately $10^{-9} M$ has been calculated by Mato *et al.* (1978d) for *D. discoideum* NC-4. In *P. violaceum* Scatchard plots of cGMP binding are not linear, suggesting the existence of more than one type of cGMP binding protein (Mato *et al.*, 1979). Two cGMP binding proteins have been separated by chromatography on Ultrogel AcA-34 from homogenates of *D. discoideum, D. rosarium,* and *P. violaceum* cells. All these species have in common a binding protein with a molecular weight of about 250,000 (Mato *et al.*, 1979). Rahmsdorff and Gerisch (1978) have also purified by sequential chromatography on DEAE-cellulose and Sephadex G-200 a cGMP binding protein with a molecular weight greater than 180,000 which eluted immediately after the void volume. None of these cGMP binding proteins seems to be associated to protein kinase stimulation when histone is used as substrate (Rahmsdorff and Gerisch, 1978; Mato *et al.*, 1979). Similar cGMP binding proteins specific for this nucleotide, with a high affinity and without definable enzymic activity, have been purified from other systems [for a review, see Goldberg and Haddox (1977)]. cGMP binding activity is two- to threefold higher in aggregation-competent cells than in growth phase cells (Rahmsdorff and Gerisch, 1978; Mato *et al.*, 1978). The high affinity and specificity make these cGMP binding proteins obvious candidates for mediators of cGMP action *in vivo* in cellular slime molds.

XI. CALCIUM ION, cGMP, AND THE CHEMOTACTIC RESPONSE

In 1875 Schulze (1875) proposed that cytoplasmic contraction was the force driving ameboid movement. About 100 years later, Wohlfarth-Bottermann (1960) and Simard-Duquense and Couillard (1962) observed the presence of actomyosin proteins in amebas, giving strong support to the contraction theory. Although this theory is nowadays generally accepted, two controversial models have been proposed to explain how contraction initiates cell movement: (1) the ectoplasmic contraction theory (Pankin, 1923; Mast, 1926), according to which contraction occurs at the back of a cell, and by raising the intracellular pressure at this region pushes the endoplasm to the front end, forming a pseudopod; and (2) the front zone contraction theory (Allen, 1961), according to which traction is generated by contraction of the endoplasm at the front end of the cell.

Current ideas on ameboid movement favor contraction at the back of a cell and relaxation at the front, instead of contraction at the front (Komnick *et al.*, 1973). Korohoda (1977) showed that in *Amoeba proteus* local relaxation of the plasma membrane—induced by applying benzene in the proximity of a cell with a microcapillary—induced pseudopod formation and cell movement toward the source of the benzene. Yet the mechanism for initiating cell movement may not be unique and, as proposed by Komnick *et al.* (1973), both possibilities may operate when necessary. Therefore, rather than trying to determine where cell contraction takes place, we limit ourselves to a review of the evidence connecting chemotactic stimulation with cellular contraction and/or relaxation.

Ameboid movement in *D. discoideum* has been described in detail by Shaffer (1965). Cytoplasmic streaming is localized in the advancing region of a cell, in contrast to the situation in *A. proteus* where streaming starts at the rear and is directed toward the front (Eckert *et al.*, 1977). As we have seen, cells have small fingerlike projections which extend in many directions and occasionally form a pseudopod (Eckert *et al.*, 1977). Actin and myosin have been purified from *D. discoideum* cells (Woolley, 1972; Clarke and Spudich, 1974). In addition, homogenates of *D. discoideum* contain a protein factor that confers Ca^{2+} sensitivity to the activation of myosin ATPase by actin (Mockrin and Spudich, 1976). Contractile actin filaments are distributed around the entire *D. discoideum* cell (Eckert *et al.*, 1977) and are visualized in association with plasma membranes (Clarke *et al.*, 1975; Eckert *et al.*, 1977). Actin and myosin are components of the plasma membrane of *D. discoideum* cells (Parish *et al.*, 1977b).

Recently, Eckert and Lazarides (1978) have localized actin in *D. discoideum* cells by immunofluorescence. Whereas vegetative cells show uniform fluorescence, in migrating aggregative cells bright fluorescence staining is found in the advancing area. Filopodia and pseudopodia are also brightly fluorescent. From these observations it is clear that cytoplasmic contraction is most probably the driving force of movement in *D. discoideum*. Rahmsdorff *et al.* (1978) found that the addition of cAMP to a suspension of *D. discoideum* cells, but not to a homogenate, induced phosphorylation of a protein with the same molecular weight as myosin. Phosphorylation of myosin regulates the actin-activated ATPase activity of myosins from a variety of smooth muscle and nonmuscle cells [for a review, see Adelstein (1978)]. Phosphorylation of the light myosin chain is catalyzed by a kinase which is not activated by cAMP. Actin-activated ATPase activity in phosphorylated platelet myosin is sevenfold higher than in the nonphosphorylated myosin. Light chain kinases of smooth and skeletal myosin and of platelets require Ca^{2+} for activity. The Ca^{2+} dependence of myosin kinases is mediated by a Ca^{2+} binding protein which appears to be identical to the Ca^{2+}-dependent regulator which is known to activate cyclic nucleotide phosphodiesterases from a variety of organisms and tissues. Whether the same mechanism applies to *D. discoideum* cells is not yet known. It would be interesting to know whether *D. discoideum* cells contain the Ca^{2+}-dependent regulator.

Calcium ions may have a universal role in controlling cell movement. They regulate chemotactic behavior in bacteria; when the intracellular Ca^{2+} concentration rises, tumbling is favored, and when the Ca^{2+} concentration decreases bacteria swim smoothly (Ordal, 1977). In *Paramecium*, Ca^{2+} also regulates the direction of movement (Eckert, 1972) and in *Chaos chaos* a Ca^{2+} current may regulate the direction of cytoplasmic streaming (Nuccitelli *et al.*, 1977). Stimulation of aggregative *D. discoideum* cells by cAMP induces a fast Ca^{2+} influx (Wick *et al.*, 1978), suggesting a function for Ca^{2+} during the chemotactic response. However, chemotaxis can take place in the presence of 1 m*M* EGTA (Mato *et al.*, 1977c) and cAMP induces changes in light scattering in the presence of 10 m*M* EDTA (Gerisch *et al.*, 1975a) or 1 m*M* EGTA (Mato *et al.*, 1977c). Mason *et al.* (1977) have reported that cell aggregation does not occur below an extracellular Ca^{2+} concentration of 10^{-6} *M*. These results contrast with those of Mato *et al.* (1977c) indicating that cell aggregation could take place in the presence of 1 m*M* EGTA. The reason for these differences is not yet clear. Whereas the experiments of Mason *et al.* were carried out on a cellophane membrane, those of Mato *et al.* utilized an agar surface. As shown previously, extracellular Ca^{2+} is not necessary for obtaining cGMP accumulation, a light scattering decrease, or chemotaxis in response to an attractant. A 10 m*M* concentration of EDTA reversibly inhibits autonomous light scattering oscillations (Gerisch *et al.*, 1975c). Yet Geller and Brenner (1978) have recently shown that it is possible to inhibit autonomous oscillations without blocking the transduction of a chemotactic signal. Therefore, these results indicate that the transductive coupling during chemotactic response in *D. discoideum* seems to be independent of extracellular Ca^{2+}. Cells of *D. discoideum* preloaded with ^{45}Ca continuously release ^{45}Ca to the extracellular medium (Wick *et al.*, 1978). Thus, an increase in cytosolic Ca^{2+} in response to cAMP stimulation might result from an increase in the calcium permeability of the plasma membrane and/or by an inhibition of the efflux of Ca^{2+}. Extrusion of Ca^{2+} is believed to be mediated by a Ca^{2+} pump which is probably equivalent to the Ca^{2+}-activated ATPase found in red blood cells (Schatzmann and Vincenzi, 1969). Homogenates of *D. discoideum* cells obtained by sonication accumulate ^{45}Ca (Mato and Marin Cao, unpublished observations). The uptake of ^{45}Ca is ATP and Mg dependent, and experiments with the ionophore A23187 indicate that Ca^{2+} accumulates inside vesicles and is neither membrane bound nor precipitated. The mitochondrial uncouplers 2,4-dinitrophenol and NaN_3 are not inhibitors of the ATP-dependent Ca^{2+} uptake in lysed *D. discoideum* cells. Addition of cAMP to a suspension of *D. discoideum* cells, but not to a homogenate, inhibits the ATP-dependent ^{45}Ca uptake (Mato and Marin Cao, unpublished observations). These results suggest that chemotactic stimulation by cAMP might induce an increase in cytoplasmic Ca^{2+} not only as a result of a stimulated influx but also from simultaneous inhibition of the efflux of Ca^{2+}. How might cAMP affect the translocation of Ca^{2+}? The addition of 1 mM S-adenosyl methionine (SAM) to a homogenate of *D. discoideum* cells causes a threefold inhibition of the ATP-dependent Ca^{2+} uptake (Figure 9). Concentrations of 0.1 mM SAM inhibit 20% and 0.01 mM SAM had no effect. When SAM is added to a homogenate of *D. discoideum* cells preloaded with ^{45}Ca, it immediately induces an efflux of radioactivity into the medium. These results suggest that SAM is, probably via methylation, a potential candidate for regulating Ca^{2+} movements in *D. discoideum*. It would be interest-

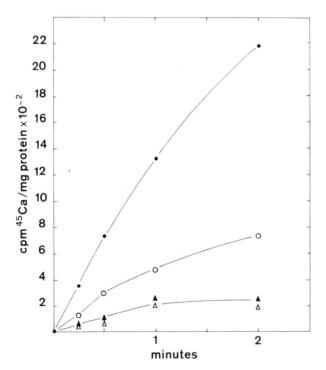

Figure 9.   Time course of ^{45}Ca uptake by a homogenate of sonicated *D. discoideum* cells. The uptake of ^{45}Ca was measured basically as described by Kendrick *et al.* (1977): ●, 1 mM ATP; ○, 1 mM ATP + 1 mM SAM; ▲, no added ATP; and △, no added ATP + 1 mM SAM (Mato and Marin Cao, unpublished data).

ing to know whether cAMP stimulation induces the methylation of specific membrane proteins. In bacteria, the methylation of several inner membrane proteins by SAM is specifically influenced by chemotactic stimulation and they play a central role during transduction of the chemotactic information into specific changes in flagellar rotation [for a review, see Koshland (1977)]. Inverted membrane vesicles isolated from bacteria accumulate Ca^{2+} in the presence of ATP (Rosen and McClees, 1974; Kobayashi *et al.*, 1978). Ca^{2+} accumulation in bacteria also seems to be regulated by SAM (Mato, unpublished observations). These results suggest a general role for SAM as regulator of Ca^{2+} metabolism.

In addition to an increase in cytosolic Ca^{2+}, cAMP induces in *D. discoideum* cells a biphasic efflux of protons (Malchow *et al.*, 1978b) and an increase in the carbon dioxide output (Gerisch *et al.*, 1977; Chung and Loe, 1978). The increase in protons does not seem to be due to cAMP release or hydrolysis, but rather to the expression of the transduction of a cAMP signal. Ca^{2+} ($2 \times 10^{-6} M$), but not Mg^{2+} ($2 \times 10^{-6} M$), decreases the proton efflux (Malchow *et al.*, 1978b). The mechanism underlying cAMP-mediated proton efflux is not known. A possibility is the transport of a weak acid as suggested by the increased carbon dioxide output upon cAMP stimulation. Brachet and Klein (1977) have given evidence which suggests an inhibitory effect by Ca^{2+} on development. The addition of the Ca^{2+} ionophore A23187 accelerates the onset of aggregation in *D. discoideum* cells, and in its presence two aggregateless mutants could develop normally. The possibility that Ca^{2+} acts via regulation of basal adenylate cyclase activity will be discussed in Section XIII.

The next question is what function, if any, does cGMP have during cellular contraction? Although cGMP was discovered in urine about 15 years ago (Ashman *et al.*, 1963), its biological function remains unknown. cGMP suppresses the rate and/or intensity of contraction in isolated cardiac cells (Krause *et al.*, 1972). In the spontaneously beating frog heart, the cGMP concentration decreases at the onset of contraction and returns to basal levels in the relaxation phase (Wollenberger *et al.*, 1973). In the ductus deferens of the rat the relaxing effect of several drugs has been correlated with their capacity to elevate cGMP levels (Schultz *et al.*, 1977). However, Diamond (1978) has presented evidence which argues against a role for cGMP as a mediator of drug-induced relaxation in smooth muscle and concludes that cGMP is not an important regulator of smooth muscle tension. In striated muscle, Ong and Steiner (1977) have shown, by immunofluorescence techniques, that cGMP is associated with myosin bands. In conclusion, although in some cases there seems to be a temporal relationship between cGMP changes and the tension stage of a cell, the question concerning the functions of cGMP is far from being answered and remains one of the most challenging problems.

Microtubules are clearly important for cell morphology, secretion, and motility. It has been suggested (Rebhun, 1967) that microtubules control cell motility by regulating the distribution of microfilaments and the cellular region of application of the force generated by contraction of these structures. cGMP has been reported to stimulate *in vivo* the assembly of microtubules in higher organisms (Weissmann *et al.*, 1975). Unfortunately very little is known about microtubule polymerization in *D. discoideum*. As expected, homogenates of *D. discoideum* cells contain tubulin, which amounts to about 3% of the total cell protein during the vegetative stage (Cappuccinelli *et al.*, 1977). Very few microtubules have been visualized in *D. discoideum* cells (Eckert *et al.*, 1977). Since high concentrations of inhibitors of microtubule action (colchicine, vinblastine, and griseofulvin) only induce a delay in cell aggregation (Cappuccinelli and Ashworth, 1976; Eckert *et al.*, 1977), it does not seem likely that microtubules are important during chemotaxis in *Dictyostelium*.

XII. cAMP CHANGES DURING CHEMOTACTIC TRANSDUCTION

Dictyostelium discoideum cells not only react chemotactically when triggered by cAMP, but also after being signaled release this nucleotide to the extracellular medium (Roos *et al.*, 1975; Shaffer, 1975). The amount of cAMP released increases with the size of the stimulus. Concentrations as low as $10^{-10} M$ cAMP can already induce the release of cAMP; saturation occurs at about $10^{-5} M$ cAMP and halfsaturation at $5 \times 10^{-8} M$ cAMP (Devreotes and Steck, 1979). The amount of cAMP released can be up to 100 times the magnitude of the input signal (Gerisch and Malchow, 1976; Devreotes and Steck, 1979). The duration of the cAMP response is proportional to that of the stimulus for up to 2–3 min and then the rate of cAMP release rapidly decreases (Devreotes and Steck, 1979). The abrupt cessation of the relay response during continuous cAMP stimulation indicates a desensitization to the stimulus. However, if the concentration of extracellular cAMP rises, a new cAMP response can be elicited. Thus, by administration of a discontinuous, geometrically rising cAMP gradient, Devreotes and Steck (1979) could elicit a continuous release of cAMP during about 40 min. These observations indicate that the relay response in *D. discoideum* cells is triggered by an increment in the extracellular cAMP concentration, adapts within minutes, and remains adapted for as long as the cAMP concentration is kept constant.

Gerisch and Wick (1975) found that suspensions of *D. discoideum* cells spontaneously synthesized and released cAMP pulses. Spontaneous oscillations of cAMP are in phase with spontaneous changes in light scattering

(Gerisch and Wick, 1975). The extracellular concentration of cAMP reaches its peak 30–40 sec later than the intracellular peak. During the increase in intracellular cAMP there is an increase in the number of cytoplasmic vesicles, and during the release of cAMP the number of vesicles showing exocytosis increases (Maeda and Gerisch, 1977). cAMP is supposedly stored in these vesicles and then released into the extracellular medium, as in the excretion of neurotransmitters. Whether these vesicles are indeed filled with cAMP is not known. Unlike cGMP, most of the cAMP is released into the medium (Gerisch and Wick, 1975). Changes in cAMP levels following exposure to a chemoattractant do not occur in all species. The attractants of *D. lacteum* and *P. violaceum* elevate the cGMP content without modifying cAMP levels (Mato and Konijn, 1977a; Wurster *et al.*, 1978).

It is not yet known whether cAMP-induced chemotaxis and cAMP release are mediated by the same or different receptors. No systematic study on the specificity of relay in relation to chemotaxis has been carried out. Nonlinear Scatchard plots of cAMP binding have been obtained by Green and Newell (1975), suggesting the existence of either two types of receptors or a negative cooperative interaction. Further investigations should clarify this ambiguity.

In addition to its intercellular function cAMP also plays an intracellular role in the induction of stalk cell formation. Thus, *D. discoideum* amebas plated on 1 mM cAMP containing agar differentiate into stalk cells (Bonner, 1970; Town *et al.*, 1976), and high concentrations of cAMP also alter morphogenesis and enzyme patterns (Nestle and Sussman, 1972). Similar cAMP effects have also been found in *P. pallidum* (Francis, 1975; Hohl *et al.*, 1977). That basal adenylate cyclase activity in *D. discoideum* also increases during differentiation (see Section XIII) further supports an intracellular function for cAMP in cellular slime molds. Recent experiments indicate that cAMP and ammonia have opposing effects during cell differentiation in *Dictyostelium*. While extracellular cAMP increases the number of aggregation centers per square centimeter (Bonner *et al.*, 1969; Konijn, 1975), ammonia decreases this number (Lonski, 1976; Thadani *et al.*, 1977). Ammonia causes a decrease in the release of cAMP by *D. mucoroides* (Thadani *et al.*, 1977) and *D. discoideum* (Schindler and Sussman, 1977a) cells. It also inhibits the accumulation of EDTA-resistant contact sites and the synthesis of extracellular and membrane-bound phosphodiesterase (Schindler and Sussman, 1977a). *Dictyostelium* cells are known to give off and accumulate ammonia during aggregation (Cohen, 1953; Gregg *et al.*, 1954; Schindler and Sussman, 1977b). The exact level of cAMP synthesis and/or release at which ammonia works is not yet known. The possible effect of ammonia on cAMP-mediated chemotaxis is not known. An interesting model for morphogenetic regulation in *D. discoideum,* based on the opposing effects of ammonia and cAMP, has been proposed by Sussman and Schindler (1978).

XIII. ADENYLATE CYCLASE

Adenylate cyclase has been measured in homogenates of *D. discoideum* cells, its activity increasing 10- to 40-fold during development (Klein, 1976; Roos *et al.,* 1977; Loomis *et al.,* 1979). The enzyme has an apparent K_m of 0.2–0.5 mM (Klein, 1976) and is inhibited by 0.1 mM Ca^{2+} (Klein, 1976; Loomis *et al.,* 1979). Mn^{2+} can overcome the inhibition by Ca^{2+} and its addition to an otherwise aggregateless mutant induced normal aggregation and development (Loomis *et al.,* 1979). The ionophore A23187 also permits the aggregation of two aggregateless mutants and accelerates the onset of aggregation in the wild type (Brachet and Klein, 1977). Treatment with the ionophore A23187 results in increased accumulation of adenylate cyclase (Loomis *et al.,* 1979). These results suggest that Ca^{2+} can block cell differentiation and that regulation of its levels is necessary for obtaining normal development. Furthermore, these results point to adenylate cyclase as having a central role during cell differentiation. The addition of a cAMP pulse to a cell suspension of sensitive *D. discoideum* amebas induces a fast, up to fivefold, activation of adenylate cyclase that decays to basal activity in about 1 min (Roos and Gerisch, 1977). These results indicate that cAMP-induced cAMP release is, at least in part, mediated by activation of adenylate cyclase. Activation is only observed with whole cells and not when cAMP is added to a cell homogenate (Roos and Gerisch, 1977). In addition, no activation is observed by mixing activated with nonactivated homogenates. These results argue against a role for Ca^{2+} during cAMP-mediated activation of adenylate cyclase. Thus, whereas Ca^{2+} seems to be important in regulating basal adenylate cyclase activity, it seems to have no effect on the stimulated enzyme. In contrast to earlier results in which adenylate cyclase was activated by 5'-AMP (Rossomando and Sussman, 1973), Klein (1976) and Roos and Gerisch (1976) found no modification of adenylate cyclase by this nucleotide, nor by NaF or GTP. Spontaneous oscillations of adenylate cyclase in phase with oscillations in the light scattering response have been observed (Roos *et al.,* 1977b; Klein *et al.,* 1977). Because cGMP accumulates before cAMP (Wurster *et al.,* 1977), the former may be involved in adenylate cyclase activation. However, no evidence is yet available about a cross-interaction between cGMP and adenylate cyclase. Three interesting theoretical models by Goldberg and Segel (1977), Cohen (1977), and Rapp and

Berridge (1977) have been developed to explain cAMP oscillations. In these models oscillations are sustained by an autocatalytic mechanism for cAMP synthesis and periodic changes in a feedback regulator of cAMP synthesis. Goldberg and Segel (1977) have proposed oscillations in the ATP levels to obtain oscillations in cAMP synthesis. Cohen (1977) explains adenylate cyclase oscillations by predicting changes in an undefined metabolite, possibly a glycolytic intermediate. In the model of Rapp and Berridge (1977), oscillations in adenylate cyclase are supported by periodic changes in Ca^{2+}. As we have seen, current evidence, however, does not support an effect of Ca^{2+} on adenylate cyclase oscillations but rather on the basal activity of the enzyme. Chung and Coe (1978) have recently reported that the addition of cAMP to *D. discoideum* NC-4 cells induces a transient ADP increase with a maximum between 40 and 100 sec, a transient decrease in ATP which coincides in time and magnitude with the ADP elevation, a decrease in pyruvate simultaneous with the ADP rise, and an increase in glucose 6-phosphate with a peak about 2 min after the ADP maximum. These results contrast with those of Geller and Brenner (1979), where no changes in ATP, GTP, glucose 1-phosphate, and glucose 6-phosphate were observed during cAMP oscillations. Roos *et al.* (1977b) could not find detectable changes in the ATP concentration during spontaneous adenylate oscillations, but their data do not exclude the possibility of a small ATP variation (less than 15%). The changes reported by Chung and Coe (1978) on ADP and ATP levels are, respectively, less than 15 and 5%. It is interesting, however, to point out that oscillations of an amplitude within this range are theoretically sufficient to maintain cAMP oscillations (Goldberg and Segel, 1977; Cohen, 1977). That the addition of 2,4-dinitrophenol inhibits autonomous oscillations (Geller and Brenner, 1978) further suggests that an oxidative metabolite might be involved in the generation of periodic activities. In conclusion, it remains an open question whether oxidative metabolites are constituents of the oscillating mechanism in *D. discoideum*. The observed changes in ATP and ADP could be the result of an increased energy consumption (i.e., contraction, cAMP release) during cAMP responses and therefore not intrinsic constituents of the cellular oscillator.

As previously seen, adenylate cyclase is developmentally regulated, its basal activity increasing up to 40-fold during differentiation. Roos *et al.* (1977a) have reported that the application of cAMP pulses speeds up the increase in adenylate cyclase activity. In contrast to these results, Klein and Darmon (1977) observed that the application of cAMP pulses did not affect the timing of adenylate cyclase differentiation. The reason for these differences is not yet clear. Most of the other components of the cAMP signal system seem to be under the control of cAMP pulses.

Thus, cAMP-binding sites (Mato *et al.*, 1977b; Roos *et al.*, 1977a; Klein and Darmon, 1977), contact sites A (Gerisch *et al.*, 1975b), and intracellular, cell surface bound and extracellular soluble phosphodiesterase (Roos *et al.*, 1977a; Klein and Darmon, 1977; Tsang and Coukell, 1977) are stimulated by the application of cAMP pulses. In addition, release of the inhibitor of extracellular soluble phosphodiesterase is switched off by the application of cAMP pulses (Klein and Darmon, 1977; Tsang and Coukell, 1977).

The existence of an extracellular soluble phosphodiesterase was first reported by Chang (1968). This enzyme shows Michaelian kinetics and is inhibited by binding to a protein inhibitor released by the cells (Riedel and Gerisch, 1971; Riedel *et al.*, 1972; Malchow *et al.*, 1975). The time course of release of the inhibitor differs among strains, yet its secretion always starts after starvation (Gerisch, 1976). It is interesting to note that *P. violaceum*, a species that does not respond to cAMP as an attractant, also releases this enzyme into the extracellular medium (Gerisch *et al.*, 1972; Bonner *et al.*, 1972). A mutant has been found which lacks cell surface and soluble phosphodiesterases and does not show spontaneous light-scattering oscillations (Brachet, personal communication). On the addition of a phosphodiesterase to a cell suspension of this mutant, light-scattering oscillations start, indicating that a possible major function of the phosphodiesterase is to participate in the generation of spontaneous activities.

XIV. CONTACT SITES A, LECTINS, AND CONCANAVALIN A RECEPTORS OF *DICTYOSTELIUM DISCOIDEUM*

In the cellular slime molds cell aggregation is the result of a chemotactic response and enhanced cellular adhesion. Two types of cell surface proteins, contact sites A and discoidins, have been related to cell-to-cell adhesion during aggregation in *D. discoideum*. Rosen and Barondes (1978) have recently written an excellent review on cell adhesion in the cellular slime molds.

During differentiation to aggregation competence, cell-specific contact sites A develop on the surface of *D. discoideum* cells (Beug and Gerisch, 1969; Beug *et al.*, 1970, 1971, 1973). Univalent antibodies fragments (Fab) against contact sites A block specific cell-to-cell adhesion without effect on the shape and motility of the cells (Beug *et al.*, 1970). Furthermore, Fab inhibited agglutination of aggregation-competent cells in a gyrated suspension (Beug *et al.*, 1973) and dissociated the pseudoplasmodium of *D. discoideum* (Beug *et al.*, 1971). About 3×10^5 contact sites A per cell have been calculated by Gerisch *et al.* (1979). These contact sites A have been recently purified and are Con A-binding glycoproteins (Huesgen and Gerisch, 1975; Eitle and Gerisch, 1977) with an apparent molecular weight by SDS-polyacrylamide gel electrophoresis of 80,000 (Müller and Gerisch, 1978). Based on chromatography on Sephadex G-200, an apparent molecular weight for solubilized contact sites A of about 130,000 has been reported (Huesgen and Gerisch, 1975). Whether these differences are due to a dimeric state of the solubilized protein is not known. Contact sites A are minor constituents of the cell surface. A number of 500 contact sites A per μm^2 surface area for a rounded cell of 11 μm has been calculated (Müller and Gerisch, 1978). The exact function of contact sites A is not known. As pointed out by Müller and Gerisch (1978), it is possible that two adjacent cells are held together by the contact sites A, but they also could regulate the activity of true adhesion sites. Two types of lectins, discoidin I and II, have been isolated from *D. discoideum* amebas (Rosen *et al.*, 1973; Frazier *et al.*, 1975). These two proteins have a molecular weight of about 100,000 and are made up of four identical subunits of 26,000 (discoidin I) or 24,000 (discoidin II) daltons (Frazier *et al.*, 1975). *Polysphondylium pallidum* cells have a different lectin called pallidin (Rosen *et al.*, 1975). Both discoidins and pallidin are present at the cell surface and inside the cell in the soluble fraction (Rosen *et al.*, 1974, 1975, 1976; Chang *et al.*, 1975; Reitherman *et al.*, 1975). Discoidin levels increase over 400-fold during differentiation (Rosen *et al.*, 1973). Agglutination experiments have provided evidence that there are cell surface receptors for discoidins and pallidin and that the agglutinability of fixed cells by added discoidins and pallidin increases during development (Rosen *et al.*, 1979; Reitherman *et al.*, 1975). These receptors for lectins have not been yet purified. Experiments where inhibition of cell adhesion by univalent antibodies against pallidin have been tested are not conclusive (Rosen *et al.*, 1976, 1977; Bozzaro and Gerisch, 1978). Therefore, as for contact sites A, the exact functions of these lectins is not yet known.

In addition to contact sites A, extracellular and membrane-bound phosphodiesterases, as well as the inhibitor of the extracellular phosphodiesterase, seem to be Con A-binding proteins (Eitle and Gerisch, 1977). Con A receptors have been visualized at the cell surface of *D. discoideum* and differ from those for wheat germ agglutinin (Molday *et al.*, 1976). Up to 6×10^7 Con A receptors per cell (Weeks, 1975) and more than 35 different molecular species binding Con A (West and McMahon, 1977) have been identified in the plasma membrane of *D. discoideum*. Con A-binding proteins are developmentally regulated (Geltosky *et al.*, 1976). At least 12 Con A-binding proteins diminish and 12 increase during differentiation (West and McMahon, 1977). The number of contact sites A accounts for less than 1% of the total number of Con A receptors (Eitle and Gerisch, 1977). The existence of a carbohydrate moiety in the contact site A molecule is well documented, yet there is no evidence involving such a carbohydrate moiety in cell-to-cell adhesion (Eitle and Gerisch, 1977). The incubation of cells with Con A induces rapidly capping (Molday *et al.,* 1976) and the inhibition of attractant-mediated changes in light scattering (Mato *et al.,* 1978c). This is accompanied by a rapid fourfold increase in membrane-bound phosphodiesterase activity within 30 min (Gillette and Filosa, 1973) and a twofold increase in the number of cAMP-binding sites at the cell surface within 10 min (Juliani and Klein, 1977), the former being probably due to the binding of the intracellular enzyme to membranes by Con A (Filosa, 1978).

Con A addition can stimulate or delay cell aggregation, depending on the concentration and length of incubation (Gillette and Filosa, 1973; Weeks and Weeks, 1975; Darmon and Klein, 1976). Recently, Hoffman and McMahon (1978) showed that five of six differences between the plasma membrane of a nonaggregating mutant with abnormal cellular interactions and that of the wild type were differences in glycoproteins. Although the mechanism underlying all these Con A actions is not known, it is interesting to see how interaction with plasma membrane glycoproteins can regulate the differentiation program.

XV. CONCLUSIONS

Although the molecular structure of an attractant depends on the organism, the structure and function of the proteins responsible for cell movement—actin and tubulin—have been carefully preserved during eukaryote evolution. Therefore, one is inclined to assume that there might be some basic principles involved in all eukaryotic chemotactic responses; information released by the receptors is translated into changes in the behavior of movement. Even more, the study of chemosensory functions in microorganisms may provide us with a model for the study of sensory functions in higher organisms (Clayton, 1953; Koshland, 1974), a possibility originally proposed by Binet (1889). Yet although it is reasonable to assume that such common principles exist in nature, it would be desirable to have a better understanding of the molecular processes that underlie sensory transduction in microorganisms as a basis for generalizations. Otherwise we risk developing concepts based purely on phenomenology.

In *D. discoideum* some of the basic elements of chemotactic transduction and cell differentiation seem to have been identified (cAMP, cGMP, Ca^{2+}, protein phosphorylation). A tentative model is shown in Figure 10. It would be interesting to determine whether chemoattractants also in-

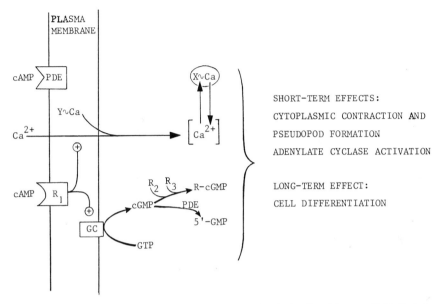

Figure 10. A model for chemosensory transduction in *D. discoideum*. cAMP binds to specific binding sites (R_1) or is hydrolyzed by a membrane-bound phosphodiesterase (PDE). The cAMP–R_1 interaction triggers an increase in the concentration of free intracellular calcium and a transient activation of guanylate cyclase (GC). The cGMP formed is either destroyed by an intracellular phosphodiesterase (PDE) or binds to specific receptors (R_2, R_3). The temporal increase in Ca^{2+}, R–cGMP, and other as-yet-unidentified factors control the short- and long-term effects of cAMP.

duce a rapid change in the cGMP level in other organisms. The activation of guanylate cyclase by cAMP in *D. discoideum* shows that stimuli working at the cell surface can have a direct action on this enzyme activity. These results provide a new mechanism, which had been assumed but not proven in other organisms, by which cGMP metabolism is regulated *in vivo*. Still there are some unsolved fundamental questions on chemotactic transduction in cellular slime molds. How many types of cAMP receptors does a cell have? What is the mechanism of desensitization and how does a cell recover from stimulation? What type of modification (e.g., phosphorylation or other covalent change) switches guanylate cyclase and adenylate cyclase activities on and off? What is the mechanism for cellular oscillations? By what means do cGMP, Ca^{2+}, and phosphorylation regulate the location and magnitude of cytoplasmic contraction? Is cGMP involved in cell differentiation? What is the intracellular function of cAMP? The usefulness of *Dictyostelium* in shedding light on basic problems in the past appears promising for future research.

REFERENCES

Adelstein, R. S. (1978). *Trends Biochem. Sci.* February 1978, 27–30.
Alcantara, F., and Monk, M. (1974). *J. Gen. Microbiol.* **85,** 321–334.
Allen, R. D. (1961). *Exp. Cell Res.* **8,** 17–31.
Ashman, D. F., Lipton, R., Melicow, M. M., and Price, T. D. (1963). *Biochem. Biophys. Res. Commun.* **11,** 130–134.
Beug, H., and Gerisch, G. (1969). *Naturwissenschaften* **56,** 374
Beug, H., Gerisch, G., Kempff, S., Riedel, V., and Cremer, G. (1970). *Exp. Cell Res.* **63,** 147–158.
Beug, H., Gerisch, G., and Müller, E. (1971). *Science* **173,** 742–743.
Beug, H., Katz, F. E., and Gerisch, G. (1973). *J. Cell Biol.* **56,** 647–658.
Binet, A. (1889). "Psychic Life of Microorganisms." Open Court, Chicago, Illinois.
Bonner, J. T. (1947). *J. Exp. Zool.* **106,** 1–26.
Bonner, J. T. (1967). "The Cellular Slime Molds," 2nd ed. Princeton Univ. Press, Princeton, New Jersey.
Bonner, J. T. (1970). *Proc. Natl. Acad. Sci. U.S.A.* **65,** 110–113.
Bonner, J. T. (1977). *Mycologia* **69,** 443–459.
Bonner, J. T., Kelso, A. P., and Gillmor, R. G. (1966). *Biol. Bull. (Woods Hole, Mass.)* **130,** 28–42.
Bonner, J. T., Barkley, D. S., Hall, E. M., Konijn, T. M., Mason, J. W., O'Keefe, G., and Wolfe, P. B. (1969). *Dev. Biol.* **20,** 72–87.
Bonner, J. T., Hall, E. M., Sachsenmaier, W., and Walker, B. K. (1970). *J. Bacteriol.* **102,** 682–687.
Bonner, J. T., Hall, E. M., Noller, S., Oleson, F. G., and Roberts, A. B. (1972). *Dev. Biol.* **29,** 402–409.
Bozzaro, S., and Gerisch, G. (1978). *J. Mol. Biol.* **120,** 265–279.
Brachet, P., and Klein, C. (1977). *Differentiation* **8,** 1–8.
Cappuccinelli, P., and Ashworth, J. M. (1976). *Exp. Cell Res.* **103,** 387–393.
Cappuccinelli, P., Hames, B. D., and Cuccureddu, R. (1977). *In* "Developments and Differentiation in the Cellular Slime Moulds" (P. Cappuccinelli and J. M. Ashworth, eds.), pp. 231–241. Elsevier, Amsterdam.
Chang, C. M., Reitherman, R. W., Rosen, S. D., and Barondes, S. H. (1975). *Exp. Cell Res.* **95,** 136–142.
Chang, Y. Y. (1968). *Science* **160,** 57–59.
Chung, W. J. K., and Coe, E. L. (1978). *Biochim. Biophys. Acta* **544,** 29–44.
Clarke, M., and Spudich, J. A. (1974). *J. Mol. Biol.* **86,** 209–222.
Clarke, M., Schatten, G., Mazia, D., and Spudich, J. A. (1975). *Proc. Natl. Acad. Sci. U.S.A.* **72,** 1758–1762.
Clayton, R. K. (1953). *Arch. Mikrobiol.* **19,** 141–165.
Cohen, A. L. (1953). *Proc. Natl. Acad. Sci. U.S.A.* **39,** 68–74.
Cohen, M. H. (1977). *J. Theor. Biol.* **69,** 57–85.
Cohen, M. H., and Robertson, A. (1971). *J. Theor. Biol.* **31,** 101–118.
Cohen, M. S. (1977). *J. Theor. Biol.* **69,** 57–85.
Darmon, M., and Klein, C. (1976). *Biochem. J.* **154,** 743–750.
Darmon, M., Brachet, P., and Pereira Da Silva, L. H. (1975). *Proc. Natl. Acad. Sci. U.S.A.* **72,** 3163–3166.
Devreotes, P. N., and Steck, T. L. (1979). *J. Cell Biol.* **80,** 300–309.

Diamond, S. (1978). *Adv. Cyc. Nucl. Res.* **9**, 327–340.

Eckert, B. S., and Lazerides, E. (1978). *J. Cell Biol.* **77**, 714–721.

Eckert, B. S., Warren, R. H., and Rubin, R. W. (1977). *J. Cell Biol.* **72**, 339–350.

Eckert, R. (1972). *Science* **176**, 473–481.

Eitle, E., and Gerisch, G. (1977). *Cell Differ.* **6**, 339–346.

Filosa, M. P. (1978). *Differentiation* **10**, 177–180.

Francis, D. (1975). *Nature (London)* **258**, 763–765.

Frazier, W. A., Rosen, S. D., Reitherman, R. W., and Barondes, S. H. (1975). *J. Biol. Chem.* **250**, 7714–7721.

Geller, J. S., and Brenner, M. (1978). *Biochem. Biophys. Res. Commun.* **81**, 814–818.

Geller, J. S., and Brenner, M. (1979). *J. Cell Phys.* **97**, 413–420.

Geltosky, J. E., Siu, C. H., and Lerner, R. A. (1976). *Cell* **8**, 391–396.

Gerisch, G. (1968). *Curr. Top. Dev. Biol.* **3**, 157–197.

Gerisch, G. (1976). *Cell Differ.* **5**, 21–25.

Gerisch, G., and Hess, B. (1974). *Proc. Natl. Acad. Sci. U.S.A.* **71**, 2118–2122.

Gerisch, G., and Malchow, D. (1976). *Adv. Cyclic Nucleotide Res.* **7**, 49–68.

Gerisch, G., and Wick, U. (1975). *Biochem. Biophys. Res. Commun.* **65**, 364–370.

Gerisch, G., Malchow, D., Riedel, V., Müller, E., and Every, M. (1972). *Nature (London), New Biol.* **235**, 90–92.

Gerisch, G., Beug, H., Malchow, D., Schwarz, H., and von Stein, A. (1974). *In* "Biology and Chemistry of Eucaryotic Cell Surfaces" (E. Y. C. Lee and E. E. Smith, eds.), Miami Winter Symposia, No. 7, pp. 49–66. Academic Press, New York.

Gerisch, G., Malchow, D., Huesgen, A., Nanjundiah, V., Roos, W., and Wick, U. (1975a). *In* "Developmental Biology" (D. McMahon and C. F. Fox, eds.), Vol. 2, pp. 76–88. Benjamin, New York.

Gerisch, G., Hülser, D., Malchow, D., and Wick, U. (1975b). *Philos. Trans. R. Soc. London, Ser. B* **272**, 181–192.

Gerisch, G., Fromm, H., Huesgen, A., and Wick, U. (1975c). *Nature (London)* **255**, 547–549.

Gerisch, G., Maeda, Y., Malchow, D., Roos, W., Wick, U., and Wurster, B. (1977). *In* "Developments and Differentiation in the Cellular Slime Moulds" (P. Cappuccinelli and J. M. Ashworth, eds.), pp. 105–124. Elsevier, Amsterdam.

Gillette, M. U., and Filosa, M. F. (1973). *Biochem. Biophys. Res. Commun.* **53**, 1159–1166.

Goldberg, N. D., and Haddox, M. K. (1977). *Annu. Rev. Biochem.* **46**, 823–896.

Goldbeter, A., and Segel, L. A. (1977). *Proc. Natl. Acad. Sci. U.S.A.* **74**, 1543–1547.

Gordon, A. S., Davies, C. G., Milfay, D., and Diamond, I. (1977). *Nature (London)* **267**, 539–540.

Green, A. A., and Newell, P. C. (1975). *Cell* **6**, 129–136.

Gregg, J. H., Hackney, A. L., and Krivanek, J. P. (1954). *Biol. Bull. (Woods Hole, Mass.)* **107**, 226–235.

Henderson, J. (1975). *J. Biol. Chem.* **250**, 4730–4736.

Hoffman, S., and McMahon, D. (1978). *J. Biol. Chem.* **253**, 278–287.

Hohl, H. R., Honegger, R., Traub, F., and Markwalder, M. (1977). *In* "Developments and Differentiation in the Cellular Slime Moulds" (P. Cappuccinelli and J. M. Ashworth, eds.), pp. 149–172. Elsevier, Amsterdam.

Huesgen, A., and Gerisch, G. (1975). *FEBS Lett.* **56**, 46–49.

Jastorff, B., Konijn, T. M., Mato, J. M., Hoppe, J., and Wagner, K. G. (1978). *Hoppe-Seyler's Z. Physiol. Chem.* **359**, 281.

Johnson, G. L., Wolfe, B. B., Harden, T. K., Molinoff, P. B., and Perkins, J. P. (1978). *J. Biol. Chem.* **253**, 1472–1480.

Juliani, M. H., and Klein, C. (1977). *Biochim. Biophys. Acta* **497**, 369–376.
Kakebeeke, P. I. J., Mato, J. M., and Konijn, T. M. (1978). *J. Bacteriol.* **133**, 403–405.
Keating, M. T., and Bonner, J. T. (1977). *J. Bacteriol.* **130**, 144–147.
Kendrick, N. C., Blaustein, M. P., Fried, R. C., and Ratzlaff, R. W. (1977). *Nature (London)* **265**, 246–248.
King, A. C., and Frazier, W. A. (1977). *Biochem. Biophys. Res. Commun.* **78**, 1093–1099.
Klein, C. (1976). *FEBS Lett.* **68**, 125–128.
Klein, C., and Darmon, M. (1977). *Nature (London)* **268**, 76–78.
Klein, C., and Juliani, M. H. (1977). *Cell* **10**, 329–335.
Klein, C., Brachet, P., and Darmon, M. (1977). *FEBS Lett.* **76**, 145–147.
Kobayashi, H., van Brunt, S., and Harold, F. M. (1978). *J. Biol. Chem.* **253**, 2085–2092.
Komnick, H., Stockem, W., Wohlfarth-Bottermann, K. E. (1973). *Int. Rev. Cytol.* **34**, 162–249.
Konijn, T. M. (1965). *Dev. Biol.* **12**, 487–497.
Konijn, T. M. (1969). *J. Bacteriol.* **99**, 503–509.
Konijn, T. M. (1970). *Experientia* **26**, 367–369.
Konijn, T. M. (1972). *Adv. Cyclic Nucleotide Res.* **1**, 17–31.
Konijn, T. M. (1973). *FEBS Lett.* **34**, 263–266.
Konijn, T. M. (1975). *In* "Primitive Sensory and Communication Systems" (M. J. Carlile, ed.), pp. 101–153. Academic Press, New York.
Konijn, T. M., and Jastorff, B. (1973). *Biochim. Biophys. Acta* **304**, 774–780.
Konijn, T. M., and Raper, K. B. (1961). *Dev. Biol.* **3**, 725–756.
Konijn, T. M., Van de Meene, J. G. C., Bonner, J. T., and Barkley, D. S. (1967). *Proc. Natl. Acad. Sci. U.S.A.* **58**, 1152–1154.
Konijn, T. M., Barkley, D. S., Chang, Y. Y., and Bonner, J. T. (1968). *Am. Nat.* **102**, 225–234.
Korohoda, W. (1977). *Cytobiologie* **14**, 338–349.
Koshland, D. E. (1974). *FEBS Lett.* **40**, 33–39.
Koshland, D. E. (1977). *Science* **196**, 1055–1063.
Krause, E. G., Halle, W., and Wollenberger, A. (1972). *Adv. Cyclic Nucleotide Res.* **1**, 301–305.
Lonski, J. (1976). *Dev. Biol.* **51**, 158–165.
Loomis, W. F. (1975). "Dictyostelium discoideum: A Developmental System." Academic Press, New York.
Maeda, Y., and Gerisch, G. (1977). *Exp. Cell Res.* **110**, 119–126.
Malchow, D., and Gerisch, G. (1974). *Proc. Natl. Acad. Sci. U.S.A.* **71**, 2423–2427.
Malchow, D., Nägele, B., Schwarz, H., and Gerisch, G. (1972). *Eur. J. Biochem.* **28**, 136–142.
Malchow, D., Fuchila, J., and Nanjundiah, V. (1975). *Biochim. Biophys. Acta* **385**, 421–428.
Malchow, D., Nanjundiah, V., and Gerisch, G. (1978a). *J. Cell Sci.* **30**, 319–330.
Malchow, D., Nanjundiah, V., Wurster, B., Eckstein, F., and Gerisch, G. (1978b). *Biochim. Biophys. Acta* **538**, 473–480.
Malkinson, A. M., Kwasniak, J., and Ashworth, J. M. (1973). *Biochem. J.* **133**, 601–603.
Mason, J. W., Rasmussen, H., and Dibella, F. (1971). *Exp. Cell Res.* **67**, 156–160.
Mast, S. O. (1926). *J. Morphol. Physiol.* **41**, 347–425.
Mato, J. M. (1978). *Biochim. Biophys. Acta* **540**, 408–411.
Mato, J. M., and Konijn, T. M. (1975a). *Biochim. Biophys. Acta* **385**, 173–179.
Mato, J. M., and Konijn, T. M. (1975b). *Dev. Biol.* **47**, 233–235.
Mato, J. M., and Konijn, T. M. (1976). *Exp. Cell Res.* **99**, 328–332.

Mato, J. M., and Konijn, T. M. (1977a). *In* "Development and Differentiation in Cellular Slime Moulds" (P. Cappuccinelli and J. M. Ashworth, eds.), pp. 93–103. Elsevier, Amsterdam.

Mato, J. M., and Konijn, T. M. (1977b). *FEBS Lett.* **75**, 173–176.

Mato, J. M., and Malchow, D. (1978). *FEBS Lett.* **90**, 119–122.

Mato, J. M., Losada, A., Nanjundiah, V., and Konijn, T. M. (1975). *Proc. Natl. Acad. Sci. U.S.A.* **72**, 4991–4993.

Mato, J. M., Van Haastert, P. J. M., Krens, F. A., and Konijn, T. M. (1977a). *Dev. Biol.* **57**, 450–453.

Mato, J. M., Krens, F. A., Van Haastert, P. J. M., and Konijn, T. M. (1977b). *Biochem. Biophys. Res. Commun.* **77**, 399–402.

Mato, J. M., Krens, F. A., Van Haastert, P. J. M., and Konijn, T. M. (1977c). *Proc. Natl. Acad. Sci. U.S.A.* **74**, 2348–2351.

Mato, J. M., Van Haastert, P. J. M., Krens, F. A., Rhijnsburger, E. H., Dobbe, F. C. P. M., and Konijn, T. M. (1977d). *FEBS Lett.* **79**, 331–336.

Mato, J. M., Jastorff, B., Morr, M., and Konijn, T. M. (1978a). *Biochim. Biophys. Acta* **544**, 309–314.

Mato, J. M., Roos, W., and Wurster, B. (1978b). *Differentiation* **10**, 129–132.

Mato, J. M., Van Haastert, P. J. M., Krens, F. A., and Konijn, T. M. (1978c). *Cell Biol. Int. Rep.* **2**, 163–170.

Mato, J. M., Woelders, H., Van Haastert, P. J. M., and Konijn, T. M. (1978d). *FEBS Lett.* **90**, 261–264.

Mockrin, S. C., and Spudich, J. A. (1976). *Proc. Natl. Acad. Sci. U.S.A.* **73**, 2321–2325.

Molday, R., Jaffe, R., and McMahon, D. (1976). *J. Cell Biol.* **71**, 314–322.

Mullens, I. A., and Newell, P. C. (1978). *Differentiation* **10**, 267–276.

Müller, K., and Gerisch, G. (1978). *Nature (London)* **274**, 445–449.

Nanjundiah, V., and Malchow, D. (1976). *J. Cell Sci.* **22**, 49–58.

Nestle, M., and Sussman, M. (1972). *Dev. Biol.* **28**, 545–554.

Newell, P. C. (1977). *In* "Microbial Interactions" (J. L. Reissig, ed.), pp. 3–57. Chapman & Hall, London.

Nuccitelli, R., Poo, M., and Jaffe, L. F. (1977). *J. Gen. Physiol.* **69**, 743–763.

Olive, L. S. (1974). "The Mycetozoans." Academic Press, New York.

Ong, S. H., and Steiner, A. L. (1977). *Science* **195**, 183–185.

Ordal, G. W. (1977). *Nature (London)* **270**, 66–67.

Pan, P., Hall, E. M., and Bonner, J. T. (1972). *Nature (London), New Biol.* **237**, 181–182.

Pan, P., Hall, E. M., and Bonner, J. T. (1975). *J. Bacteriol.* **122**, 185–191.

Pankin, C. F. A. (1923). *J. Mar. Biol. Assoc. U.K.* **13**, 24–69.

Pannbacker, R. G., and Bravard, L. J. (1972). *Science* **175**, 1014–1016.

Parish, R. W., Müller, U., and Schmidlin, S. (1977a). *FEBS Lett.* **79**, 393–395.

Parish, R. W., Schmidlin, S., and Müller, U. (1977b). *In* "Developments and Differentiation in the Cellular Slime Moulds" (P. Cappuccinelli and J. M. Ashworth, eds.), pp. 85–92. Elsevier, Amsterdam.

Parnas, H., and Segel, L. A. (1977). *J. Cell Sci.* **25**, 191–204.

Perekalin, D. (1977). *Arch. Microbiol.* **115**, 333–337.

Posternak, T., Sutherland, E. W., and Henion, W. F. (1962). *Biochim. Biophys. Acta* **65**, 558–560.

Rahmsdorff, H. J., and Gerisch, G. (1978). *Cell Diff.* **7**, 249–258.

Rahmsdorff, H. J., and Pai, S. H. (1979). *Biochim. Biophys. Acta.* (in press).

Rahmsdorff, H. J., Malchow, D., and Gerisch, G. (1978). *FEBS Lett.* **88**, 322–326.

Raper, K. B. (1935). *J. Agric. Res.* **50**, 135–147.

Raper, K. B. (1973). *In* "The Fungi" (G. C. Ainsworth, F. K. Sparrow, and A. S. Sussman, eds.), Vol. 4B, pp. 9–36. Academic Press, New York.

Rapp, P. E., and Berridge, M. J. (1977). *J. Theor. Biol.* **66**, 497–525.

Rebhun, L. I. (1967). *J. Gen. Physiol.* **50**, 223–239.

Reitherman, R. W., Rosen, S. D., Frazier, W. A., and Barondes, S. H. (1975). *Proc. Natl. Acad. Sci. U.S.A.* **72**, 3541–3545.

Riedel, V., and Gerisch, G. (1971). *Biochem. Biophys. Res. Commun.* **42**, 119–123.

Riedel, V., Malchow, D., Gerisch, G., and Nägele, B. (1972). *Biochem. Biophys. Res. Commun.* **46**, 279–287.

Robertson, A., Drage, D. J., and Cohen, M. H. (1972). *Science* **175**, 333–335.

Roos, W., and Gerisch, G. (1976). *FEBS Lett.* **68**, 170–172.

Roos, W., Nanjundiah, V., Malchow, D., and Gerisch, G. (1975). *FEBS Lett.* **53**, 139–142.

Roos, W., Malchow, D., and Gerisch, G. (1977a). *Cell Differ.* **6**, 229–240.

Roos, W., Scheidegger, C., and Gerisch, G. (1977b). *Nature (London)* **266**, 259–261.

Rosen, B. P., and McClees, J. S. (1974). *Proc. Nat. Acad. Sci. U.S.A.* **71**, 5042–5046.

Rosen, S. D., and Barondes, S. H. (1978). *In* "Specificity of Embryological Interactions. Receptors and Recognition" (R. Gamod, ed.), Ser. B, Vol. 4, pp. 233–264. Chapman and Hall, London.

Rosen, S. D., Kafka, J. A., Simpson, D. L., and Barondes, S. H. (1973). *Proc. Natl. Acad. Sci. U.S.A.* **70**, 2554–2557.

Rosen, S. D., Simpson, D. L., Rose, J. E., and Barondes, S. H. (1974). *Nature (London)* **252**, 149–151.

Rosen, S. D., Reitherman, R. W., and Barondes, S. H. (1975). *Exp. Cell Res.* **95**, 159–166.

Rosen, S. D., Haywood, D. L., and Barondes, S. H. (1976). *Nature (London)* **263**, 425–427.

Rosen, S. D., Chang, C. H., and Barondes, S. H. (1977). *Dev. Biol.* **61**, 202–223.

Rossomando, E. F., and Sussman, M. (1973). *Proc. Natl. Acad. Sci. U.S.A.* **70**, 1254–1257.

Runyon, E. H. (1942). *Collect. Net* **17**, 88.

Sampson, J. (1977). *Cell* **11**, 173–180.

Samuel, E. W. (1961). *Dev. Biol.* **3**, 317–336.

Schatzmann, H. J., and Vincent, F. F. (1969). *J. Physiol.* **201**, 369–395.

Schindler, J., and Sussman, M. (1977a). *Biochem. Biophys. Res. Commun.* **79**, 611–617.

Schindler, J., and Sussman, M. (1977b). *J. Mol. Biol.* **116**, 161–169.

Schultz, G., and Hardman, J. G. (1975). *Adv. Cyclic Nucleotide Res.* **5**, 339–351.

Schultz, K. D., Schultz, K., and Schultz, G. (1977). *Nature (London)* **265**, 750–751.

Schulze, F. E. (1875). *Arch. Mikrosk. Anat.* **87**, 389–471.

Shaffer, B. M. (1965). *Exp. Cell Res.* **37**, 12–25.

Shaffer, B. M. (1975). *Nature (London)* **255**, 549–552.

Simard-Duquesne, N., and Couillard, P. (1962). *Exp. Cell Res.* **28**, 92–98.

Sussman, M., and Schindler, J. (1978). *Differentiation* **10**, 1–5.

Teichberg, V. I., Sobel, A., and Changeux, J. P. (1977). *Nature (London)* **267**, 540–542.

Thadani, V., Pan, P., and Bonner, J. T. (1977). *Exp. Cell Res.* **108**, 75–78.

Town, C. D., Gross, J. D., and Kay, R. R. (1976). *Nature (London)* **262**, 717–719.

Tsang, A. S., and Coukell, M. B. (1977). *Cell Differ.* **6**, 75–84.

Veron, M., and Patte, J. C. (1978). *Dev. Biol.* **63**, 370–376.

Ward, A., and Brenner, M. (1977). *Life Sci.* **21**, 997–1008.

Weeks, C., and Weeks, G. (1975). *Exp. Cell Res.* **92**, 372–382.

Weeks, G. (1975). *J. Biol. Chem.* **250**, 6706–6710.

Weinstein, B. I., and Koritz, S. B. (1973). **34**, 159–162.

Weissmann, G., Goldstein, I., Hoffstein, S., and Tsung, P. (1975). *Ann. N.Y. Acad. Sci.* **253**, 750–762.

West, C. M., and McMahon, D. (1977). *J. Cell Biol.* **74,** 264–273.

Wick, U., Malchow, D., and Gerisch, G. (1978). *Cell Biol. Int. Rep.* **2,** 71–79.

Wohlfarth-Bottermann, K. E. (1960). *Protoplasma* **52,** 58–107.

Wollenberger, A., Babskii, E. B., Krause, E. G., Genz, S., Blohm, D., and Bogdanova, E. V. (1973). *Biochem. Biophys. Res. Commun.* **55,** 446–452.

Woolley, D. E. (1972). *Arch. Biochem. Biophys.* **150,** 519–530.

Wurster, B., and Schubiger, K. (1977). *J. Cell Sci.* **27,** 105–114.

Wurster, W., Pan, P., Tyan, G. G., and Bonner, J. T. (1976). *Proc. Natl. Acad. Sci. U.S.A.* **73,** 795–799.

Wurster, B., Schubiger, K., Wick, U., and Gerisch, G. (1977). *FEBS Lett.* **76,** 141–144.

Wurster, B., Bozzaro, S., and Gerisch, G. (1978). *Cell Biol. Int. Rep.* **2,** 61–69.

The Genetics of Swimming and Mating Behavior in *Paramecium*

6

DONALD L. CRONKITE

The stereotyped behavior of *Paramecium* invites analysis. As pointed out by Jennings (1906), *Paramecium* swimming behavior can be broken down into a few simple components. Another aspect of behavior in *Paramecium*, mating, can be looked upon as a modification of swimming behavior, especially when one focuses on mechanisms of control.

Among the more powerful tools of analysis available to modern students of cell physiology and behavior is the tool of "genetic dissection" (Benzer, 1967). By judicious use of mutations which interfere with a process under investigation, one can map out steps underlying the process. In the analysis of behavior, genes which block or modify particular behavior patterns may have associated modifications of physiology or anatomy which will give insight into the control of behavior at a profound level. This is especially important in *Paramecium* since techniques using drugs

221

which selectively block aspects of behavioral physiology have not been very successful (Eckert, 1977).

Paramecium, especially the group of sibling species called *Paramecium aurelia* (Sonneborn, 1975), is well known genetically. Culture is fairly easy in either bacterized or axenic medium, and the life cycle is open to manipulation (Sonneborn, 1970a). There is a strong tradition of behavior analysis founded upon Jenning's (1906) pioneering work, and there is a developing body of knowledge about the genetics of both swimming and mating behavior in *Paramecium*.

The aim of this chapter is to complement those reviews of the physiology of behavior already available (Eckert *et al.*, 1976; Byrne and Byrne, 1978b) and another chapter by Naitoh that will appear in a future volume of this treatise. The physiology of mating has been reviewed (Hiwatashi, 1969; Miyake, 1974, 1978), and Miyake includes another chapter in this treatise (Volume 4). It is the intent of this chapter to focus rather narrowly on the genetic analysis of swimming and mating behavior. The techniques brought to bear in the genetic study of *Paramecium* behavior should serve as a model of the kind of attack possible on any number of questions about the biochemistry and physiology of protozoa. In addition, there has been no recent attempt to unite our understanding of mating behavior and swimming behavior. It is now becoming possible to see an overlap in the mechanisms of control of these two aspects of *Paramecium* physiology, and this chapter explores that overlap.

I. THE COMPONENTS OF SWIMMING BEHAVIOR

A. The Action System

Jennings (1906) provides a systematic introduction to behavior in *Paramecium*, in which swimming behavior of *Paramecium* is broken into a set of actions which are combined into a coordinated "action system." Contemporary genetic analysis has been aimed at further defining parts of this action system through mutational lesions (see, e.g., Kung *et al.*, 1975).

The action system includes characteristic patterns of forward and reverse swimming, both based upon the shape of the cell, the distribution of cilia on the cell, and the direction and strength of beat of these cilia. *Paramecium* cells are not symmetrical. The oral groove on one side of the animal gives the cell a twist. In addition, the cilia are not evenly distributed over the surface. Cilia on the oral surfaces are arranged differently from those of the aboral surface (see, e.g., Jurand and Selman, 1969; Sonneborn, 1970b). In forward movement the slightly right-oblique, posteriorly directed effective stroke of the cilia rotates the cell about its long axis as it moves forward (Jennings, 1906). The cilia on the oral side are greater in number and give a stronger effective stroke than the aboral ones. Thus, as in a row boat with one oar working more strongly than the other, the cell turns in the direction away from its oral surface as it moves forward. Circular motion is avoided by the rotation about the long axis, which constantly changes the direction of the aboral surface. The result is an open left-handed spiral as the characteristic forward movement. Detailed descriptions of ciliary beat are now available using scanning electron microscopy on instantaneously fixed cells (Tamm, 1972; Tamm *et al.*, 1975) and light microscopy coupled with flash photography (Machemer, 1972). These descriptions greatly enrich our understanding of swimming behavior without changing the general outline of Jennings.

When stimulated mechanically at the posterior end, *Paramecium* swims forward with increased speed (Naitoh and Eckert, 1969). Anterior mechanical stimulation or chemical stimulation results in the "avoiding reaction" (Jennings, 1906), in which the effective stroke of the cilia is reversed, and the cell swims backward in a tight right-handed spiral. At the end of a strong avoiding reaction lasting 0.5–1 sec, the cell ceases backward movement and whirls in place, often moving the anterior end in a wide circle while holding the posterior end still. Forward movement is eventually resumed. In weak avoiding reactions stationary whirling may be seen without backward swimming.

Dryl (1961) and Grebecki (1965) emphasize that the avoiding reaction may take any of several forms. Sometimes the cells display long continuous episodes of backward swimming (continuous ciliary reversal or CCR). At other times there will be short episodes of backing alternating with short forward movements (periodic ciliary reversal or PCR). If only some of the cilia reverse their beat, the result will be a cell which circles or whirls in place, as seen at the end of strong avoiding reactions (partial ciliary reversal or PaCR).

A final component of the action system is just holding still. Jennings discusses in detail the stopping of *Paramecium* when it comes into contact with a surface.

B. Underlying Physiology

The physiological bases of the action system are now becoming clear. Reviews cited earlier should be consulted for details. A bare outline will be presented here.

There is a correlation between the behavior of *Paramecium* and the electrical properties of the cell surface membrane as measured with intracellular electrodes. Forward swimming is correlated with a rather constant negative resting potential across the membrane whereas the reversal of ciliary beat is correlated with a rapid depolarization of the membrane seen as an action potential (Kinosita *et al.*, 1964a,b, 1965). The rapid upward rise of the action potential is followed by a return to resting levels, and this return is correlated with a resumption of forward swimming.

In the upstroke of the action potential in *Paramecium* the current is carried inward by Ca^{2+} (Eckert, 1972). Naitoh and Kaneko (1972, 1973) used Triton X-100-extracted models of *Paramecium caudatum* to show that cilia reverse their direction of beat when the internal concentration of Ca^{2+} exceeds 10^{-6} M. Browning and Nelson (1976) demonstrated an energy using pump which actively excludes Ca^{2+} from cells.

Thus the outline of events involved in backward swimming begins when mechanical, chemical, or electrical stimulation results in depolarization of the cell surface membrane. The calcium conductance of the membrane increases, and Ca^{2+} rushes into the cell. When the internal Ca^{2+} concentration exceeds $10^{-6} M$, the cilia reverse their beat. Meanwhile the active pump excludes Ca^{2+} from the cell so that soon the internal Ca^{2+} concentration drops below 10^{-6} M, and the cells return to forward movement. Further details of this outline will be filled in as specific mutants of swimming behavior are discussed.

Gene action plays a role in synthesizing and organizing the structures which are involved in the action system and in setting up the molecular arrangements of receptors and effectors which react to stimulation. It is thus possible to interfere mutationally with the action system. We see mutants which swim at speeds different from those of normal cells, which display different patterns of forward or reverse movement, which lack the ability to back up when stimulated by all agents or a limited number of them, which show exaggerated avoiding reactions, or which fail to swim at all. Those which have been studied in some detail in *Paramecium tetraurelia* and *P. caudatum* form the major subject matter of this chapter.

Mating behavior in *Paramecium* occurs among swimming cells. It is useful to think of early mating behavior as a modified swimming behavior, which, in fact, Jennings did. Mutations of early mating have been studied and have been found in some cases to result in changes in the action system, thus suggesting a similarity in control mechanisms of the two types of behavior.

II. GENETIC DISSECTION OF SWIMMING BEHAVIOR

A. Modifications of Cell Shape

Early mutational interference with swimming behavior involved changes in cell shape. Jennings (1906) had emphasized the role of the placement of cilia on a cell of particular shape in generating the kind of normal swimming pattern seen in *Paramecium*. Thus it is to be expected that dramatic changes in cell shape or the positioning of the cilia would alter swimming behavior. Jennings (1908, 1913) and Stocking (1915) describe numerous cases of cell deformity, some arising after conjugation within cultures. Deformities could be made more or less extreme by selection, and many of the more extreme alterations resulted in cells which really could not swim at all and sank to the bottom of the culture vessel.

Dawson (1926, 1928) gives an extensive account of a line of *P. aurelia* with an anterior truncation of variable severity. The abnormality was inherited through 840 asexual generations, but, as with the abnormalities reported by Jennings and Stocking, no genetic analysis using progeny of controlled mating was done.

Sonneborn's (1937) report of mating types in *P. aurelia* allowed an analysis of such lines. Beisson and Sonneborn (1965) analyzed a "twisty" mutant of *P. tetraurelia* and discovered that it was the result of a heritable change in the pattern of distribution of cilia on the cell. No nuclear mutation was involved. Twisty cell lines all have several rows of cilia (kineties) which have been rotated 180° from normal and, as shown by Tamm *et al.* (1975), which beat in the opposite direction (Figure 1). Thus the cells swim abnormally in both the forward and reverse directions.

An important result of the studies of Beisson and Sonneborn and Tamm *et al.* is the conclusion that the reference point for determining direction of ciliary beat is not the whole organism. Individual kineties or perhaps even individual cilia use the orientation of surrounding cortical structures as guides for the direction of beat.

B. Concentrated Attack—Methodology

Most of the studies of the effect of cell shape and distribution of cilia on swimming behavior were carried out for reasons other than dissection of behavior. There has been no systematic look at the effect of morphological abnormalities on components of swimming behavior. A concentrated attack on the genetics of swimming behavior is directed at the problems of ionic control of ciliary beat (Kung, 1971a,b; Kung *et al.*, 1975).

Reliable methods are available for producing mutations at high frequency, for increasing the proportion of specific mutations in the population by selection, and for genetic analysis. These methods are not unlike those used for other organisms, but since they are based on peculiarities of *Paramecium* biology, a brief discussion will be of value.

1. Mutagenesis

A sufficiently high rate of mutation is necessary if there is to be a reasonable chance of recovering specific mutations by screening. The

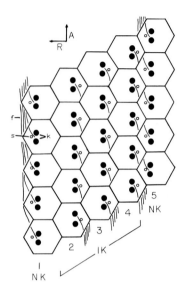

Figure 1.   Diagram of the structure and arrangement of cortical units on the surface of *P. aurelia* with inverted kineties. Part of two normally oriented rows of cortical units (NK, 1 and 5). In describing the asymmetry of cortical units, "to the right" (R arrow) means in the clockwise direction around the cell when facing its anterior end (A arrow); "to the left" refers to the anticlockwise direction. A normally oriented cortical unit (rows 1 and 5) includes one or two kinetosomes (k) lying slightly to the right of the unit midline (only two-kinetosome units are shown here; units with one kinetosome lack the anterior one), a parasomal sac(s) to the right of the kinetosomes and a kinetodesmal fiber (f) emerging to the right and extending anteriorly from the posterior kinetosome (when two are present). Inverted cortical units contain the same structures as normally oriented ones, but with reversed polarity (rows 2–4): the kinetosomes and parasomal sac lie to the left of the unit midline and the kinetodesmal fibers emerge to the left and run posteriorly. A group of inverted kineties is bounded on its right side by an unusually wide space between oppositely pointing units in adjoining inverted and normal rows (1 and 2, see Figure 3a), and, on its left side, by an unusually narrow space where the visible structures point toward each other and are closer together (rows 4 and 5). (From Tamm *et al.*, 1975, reproduced by permission.)

mutagen used almost exclusively for studies described in this review is *N*-methyl-*N'*-nitro-*N*-nitrosoguanidine (MNNG). Details of the mutagenic action of MNNG on *Paramecium* may be found in Kimball (1970). Behavioral mutants have also been obtained (Kung, personal communication) with X rays and the chemical mutagen ICR 170. Refer to Sonneborn (1970a), Kung (1971b), and Byrne (1973) for more detailed discussion of the use of mutagens.

2. Achieving Homozygosity

Many mutations produced during mutagenesis are recessive. In diploids such as *Paramecium* this means that an inbreeding program is necessary to make mutations homozygous. In members of the *P. aurelia* group of sibling species homozygosity is simple to achieve since cells undergo autogamy, a form of self-fertilization. Like most ciliates, *Paramecium* has both a macronucleus and one or more diploid micronuclei. At conjugation or autogamy the macronucleus eventually disintegrates to be replaced by a postfertilization product of a micronucleus. It is the micronucleus which undergoes meiosis and fertilization, and the peculiar maneuvers of the micronuclei during autogamy result in cells homozygous for all of their genes. Details of the cytological processes and genetic consequences of autogamy may be found in Sonneborn (1947) and Beale (1954).

Autogamy occurs when well-fed cells of requisite age begin to starve. Thus the common practice is to prepare mass cultures of well-fed cells, treat them with mutagen while they are still dividing, wash out the mutagen, and let the cells starve. Autogamy will occur, and screening for mutations of behavior can be carried out in postautogamous cells, homozygous for all genes including any new mutations.

Unfortunately, *P. caudatum* does not ordinarily undergo autogamy. Thus, a more complex program of inbreeding is used to achieve homozygosity (Takahashi and Naitoh, 1978; Takahashi, 1979). After treatment of cells with MNNG, individuals were isolated and the clones were grown in isolation for 12–15 fissions. Stability of mating type expression is often lost in old clones of *P. caudatum* so that mating within a clone (selfing) can be induced in mass cultures (Hiwatashi, 1960; Myohara and Hiwatashi, 1975). When selfing occurred in isolated, mutagen-treated clones, they were again fed and then screened. This laborious process has procured several interesting behavioral mutants. Fortunately, the induction of autogamy by chemical means has been perfected in *P. caudatum* (Tsukii and Hiwatashi, 1977). A more rapid development of mutagenic studies in this physiologically well-known species is to be expected in the future.

3. Screening

Picking a mutant with a particular behavioral phenotype from a population of cells having various mutations or no mutations at all requires a screening method. A commonly used method has been to subject *Paramecium* to two competing stimuli to obtain mutants insensitive to the prevailing stimulus or oversensitive to it. Details of such screening methods are given in sections dealing with specific genes. Another method used to obtain mutants of specific kinds has been based on looking for cells which swim the wrong way in an electric field or which fail to swim at all (Van Houten *et al.*, 1977; Takahashi and Naitoh, 1978). Other screening techniques are described in appropriate sections.

4. Genetic Analysis

Once one obtains variants of swimming behavior, genetic analysis is needed to determine the contribution of genes to the variation. As demonstrated by Beisson and Sonneborn (1965), there is no reason to assume that behavioral variation inherited asexually will have a nuclear gene basis. Since mating types of *Paramecium* exist, crosses between stocks, inbreeding by selfing, and inbreeding by autogamy can all be carried out to determine if the character is genetic or otherwise controlled. Since fairly high concentrations of mutagens are often used to obtain mutations, it is also necessary to determine if multiple mutations or a single gene is contributing to a mutant phenotype. Details of the theory and practice of *Paramecium* genetic analysis may be found in Beale (1954) and Sonneborn (1950, 1970a).

C. Paramecium tetraurelia

1. Pawn Mutants

The most extensively studied behavioral mutants of *P. tetraurelia* have been those of the class called pawn (Kung, 1971a,b) which, like the chess piece, only moved forward. Neither mechanical nor chemical stimulation will bring about avoiding reactions in these cells. It is understandable that these should be the most well-studied mutants. Mutants of shape alter behavior, but do not give insights into molecular issues of interest to cell physiologists. Other behavior mutants such as paranoiac and fast (Kung, 1971a,b) display exaggerations of normal behavior which are by their nature more difficult to study than pawn, which often clearly lacks a normal function (Table I).

At least 130 lines of pawn have been isolated (Chang and Kung, 1973b; Chang *et al.*, 1974; Schein, 1976b), of which at least 50 are clearly independent mutations. Many of the mutants are "leaky," showing some slight wild-type behavior, and some are temperature sensitive, showing pawn behavior only at elevated temperatures. The large number of pawn lines is attributable in part to the ease of their selection.

Kung's (1971b) method of screening is derived from the observations of Jennings (1906) on the behavior of *Paramecium* subjected to competing stimulus. *Paramecium* swims forward against the force of gravity (i.e., "upward"). When stimulated with solutions high in Na^+ or K^+ and low in Ca^{2+}, cells respond with avoiding reactions. The ion-stimulated avoiding reactions prevail over the upward swimming stimulated by the pull of gravity. But in pawn cells no avoiding reactions occur. In a population in the bottom of a column filled with a stimulating solution, wild-type cells

Table I Mutations of Swimming Behavior in *Paramecium*

Mutant class[a]	Swimming behavior	Physiological correlations	References
P. tetraurelia			
Twisty (cortical mutant)	Wide forward spiral or swims in a circle. Backward swimming also "twisty"	Several ciliary rows reversed in anterior–posterior polarity	Beisson and Sonneborn (1965); Tamm *et al.* (1975)
Pawn[b] (*pwA,pwB,pwC*)	Avoiding reaction absent or greatly reduced	No Ca-activated change in membrane potential. Little stimulated Ca^{2+} uptake	Kung (1971a,b); Chang and Kung (1973a,b); Chang *et al.* (1974); Schein (1976b); Schein *et al.* (1976); Browning *et al.* (1976); Oertel *et al.* (1977)
Paranoiac[b] (*PaA,paB,PaC,PaD,fna*)[c]	Prolonged backward swimming in Na solutions	Extended membrane depolarization in Na solutions. Stimulated K loss and Na uptake greater than wild type	Kung (1971a,b); Satow *et al.* (1976); Hansma and Kung (1976); Van Houten *et al.* (1977); Byrne and Byrne (1978a)
Fast-2 (*fna*)	No avoiding reaction stimulated by Na. Rapid forward swimming	Possible increase in K permeability relative to Ca and Na	Kung (1971a,b); Satow and Kung (1976a)
TEA⁺-insensitive (*teaA,teaB*)	No avoiding reaction stimulated by tetraethylammonium ion	K permeability increased	Chang and Kung (1976); Satow and Kung (1976c)
Fast-1 (*fA,fb*)	Rapid forward swimming	—	Kung (1971a,b); Cooper (1965)
Atalanta (*ata*)	No avoiding reaction. Stop-and-go forward movement	Probably defective in ciliary motile system	Kung *et al.* (1975); Kung (personal communication)
Spinner (*sp*)	Spin in place when stimulated chemically or mechanically	—	Schein (1976b); Kung *et al.* (1975)
Stacatto (*st*)	Spontaneous backward dashes	—	Kung *et al.* (1975); Kung (personal communication)

(Continued)

Table I (*Continued*)

Mutant class[a]	Swimming behavior	Physiological correlations	References
P. tetraurelia			
Chemokinesis-defective (*che*⁻)	Do not accumulate in an attractant	Increased hyperpolarization upon chemical stimulation	Van Houten (1977)
Sluggish (*sl*)	Very slow forward movement. Often stopped	—	Kung *et al.* (1975)
P. caudatum			
CNR (*cnrA,cnrB,cnrC*)	Avoiding reactions absent or greatly reduced	No Ca-activated change in membrane potential	Takahashi and Naitoh (1978); Takahashi (1979)
K-sensitive (probably dominant)	Prolonged backward swimming in K solutions	Increased K permeability	Takahashi and Naitoh (1978); Takahashi (1979)
Sluggish (*sl*)	Very slow forward movement	Probably defective in ciliary motile system	Takahashi and Naitoh (1978)
Temperature shock behavioral	Prolonged backward swimming when exposed to rapid elevation of temperature	—	Takahashi (1979)
Frequently reverse	High frequency of spontaneous avoiding reactions prolonged backing in K or Ba solutions.	—	Takahashi (1979)
Ba-sensitive	Prolonged backward swimming in Ba solutions	—	Takahashi (1979)
Spinner	Spin in place when stimulated chemically or mechanically	—	Takahashi (1979)
Zig-zag swimmer	Stop-and-go forward movement	—	Takahashi (1979)
TEA⁺-insensitive	No avoiding reaction stimulated by tetraethyl-ammonium ion	—	Takahashi (1979)

[a] Symbols in parentheses designate known loci. Where no symbols appear, genetic data are incomplete or lacking.
[b] Includes temperature-sensitive alleles.
[c] *fna* and *fna^p* are alleles of one locus.

will remain at the bottom doing avoiding reactions while pawn cells will swim to the top, where they can be recovered.

The column screening method was improved by Chang and Kung (1973a), who adapted cells to sucrose solution before introducing them into the bottom of the column. The denser solution containing *Paramecium* could be introduced into the bottom of the column without extensive mixing of cells with the stimulating solution. This reduced the number of wild-type cells that accidentally found their way to the top of the column.

Schein (1976b) has isolated pawn by a "nonbehavioral" screening technique. His was based on physiological findings about pawn to be reviewed later in this section. Pawns have altered calcium channels, preventing normal stimulated influx of Ca^{2+}. Barium ion is known to enter the calcium channel more readily than Ca^{2+}, and when it does it paralyzes the cell. Thus Schein looked for barium-resistant *Paramecium* by treating mutagenized cells with critical concentrations of Ba^{2+} and rescuing swimming survivors. In this way Schein isolated seven different pawn lines. Recent preliminary experiments (Ling, personal communication) indicate that barium-resistant lines will include a number of other behavioral phenotypes as well. Both Chang *et al.* (1974) and Schein (1976b) report isolation of pawns by watching for unusual behavior during movement of cells in an electric field.

Genetic analysis (Chang and Kung, 1973b; Chang *et al.*, 1974; Schein, 1976b) identifies three loci, each of which results in pawn behavior. The loci, designated *pwA*, *pwB*, and *pwC*, are unlinked, and the *pawn* alleles are recessive to wild type.

The three loci of *pawn* can be distinguished in several ways. The mutants of *pwC* isolated to date by Chang and Kung (1973b) and Schein (1976b) are temperature sensitive (*ts-pawn*). They display apparently normal backward swimming when cultured at 23°C, but if grown for 4–12 hr at 32°C, the cells become pawn. Several hours after returning growing cultures to 23°C, normal behavior is restored (Satow et al., 1974). Some *pwA* alleles are temperature sensitive, but no known *pwB* are. Almost all *pwA* are leaky. Some slow their forward movement, whirl, or manifest a brief avoiding reaction when introduced into stimulating solutions. Almost all of the pwA mutants sometimes display a weak avoiding reaction when they encounter walls of culture vessels. Mutants of *pwB*, on the other hand, are usually not leaky. The leakiness of cultures of any pawn genotype is greatly reduced by growing them at 32°C (Chang *et al.*, 1974), suggesting that ts-pawns are extreme examples of leaky phenotypes.

Loci *pwA* and *pwB* can be distinguished by resistance to killing in barium solutions (Schein, 1976b). For two stocks showing the same degree of pawn expression, the pwB allele will be more resistant than pwA to barium killing.

F2 lines homozygous for two pawn mutations provide interesting information. Double mutants of nonleaky pawns are nonleaky (Chang et al., 1974). Double mutants of two leaky pawns are also usually nonleaky (Chang et al., 1974; Schein, 1976b). A double mutant of two ts-pawns is unconditionally pawn (Chang and Kung, 1973b). This suggests that ts-pawn gene products do not function normally even at 23°C.

Some unusual genetic results were found (Chang and Kung, 1973b) when a temperature-sensitive pwC homozygote was crossed to an unconditional pwA homozygote. Since the genes are not allelic, the F1 was wild type as expected. In the F2 a body deformation marker gene segregated as expected when technical difficulties have not arisen. Yet despite evidence for a good F2, the segregation of the pawn character was unexpected. If two recessive genes assorted independently, and if the unconditional pwA masked the expression of pwC, a ratio of 2 unconditional pawn : 1 ts-pawn : 1 wild type is expected. The actual ratio was 79 unconditional pawn : 12 ts-pawn : 86 wild type, a significant deviation with a decided excess of wild type at the expense of ts-pawn.

An excess of wild type would be expected if the F2 were only partially an F2, F1 cells being included through failure of autogamy in a significant number of F1 cells or through macronuclear regeneration after autogamy. Such events probably do not explain this result, first because of the good 1 : 1 segregation of the marker gene and second because the wild type increase results in differential decrease of the two pawn classes. There is a possibility of some further genetic or epigenetic control of pawn phenotype beyond genes pwA, pwB, and pwC, which should be of interest for further analysis.

A failure to swim backward when stimulated could be due either to a lesion in the mechanism which increases calcium conductance following depolarization of the cell surface membrane (Eckert, 1972) or to a lesion which destroys the ability of cilia to respond to raised internal Ca^{2+} concentration with a reversal of beat. Three kinds of evidence establish that pawn lesions are in the mechanism of Ca^{2+} entry (the "calcium activation system") and that the ability of cilia to respond to raised Ca^{2+} concentration is unimpaired.

Naitoh and Kaneko (1972, 1973) produced "models" of P. caudatum by extracting cells with the detergent Triton X-100. This treatment destroys surface membrane integrity so that the concentration of ions in the external solution is approximated by internal ion concentrations. Extracted cells will swim forward if supplied with Mg^{2+}, ATP, and EGTA, and will swim backward if, in addition, Ca^{2+} is added in excess of 10^{-6} M. Kung and Naitoh (1973) made Triton-extracted models of both the wild type and *pwB* cells. Both swam forward in Mg–ATP–EGTA, and both swam backward when sufficient Ca^{2+} was supplied. Thus, destruction of the membrane components of calcium activation demonstrates that the cilia of pawns are still able to respond to raised internal Ca^{2+} concentrations and suggests that the pawn lesion is in the membrane components. In a similar way, Schein (1976b) showed that chlorpromazine, which renders cells permeable to Ca^{2+}, also restores the ability to back up to all three types of pawn.

Browning *et al.* (1976) examined the uptake of radioactive $^{45}Ca^{2+}$ by mutant and wild-type *P. tetraurelia*. Cells were incubated at 0°C to inhibit an active pump which rapidly removes entering Ca^{2+} from cells. When stimulated by K^+ or Na^+, wild-type cells accumulated $^{45}Ca^{2+}$ at a rate five to ten times that of unstimulated cells. Unconditional *pwB* showed no increase in calcium uptake, and *ts-pawns* (either *pwA* or *pwC*) could be stimulated to accumulate $^{45}Ca^{2+}$ if the cells had been grown at 23°C, but not if the cells were grown at 32°C.

Kung and Eckert (1972) provided the first electrophysiological evidence for a failure in pawn of the system of calcium activation. Cells of *pwB* or wild type were impaled on two microelectrodes. One delivered a stimulating pulse, and the other recorded the potential difference between the inside and outside of the cell and the current across the cell membrane. In K^+-containing solutions long depolarizing currents from the stimulating electrode produced in wild type a rapidly rising regenerative response graded to the strength of the stimulus, a characteristic of the increased calcium conductance activated by depolarization. Pawns showed no such response.

In barium-containing solutions *P. caudatum* responded to current pulses with all-or-none action potentials (Naitoh and Eckert, 1968). Wild-type *P. tetraurelia* responded in the same way in the study of Kung and Eckert. The membrane became sensitive to depolarization in the barium solution, action potentials had a distinct (and low) threshold stimulus, and trains of action potentials often resulted when the threshold was exceeded. In contrast, cells homozygous for *pwB* showed no action potentials even when stimulating pulses were used which are 20 times the wild-type threshold.

Passive properties of the wild-type and pawn membranes are quite similar. Both had resting potentials near −24 mV; the input resistances of the two stocks were not significantly different; both responded to hyperpolarizing stimuli in the same way. The electrophysiological evidence of Kung and Eckert coupled with the results of experiments using Triton-extracted cells and chlorpromazine-treated cells and the studies of $^{45}Ca^{2+}$ uptake comprise strong evidence that *pwB* produces a defect in the system of calcium activation. Further studies bear this out and fill in the picture for other *pawn* loci.

The *pwB* gene also blocks calcium activation by chemical stimulation (Satow and Kung, 1974). When Ba^{2+} was introduced into the solution bathing wild-type cells, a train of action potentials was generated. The membrane of cells homozygous for *pwB* never fired in Ba^{2+} solution. Satow and Kung found a characteristic response of wild-type cells when a bath lacking Na^+ was replaced with a Na^+-containing solution. There is a train of depolarizations of characteristic shape, each depolarization in the train lasting about 0.5 sec. Each cycle included a rapid depolarization from resting level terminating in a spike, a brief oscillation at the depolarized level, and a repolarization to the resting level. Cells of *pwB* also depolarized in the Na^+ solution, but there was no spike, and behavioral observations showed that this electrical response correlated with no ciliary reversal.

Ts-pawns of *pwA* or *pwC* grown at the restrictive temperature of 32°C show no action potential when stimulated with Ba^{2+} (Satow *et al.,* 1974) or with an injected depolarizing current (Satow and Kung, 1976b). Thus at 32°C ts-pawns look like unconditional pawns. The electrophysiological studies also confirm the suggestion from genetic studies that even at 23°C ts-pawns are not fully normal.

In parallel with the genetic findings (Chang and Kung, 1973b), the barium-induced action potentials seen in wild type at 23°C are also seen in *pwA* or *pwC*. But a double homozygote of the two *ts-pawn* loci generated no barium-induced action potentials even at 23°C (Satow *et al.,* 1974). There was a slight difference between the barium response of wild type and ts-pawn single mutants, but these differences were more clearly delineated by the study of electrically evoked action potentials (Satow and Kung, 1976b).

At 23°C, electrically evoked action potentials are not the same for cells homozygous for *ts-pawn* genes and for wild type. There is an action potential, but the peak is lower for a given stimulus strength, and the maximal rate of rise is slower in the mutants (Satow and Kung, 1976b). With strong currents the maximal rate of rise of wild type is about twice that of mutants, and the excitation threshold is higher in the mutants. This difference in peak level and rate of rise is seen when cells are stimulated in bathing solutions containing either K^+ or Ba^{2+}.

Tetraethylammonium (TEA^+), which blocks potassium conductance in *Paramecium* (Friedman and Eckert, 1973), was added to the potassium-containing bath solution, and evoked potentials were again measured. Satow and Kung found that the maximal rate of rise and the peak potential for both mutant and wild type were raised in the presence of TEA^+, presumably because the short-circuiting effect of outward potassium flow had been reduced. Under these conditions the peak potential of mutants rises to very nearly the level of wild type, but the maximal rate of rise in the mutants is still about half that of wild type. Passive membrane properties such as resting potential and membrane resistance are not different in mutant and wild type at either 23° or 32°C.

These results favor the presence of a defect in calcium activation of ts-pawn even at 23°C, at which temperature the behavioral phenotype is wild type. The addition of TEA$^+$ does not correct the defect in the mutants reflected in a lower maximal rate of rise of the action potential, which makes it unlikely that the lesion of ts-pawns at 23°C is in a later activation of potassium conductance.

Schein *et al.* (1976) compared quantitatively the electrical properties of membranes of both leaky and extreme pawns at all three loci. This is the only study which provides information using a single method on all loci and on more than one allele of *pwA* and *pwB*. Active inward current was calculated after current of known magnitude was applied and the voltage and time derivative of voltage were recorded. A graphic subtraction of $C\ dV/dt$ from the applied current gave the total current across the membrane due to ionic flow (I_{ionic}). Then the passive ionic current was determined and subtracted from I_{ionic} to give the active ionic current across the membrane. The active inward current was plotted against time for each of the genotypes used.

The active inward current for all of *pawn* lines was less than that of wild type (Table II). Schein's (1976b) *pwC* is excitable and shows nearly normal avoiding reactions, yet its active inward current is reduced to 50%

Table II Summary of Active Electrical Properties of Pawns[a]

Cell type	Maximum active inward current (nA)	Presence of delayed rectification	Absolute activation voltage of anomalous rectification (mV)
Wild type	2.8	+	−40
pwnA(214)	0.13	+	−40
pwnA(414)	<0.1	+	−40
pwnA(419)	<0.1	+	−40
pwnB(314)	0.7	+	−57
pwnB(100)	<0.1	+	−72
pwnC(320)	1.3	+	−40

[a] Reprinted from Schein *et al.* (1976), by permission of the Company of Biologists.

that of wild type. The nonleaky *pwA* of Schein has no significant active inward current. A leaky *pwA* had an active inward current about 5% that of wild type. A very leaky *pwB* line showed active inward current reduced to 25% of wild type whereas two nonleaky *pwB* had small but definite active inward currents.

In addition to active inward current, Schein *et al.* looked at anomalous rectification (Naitoh and Eckert, 1968), a delayed increase in membrane conductance following application of hyperpolarizing currents. Anomalous rectification of *pwA* and *pwC* was like that of wild type, first detected when the membrane is hyperpolarized 15 mV more negative than resting potential. Greater hyperpolarization is needed to see anomalous rectification in *pwB* (35 mV more negative in the leaky *pwB* and 52 mV more negative in an extreme *pwB*). Thus a physiological difference is seen which differentiates one *pawn* locus from the others.

Double mutants studied by Schein *et al.* lacked active inward current. Even the combination of the very leaky *pwC* and *pwB* genes lacked detectable active inward current. This double mutant had the anomalous rectification characteristics of the *pwB* parent.

Electrical properties of the membrane of pawns have been correlated with behavioral characteristics of the mutants so as to make it almost certain that pawn is deficient in calcium activation. The difference in characteristics of anomalous rectification between *pwB* and the other loci suggests that *pwB* may control a different component of the calcium-activation system, although the physiological significance of anomalous rectification is not known. Schein *et al.* suggest that *pwA* may control a mechanism for opening the calcium channel in response to depolarization while *pwB* may control a component of the structure of the channel itself.

The demonstration that *pawn* genes block calcium activation allows further probing of the nature of the calcium-activation system. Oertel *et al.* (1977) used *pwB* as a selective blocker of the calcium channel to separate the calcium current from other contributions to the total current observed upon electrical stimulation. A voltage clamp was used to give depolarizing pulses to cells, and the resulting current across the membrane was measured. In wild-type cells there is a transient inward current for about 5 msec following stimulation and then a steady state outward current as long as the stimulus is applied. The level of the steady state outward current increases with increased voltage of the step depolarization. In pawn a steady state outward current not unlike that of wild type appears, but there is no transient inward current.

Using a graphic subtraction of the currents of pawn and wild type, Oertel *et al.* produced a graph of the time course of the calcium current alone (Figure 2). Since the calcium channel is blocked in pawn, the differ-

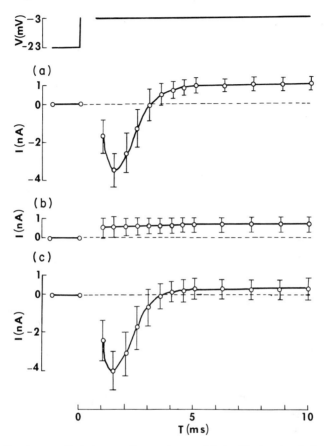

Figure 2. Average time course of currents associated with step depolarizations of 20 mV in wild type (a) and pawn (b). Mean current magnitudes ± SD were measured at various times from four wild-type cells and five *pawn* cells. In (c) is shown the difference between current measured in wild type (a) and *pawn* (b) ± SD of the difference of the means. Measurements were not made in the first 1.0 msec after the step depolarization because the voltage was not well controlled during this time. (From Oertel *et al.*, 1977, reproduced by permission of *Nature* (*London*) and MacMillan Journals Limited.)

ence in the graphs must be the missing calcium current. When wild-type cells were bathed in a medium in which 99% of the Ca^{2+} was replaced by Mg^{2+}, the transient inward current disappeared, and only the steady state outward current was seen, confirming that the transient inward current is a calcium current. Surprisingly, the results of the graphic subtraction of wild type from pawn suggests that the calcium current does not persist during the entire time of depolarization but rapidly inactivates.

To eliminate the possibility that the loss of inward current might be due
to short circuiting by a strong activated potassium current, Oertel *et al.*
studied the effect of hyperpolarization on the inward and outward currents. If there is an outwardly directed potassium current, hyperpolarization of the membrane below the equilibrium potential for potassium (E_K)
should shift the potassium current inward. If an outward potassium current were canceling the inward calcium current, a depolarization followed
by a hyperpolarization more negative than E_K would first activate the
calcium current and the potassium current and then shift the potassium
current inward to produce a large inward current. A hyperpolarization
alone would not show a high inward current since neither of the activated
currents would be functioning. However, if the calcium current had inactivated, either depolarization followed by hyperpolarization or simple hyperpolarization should give the same result, a low inward current. The
result actually observed was the result expected if the calcium current
inactivated. Rapid inactivation of the inward calcium current has now
been demonstrated by other means in *P. caudatum* (Brehm and Eckert,
1978).

The observation of rapid inactivation of the calcium current appears to
be in conflict with another observation (Machemer and Eckert, 1975). The
short inward calcium current observed by Oertel *et al.* lasts but 5 msec.
Yet cilia reverse their beat under step depolarization and continue to beat
in reverse direction throughout depolarizations even of several seconds. If
Ca^{2+} has ceased to flow into the cell after 5 msec, how is the internal Ca^{2+}
concentration maintained to activate ciliary reversal? Reversal of ciliary
beat is a function of internal Ca^{2+} concentration, not calcium current, and
it is possible that a depolarization-related mechanism retains Ca^{2+} which
flows into the cell even after the calcium current has ceased. In any case,
that such a contradiction should arise attests to the value of the method of
genetic dissection. This interesting problem would not otherwise have
arisen. It should stimulate study by biochemical means of the levels of
internal Ca^{2+} during depolarization.

Berger (1976) used pawn for quite another kind of analysis. He was
interested in the phenomenon of ''phenomic lag'' (Sonneborn, 1954). After
conjugation, parental phenotypes sometimes persist for several fissions
even when progeny genotypes are different. Pawn can be used for quantitative study of phenomic lag. By transferring cells into stimulating solutions and noting whether or not the cells swim backward, one can rapidly
determine phenotypes without destroying the cells. Cells may be returned
to culture fluid for later tests, and in this way the same line can be tested
after each fission. The transition from wild type to pawn was usually
accomplished by inducing autogamy in *pwA*/+ heterozygotes, after which

half of the cells will be homozygous *pwA*. The exautogamous *pwA* homozygotes nevertheless retain fragments of the old *pwA*/+ macronucleus, which are diluted out in subsequent divisions. Transition from pawn to wild type was accomplished by mating pawn and wild type and studying the exconjugants of pawn parentage.

In the transition from wild type to pawn, well-fed exconjugants or exautogamonts remained wild type for a median 7.0–7.6 fissions, although there was great variability within clones. When cells were starved for 4 days after autogamy and then fed, the median lag was 3.3 fissions. The difference between the results of well-fed and starved exautogamonts is probably due to the difference in persistence of the heterozygous fragments of the old macronucleus. These persist for as long as 11 fissions in well-fed cells (Berger, 1973) but are autolyzed rapidly in starved cells (Berger, 1974). Evidence that the fragments were a main contributor to phenomic lag was the finding (Berger, 1976) that there was an association between the persistence of wild-type phenotype and the persistence of fragments. After eight to nine fissions, when most cells were wild type, Berger stained all of the cells of an exautogamous clone after each cell had been tested for behavior. There was a strong association between wild-type phenotype and possession of fragments.

In the transition from pawn to wild type, cells became phenotypically wild in as little as 2 hr after conjugation. Some cells, in fact, became wild type before conjugation was over. All cells were wild type within two fissions after conjugation. The early change from pawn to wild type is probably due to cytoplasmic exchange between the conjugants. Berger (1976) mixed mating reactive cells labeled with radioactive leucine with cells of complementary type not labeled and found that a significant amount of label crossed into the unlabeled mate by 3 hr after complementary mating types had been mixed. To determine the time that new nuclear activity contributed to wild-type phenotype, heterokaryons were constructed with a *pwA*/*pwA* macronucleus and a +/*pwA* micronucleus. When such cells undergo autogamy, half will become +/+. Since they had no mate, these cells can be used to determine phenomic lag without cytoplasmic exchange. Transition from pawn to wild type occurred by the end of the first cell cycle or the beginning of the second after autogamy, still much more rapid than in the transition from wild type to pawn.

Schein (1976a) followed electrophysiological characteristics of cells undergoing the transition from wild type to pawn. He tested the excitability of exautogamous homozygous pawn cells from *pw*/+ parents. Both *pwA* and *pwB* were studied. In addition to electrophysiological tests, the behavior of the cells was tested in a stimulating solution containing a low Ba^{2+} concentration which induced frequent reversals in wild type, and resistance was tested to a paralyzing solution with a high concentration of Ba^{2+} which paralyzed wild-type cells in 15 sec. Pawns were not readily paralyzed by the high Ba^{2+} solution.

With the paralyzing solution, pawn phenotype could be detected even after the first fission following autogamy, even though the cells were still wild type in response to stimulating solution. By the fourth fission after autogamy the stimulating solution elicited only a weak reversal response, and by the fifth fission the cells were extreme pawns. The same kinetics of lag was seen for *pwA* and *pwB*. Berger only used a stimulating solution in his studies and found somewhat longer lags than Schein. In experiments using cells starved after autogamy, however, the lag found by Berger (3.3 fissions) and that of Schein (4 fissions) were similar.

Schein measured active inward calcium current by the method of Schein *et al.* (1976) during phenotypic transition. The heterozygotes were fully wild type in inward calcium current. At succeeding fissions after autogamy there was a progressive halving of inward calcium current with each fission until at last a fully pawn condition was reached in which no inward calcium current was seen.

Schein used his results on measurements of electrical properties of membranes during phenotypic transition to estimate the stability of the calcium channel. Components of channels under the control of *pwA* and *pwB* might be very stable and thus disappear during phenotypic transition only at the rate at which they are diluted during fission. Or the channel components could be rather unstable, in which case they would disappear much more rapidly than expected for simple dilution. In fact, the level of excitability halves with each fission in rapidly growing cells, as expected of a stable channel. When cells were grown slowly enough during phenotypic transition, channels decayed somewhat more rapidly than they would have by simple dilution, and Schein calculated the lifetime of channel components under the control of *pwA* to be 5.5–8 days and of *pwB* to be 5.5–9 days. The channel components are, therefore, rather stable entities.

The rapid replacement of pawn phenotype by wild type through cytoplasmic exchange of soluble substances (Berger, 1976) is hard to reconcile with the stability of channels. Presumably the channels are composed of macromolecules integrated into the surface membrane. Microinjection of wild-type cytoplasm into pawn cells might be used to discover what diffusible substance repairs pawn gates so rapidly.

2. Paranoiac Mutants

Compared with the information available about pawn, relatively little is known about other mutants of swimming behavior. However, the class of paranoiacs has been studied and has the distinction of being the first class of mutants for which an ultrastructural correlate of behavioral and physiological characteristics was discovered.

Whereas pawn is characterized by the loss of a part of normal swimming behavior, paranoiac mutants are exaggerated in behavior. They overreact to stimulation by Na^+ with lengthy continuous ciliary reversal (Kung, 1971a; Van Houten *et al.*, 1977). Spontaneous CCR in culture fluid also occurs more often than in wild-type cells.

Paranoiac lines were selected in one of two ways (Van Houten *et al.*, 1977). One method involved a reversal of the column used to screen for pawn. The column was filled with a solution not expected to elicit avoiding reactions in wild-type cells (high in Ca^{2+} relative to the concentration of other cations). Mutagen-treated cells were introduced to the top of the column, and collections of possible mutant cells were removed from the bottom, presumably after they backed up. No tests, however, have been done on this method to determine if it actually does enrich for mutants.

A galvanotactic screening method exploited the tendency of *Paramecium* to swim forward toward the cathode in an electric field. Paranoiac cells swam backward to the anode. After groups of mutagen-treated cells were subjected to an electric field, cells were removed that did not migrate or that migrated toward the anode, and these cells were examined as possible paranoiacs. Again, the effectiveness of the method in truly enriching for mutants has not been determined.

All seven lines of paranoiac now reported have as their bases single nuclear genes (Kung, 1971b; Van Houten *et al.*, 1977) at one of five different loci. At two of the loci the *paranoiac* allele is recessive to wild type (*paB* and *fna*[p]). At the other loci the *paranoiac* allele is either dominant or codominant to wild type (*PaA, PaC, PaD*). Two alleles of *PaA* are known, and *PaA* is closely linked (3 map units) with *PaC*. A ts-paranoiac is also known (Kung *et al.*, 1975).

Surprisingly, *fna*[p] is allelic to *fast-2* (Van Houten *et al.*, 1977), yet the phenotype is completely different. Gene *fna*[p] results in overreaction to Na^+ and spontaneous CCR like *PaA*, but the main characteristic of *fast-2* is complete lack of response to Na^+ (Kung, 1971a), probably because of a change in potassium permeability (Satow and Kung, 1976a). Alleles of a single locus would presumably affect the same product, so that it is surprising to see that the F1 of a cross between *fast-2* and *fna*[p] with no wild-type alleles at the locus is wild type. This case might provide an interesting study of intragenic complementation in *Paramecium*.

Pawn mutants are epistatic to paranoiac, i.e., in double homozygotes of pawn and paranoiac the phenotype is pawn (Kung, 1971b; Van Houten, 1977). Pawn blocks the upstroke of the action potential correlated with

backward swimming whereas paranoiac prevents the return of the depolarized membrane to its resting state (Satow and Kung, 1974). One
would expect that mutations acting earlier in a process would mask the
expression of genes acting later.

The genetic analysis of paranoiac includes an interesting nonconformity
inviting further study. Kung (1971b) reports that many of his crosses
involving paranoiac had a slight excess of mutants when a 1 : 1 ratio was
expected. That this may be more than chance deviation is indicated by the
occurrence of similar results in Van Houten *et al.* (1977). Most striking is
the report of Van Houten *et al.* of crosses between *PaC* and two different
pawn genes, *pwA* and *pwB*. Because of the epistasis of *pw* over *Pa,* a ratio
of *pawn* : *paranoiac* : wild type of 2 : 1 : 1 was expected. But in both cases a
clear excess of *paranoiac* resulted in ratios much more like 2 : 2 : 1. It is
difficult to explain this result as selective death in any single class, nor is it
possible to account for it by any of the usual problems in *Paramecium*
crosses such as macronuclear regeneration or unilateral fertilization. Such
mechanisms would result in an increase in wild type, although the con-
trary occurred. Further attention to this problem could produce as yet
unnoticed information about the interaction of *pw* and *Pa.*

The behavioral phenotype of overreaction to Na^+ has been correlated
with electrophysiological, biochemical, and ultrastructural characteristics
of paranoiac. A summary of findings in each area will round out our view
of paranoiac.

In wild-type cells there is a distinct pattern of active electrogenesis
when a bathing solution containing K^+ is replaced by one containing Na^+
(Satow and Kung, 1974; Satow *et al.,* 1976). A complete cycle of the
pattern takes about 0.5 sec and includes a rapid initial depolarization from
the resting level ending in a spike, a brief oscillation at the depolarized
level, and a repolarization to the resting level or slightly below it. This
cycle recurs in trains, with the later cycles being slightly damped.

When paranoiac is subjected to the same conditions, Satow and Kung
and Satow *et al.* found a strikingly modified pattern. The upstroke and the
spike look like wild type, but once depolarized the membrane remains at
that level for 2 to 60 sec. Thus the short jerks of periodic ciliary reversal in
wild type and the extended continuous ciliary reversal of paranoiac in Na^+
are correlated with differences in the pattern of active electrogenesis.

When depolarizing current is injected across the surface membrane of
cells bathed in Na^+ solution, similar differences between wild type and
paranoiac can be seen (Satow *et al.,* 1976). During the period of the injected current, paranoiac and wild type both undergo a sharp upstroke of
potential culminating in a spike and followed by a depolarized plateau.
When the injected current is shut off, the wild-type membrane depolarizes

to the resting level. However, the paranoiac membrane remains at the depolarized level for many seconds. During the extended period of depolarization, the resistance of the membrane to inward injected current is markedly decreased relative to the resting membrane. The strength of the injected current need not be large to trigger the long depolarization of the paranoiac membrane.

Kung (1971a) and Hansma and Kung (1976) showed that the duration of backward swimming of paranoiacs increased with increasing concentration of Na^+ in the stimulating solution within the range of 0–20 mM Na^+. This behavioral observation is also correlated with electrophysiological data (Satow *et al.*, 1976). Extended depolarized plateaus were evoked electrically. As the Na^+ concentration in the bath increased, so did the duration and potential level of the plateau. The resting level to which the paranoiac membrane eventually returned also became less negative as Na^+ concentration increased.

Biochemical observations are in accord with and illuminate the electrophysiological data. In wild-type cells, uptake of $^{45}Ca^{2+}$ can be stimulated with K^+ or Na^+, with K^+ more effective than Na^+ (Browning *et al.*, 1976). However, in certain paranoiac lines Na^+ is as effective (*PaA*) or more effective (*PaC*) in stimulating calcium uptake. The flow of Ca^{2+} into cells that occurs upon membrane depolarization (Eckert, 1972) might be expected to continue for a longer time in cells which remain depolarized longer as paranoiacs will in Na^+ solutions.

Both K^+ and Na^+ movements have been studied in paranoiac in Na^+ solution (Hansma and Kung, 1976; Satow *et al.*, 1976). Over the range of 0–20 mM external Na^+ the concentration of K^+ in wild-type cells remains a nearly constant 12 mM. But in three paranoiac stocks studied K^+ leaks out of the cell in amounts reaching up to three-fourths the internal K^+ at high concentrations of Na^+ (Hansma and Kung, 1976). Counts of cells before and after the experiments showed that cell lysis could not account for the loss. It is interesting that the different mutants showed different kinetics of K^+ loss. Stocks homozygous for *fna*[p] lost K^+ at a rate slower than *PaA* or *PaC*. The duration of backward swimming of *fna*[p] in Na^+ is also less than that of *PaA* and *PaC*.

Mutant *fna*[p] and wild type were compared in their regulation of K^+ content in K^+ solution (Hansma and Kung, 1976). No significant differences were seen. Both paranoiac and wild type gained K^+ as the external K^+ concentration increased. Thus paranoiac again reacts differently from wild type specifically in Na^+ solutions.

Fluxes of Na^+ were studied using $^{22}Na^+$ (Hansma and Kung, 1976; Satow *et al.*, 1976). As the concentration of external Na^+ increased, internal Na^+ increased in both mutant and wild type. However, the initial rate of Na^+ uptake in fna^p and PaC was significantly higher than in wild type so that final concentrations of sodium were higher in *paranoiac*. Satow *et al.* (1976) showed that during periods of extended depolarization in stimulated *paranoiacs* the conductance of the membrane increased. The flow of K^+ out of *paranoiac* cells (balanced by a flow of Na^+ into the cells) is consistent with the observation of heightened conductance of the membrane. This increase in flow of monovalent cations across the *paranoiac* membrane is another aspect of the phenotype of *paranoiac*. As the concentration of Na^+ is increased, the cells swim backward longer, produce a longer plateau of depolarization, and lose more K^+ while gaining more Na^+. The different loci, *PaA, PaC,* and *fna^p*, show different kinetics of K^+ loss, respond with different intensities of backward swimming when Na^+ concentration is increased (Hansma and Kung, 1976), and respond differently to the stimulation of calcium uptake by potassium and sodium (Browning *et al.*, 1976). All *paranoiac* genes apparently block the ability of depolarized membranes to return readily to resting potential. But the differences in phenotypes at the biochemical level suggest that the different loci might be blocking the return in different ways.

In addition to the suggestive electrophysiological and biochemical data available for paranoiac, an important recent finding has been the discovery of differences in ultrastructure between paranoiac and wild type (Byrne and Byrne, 1978a). A very regular series of particles is found at the base of freeze-fractured cilia of wild-type *Paramecium* examined with the electron microscope. This array, called the ciliary granule plaque (Plattner, 1975), is located within the ciliary membrane and consists of three vertical columns of 10-nm particles in variable numbers of horizontal rows. In the work of Byrne and Byrne the number of rows was 3 to 7 with a mean of 4.84, and all cilia encountered had plaques (Figure 3a). There are nine plaques at the base of each cilium (Plattner, 1975).

Plattner found calcium deposits associated with ciliary granule plaques and suggested a role for the plaques in regulation of ciliary activity. Polycationic ferritin will also bind selectively at the region of the plaques, demonstrating that the plaques have a high net negative charge (Dute and Kung, 1977).

In the mutant PaA^1, an allele of *PaA*, cilia were often found by Byrne and Byrne that lacked plaques or that had irregular plaques (Figure 3b). Of 40 cilia examined, 16 had no plaques at all. The loss and disarray of plaques in the PaA^1 mutant were not altered by changes in growth conditions. Another *PaA* allele was also examined and found to have more irregular plaques than wild type. The differences, however, were not as extreme as in PaA^1, and no cases of completely missing plaques were encountered.

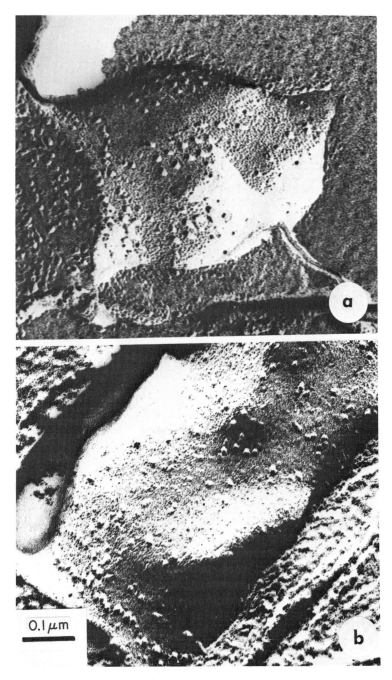

Figure 3. Ciliary granule plaques of wild type (a) and *paranoiac* PaA[1] (b) of *P. tetraurelia*. In (a) the characteristic wild-type array of ciliary granules into three columns can be clearly seen. In (b) no ordered array is evident, although some granules are distributed about the membrane. (Photo courtesy of Bruce Byrne and Barbara Byrne, Wells College.)

It is tempting to look on these plaque abnormalities as the structural basis of the paranoiac behavioral and physiological characteristics. Do the particles of the plaque in regular array play a role in regulating the depolarization of the membrane? It is possible that the plaques play such a role, but it is also possible that they represent another effect of the *PaA* gene unrelated to the paranoiac phenotype or even a secondary effect of the primary cause of paranoiac behavior. However, as the first ultrastructural correlates of membrane function in *Paramecium,* the irregular and missing plaques should be the beginning of an eventual unified understanding of swimming behavior from the point of view of electrophysiology, biochemistry, and anatomy.

3. The Fast Mutants

Two types of "fast" mutants have been discovered (Kung, 1971a,b), both having as one characteristic increased forward speed when agitated. Cells introduced into culture medium from micropipets at first swim rapidly, showing little or no avoiding reaction when encountering the side of the culture dish. As the cells slow down, they begin to display avoiding reactions. The two types of fast are distinguished by their reactions to solutions of Na^+ low in Ca^{2+} (Kung, 1971a). Fast-1 reacts as wild type does, giving repeated avoiding reactions. Fast-2 does not avoid sodium solutions, although it acts like wild type in solutions high in K^+ or Ba^{2+} and low in Ca^{2+}.

Both kinds of mutants were selected using a column like that used for pawn, containing a stimulating solution high in Na^+. Of course, the fast-2 acts like pawn in such a column. Fifteen lines of fast-1 and 10 lines of fast-2 were isolated in the first experiments of Kung (1971b).

Fast-1 cells have a fast forward swimming speed of 1.39–1.43 mm/sec whereas wild type swims at 0.63 mm/sec. Fast-2 swims at 1.17–1.22 mm/sec (Kung, 1971a). A fast mutant of Cooper (1965) swam at about the speed of fast-1 and like some fast-1 was unable to mate.

Among those fast-1 lines able to mate, two loci were recovered, *fA* and *fB*. Mutant alleles of both loci are recessive to wild type and are unlinked to each other and to other genes tested. No further information is available in the published literature on *fast-1*.

Of the five lines of fast-2 studied by Kung (1971b), all carried a mutation at the same locus, *fna*. This is an allele of the same locus as *fna*[p] which was reviewed in the section on paranoiac.

Electrophysiological analysis of *fna* has been carried out (Satow and Kung, 1974, 1976a). Cells were adapted to a potassium-containing bath and then the potassium solution was replaced with a sodium solution in which there was no K^+. In such a solution wild-type cells displayed an avoiding reaction and fast-2 cells did not. Furthermore, the wild-type cells monitored after change of bath from one containing K^+ to one containing Na^+ displayed a characteristic depolarization and action potential while fast-2 did not. Under the same conditions pawn showed a depolarization without the spike seen in wild type, and paranoiac had a prolonged plateau following the spike (Satow and Kung, 1974). The fast-2 pattern was very different. The membrane potential oscillated about the resting potential and showed sporadic spikes of hyperpolarization. Over the time of such experiments the resting potential of fast-2 drifted in a more negative direction (Satow and Kung, 1976a).

Satow and Kung showed that wild-type cells artificially hyperpolarized by as much as these fast-2 cells also failed to show sodium-triggered action potentials, although both mutant and wild-type membranes could be depolarized when in this state by injecting depolarizing current. Fast-2, but not wild type, also shifted its resting potential in a negative direction when bathed in a solution containing only Ca^{2+}.

The resting level of fast-2 membranes in a potassium-free solution was 20–30 mV more negative than wild type. A plot of resting potential versus external K^+ concentration showed that, as K^+ concentration increased, the resting potential became more like that of wild type. When 3 mM Ca^{2+} was the only ion besides K^+ in the bath, wild-type and mutant membranes had the same resting potential above 4 mM K^+. The addition of Na^+ as well as Ca^{2+} resulted in lower resting potentials for fast-2 membranes over a greater range of K^+ concentrations. When TEA^+ was included in the K^+-free bathing solution, the resting potential did not become more negative than that of wild type, and the active electrogenesis in response to sodium was also like that of wild type. The resting membrane resistance of fast-2 and wild type did not differ in potassium-free solution or when K^+ concentration was 4 mM. At 16 mM external K^+, the resistance of the mutant was significantly less than that of wild type (37 $M\Omega$ for wild type; 17 $M\Omega$ for fast-2).

All these data favor the hypothesis that fast-2 has a higher permeability to K^+ relative to the permeability of Na^+ and Ca^{2+} than does wild type. Since fast-2 shows barium spikes and electrically evoked action potentials, the problem is clearly not that of pawn. The calcium activation system will work. The proposed greater permeability would account for the more negative resting potential in potassium-free and low-potassium solutions and the lowered membrane resistance in high potassium solutions. These phenomena are not seen in the presence of TEA^+, an ion known to decrease potassium permeability in *Paramecium* (Friedman and

Eckert, 1973; Satow and Kung, 1976c). Fast-2 is not the only mutant which alters potassium conductance, and we now turn to another, TEA$^+$-insensitive.

4. The TEA$^+$-Insensitive Mutant

Knowledge of the physiology of the excitable membrane of *Paramecium* is now advanced enough so that screening techniques for particular kinds of membrane lesions can sometimes be designed based on physiological knowledge. The TEA$^+$-insensitive mutant is an example of such a situation. Friedman and Eckert (1973) showed that the TEA$^+$ blocked the efflux of K$^+$ normally associated with depolarization of the *Paramecium* membrane. The potassium efflux normally short-circuits the calcium influx, thus reducing the action potential (Naitoh *et al.*, 1972). TEA$^+$ thus enhances the calcium action potential, rendering the membrane more excitable. Wild-type *P. tetraurelia* introduced into 10 mM TEA$^+$ solutions low in Ca$^+$ exhibit avoiding reactions.

To look for mutants with an enhanced potassium conductance, Chang and Kung (1976) made use of the column employed for *pawn* screening, modified to take into account the effect of TEA$^+$. Columns were filled with solutions high in Na$^+$ and low in Ca^{2+} with 5 mM TEA$^+$. Only cells which do not avoid TEA$^+$ solution should rise to the top. Presumably some of these cells would be *pawn*, but some might have an increased potassium conductance. In such cells potassium might short-circuit the calcium action potential even in TEA$^+$ solution, thus failing to avoid the solution.

Indeed, 26 pawns were isolated from the columns, and 1 TEA$^+$-insensitive mutant was found (Chang and Kung, 1976). The difference in proportions of pawn and TEA$^+$-insensitive mutants isolated from the column could exist for any of several reasons. *Pawn* genes might be more numerous or more mutable than *TEA$^+$-insensitive* genes. On the other hand, many cells with only slightly enhanced potassium conductance might be enough affected by the TEA$^+$ to give avoiding reactions. The selection procedure should only isolate fairly extreme cases of increased potassium conductance. TEA$^+$ insensitivity was due to a single recessive gene *teaA* (Chang and Kung, 1976). This gene is unlinked to *pwA*, *pwB*, or *pwC*, to *fna*, or to *PaA*, all of which are epistatic to it.

Electrophysiological investigations of *teaA* (Satow and Kung, 1976a) suggest an increased potassium conductance in the mutant. Small or moderate depolarizing currents yielded smaller and slower action potentials in the mutant than in the wild type, as expected if increased potassium efflux was short-circuiting calcium influx. Smaller action potentials would also occur if the calcium activation system were impaired. However, large depolarizing currents produce very similar action potentials in mutant and wild type, and there was a relatively normal barium spike produced by TEA⁺-insensitive, both of which demonstrate that this is not another pawn.

Electrically evoked action potentials had steeper downstrokes in the mutant, and the downstroke usually undershot the resting level, again expected if potassium efflux increased. Large hyperpolarizing currents are carried by cations in the bathing solution. Satow and Kung found that the resistances of mutant and wild type, when given large hyperpolarizing currents were nearly the same in Ca^{2+}, TEA^+, or Ba^{2+} solutions. But in potassium solutions the resistance of the mutant membrane was significantly lower than the wild type, further evidence for a change specifically in potassium conductance.

Friedman and Eckert (1973) showed that internally applied TEA^+ blocked the efflux of K^+ from *Paramecium*. Satow and Kung showed that high potassium conductance in *teaA* could be blocked by internal application of TEA^+ or by Ba^{2+}.

The relation between fast-2 and TEA⁺-insensitive is unclear. TEA⁺-insensitive weakly avoids sodium solutions whereas fast-2 fails to avoid them. Fast-2 is not insensitive to TEA^+. Depolarizing currents stimulate normal looking action potentials in fast-2, but stimulate slower action potentials of lower magnitude in TEA⁺-insensitive. The two unlinked loci must have effects on different components of the surface membrane, each of which results in increased potassium conductance. Unfortunately our knowledge of K^+ metabolism in *Paramecium* is limited so as to prevent a clear understanding of the differences between the mutants. If there is a calcium-activated potassium conductance in *Paramecium* (Eckert *et al.*, 1976; Satow, 1978), one of the mutants might affect the resting conductance while the other might represent a lesion in the activated system. Satow and Kung (1976c) suggest that *teaA* may result in an activated system which is always partially on, thus increasing potassium conductance.

5. The Potassium-Resistant Mutants

Schein's (1976b) demonstration that pawn mutants could be isolated as barium-resistant cells has prompted the use of "nonbehavioral screening" in the search for other behavioral mutants. Shusterman *et al.* (1978) have recovered mutants which survive concentrations of potassium lethal to wild type cells. Among 27 independent mutations isolated in this way, at least three unlinked loci have been identified.

Two classes of potassium-resistant mutants have been found, one of which survives high concentrations of KCl (up to 80 mM) added to the culture medium and one of which survives intermediate concentrations (around 40 mM). Either the high or intermediate concentrations kill wild-type cells. The two classes of mutants may also be distinguished by resistance to rubidium ions, but the mutants are as sensitive as wild type to Na^+, Ba^{2+}, and Ca^{2+}.

When mutants are grown in normal culture medium, their swimming behavior is indistinguishable from that of wild type. They show the same range of stimulated behavior patterns in barium or sodium solutions. But the potassium mutants qualify as behavior mutants because of differences which arise after growth for 2 days in culture medium enriched with 15 mM KCl. In such circumstances, wild type loses its ability to react to barium. The characteristic periodic ciliary reversal associated with barium stimulation is not lost in the potassium-resistant mutants. Those mutants most resistant to potassium respond with vigorous ciliary reversal, and the intermediate class responds with alternating backward and forward swimming. Prolonged growth of mutants in 15 mM KCl or growth in increased potassium concentrations results in loss of barium response even in the mutants.

Raising the concentration of calcium in potassium-enriched medium raises the level of potassium which kills wild-type cells and also preserves their ability to swim backward in barium solutions. Shusterman *et al.* (1978) speculate that over the range of low potassium concentrations ordinarily encountered in fresh water *Paramecium* is able to regulate permeability to potassium. This regulation maintains the equilibrium potential across the membrane and thus permits forward swimming even as external potassium concentrations fluctuate. The regulation of potassium permeability may have been lost in these mutants. Electrophysiological and biochemical data on these mutants should illuminate this interesting possibility.

6. Chemokinesis Mutant

Mutants of chemokinesis, which might include cells altered in any number of aspects of the action system, can be screened in a countercurrent apparatus in which two solutions of different density, one containing an attractant or a repellent, flow past one another in opposite directions in a tube (Van Houten *et al.*, 1975). Paramecia separate themselves in the two solutions according to their relative attraction or repulsion to the cells. Thus more complex behavior may now be studied by genetic dissection.

The genetics and physiology of chemokinesis in protozoa including *Paramecium* will be reviewed by Van Houten *et al.* in another volume of this treatise. However, one mutant of chemokinesis in *P. tetraurelia* has been described and should be mentioned in the context of this chapter. Van Houten (1977) isolated a mutant that was repelled by sodium acetate.

Wild-type cells are attracted to the substance, apparently by decreasing the frequency of avoiding reactions when coming into contact with sodium acetate. The mutant also decreases the frequency of avoiding reactions in sodium acetate, but in addition increases forward velocity significantly so that it swims forward, usually away from the sodium acetate. Thus we have another fast mutant, this time one with increased hyperpolarization of the membrane upon chemical stimulation (Van Houten, 1977). Electrophysiological data on this mutant have not yet been published.

7. Other Mutations of Swimming Behavior

Kung and his associates have now obtained over 350 lines of mutants of swimming behavior that are mapped on over 20 genes (Kung, 1976) (Table I). In addition to those mutants already discussed, several others are known only superficially and may be found listed in Kung *et al.* (1975). These other mutants include spinner, which spontaneously spins upon mechanical stimulation of the anterior end or upon stimulation by sodium solutions; staccato, which displays frequent short avoiding reactions in culture medium; sluggish, which swims slowly or not at all, and gives avoiding reactions when stimulated by Ba^{2+} or Na^+; and atalanta, which stops transiently in culture fluid and shows barely discernible backing in Ba^{2+} solution. Further analysis of these mutants should enrich our knowledge of the control and coordination of components of the action system.

Search for other mutants of swimming behavior should be continued. Extension of Schein's (1976b) concept of nonbehavioral selection for behavior mutants becomes more possible as more information about the physiology of behavior makes possible the construction of screening techniques. Byrne and Byrne (1978b) discuss other screening methods designed to obtain mutants needed to answer specific problems. In addition, the use of related ciliates known genetically should make possible the characterization of new aspects of behavior.

D. Paramecium caudatum

Paramecium caudatum is larger than members of the *P. aurelia* complex of species and has been used for many of the physiological studies of behavior by Kinosita and his associates, Eckert, Naitoh, and others. However, genetic studies of *P. caudatum* are more difficult than these on the *P. aurelia* species. There is no natural autogamy, and there is often a long immature period following conjugation and a high mortality rate following inbreeding. Yet many of the difficulties were overcome by Takahashi (Takahashi and Naitoh, 1978; Takahashi, 1979), who produced a number of interesting mutants of swimming behavior in syngen 3 of *P.*

caudatum (Table I). The breeding techniques for this study are discussed in Takahashi (1978).

Strain KyK 201 of *P. caudatum* was chosen for mutagenesis because of a high rate of selfing in mating type VI and a high rate of survival in selfing exconjugants. Screening was in some cases by the gravity column method of Kung and sometimes by galvanotaxis. An important difference between Kung's gravity columns and Takahashi's was the use by Takahashi of high K^+ solutions for stimulation rather than high Na^+ solutions. For galvanotaxis, populations of mutagenized cells which had been selfed were subjected to an electric field, and any cell swimming the wrong direction, not swimming at all, or in some other way displaying unusual behavior was removed and cloned for further study. As with *P. tetraurelia,* the efficiency of the screening methods has not been established.

1. CNR Mutants

Mutants much like *pawn* of *P. tetraurelia* were recovered. In eight different experiments 650,000 cells were mutagenized. Of these, 23 were isolated with unusual behavior; 10 have been studied genetically, and of the 10, 7 are *pawn*like (Takahashi, 1978). Takahashi called the *pawn*-like cells CNR (*c*audatum-*n*on-*r*eversing).

CNR cells fail to swim backward when stimulated mechanically or chemically. They do not produce an action potential when stimulated by depolarizing current (Takahashi and Naitoh, 1978). Triton-extracted models of CNR do reverse when Ca^{2+} is added to the reactivation medium. Both wild type and CNR increase the frequency of ciliary beat in response to mechanical stimulation of the posterior end or to hyperpolarizing current. Thus CNR looks very much like pawn with a lesion in the calcium activation system.

Genetic analysis was hampered by low survival after crossing to wild-type strains. Crosses often had to be done to several wild stocks to find one combination with sufficient F1 to yield good data (Takahashi, 1979). Selfing is common when complementary mating types are mixed in *P. caudatum.* In mating CNR to wild type, selfing pairs were eliminated by looking at the behavior of pairs in potassium solutions and eliminating pairs in which both members behaved the same. If both members of a pair have the same behavioral phenotype, they must both be of the same parental type, and the pair cannot produce a true cross. F2 clones were produced by selfing of F1 or by crossing two CNR of complementary mating type. F2 survival, especially when selfers were used, was often low.

CNR is due, in every case analyzed, to an allele recessive to wild type. Three complementation groups were found, and Takahashi therefore assigns CNR alleles to three loci, *cnrA, crnB,* and *cnrC.*

Paramecium caudatum, because it does not normally undergo autogamy, is excellent material for the study of the phenotype of heterozygotes. Takahashi produced a number of multiple heterozygotes and homozygotes and measured the duration of potassium-stimulated backward swimming. She found a statistically significant shortening of the duration of backward swimming of single heterozygotes when compared to wild type and of double heterozygotes compared to single heterozygotes. The *cnr* loci are not fully recessive but have some effect in the heterozygous condition. It is interesting to compare this result with the finding of Schein (1976a) that the electrophysiological properties of heterozygous *pwA*/+ of *P. tetraurelia* are normal.

Wild-type *P. caudatum* show repeated short avoiding reactions in barium solutions (Naitoh and Eckert, 1968), and the cells die in barium solutions. Takahashi tested the swimming response and survival of single and double homozygotes in Ba^{2+}. Clones of *cnrB* are most sensitive to barium killing of the single mutants, but all are more resistant to barium killing than wild-type cells. Clones of double homozygotes are more resistant to barium killing than single mutants, and double homozygotes lack the leakiness characteristic of single homozygote parents.

Following the methods of Berger (1976), Takahashi and Hiwatashi (personal communication) looked at expression of the mutant character when CNR was mated to wild type. In matings of *cnrC* homozygotes to wild type the exconjugant of *cnr* parentage became wild type very early because of cytoplasmic exchange, as Berger (1976) reported for *P. tetraurelia* *pwA*. When *cnrA* or *cnrB* was used, no change of phenotype was observed during conjugation and early postconjugation. When either *cnrA cnrC* and *cnrB cnrC* double homozygotes were used, the result was the same as for the *cnrA* or *cnrB* single homozygotes. No early phenotypic change was seen. These results encourage examination of other *pw* loci in *P. tetraurelia* to see if similar results are obtained.

An interesting physiological finding about CNR (Takahashi and Naitoh, 1978) was that, while the calcium action potential was absent from CNR, a small depolarization was seen when the anterior end was stimulated mechanically. The mechanoreceptor potential seems to be uncoupled from the calcium activation potential in the mutant, emphasizing the separate mechanisms of these two processes, and encouraging the search for mechanoreceptor mutants in *Paramecium*.

2. The Potassium-Sensitive Mutant

A class of mutant found in *P. caudatum* but not identified in *P. tetraurelia* is postassium-sensitive (Takahashi and Naitoh, 1978; Takahashi, 1979). These cells swim backward about five times as long as wild type when introduced into potassium-rich medium. Paranoiac of *P. tetraurelia* reacts in a similar way in sodium-rich solutions. That these different mutants of ion sensitivity could be found in these two species is probably due to differences in culture techniques. The fresh lettuce juice medium used for culture of *P. caudatum* (Hiwatashi, 1968) is made up in Dryl's (1959) solution, a medium rich in sodium. Takahashi reports that cells cultured in this medium are rather insensitive to stimulation by Na$^+$. But Kung (1971a,b) used Dryl's solution and other solutions high in sodium for stimulation tests and for screening of mutants. *Paramecium tetraurelia* is routinely cultured in Cerophyl medium (Sonneborn, 1970a), a medium containing less sodium. Thus it would not be expected to find sodium-sensitive mutants under the culture conditions of Takahashi.

The potassium-sensitive mutant has a higher permeability to K$^+$ than wild type, as reflected by a more negative resting potential at low concentrations of external K$^+$ (Takahashi and Naitoh, 1978). It would be of interest to culture this strain under the same conditions as *P. tetraurelia* to compare its physiological characteristics with those of *fast-2* and *TEA$^+$-insensitive*. In fact, for comparison of any of the *P. caudatum* mutants with *P. tetraurelia* it is crucial to find common culture conditions for both species.

Details of the genetics of the potassium-sensitive mutant remain to be worked out, but F1 of a cross of potassium-sensitive to wild type were found in a ratio of 1:1 potassium-sensitive:wild. Thus potassium-sensitive may be due to a dominant gene, and the original stock may have been heterozygous.

3. Slow Swimmer Mutant

The third mutant for which there is appreciable information is slow swimmer, due to a recessive gene designated *sl*. One mutation has been recovered (Takahashi and Naitoh, 1978). Triton-extracted models of slow swimmer act like live cells, swimming more slowly than wild type. Electrophysiological characteristics of slow swimmer look like wild type. Probably the mutant is defective in its ciliary motile system. Kung *et al.* (1975) report a "sluggish" mutant in *P. tetraurelia* of similar phenotype.

4. Other Mutants

Takahashi (1979) has isolated several other stocks with altered swimming behavior for which genetic data are lacking or incomplete (Table I). "Temperature shock behavioral" responds to rapid elevation in temperature with a more prolonged avoiding reaction than does wild type. "Frequently reverse" displays spontaneous avoiding reactions much more frequently than wild type, reminiscent of paranoiac. In addition, frequently reverse undergoes prolonged backward swimming in potassium-

or barium-rich solutions. "Ba^{2+}-sensitive" undergoes prolonged continuous ciliary reversal before displaying the train of short ciliary reversals characteristic of wild type in Ba^{2+}. When spinner encounters the edge of a culture dish, it spins rapidly around its long axis rather than backing. "Zig-zag swimmer" shows transient and frequent stopping during forward swimming, not unlike atalanta in *P. tetraurelia*. However, whereas atalanta is probably altered in its ciliary motility system (Kung, personal communication), zig-zag swimmer probably is not. Triton-extracted models of zig-zag swimmer act like wild type. Finally, there is a "TEA$^+$-insensitive" mutant which, like that mutant in *P. tetraurelia*, shows no avoiding reaction in TEA$^+$ solution.

Further genetic and physiological studies of these mutants are anticipated. Improvements in genetics of *P. caudatum* now being developed (Tsukii and Hiwatashi, 1977 and personal communication) and the large literature on the physiology of *P. caudatum* indicate that these mutants promise important advances in our understanding of swimming behavior in *Paramecium*.

E. *Tetrahymena*

Although well-known biochemically, the behavior genetics of *Tetrahymena* is yet to be developed. This may be because of the small size of the cells. However, some behavior mutants are known (McCoy, 1977). The phenotypes of the mutants include a temperature-sensitive motility mutant, a nonconditional motility mutant, and a number of "monsters" and other shape mutants which probably are altered in swimming behavior.

III. MATING BEHAVIOR

A. Outline of Mating Events

1. *Immaturity*

Clones of ciliates pass through stages of development. The stages differ in detail for different species, and the time of onset of the stages also differs (see summary in Bleyman, 1971). However, the general features of the stages of clonal history are the same. Duration of stages is usually measured in numbers of fissions, although some stocks of *P. caudatum* and *Paramecium multimicronucleatum* use a combination of fissions and clock time in determining the length of stages (Takagi, 1970). The clonal life history begins with conjugation, and the first stage after conjugation is called immaturity, during which time cells are capable of neither conjugation nor autogamy. None of the aspects of mating behavior to be described in the next sections is seen during immaturity. According to Bleyman's (1971) review, species of the *P. aurelia* complex may lack immaturity, and in *P. caudatum* and *P. multimicronucleatum* maturity may last 40–90 fissions. In *Tetrahymena thermophila* (formerly *T. pyriformis*, syngen 1, Nanney and McCoy, 1976), 40–120 fissions ensue during immaturity. Differences between species of the *P. aurelia* complex may be found in Sonneborn (1957).

2. Clumping and Conjugation

The end of immaturity comes when mating behavior is seen. The transition from immaturity to maturity is somewhat gradual, with all of the cells in a population becoming mature over a time of several fissions in *P. primaurelia* Siegel 1961) and in *P. multimicronucleatum* and *P. caudatum* (Takagi, 1971). During the transition in certain stocks of *P. caudatum* three different characteristics of cells which are mature appear sequentially. First, cells are able to undergo mating reactions, then they can both undergo mating reactions and form holdfast unions, and finally both of the first two plus paroral unions became possible (Takagi, 1971). This sequential appearance of mating characteristics is not seen in all stocks of *P. caudatum* (Hiwatashi, personal communication).

Mating behavior begins after cells develop complementary "mating substances" on their cilia. Such cells are said to be "mating reactive." The mating substances are almost certainly proteins (Metz, 1954; Kitamura and Hiwatashi, 1978) which are transported to the surface of the cell from the interior (Takahashi and Hiwatashi, 1974). In order for *Paramecium* to develop mating substances, cells must, in addition to being mature, be neither too starved nor too well fed (Sonneborn, 1937, 1939). When mature, starving cells of complementary mating type are mixed together, the first step in mating behavior is the "mating reaction," in which cells of complementary type adhere by their cilia in clumps of variable size. The clumping is presumably the result of specific interactions between mating-type substances of complementary type.

As with swimming behavior, clumping can be described in terms of the shape of the cell and the distribution of cilia on the cell. Most, if not all, of the mating-reactive cilia are on the oral side of the cell. This was demonstrated by Hiwatashi (1961), who cut mating-reactive *P. caudatum* into oral and aboral halves and found that only the oral half and the anterior tip of the aboral half were mating reactive. In *P. multimicronucleatum* (Miyake, 1964) and *Paramecium bursaria* (Cohen and Siegel, 1963), mating-reactive cilia isolated from one mating type agglutinated only on the oral side of mating-reactive cells of complementary mating type.

Jennings (1906) points out that localization of mating reactivity on the oral surface makes sense in terms of swimming behavior. There is a stronger current of water moving on the oral side so that two cells which come close to each other by their oral sides will be drawn toward each other, but cells coming close by aboral sides will be drawn away from each other. Wild-type paramecia, having joined together in mating reactions, sometimes show avoiding reactions and pull away from the clump.

An ordered series of events follows the mating reaction. Time schedules for these events may be found for *P. aurelia* of undesignated species in Jurand and Selman (1969) and for *P. caudatum* syngen 3 in Mikami and Hiwatashi (1975). Both external changes and changes in the nuclei occur during conjugation. The nuclear events, although of course the main feature of conjugation, are not at issue in this chapter and will not be described in detail. However, the external events are of some interest, and constitute that which can be called "mating behavior" in *Paramecium*.

In *P. caudatum* cilia begin to degenerate on the anterior end of the oral surface so that a deciliated region grows in a posterior direction as mating proceeds. This has been observed in *P. caudatum* with the light microscope (Hiwatashi, 1961), and excellent scanning electron microscopic observations have been made by Watanabe (1978) (Figure 4). Miyake (1966) also observed the loss of cilia in *P. multimicronucleatum* using the light microscope. In 40–60 min after mixing complementary mating types, cells adhere at the anterior deciliated tips in a "hold fast union." The clumps of cells in the mating reaction break down into individual hold fast pairs. Between 1 and 2 hr after mixing complementary mating types, the members of a pair adhere more intimately with a second union of deciliated surfaces just posterior to the gullet. This is called the "paroral union." In *P. aurelia* the cells separate well before the micronuclei undergo the first postzygotic division, about 5–6 hr after mixing of complementary mating types (Jurand and Selman, 1969). *Paramecium caudatum* remains united in pairs until the second postzygotic division, separating after 15 hr.

Pairs of cells swim in an open left-handed spiral much as single cells do (Jennings, 1906). Pairs give an avoiding reaction when stimulated. In pairs of which one member cannot reverse because it is a *pawn* (*P. tetraurelia*) or CNR (*P. caudatum*), behavior is as expected if each member of the pair were reacting independently. When stimulated with chemicals, the pair swims backward in a very wide spiral or swims in circles (Berger, 1976; Takahashi, 1979).

Conjugation among mating-reactive cells of a single mating type may be induced by chemical solutions in which the concentration of Ca^{2+} is low

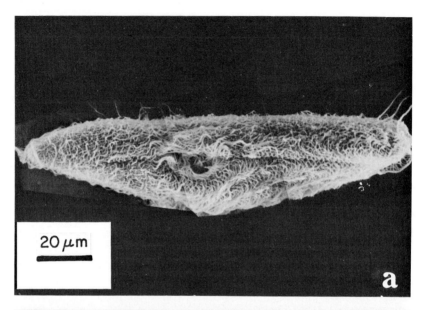

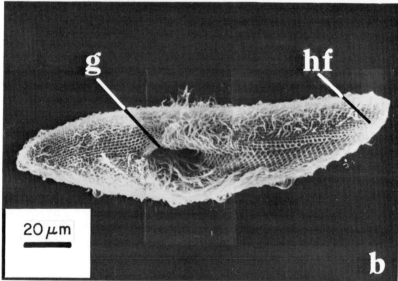

Figure 4. Scanning electron micrograph of oral surface of *P. caudatum* not involved in mating (a) and 90 min after the beginning of a mating reaction (b). In (b) cilia have been lost all along the midline of the oral surface, but especially in the anterior holdfast region (hf). The opening of the gullet is marked g. (Photo courtesy of T. Watanabe, Tohoku University.)

relative to that of any of a number of other cations including K$^+$, Na$^+$, Mn^{2+}, Mg^{2+}, and NH$_4^+$ (Miyake, 1958, 1968a; Miyake and Okamoto, 1958; Hiwatashi, 1959; Hiwatashi and Kasuga, 1960). Not all inducing solutions are effective on all species. A history of the development of these techniques may be found in Miyake (1968a) and Hiwatashi (1969).

In chemically induced conjugation the cells must be mating reactive, but interaction of complementary mating types does not occur. No mating reaction is seen, but after 40–60 min holdfast pairs appear in the culture, and the other events of conjugation appear normal. During chemical induction of mating in *P. caudatum* cilia are lost in much the same pattern as when normal mating occurs (Watanabe, 1978).

An interesting swimming pattern has been reported for cells subjected to chemically induced mating (Miyake, 1958; Tsukii and Hiwatashi, 1978), and this pattern has been analyzed by Cronkite (1972). Cells settle to the bottom of culture vessels and whirl their anterior ends in wide circles while holding the posterior end stationary. Cronkite described this behavior by counting the number of cells of *Paramecium octaurelia* involved in these movements which were touching the bottom of the culture depression during chemical induction. Figure 5 shows the results for wild-type cells. After an initially large number of cells whirled at 5 min, most cells began to swim normally again, reflected by a low number of whirlers at 20 min. Then cells again settled to the bottom and reached a second peak of

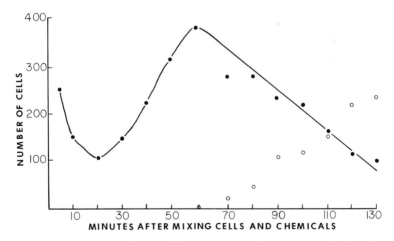

Figure 5. Stationary whirling of wild-type *P. octaurelia* (Stock 31) during chemical induction of mating. At Time 0 mating-reactive cells of one mating type were mixed with 20 mM KCl and 1 mg% acriflavine. Closed circles represent cells whirling in place; open circles indicate cells in holdfast union. All points are means of five identically designed experiments. (From Cronkite, 1972.)

stationary whirling at 60 min, just when the first holdfast pairs began to form. The number of stationary whirlers then decreased at about the same rate as the number of pairs increased.

The large number of cells whirling in partial ciliary reversal at the first observation 5 min after mixing with chemicals was not unexpected. The high concentration of K^+ relative to Ca^{2+} should stimulate such behavior. Yet the second peak of stationary whirling is an event not described by students of swimming behavior. The second peak was only seen in cultures of mating-reactive cells (Cronkite, 1972). It would be of interest to look for electrophysiological correlates which might account for the mating-reactive-specific behavior described here. It is also possible that the loss of cilia in early mating alters the swimming pattern.

Mating behavior in *Tetrahymena* is in many ways quite unlike that of *Paramecium,* and the specifics of mating are not within the scope of this chapter. However, recent studies of early events of conjugation using the scanning electron microscope (Wolfe and Grimes, 1979) emphasize that the loss of cilia preceding pair formation follows a pattern similar to that of *Paramecium.* In "tip transformation" the cilia are lost from the anterior end of the oral surface. Then the ridged surface that remains smooths out, and cells of complementary mating type join in the deciliated region.

3. Autogamy

Species of *P. aurelia* follow the period of maturity with a period of "senescence," in which autogamy as well as mating reactivity occurs upon starvation. The nuclear events of autogamy are like those of conjugation except that the result is self-fertilization rather than exchange of nuclei. Sonneborn and Dippell (personal communication) see no loss of cilia during normal autogamy of *P. tetraurelia.* However, they were able to induce nuclear reorganization in single cells using a chemical solution like that used for induction of mating among cells of a single mating type. In such cases cilia loss was seen as in normal mating.

Although autogamy in *P. caudatum* and *P. multimicronucleatum* cannot be induced by starvation alone, chemical induction of autogamy is possible using the same medium as is used in chemical induction of conjugation plus a protease such as papain or ficin (Miyake, 1968b; Tsukii and Hiwatashi, 1977). Tsukii and Hiwatashi used genetic markers to show that, for *P. caudatum* at least, true autogamy occurred, giving the expected phenotypic ratios. As in normal conjugation and autogamy, cells undergoing chemical induction of autogamy dedifferentiate their gullet region during autogamy and replace it after the process is over. Cells swim as they do in chemical induction of mating (Tsukii and Hiwatashi, 1978) and they lose cilia as in normal mating (Watanabe, 1978).

Presumably the precise pattern of development of the clonal life history through immaturity, maturity, and senescence is under genetic control, as is the ordered sequence of events of mating behavior. How does a cell count fissions? What triggers the set of events of conjugation? How are chemical induction of mating and induction of mating with interaction of mating types related? These questions are subject to a variety of different kinds of attacks. The genetic dissection of the control of mating behavior is not as well developed as is the genetic dissection of swimming behavior. But a beginning has been made with many suggestive insights already forthcoming, and the intimate interweaving of swimming behavior and mating behavior already noted in this outline of mating events allows us to exploit some of the work on genetics of swimming behavior in an attack upon mating behavior.

B. Genetics of Mating Reactivity, Conjugation, and Autogamy

A number of mutations affecting mating type determination have been described (Butzel, 1955; Taub, 1966; Byrne, 1973; Brygoo, 1977; Hiwatashi, 1968; Hiwatashi and Myohara, 1976), and "early mature" mutants with shortened immature periods are known in *T. thermophila* (Bleyman and Simon, 1968; Bleyman, 1971, 1972) and *P. caudatum* (Myohara and Hiwatashi, 1978). However, only a few mutations of the process of mating itself have been reported. Such mutations are not readily amenable to genetic analysis if they interfere unconditionally with mating. Only when mating can be conditionally blocked, allowing it under some restricted set of conditions, is a thorough analysis possible. Nevertheless hereditary blocks of mating have provided important information about the initiation of mating and autogamy even when complete genetic information was not available (Table III).

A line of *P. tetraurelia* was recovered (Cooper, 1965) which did not become mating reactive under any conditions studied. In a sense the stock is a permanently immature one. Yet the cells undergo autogamy after the requisite number of fissions, indicating that the block is restricted specifically to the production of mating substances or to their transport to surface sites. Chemical induction of mating has been unsuccessful when tried on well-fed, moderately starved, and starved cells (Cronkite, unpublished observations). These cells swim forward faster than wild type (Cooper, 1965). The fast behavior resembles that of fast-1 of Kung (1971a), since the cells show avoiding reactions in either potassium or sodium solutions (Cronkite, unpublished observations). Kung (1971a) has found some mutants that behave like fast-1 but which fail to mate. Takahashi and Hiwatashi (1974) reported that K^+ played a role in the transport of mating

Table III Mutants of Mating Behavior in *P. aurelia*

Mutant	Genetic basis	Mating reactivity	Hold-fast unions	Chemical induction	Autogamy	References
Cooper's fast	Inherited asexually	−	−	−	+	Cooper (1965); Cronkite (unpublished)
Fast-1	Inherited asexually[a]	−	−	Not tested	+	Kung (1971a,b)
Cm	Inherited asexually	+	−	−	+	Metz and Foley (1949); Miyake (1968a)
Stacatto	Inherited asexually[a]	+	−	−	+	Kung *et al.* (1975); Cronkite (unpublished)
kau-1	Nuclear recessive	+	+	−	+	Cronkite (1974, 1975)
kau-2	Nuclear recessive	+	+	−	+	Cronkite (1974, 1975)
pwB	Nuclear recessive	+	+	−	+	Kung (1971a,b); Cronkite (1976)
pwC						
32°C	Nuclear recessive	+	+	−	+	Satow *et al.* (1974);
23°C		+	+	+	+	Cronkite (1976)

[a] Some Fast-1 and stacatto mutants will mate, are nuclear recessives, and act like the wild type in all categories of the table.

substances to the surface of cells from their site of synthesis inside the cell. It would be interesting to study the potassium permeability of Cooper's fast mutant, especially since it has a fast behavior coupled with the inability to mate. A lesion in potassium transport could conceivably explain both the swimming behavior and the lack of mating behavior of these cells.

Another set of stocks of *Paramecium primaurelia* and *P. tetraurelia* (Sonneborn, 1942; Metz and Foley, 1949) produce mating substances and are thus capable of mating reactions. But the reacting cultures fail to form holdfast pairs or in any other way go beyond the stage of clumping. These stocks, called "Cannot Mate" or CM stocks, are nevertheless capable of autogamy. Kung (personal communication) reports that some "stacatto" stocks, like CM, give mating reactions but fail to form pairs.

Metz and Foley (1949) used a CM stock in an analysis of the initiation of mating. Metz (1947) had found that living animals of one mating type could clump with killed animals of complementary type and would then form so-called pseudoselfing pairs of two living cells. These pairs consisted of cells of the same mating type. Pseudoselfing pairs were capable of the entire sequence of mating events, including meiosis of the micronuclei and macronuclear breakdown. When live wild-type cells were mixed with killed CM of complementary type, these cells formed pseudoselfing pairs. But when live CM were mixed with dead wild-type cells, only clumping occurred. Thus CM animals are incapable of forming holdfast unions, yet induce these unions in normal cells. The initial interaction of mating substances during the mating reaction is apparently sufficient to initiate all of the sequence of steps characteristic of mating. That both Cooper's fast and the CM stocks are capable of autogamy indicates a separate route of initiation for that process and conjugation. These conclusions are not clearly applicable to *Paramecium* species other than *P. tetraurelia*. In *P. caudatum* (Hiwatashi, 1955) the mating reaction is not sufficient to induce all of the remaining events of conjugation.

More detailed genetic analysis of mating behavior became possible when mutants were discovered in *P. octaurelia* (Cronkite, 1974, 1975) that were blocked in chemical induction of mating but that mated normally when mixed with cells of complementary mating type. Three genes have so far been identified, two of them unlinked recessive genes (*kau-1* and *kau-2*), either of which will block chemical induction by solutions containing KCl and acriflavine or $MgCl_2$ and acriflavine. The third gene is a dominant suppressor [*Su(kau-2)*] which restores partial wild-type characteristics to cells homozygous for *kau-2* (Cronkite, 1975).

The analysis of these mutants emphasizes the overlap between control of swimming behavior and mating behavior. Genes *kau-1* and *kau-2*, which block chemically induced mating, also interfere with the pattern of stationary whirling associated with chemical induction (Figure 6). When cells are homozygous for *kau-1,* neither the first nor second peak of stationary whirling occurs. Cells homozygous for *kau-2* do show the first peak of whirling, but no second peak is seen. These cells reach a low at 20 min just as in the wild type, but remain low for the rest of the experiment.

The conditions necessary for chemical induction parallel those which stimulate avoiding reactions in *Paramecium,* low concentration of Ca^{2+} relative to some other cation such as K^+ or Na^+. That partial ciliary reversal or whirling is lacking in *kau-1* cells and that the second peak of whirling is lacking in *kau-2* cells suggests that uninducible mutants may have a lesion in the calcium activation mechanism. Two additional lines of evidence strengthen this possibility.

The duration of backward swimming is related to the flow of Ca^{2+} across the cell surface membrane following depolarization (Eckert, 1972). Since a critical internal concentration of Ca^{2+} is needed for ciliary reversal (Naitoh and Kaneko, 1972, 1973), an impaired calcium conductance should result in a shorter duration of ciliary reversal. Cronkite (1976) measured the duration of backward swimming of mutant and wild-type *P. octaurelia* in components of a chemical induction medium effective for that species (20 m*M* KCl or 1 mg% acriflavine or both). The results, summarized in Table IV, show that both *kau-1* and *kau-2* act as "acriflavine pawns," eliminating backward swimming stimulated by 1 mg%

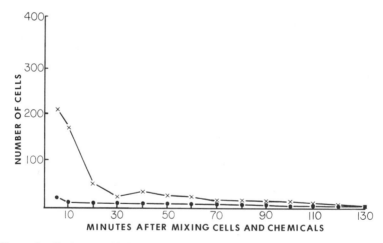

Figure 6. Stationary whirling in mutants of *P. octaurelia* homozygous for *kau-1* (closed circles) and *kau-2* (X). At Time 0 mating-reactive cells of one mating type were mixed with 20 m*M* KCl and 1 mg% acriflavine. Points are means of five identically designed experiments.

Table IV Mean Duration of Ciliary Reversal for Wild Type and *kau* Mutants[a,b]

Test solution	Wild type	*kau-1*	*kau-2*
20 m*M* KCl	26.7	27.9[c]	24.3
20 m*M* KCl + 1 mg% acriflavine	45.7	35.1	34.1
1 mg% acriflavine	6.2	—[d]	—[d]

[a] Mean duration in seconds.

[b] Reprinted from Cronkite (1975), by permission of *Genetics* and the Genetics Society of America.

[c] A line drawn below two or more means indicates that those means are not significantly different at 95% confidence level. Those not underlined are significantly different from the other measurements in the row (Student–Newman–Keul multiple range test was applied).

[d] Of 1050 cells of each stock tested none swam backward.

acriflavine and decreasing the duration of backward swimming in KCl + acriflavine.

Little is known about the behavioral physiology of *P. octaurelia*, nor is there information on the effect of acriflavine on swimming behavior. Mutants of *P. tetraurelia* of known physiological effect were studied for their chemical inducibility (Cronkite, 1976). Mutants homozygous for *pawn* (*pwB, pwC*), *fast-2* (*fna*), and *atalanta* were cultured for high mating reactivity and mixed with 20 m*M* KCl + 1 mg% acriflavine. Wild type, fast-2, atalanta, and ts-pawn (*pwC*) at 23°C were all capable of chemical induction of mating. Cells of *pwB* were never successfully induced to mate with chemicals nor were cells of *pwC* at 32°C (Table V). At 32°C chemical induction of mating in wild type was possible. Chemical induction occurred, in other words, in all of those lines in which calcium activation was not blocked. The lesion in atalanta that prevents it from swimming backward is not in the calcium activation system, but probably in the ability of the cilia to respond to raised levels of internal Ca^{2+} (Kung, personal communication). Atalanta was used as a control to eliminate the possibility that backward swimming per se and not calcium activation was the necessary event for chemical induction. So calcium activation probably is a necessary step in chemically induced mating. Takahashi's report (personal communication) that chemical induction is possible for some CNR in *P. caudatum* makes the conclusion less than certain, however. Other *pw* loci, including leaky ones, should be studied for susceptibility to chemical induction.

Normal mating is not impaired in any of the mutants incapable of chemical induction. We still must ask what the relationship is of chemical induction to induction of mating by interaction of mating types and what

Table V Chemical Induction of Various Genotypes of *P. tetraurelia*[a]

Stock	Temper- ature (°C)	Active membrane depolar- ization[b]	Backward swim- ming behavior	Chemical induction of mating
51s (wild type	23	+	+	+
	32	+	+	+
d4-94 (pawn A)	23	−	−	−
7-2-34 (ts pawn C)	23	+	+	+
	32	−	−	−
d4-91 (fast-2)	23	+	+	+
10-3-35 (atalanta)	23	+	−	+

[a] Reprinted from Cronkite (1976), by permission of *Journal of Protozoology* and The Society of Protozoologists.

[b] Data on membrane depolarization are from Kung *et al.* (1975) and Satow *et al.* (1974). The presence of active membrane depolarization is taken to indicate that a transient inward flow of Ca^{2+} across the cell membrane is possible when cells are mixed with solutions high in the concentration of K^+ relative to the concentration of Ca^{2+} (Eckert, 1972).

role, if any, is played by calcium activation in normal mating. No further genetic evidence bears on these issues, but some other analyses shed light on the matter when considered with the genetic studies.

In vegetative cells of *P. caudatum* the micronucleus sits in a pocket of the macronucleus. However, when mating is induced by interaction of mating types or by chemicals the micronucleus moves out of the pocket and into the cytoplasm (Fujishima and Hiwatashi, 1977). This "early micronuclear migration" (EMM) is the earliest event so far described which follows the initial mating reaction or the mixing of mating reactive cells with chemicals (Figure 7). Among cells which will show very high frequency of conjugation, 30% of the cells show EMM by 10 min after the beginning of the induction of mating, and by 30 min nearly 90% can show EMM. Immature cells related to those mature cells could not be induced to mate either by interaction of mating substances or by chemical induction, and no EMM was seen in these cells.

Because calcium activation is implicated as a necessary step in chemical induction of mating, the effect of changing calcium concentration on EMM has been studied (Cronkite, unpublished observations). Mating-reactive *P. caudatum* were cooled in an ice bath for 60 min to allow Ca^{2+} accumulation (Browning and Nelson, 1976; Browning *et al.*, 1976) in a salt solution which stimulates backward swimming and maintains strong mating reactivity (1.5 mM NaCl, 1.55 mM KCl, 0.1 mM $MgCl_2$, 0.01 mM $CaCl_2$, 10 mM Tris, pH 7.2). After 60 min the cells were rapidly warmed to room temperature, and at intervals after warming samples were fixed and

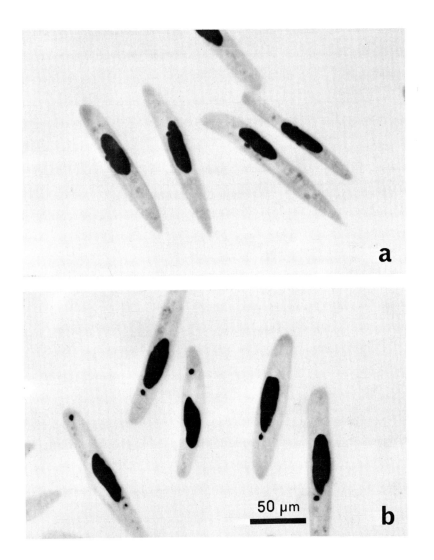

Figure 7. Early micronuclear migration in *P. caudatum*. Cells in stationary phase (a) have micronucleus in a pocket of macronucleus. Cells 30 min after beginning of mating reaction (b) have micronucleus in cytoplasm. Pictures are of cells fixed with Carnoy solution and stained with Feulgen's stain (From Fujishima and Hiwatashi, 1977, reproduced by permission of *J. Exp. Zool.* and Wistar Press.)

stained with Feulgen's stain to look at EMM. The results given in Table VI indicate that the cold treatment does not increase the frequency of EMM, but that cells that have been warmed after cold treatment undergo EMM at frequencies higher than those of controls. Two interpretations

Table VI Percentage of Early Micronuclear Migration after
Cold Treatment of Mating-Reactive *P. caudatum*[a]

Time after treatment (min)	Control	Cold-treated
0	3.7 ± 1.1	5.7 ± 1.7
15	3.3 ± 1.2	21.5 ± 7.8
30	3.2 ± 0.6	16.2 ± 5.6
60	3.7 ± 1.7	12.5 ± 4.9
90	4.3 ± 2.2	14.9 ± 2.1
120	3.9 ± 1.2	9.7 ± 2.6

[a] Figures are means $\pm$ 1 SD for three identically designed experiments. Mating reactive cells were kept at 0°C for 60 min and were then rapidly warmed to 24°C at Time 0. Samples of cells were fixed with Carnoy's solution and stained with Feulgen's stain to observe early micronuclear migration. From Cronkite (unpublished observations).

are possible. While in the cold, energy may not be available to move the micronucleus. Return to room temperature would then mobilize an already triggered EMM. Alternatively, the accumulation of Ca^{2+} during cold treatment may not trigger EMM. Instead, the increase of Ca^{2+} concentration followed by the decrease when cells return to room temperature may be the triggering event. Further experiments are underway to clarify this point.

Cells were also treated with the calcium ionophore A23187 (gift of Dr. R. L. Hamill, Lilly Research Laboratories, Indianapolis, Indiana) after they had been washed in the same solution used in cold treatment. Treatment with ionophore increases significantly the frequency of EMM in mating-reactive cells. Interestingly, immature cells related to the mature cells used in this study do not respond either to cold treatment or to ionophore treatment with an increase in the frequency of EMM. Immaturity, which is seen operationally as a lack of mating substances on the surface of cells, is probably a more profound difference in cellular physiology and architecture. The mechanism of movement of the micronucleus could only be very indirectly related to the appearance of mating substances on the cell surface.

Conjugation induced either by interaction of mating substances or by chemicals is blocked by concanavalin A (Tsukii and Hiwatashi, 1978). Neither the mating reaction nor the whirling behavior seen in chemical induction is inhibited by concanavalin A, and α-methyl-D-mannoside prevents Con A inhibition. There is a 30-min sensitive period for Con A inhibition at the beginning of the induction of mating, after which the lectin has no effect. There must be essential changes in the membrane in early mating which are inhibited when Con A binds to specific receptors on the surface. Further studies of this point should be fruitful, and contribute to a general understanding of the mechanism of initiation of mating.

C. Conclusions

The genetic and physiologic investigations of the initiation of mating reveal certain facts which must be taken into account in any understanding of mechanisms. The results of studies of mutants of mating behavior show clearly that knowledge of ion transport which has contributed so much to our understanding of the control of swimming behavior will have to be brought to bear in any attempt to explain mating behavior. But such knowledge as it now exists is not sufficient to account for all of the facts about mating.

A necessary condition for mating is maturity, seen as the ability to produce mating substances, but including the ability to perform other cell functions not observed in immature cells such as EMM. A second necessary condition is semistarvation. Little information on variations of ion transport phenomena with nutritional state or clonal history is available. If fluxes of ions play a role in controlling mating behavior, then studies of such variations could be of great value.

At least in species of the *P. aurelia* complex, a necessary condition for chemical induction of mating is a flux of Ca^{2+} into the cell at the very beginning of induction. The flux is almost certainly not a necessary condition of mating induced by interaction of mating substances since *pawn* cells mate quite well. It is possible that some internal cellular state ordinarily altered when mating substances interact is also altered by temporary increases in internal calcium concentration. For instance, the triggering event for the subsequent mating events may be the release of an inhibitor inside the cell, with the release occurring in either of two ways.

The finding by Takahashi that CNR of *P. caudatum* are sometimes chemically inducible suggests that initiation of mating in that species may proceed in a manner different from that in *P. aurelia*. Hiwatashi's (1955) finding that interaction of mating substances is not sufficient to trigger mating in *P. caudatum* suggests this as well. Differences in culture conditions have been cited in this chapter as one source of observed physiological differences in the two species, but there may be differences of a more profound nature between the two species. Further comparative studies of *P. aurelia* and *P. caudatum* should provide insight into the general problems of control of mating behavior. In addition, studies of *P. bursaria*, which has not been successfully induced to mate chemically, should provide yet other dimensions to our knowledge of mating behavior.

Autogamy in *P. tetraurelia* is not dependent on the same triggering events as mating. Cooper's fast mutant undergoes autogamy, demonstrating that the appearance of mating substances on the cell surface is not necessary for induction of autogamy. Even chemical induction of mating requires the presence of these substances. In fact, Cooper's fast mutant is not chemically inducible. The appearance of autogamy in CM mutants shows that steps after the production of mating substances are also not required in inducing autogamy. And the appearance of autogamy in kau mutants shows that chemical induction of mating also proceeds along a pathway at least initially unlike the pathway of induction of autogamy. No mutants of autogamy have been reported. Such mutants would be of great value in seeing how independent of other processes autogamy actually is.

The powerful technique of genetic dissection which has been so successful in helping to unravel the control of swimming behavior has yet to be applied in any systematic way to a study of mating behavior. *Paramecium tetraurelia* is now sufficiently well known genetically that screening techniques can be devised to detect mutants of mating. Temperature-sensitive mating mutants would be of value. In addition, the intimate connection between mating behavior and swimming behavior suggests that further pursuit of the details of ion transport phenomena will also yield information of importance in understanding the control of mating behavior.

ACKNOWLEDGMENTS

I thank Ching Kung, T. M. Sonneborn, Ruth Dippell, Mihoko Takahashi, Koichi Hiwatashi, and Jason Wolfe for invaluable sharing of unpublished work and work in press. In addition, critical reading of the chapter in various stages of progress was done by Barbara Byrne, Bruce Byrne, Ching Kung, Akio Kitamura, and Koichi Hiwatashi, and I gratefully acknowledge their assistance.

Unpublished data of DLC were obtained during support by an Exchange of Persons Grant from the Japan Society for the Promotion of Science and by Grant No. 154204 to K. Hiwatashi from the Ministry of Education in Japan.

REFERENCES

Beale, G. H. (1954). "The Genetics of *Paramecium aurelia*." Cambridge Univ. Press, London and New York.
Beisson, J., and Sonneborn, T. M. (1965). *Proc. Natl. Acad. Sci. U.S.A.* **53**, 275–282.
Benzer, S. (1967). *Proc. Natl. Acad. Sci. U.S.A.* **58**, 1112–1119.

Berger, J. D. (1973). *Chromosoma* **42**, 247–268.
Berger, J. D. (1974). *J. Protozool.* **21**, 145–152.
Berger, J. D. (1976). *Genet. Res.* **27**, 123–134.
Bleyman, L. K. (1971). *In* "Developmental Aspects of the Cell Cycle" (I. L. Cameron, G. M. Padilla, and A. M. Zimmerman, eds.), pp. 67–91. Academic Press, New York.
Bleyman, L. K. (1972). *Genetics* **71**, 55–56.
Bleyman, L. K., and Simon, E. M. (1968). *Genet. Res.* **10**, 319–321.
Brehm, P., and Eckert, R. (1978). *Science* **202**, 1203–1206.
Browning, J. L., and Nelson, D. L. (1976). *Biochim. Biophys. Acta* **448**, 338–351.
Browning, J. L., Nelson, D. L., and Hansma, H. G. (1976). *Nature (London)* **259**, 491–494.
Brygoo, Y. (1977). *Genetics* **87**, 633–653.
Butzel, H. M., Jr. (1955). *Genetics* **40**, 321–330.
Byrne, B. C. (1973). *Genetics* **74**, 63–80.
Byrne, B. J., and Byrne, B. C. (1978a). *Science* **197**, 1091–1093.
Byrne, B. J., and Byrne, B. C. (1978b). *Crit. Rev. Microbiol.* **6**, 53–108.
Chang, S.-Y., and Kung, C. (1973a). *Science* **180**, 1197–1199.
Chang, S.-Y., and Kung, C. (1973b). *Genetics* **75**, 49–59.
Chang, S.-Y., and Kung, C. (1976). *Genet. Res.* **27**, 97–107.
Chang, S.-Y., Van Houten, J., Robles, L. J., Lui, S. S., and Kung, C. (1974). *Genet. Res.* **23**, 165–173.
Cohen, L. W., and Siegel, R. W. (1963). *Genet. Res.* **4**, 143–150.
Cooper, J. (1965). *J. Protozool.* **12**, 381–384.
Cronkite, D. L. (1972). Ph.D. Thesis, Indiana Univ., Bloomington.
Cronkite, D. L. (1974). *Genetics* **76**, 703–714.
Cronkite, D. L. (1975). *Genetics* **80**, 13–21.
Cronkite, D. L. (1976). *J. Protozool.* **23**, 431–433.
Dawson, J. A. (1926). *J. Exp. Zool.* **44**, 133–157.
Dawson, J. A. (1928). *Science* **68**, 258.
Dryl, S. (1959). *J. Protozool.* **6**, Suppl., p. 96.
Dryl, S. (1961). *J. Protozool.* **8**, Suppl., p. 16.
Dute, R., and Kung, C. (1977). *J. Cell Biol.* **75**, 211a.
Eckert, R. (1972). *Science* **176**, 473–481.
Eckert, R. (1977). *Nature (London)* **268**, 104–105.
Eckert, R., Naitoh, Y., and Machemer, H. (1976). *In* "Calcium in Biological Systems" (C. J. Duncan, ed.), pp. 233–255. Cambridge Univ. Press, London and New York.
Friedman, K., and Eckert, R. (1973). *Comp. Biochem. Physiol. A* **45**, 101–114.
Fujishima, M., and Hiwatashi, K. (1977). *J. Exp. Zool.* **201**, 127–134.
Grebecki, A. (1965). *Acta Protozool.* **3**, 275–289.
Hansma, H. G., and Kung, C. (1976). *Biochim. Biophys. Acta* **436**, 128–139.
Hiwatashi, K. (1955). *Sci. Rep. Tohoku Univ., Ser. 4* **21**, 207–218.
Hiwatashi, K. (1959). *Sci. Rep. Tohoku Univ., Ser. 4* **25**, 81–90.
Hiwatashi, K. (1960). *Jpn. J. Genet.* **35**, 213–221.
Hiwatashi, K. (1961). *Sci. Rep. Tohoku Univ., Ser. 4* **27**, 93–99.
Hiwatashi, K. (1968). *Genetics* **58**, 373–386.
Hiwatashi, K. (1969). *In* "Fertilization" (C. B. Metz and A. Monroy, eds.), Vol. 2, pp. 255–293. Academic Press, New York.
Hiwatashi, K., and Kasuga, T. (1960). *J. Protozool.* **7**, Suppl., 20–21.
Hiwatashi, K., and Myohara, K. (1976). *Genet. Res.* **27**, 135–141.
Jennings, H. S. (1906). "Behavior of the Lower Organisms." Indiana Univ. Press, Bloomington.

272   Donald L. Cronkite

Jennings, H. S. (1908). *J. Exp. Zool.* **34**, 339–384.
Jennings, H. S. (1913). *J. Exp. Zool.* **14**, 279–391.
Jurand, A., and Selman, G. G. (1969). "The Anatomy of *Paramecium aurelia*." St. Martins Press, London.
Kimball, R. F. (1970). *Mutat. Res.* **9**, 261–271.
Kinosita, H., Dryl, S., and Naitoh, Y. (1964a). *J. Fac. Sci. Univ. Tokyo, Sect. 4* **10**, 291–301.
Kinosita, H., Dryl, S., and Naitoh, Y. (1964b). *J. Fac. Sci. Univ. Tokyo, Sect. 4* **10**, 303–309.
Kinosita, H., Murakami, A., and Yasuda, M. (1965). *J. Fac. Sci. Univ. Tokyo, Sect. 4* **10**, 421–425.
Kitamura, A., and Hiwatashi, K. (1978). *J. Exp. Zool.* **203**, 99–108.
Kung, C. (1971a). *Z. Vgl. Physiol.* **71**, 142–164.
Kung, C. (1971b). *Genetics* **69**, 29–45.
Kung, C. (1976). *In* "Cell Motility" (R. D. Goldman, T. D. Pollard, and J. L. Rosenbaum, eds.), pp. 941–948. Cold Spring Harbor Lab., Cold Spring Harbor, New York.
Kung, C., and Eckert, R. (1972). *Proc. Natl. Acad. Sci. U.S.A.* **69**, 93–97.
Kung, C., and Naitoh, Y. (1973). *Science* **179**, 195–196.
Kung, C., Chang, S.-Y., Satow, Y., Van Houten, J., and Hansma, H. (1975). *Science* **188**, 898–904.
McCoy, J. W. (1977). *Genetics* **87**, 421–439.
Machemer, H. (1972). *J. Exp. Biol.* **57**, 239–259.
Machemer, H., and Eckert, R. (1975). *J. Comp. Physiol.* **104**, 247–260.
Metz, C. B. (1947). *J. Exp. Zool.* **105**, 115–140.
Metz, C. B. (1954). *In* "Sex in Microorganisms" (D. H. Wenrich, ed.), pp. 284–334. Am. Assoc. Adv. Sci., Washington, D.C.
Metz, C. B., and Foley, M. R. (1949). *J. Exp. Zool.* **112**, 505–528.
Mikami, K., and Hiwatashi, K. (1975). *J. Protozool.* **22**, 536–540.
Miyake, A. (1958). *J. Inst. Polytech., Osaka City Univ., Ser. D* **9**, 251–296.
Miyake, A. (1964). *Science* **146**, 1583–1585.
Miyake, A. (1966). *J. Protozool.* **13**, Suppl., p. 28.
Miyake, A. (1968a). *J. Exp. Zool.* **167**, 359–380.
Miyake, A. (1968b). *Proc. Int. Congr. Genet., 12th, Tokyo* **1**, 72.
Miyake, A. (1974). *Curr. Top. Microbiol. Immunol.* **64**, 49–77.
Miyake, A. (1978). *Curr. Top. Dev. Biol.* **12**, 37–82.
Miyake, A., and Okamoto, C. (1958). *Physiol. Ecol.* **8**, 9–11.
Myohara, K., and Hiwatashi, K. (1975). *Jpn. J. Genet.* **50**, 133–139.
Myohara, K., and Hiwatashi, K. (1978). *Genetics* **90**, 227–241.
Naitoh, Y., and Eckert, R. (1968). *Z. Vgl. Physiol.* **61**, 453–472.
Naitoh, Y., and Eckert, R. (1969). *Science* **164**, 963–965.
Naitoh, Y., and Kaneko, H. (1972). *Science* **176**, 523–524.
Naitoh, Y., and Kaneko, H. (1973). *J. Exp. Biol.* **58**, 657–676.
Naitoh, Y., Eckert, R., and Friedman, K. (1972). *J. Exp. Biol.* **56**, 667–681.
Nanney, D. L., and McCoy, J. W. (1976). *Trans. Am. Microsc. Soc.* **95**, 664–682.
Oertel, D., Schein, S. J., and Kung, C. (1977). *Nature (London)* **268**, 120–124.
Plattner, H. (1975). *J. Cell Sci.* **18**, 257–269.
Satow, Y. (1978). *J. Neurobiol.* **9**, 81–91.
Satow, Y., and Kung, C. (1974). *Nature (London)* **247**, 69–71.
Satow, Y., and Kung, C. (1976a). *J. Neurobiol.* **7**, 325–338.
Satow, Y., and Kung, C. (1976b). *J. Membr. Biol.* **28**, 277–294.
Satow, Y., and Kung, C. (1976c). *J. Exp. Biol.* **65**, 51–63.
Satow, Y., Chang, S.-Y., and Kung, C. (1974). *Proc. Natl. Acad. Sci. U.S.A.* **71**, 2703–2706.

Satow, Y., Hansma, H. G., and Kung, C. (1976). *Comp. Biochem. Physiol. A* **54**, 323–329.
Schein, S. J. (1976a). *J. Exp. Biol.* **65**, 725–736.
Schein, S. J. (1976b). *Genetics* **84**, 453–468.
Schein, S. J., Bennett, M. V. L., and Katz, G. M. (1976). *J. Exp. Biol.* **65**, 699–724.
Shusterman, C. L., Thiede, E. W., and Kung, C. (1978). *Proc. Natl. Acad. Sci. U.S.A.* **75**, 5645–5649.
Siegel, R. W. (1961). *Exp. Cell Res.* **24**, 6–20.
Sonneborn, T. M. (1937). *Proc. Natl. Acad. Sci. U.S.A.* **23**, 378–385.
Sonneborn, T. M. (1938). *Proc. Am. Philos. Soc.* **79**, 411–434.
Sonneborn, T. M. (1942). *Cold Spring Harbor Symp. Quant. Biol.* **10**, 111–124.
Sonneborn, T. M. (1947). *Adv. Genet.* **1**, 263–358.
Sonneborn, T. M. (1950). *J. Exp. Zool.* **113**, 87–148.
Sonneborn, T. M. (1954). *Caryologia* **6**, Suppl., 307–325.
Sonneborn, T. M. (1957). *In* "The Species Problem" (E. Mayer, ed.), pp. 155–324. Am. Assoc. Adv. Sci., Washington, D.C.
Sonneborn, T. M. (1970a). *Methods Cell Physiol.* **4**, 241–339.
Sonneborn, T. M. (1970b). *Proc. R. Soc. London, Ser. B* **176**, 347–366.
Sonneborn, T. M. (1975). *Trans. Am. Microsc. Soc.* **94**, 155–178.
Stocking, R. J. (1915). *J. Exp. Zool.* **19**, 387–449.
Takagi, Y. (1970). *Jpn. J. Genet.* **45**, 11–21.
Takagi, Y. (1971). *Jpn. J. Genet.* **46**, 83–91.
Takahashi, M. (1979). *Genetics* (in press).
Takahashi, M., and Hiwatashi, K. (1974). *Exp. Cell Res.* **85**, 23–30.
Takahashi, M., and Naitoh, Y. (1978). *Nature (London)* **271**, 656–659.
Tamm, S. L. (1972). *J. Cell Biol.* **55**, 250–255.
Tamm, S. L., Sonneborn, T. M., and Dippell, R. V. (1975). *J. Cell Biol.* **64**, 98–112.
Taub, S. R. (1963). *Genetics* **48**, 815–834.
Tsukii, Y., and Hiwatashi, K. (1977). *Jpn. J. Genet.* **52**, 483.
Tsukii, Y., and Hiwatashi, K. (1978). *J. Exp. Zool.* **205**, 439–446.
Van Houten, J. (1977). *Science* **198**, 746–748.
Van Houten, J., Hansma, H., and Kung, C. (1975). *J. Comp. Physiol.* **104**, 211–223.
Van Houten, J., Chang, S.-Y., and Kung, C. (1977). *Genetics* **86**, 113–120.
Watanabe, T. (1978). *J. Cell Sci.* **32**, 55–66.
Wolfe, J., and Grimes, G. W. (1979). *J. Protozool.* (in press).

Lipids and Membrane Organization in *Tetrahymena*

7

YOSHINORI NOZAWA AND GUY A. THOMPSON, JR.

BIOCHEMISTRY AND PHYSIOLOGY OF PROTOZOA
SECOND EDITION, VOL. 2

I. INTRODUCTION

Tetrahymena pyriformis, a free-living, ciliated protozoan, has proved to be a valuable system for various studies in cell biology. Considerable information about the structure and function of the cell has been reviewed in two recent monographs (Hill, 1972; Elliot, 1973). Especially in recent years, when much attention has been focused on the dynamic aspects of membrane-associated phenomena, this organism as a model eukaryotic cell has contributed to a better understanding of membrane biology at the molecular level. In this chapter we will review some of these aspects, principally based on data which have been obtained with *T. pyriformis* in our laboratories.

II. ISOLATION AND STRUCTURES OF MEMBRANE COMPONENTS

The *T. pyriformis* cell has most of the typical organelles found in usual eukaryotic cells and also contains several less typical membranous structures such as cilia, food vacuoles, oral apparatus, and mucocysts. Some of these are shown in Figure 1. Many of these cell components can be isolated by cell fractionation. For convenience, the isolation methods are divided into two main groups: (1) systematic isolation methods by which several different membrane components can be isolated from the same batch of cells, and (2) nonsystematic isolation by which individual subcellular organelles are separately isolated (Nozawa, 1975).

A. Systematic Isolation Methods

The first attempt at systematic subcellular fractionation of *Tetrahymena* by use of zonal centrifugation was made by Müller *et al.* (1968). In this method, cell disruption is achieved by passage of a chilled cell suspension in $0.25\,M$ sucrose through a fritted-glass filter (pore size, $10–15\,\mu\mathrm{m}$) under mild suction. Then the homogenate is fractionated by zonal differential sedimentation through discontinuous sucrose gradients (0.3, 0.35, 0.4, 0.45, 0.5, and $0.55\,M$). Thus mitochondria, peroxisomes, and lysosomes are separated. A similar procedure has been employed by Lloyd's group (Lloyd *et al.,* 1971), except that preparation of the medium and cell disruption are carried out in a glass homogenizer. Nozawa and Thompson (1971a) have developed a useful procedure for isolating various components, including surface membrane fractions such as cilia and pellicles. A scheme for the overall procedure is shown in Figure 2. The harvested cells are suspended in cold $0.2\,M$ phosphate buffer (pH 7.2) containing $0.1\,M$

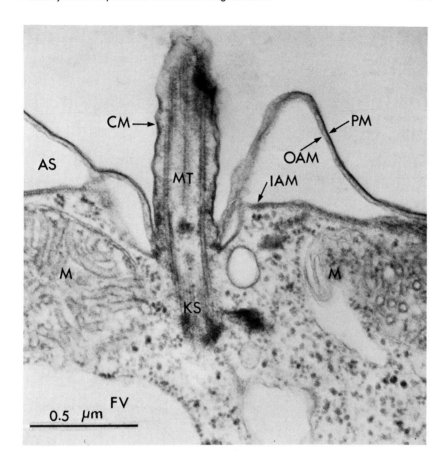

Figure 1.   General view of an ultrathin-sectioned *Tetrahymena* cell (Nozawa, 1975). PM, plasma membrane; CM, ciliary membrane; MT, microtubules; OAM, outer alveolar membrane; IAM, inner alveolar membrane; M, mitochondria; KS, kinetosome; FV, food vacuole; and AS, alveolar sac.

NaCl and 3 mM 2Na · EDTA and are then centrifuged at 100 g for 5 min. This cooling step thoroughly shrinks the cells, a prerequisite for obtaining highly purified membrane fractions. Ciliary beating is stopped in this condition. The shrunken cells are resuspended in 12–15 ml of the same cold phosphate buffer to make a suspension of approximately 6 × 10^6 to 8 × 10^6 cells per milliliter and are gently homogenized by hand (four to six strokes) in a loosely fitting glass homogenizer until most of the cilia are detached. The homogenate is then centrifuged at 1020 g for 5 min to sediment the deciliated cells (Figure 3a). The resulting supernatant is re-

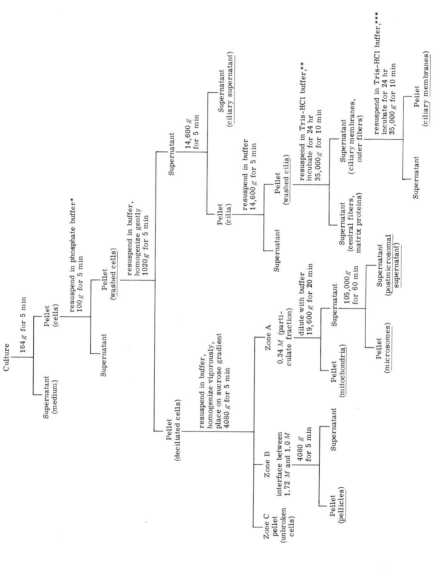

Figure 2. Scheme for the systematic isolation of subcellular membrane fractions from *Tetrahymena*:* 0.2 *M* phosphate buffer (3 m*M* EDTA, 0.1 *M* NaCl, pH 7.2). ** 1 m*M* Tris–HCl buffer (0.1 m*M* EDTA, pH 8.3). *** 10 m*M* Tris–HCl buffer (1 m*M* EDTA, 0.6 *M* KCl, pH 8.3). (From Nozawa and Thompson, 1971b, reproduced by permission.)

centrifuged at 14,600 g for 5 min, yielding a small white pellet of cilia. The deciliated cells from the 1020 g centrifugation are suspended in 4–5 ml of phosphate buffer and are homogenized vigorously by hand in a tightly fitting glass homogenizer until microscopic observation reveals only hollow pellicles remaining intact from most cells. The homogenate is layered on a discontinuous buffered sucrose gradient (0.34 M, 10 ml; 1.0 M, 15 ml; 1.72 M, 15 ml) and is centrifuged at 4080 g for 5 min. Three main zones are separated: zone A, a top layer containing mitochondria and microsomes; zone B, an interface band consisting principally of pellicles between the 1.0 and 1.72 M interface, and zone C, a bottom layer of unbroken cells. The small particulate (zone A) and pellicle (zone B) fractions are removed with a syringe. The pellicle fraction is diluted with buffer and centrifuged at 4080 g for 5 min to sediment pellicles (Figure 3c). For further purification, gentle homogenization and washing are repeated. The small particulate fraction (mainly mitochondria and microsomes) is diluted with buffer and spun at 19,600 g for 20 min to sediment mitochondria. The resulting supernatant is further centrifuged at 105,000 g for 60 min to spin down microsomes (Figure 3g). Thus five different subcellular fractions, cilia, pellicles, mitochondria, microsomes, and postmicrosomal supernatant, can be isolated in a highly purified state from the same batch of cells.

B. Nonsystematic Isolation Methods

A variety of methods have been employed to isolate individual subcellular components, especially cilia, nuclei, and pellicles. These methods are reviewed in detail in two recent publications (Nozawa, 1975; Everhart, 1972). Since Child (1959) first isolated cilia, several techniques have become available for the isolation of these appendages. They can be detached by mild shearing forces after assorted different chemical treatments. For example, ethanol (Child, 1959; Watson and Hopkins, 1962), glycerol (Gibbons, 1965), EDTA-CaCl$_2$ (Rosenbaum and Carlson, 1969), dibucaine (Thompson *et al.*, 1974), or high phosphate (Nozawa and Thompson, 1971a) may be used. Cells which are deciliated by dibucaine can regenerate cilia after transfer to a fresh medium containing no dibucaine. Ciliary membranes are separated from axonemes by incubation in 1 mM Tris-HCl buffer containing 0.1 mM EDTA (pH 8.3), and then 10 mM Tris-HCl buffer containing 1 mM EDTA and 0.6 M KCl (Figure 2) (Renaud *et al.*, 1968).

For isolating pellicles, the cells are treated with digitonin (Seaman, 1960), Tween 80 (Hartman *et al.*, 1972), or high phosphate (Nozawa and Thompson, 1971a), and are homogenized until almost all of the cells are disrupted. Then the homogenate is centrifuged to form a pellet of pellicle "ghosts."

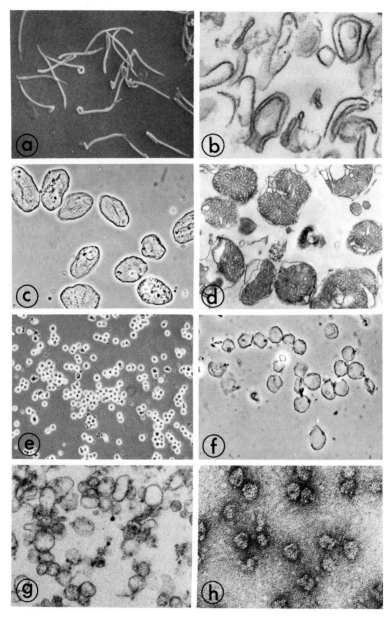

Figure 3. Features of several subcellular components from *Tetrahymena* (a) Cilia (scanning electron microscopy). Magnification: ×3800. (b) Ciliary membranes (ultrathin-sectioned). Magnification: ×54,000. (c) Pellicles (phase contrast microscopy). Magnification: ×350. (d) Mitochondria (ultrathin-sectioned). Magnification: ×13,500. (e) Macronuclei (phase contrast microscopy). Magnification: ×270. (f) Nuclear membranes (phase contrast microscopy). Magnification: ×680. (g) Microsomes (ultrathin-sectioned). Magnification: ×31,000. (h) Cytoribosomes (negative-stained, kindly provided by Dr. Jean-Jacques Curgy). Magnification: ×160,000. (From Nozawa, 1977, reproduced by permission.)

Many methods for the isolation of macronuclei have been described. There are two major categories: with and without detergents. Nozawa *et al.* (1973) have isolated macronuclei according to a modification of Gorovsky's method using *n*-octanol and gum arabic (Gorovsky, 1970). The cells are suspended in 0.1 M sucrose containing 1.5 mM MgCl$_2$ and 4% gum arabic (pH 6.7) and are washed by centrifugation. The washed cells are resuspended in the same medium, except that it contains 24 mM *n*-octanol plus 0.01% spermidine, and are homogenized gently. Then 3 volumes of the MgCl$_2$–sucrose solution are immediately added to the cell lysate to dilute the *n*-octanol concentration. The diluted cell lysate is centrifuged at 365 g for 5 min, and the pellet, resuspended in MgCl$_2$–sucrose solution, is layered over a 1.0–1.5 M discontinuous sucrose gradient and is centrifuged at 10,400 g for 5 min. Highly purified macronuclei are thereby obtained. To separate nuclear membranes from isolated macronuclei, 2–3 ml of 0.2 M phosphate buffer in 0.25 M sucrose containing 1 M NaCl (pH 7.2) is added to the pellet of macronuclei. This hypertonic shock ruptures the macronuclei, leaving intact nuclear envelope ghosts. Then the nuclear envelope suspension is centrifuged at 3000 g for 30 min through 0.2 M sucrose in phosphate buffer. The top layer, which contains almost all of the nuclear membranes, is overlaid on 1.6 M sucrose in buffer and spun at 3000 g for 20 min. The top layer from this centrifugation is diluted with buffer and centrifuged at 10,000 g for 10 min to spin down the nuclear membranes (Nozawa *et al.*, 1973).

Curgy *et al.* (1974) have isolated cytoribosomes and mitoribosomes. Cells are broken in a glass homogenizer and centrifuged at 500–800 g for 6 min. The supernatant remaining after centrifugation of the homogenate at 5000 g for 6 min is spun at 56,000 g for 20–30 min. Triton X-100 is added to the resulting supernatant (0.02 ml of 1.4% Triton X-100 per milliliter). The mixture is layered on 1.5 M sucrose in TMK buffer containing 10 mM Tris–HCl, 10 mM MgCl$_2$, 100 mM KCl, and 0.2 M sucrose and is centrifuged at 140,000 g for 150 min to pellet cytoribosomes. For mitoribosome isolation, the mitochondrial pellet is resuspended in TMK buffer and lysed with Triton X-100 and sodium deoxycholate at final concentrations of 2.5 and 0.4%, respectively, and maintained for 30 min. The lysate is layered on a linear sucrose gradient (0.3–1.4 M) in TMK buffer. Centrifugation at 35,000 rpm for 180–210 min is carried out to obtain mitoribosomes.

C. Structures of Isolated Subcellular Organelles

Several subcellular components isolated by either systematic or nonsystematic procedures are shown in Figure 3. The ciliary membrane, which is continuous with the outermost membrane of the pellicle, has a typical unit membrane integrity. The isolated pellicles appear to be almost free of cytoplasmic contaminants, as examined by thin-section electron micros-copy. The pellicle ghost consists of the outermost, plasma membrane and the outer and inner alveolar membranes. The latter two membranes are separated from each other by a space referred to as the alveolar sac. The outer face of the outer alveolar membrane shows a conspicuous paral-lelism with the plasma membrane, with a separation space of about 200 Å. Separation of the plasma membrane from the alveolar membranes has been attempted without any success, because of the tight bridges between these two membranes at many sites (Franke *et al.*, 1971). Mitochondria vary in both shape and size. Therefore, there are some difficulties in isolating a pure mitochondrial fraction.

The nuclear membranes separated from macronuclei are largely uni-form in both size and shape. The membrane integrity characteristic of the nuclear membrane is well preserved, with a bilaminar structure of the outer and inner membranes. Use of negative staining reveals nuclear pore complexes at many sites on the membrane (Nozawa *et al.*, 1973). The microsomal fraction is similar in appearance to microsomes from other cell types.

Recently, for better understanding of the intramembranous ultrastruc-ture and dynamic aspects of the biological membranes, freeze–fracture electron microscopy has proved to be of great value. The fractured faces of the ciliary membrane contain many membrane-intercalated particles (70 ± 20 Å) on the protoplasmic fracture (PF) face and few on the exo-plasmic fracture (EF) face. The EF face of the plasma membrane appears smooth with only some randomly distributed 70 ± 20 Å particles, al-though the PF face is studded with a lot of particles. As for the fractured faces of the alveolar membranes, the PF faces of the outer and the inner alveolar membranes bear many larger particles (115 ± 25 Å) than those of the plasma membrane, and the EF faces, in contrast, have a much smaller number of particles. This observation is compatible with that of the intact pellicle observed in fixed whole cells (Wunderlich and Speth, 1972). The microsomal fraction contains many vesicles of various sizes. Most of the convex faces have a large number of randomly distributed particles, and some contain very few particles, indicating that most microsomal vesicles are right side out with some inside out vesicles.

III. LIPID DISTRIBUTION IN VARIOUS MEMBRANES

It is generally accepted that the membranes of different subcellular organelles with different functions each have their own characteristic lipid composition. Such dissimilarities may provide a useful clue about how membrane lipids are associated with specific functions. As methods for isolation of subcellular organelles from *Tetrahymena* have been established, this organism is a useful cell system for demonstrating differences in intracellular lipid distribution.

A. Phospholipids

Numerous studies of phospholipids of *Tetrahymena* have been conducted and some properties have been described by Taketomi (1961). In 1967, Thompson presented the first detailed analysis using column chromatography, by which the principal components were found to be phosphatidylethanolamine (PE) and phosphatidylcholine (PC), which accounted for 60 and 20 mole%, respectively. It was also shown that the PC fraction contained to a great extent the glyceryl ether analog and that the PE fraction was rich in bound glyceryl ethers as well as 2-aminoethylphosphonic acid (AEP). Upon further analysis each of these phospholipids proved to be a mixture various compounds. The choline-containing phospholipids contained both diacyl and monoalkyl monoacyl types, whereas the ethanolamine-containing lipids consisted of three different types: diacyl–PE, diacyl–AEP, and monoalkyl monoacyl–AEP (Berger *et al.*, 1972; Liang and Rosenberg, 1966; Thompson, 1967; Smith and Law, 1970). The ether-linked alkylglyceryl moiety was represented almost entirely by 1-*O*-hexadecylglycerol, chimyl alcohol. The monoalkyl monoacyl–AEP lipid (**I**) seemed to be considerably resistant against en-

$$
\begin{array}{l}
H_2C-O-(CH_2)_{15}CH_3 \\
\qquad\qquad\quad O \\
\qquad\qquad\quad \| \\
HC-O-C-R \\
\qquad\qquad\quad O \\
\qquad\qquad\quad \| \\
H_2C-O-P-CH_2CH_2NH_2 \\
\qquad\qquad\quad OH
\end{array}
$$

(I)

zymatic and chemical digestion. Also, 2-aminoethylphosphonolipids are almost exclusively occupied at the C-2 position with linoleic ($C_{18:2\Delta^{6,9}}$) and γ-linolenic ($C_{18:3\Delta^{6,9,12}}$) acids (Berger *et al.*, 1972; Nozawa *et al.*, 1975).

Other minor components include cardiolipin and sphingolipids (Carter and Gaver, 1967; Berger *et al.*, 1972). More details about the lipid composition of *Tetrahymena* may be obtained in several reviews (Thompson and Nozawa, 1971; Hill, 1972; Holz and Conner, 1973).

The intracellular distribution of phospholipids is summarized in Table I. There are striking variations in the localization of certain phospholipids. The most remarkable pattern is seen in the ciliary membrane, which contains a particularly high level of 2-aminoethylphosphonolipids (Smith *et al.*, 1970; Nozawa and Thompson, 1971a; Jonah and Erwin, 1971; Kennedy and Thompson, 1970). A similar lipid pattern is also observed in another surface membrane, the pellicle. However, one would expect that the outermost membrane of the three pellicular membranes should be identical in the phospholipid content with the ciliary membrane, because this membrane is continuous with the pellicular outermost membrane. In fact, a figure close to the observed value can be obtained by calculation based on the assumption that the lipid composition of the two (outer and inner) alveolar membranes is identical with that of microsomes (Nozawa and Thompson, 1971a) and that of the outermost membrane is identical to cilia. The internal membrane fractions—nuclear envelope, microsomes, and mitochondria—are more or less similar to each other in the lipid distribution.

The glyceryl ether-containing phospholipids are highly localized in cilia and pellicles.

B. Neutral Lipids

Tetrahymena cells do not produce any typical sterols such as cholesterol, ergosterol, and stigmasterol but contain, instead, the pentacyclic triterpenoid tetrahymanol (**II**). This is the principal neutral lipid of growing cells, and its distribution among various membrane fractions is also presented in Table I. The surface membranes, especially cilia, contain large amounts of this unusual lipid, the function of which is as yet unresolved. Diplopterol, an analog of tetrahymanol, is also present as a minor component.

(II)

As for the fatty acid distribution, almost all of the fatty acids are bound to the phospholipid backbones in membranes. Numerous studies of *Tetrahymena* fatty acids have been conducted (reviewed in Erwin, 1973) since Erwin and Bloch (1963) first reported the fatty acid composition of

Table 1 Phospholipid Composition of Various Membrane Fractions from *T. pyriformis* WH-14 Cells[a]

Membrane fractions	C-P bond (% of lipid phosphorus)	Total phospholipids (mole%)[b]						Tetrahymanol (moles/mole lipid phosphorus)	Glyceryl ethers (moles/100 moles of lipid phosphorus)
		LysoPC	PC	LysoAEPL and lysoPE	PE	AEPL	CL		
Whole cells	29	2	33	0	37	23	5	0.057	29.7
Cilia	67	1	28	9	11	47	1	0.30	52.6
Ciliary supernatant	44	8	19	13	16	35	1	0.16	23.1
Pellicles	42	5	25	3	34	30	2	0.084	32.8
Mitochondria	26	2	35	0	35	18	10	0.048	24.7
Nuclear membranes	—	6	31	6	26	23	3	0.036	—
Microsomes	33	1	35	3	34	23	1	0.041	18.3
Postmicrosomal supernatant	26	5	34	4	30	22	2	0.016	27.4

[a] Data from Nozawa and Thompson (1971a) and Thompson *et al.* (1971).
[b] PC, phosphatidylcholines; PE, phosphatidylethanolamines; AEPL, 2-aminoethylphosphonolipid; and CL, cardiolipin.

several species. Recently, a detailed biosynthetic pathway for unsaturated fatty acids has been proposed by Koroly and Conner (1976), indicating that there are separate pathways leading from stearate and the palmitate. The novel palmitate route yields an unusual fatty acid, cilienic acid $(C_{18:2\Delta^{6,11}})$, as the terminal product. It is of interest to note that this cilienic acid is primarily bound to the 2-aminoethylphosphonolipids (Ferguson *et al.*, 1975; Fukushima *et al.*, 1976a) which are enriched in the surface membranes. When the distribution of fatty acids in various subcellular fractions is determined (Table II), the most striking difference is seen again in the ciliary membranes. They contain considerably larger amounts of γ-linolenic acid than any other membrane fractions. The fatty acyl chain composition of *Tetrahymena* membrane phospholipids changes depending on the growth temperature, as has been found in many microorganisms. More details will be discussed in Sections VI and VII.

IV. PHYSICOCHEMICAL PROPERTIES OF MEMBRANES IN *TETRAHYMENA*

Since Singer and Nicolson (1972) proposed the fluid mosaic membrane model, the concept that the biological membrane is a kind of two-dimensional solution of proteins embedded in a fluid lipid matrix has greatly attracted membrane biologists. It is now well accepted that the dynamic structure or fluidity of biological membranes is closely related to its functions (Singer, 1974). Some studies of this type have been done to characterize *Tetrahymena* membranes.

A. Physical Techniques

In recent years a number of physical techniques have become available for investigating the organization of the membrane components. These techniques include electron spin resonance (ESR), nuclear magnetic resonance (NMR), X-ray diffraction, laser Raman spectroscopy, differential scanning calorimetry (DSC), and fluorescence labeling.

We shall now describe some properties of various subcellular organelles from *Tetrahymena*. The use of spin labels to probe biological membranes has yielded much important information. However, despite the high sensitivity to the molecular motion, because the information available to this technique is limited to the local environment where the spin labels reside, one must be cautious in interpreting ESR data. The order parameters (S) calculated from ESR spectra can be used for the estimation of membrane fluidity. Figure 4 shows the temperature dependence of the order param-

Table II Fatty Acid Composition in Lipids of Whole Cells and Subcellular Fractions from *T. pyriformis* WH-14 Cells[a]

Fatty acids	Whole cells	Cilia	Pellicles	Mitochondria	Microsomes	Postmicrosomal supernatant
$C_{12:0}$	1.9	0.8	3.1	1.2	1.5	2.0
$C_{14:0}$	5.2	3.1	6.5	4.0	5.2	6.4
$C_{16:0}$	7.9	5.7	10.6	7.3	7.1	8.6
$C_{16:1,\Delta^9}$	10.8	5.5	11.4	9.2	11.9	13.6
$C_{18:0}$	0.6	0.8	0.5	0.8	0.8	1.6
$C_{18:1,\Delta^9}$	8.6	6.1	8.6	7.1	7.9	11.5
$C_{18:2,\Delta^{9,12}}$	19.8	16.6	16.8	21.4	21.0	18.0
$C_{18:3,\Delta^{6,9,12}}$	33.6	43.2	29.2	38.3	34.1	26.0
Unsaturated/saturated	4.7	6.9	3.3	5.7	5.4	3.7

[a] Data from Nozawa *et al.* (1974).

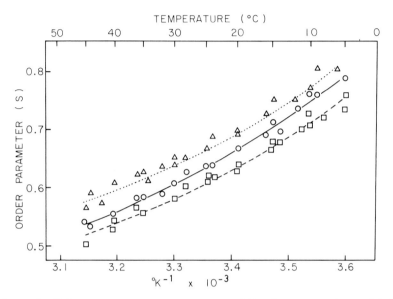

Figure 4. Order parameters versus temperature of *Tetrahymena* membranes labeled with 5-nitroxystearate. Cilia, △; pellicles, ○; microsomes, □. (From Nozawa *et al.*, 1974, reproduced by permission.)

eters of several membranes, demonstrating that the S value (S) differs from one membrane to another—larger at the higher temperatures than at the lower ones. The membranes are more fluid in the order, microsomes > pellicles > cilia (Nozawa *et al.*, 1974). It would be of great interest to establish that various membranes with different biological functions each have their own characteristic fluidity. On the other hand, the functionally distinct membranes within a particular cell have been clearly shown to have different lipid compositions. As described previously (Section III), the localization of certain lipids is striking in particular *Tetrahymena* membrane fractions, which perhaps reflects a diversity in the physical state of the membranes. Taking into consideration that the content of the sterol-like triterpenoid tetrahymanol increases in the order cilia > pellicles > microsomes, one would expect that this specific lipid might be involved in stabilizing membrane fluidity. Our preliminary data have suggested its dual effects as observed with cholesterol, namely, becoming rigid above and fluid below the transition temperature (Oldfield and Chapman, 1972). Compared to dispersions of phospholipids alone, dispersions of phospholipids with tetrahymanol become more fluid at temperatures below the apparent transition point, 27°C, and less fluid at temperatures above that point (Nozawa *et al.*, 1974). Another useful technique for probing the physical state of membranes is fluorescence labeling. For this purpose lipophilic fluorescent probes, e.g., 1,6-diphenyl-1,3,5-hexatriene (DPH), perylene, and *N*-phenyl-1-naphthylamine, are incorporated into the hydrophobic region of the membrane lipids. The fluorescence polarization analysis can give rise to useful information concerning the molecular dynamics and physical state of the lipid molecules in the vicinity of the fluorescent probe (Shinitzky and Inbar, 1974).

With the aid of the fluorescent hydrocarbon probe, 1,6-diphenyl-1,3,5-hexatriene, which has proved to be a highly sensitive probe and therefore has been most commonly employed, the fluidity or microviscosity was measured with isolated membranes of *Tetrahymena*. The degrees of fluorescence polarization are plotted as a function of temperature (Figure 5). There are pronounced differences in the microviscosity of various

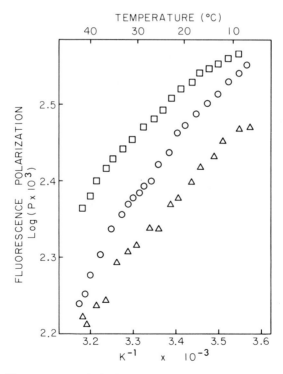

Figure 5. Fluorescence polarization versus temperature of *Tetrahymena* membranes labeled with 1,6-diphenyl-1,3,5-hexatriene. Cilia, □; pellicles, ○; microsomes, △. (From Shimonaka *et al.*, 1978, reproduced by permission.)

membrane fractions; these results are compatible with data from electron spin resonance (Nozawa *et al.,* 1974). The microviscosity of the ciliary membranes is markedly greater than the microsomal and pellicular membranes. Such diversity in the microviscosity reflects the different lipid compositions of the three membrane fractions. The ciliary membranes contain a considerably higher level of tetrahymanol and 2-amino-ethylphosphonolipid, whereas the more fluid microsomes have the lowest content of these two lipids together with the greatest degree of unsaturation in fatty acyl chains.

Wunderlich *et al.* (1975) have examined the fluorescence intensity of *Tetrahymena* microsomes with the use of an amphiphilic fluorescent probe, 8-anilino-1-naphthalenesulfonate, and have shown that there is a discontinuity between 15° and 20°C. The discontinuity was also observed by ^{1}H NMR. Moreover, they have suggested that such thermotropic membrane alterations may not be caused by typical phase transitions, but rather by the quasi-crystalline cluster formation. Recently, it has been suggested that the fluorescent probes may act as a membrane perturbant and affect the dynamic structure in the vicinity of the embedded probe molecules (Krishnan and Balaram, 1975). Therefore as in the case of ESR, care should be taken to interpret observations obtained by fluorescence labeling.

Since a method has been just recently established for obtaining highly ^{13}C-enriched membranes and phospholipids from *Tetrahymena* (Nwanze *et al.,* 1977), more detailed features about the molecular dynamics may be discovered by using ^{13}C NMR in the near future.

B. Freeze–Fracture Electron Microscopy

For better understanding the dynamic state of biological membranes, it is of great value to obtain more detailed information regarding their ultrastructure by freeze–fracture electron microscopy, since changes in the distribution of membrane-intercalated particles can be interpreted as evidence for membrane mobility or fluidity. Although the aforementioned physicochemical techniques require the preparation of isolated membrane fractions, by which some artifacts may occur, this electron microscope approach can provide useful, direct information about the state of various membranes without the isolation procedure. This technique has proved to be very valuable since it allows observation of protein particles associated with large areas of the hydrophobic matrices of membranes exposed by fracturing.

Speth and Wunderlich (1973) have first shown by freeze–fracture electron microscopy that *Tetrahymena* cells exhibit changes in the distribution of membrane-intercalated particles when shifted to lower temperatures. At 39.5°C, where *T. pyriformis* NT-I cells had been grown, all fractured faces of various membranes show a random distribution of membrane particles. When cells were cooled down to lower temperatures, however, these particles began to move by lateral diffusion within the plane of certain membranes and to form aggregates or clusters of membrane particles. Figure 6 illustrates the particle distribution pattern of the outer alveolar membrane, which is most sensitive to temperature changes. Thus one would speculate that this membrane might act as a kind of sensor for monitoring the signals transferred from outside the cell. The fractured face (PF) of this membrane of cells cooled to 15°C reveals large areas devoid of particles and also linear arrays composed of membrane particles.

To use freeze–fracture electron microscopy for monitoring the dynamic features of membranes, the degree of particle aggregation should be quantified. Several methods have been applied to certain types of cells for quantitative measurements on the translocation of the particles (Ojakian and Satir, 1974; Elgsaeter and Branton, 1974). In a previous paper, we have described a parameter called the particle density index (PDI) for monitoring quantitatively the particle distribution in *Tetrahymena* (Martin *et al.*, 1976). The protoplasmic fracture (PF) face of the outer alveolar membrane was used to determine the PDI, which can be obtained by the following equation based on the particle number per square micrometer of the fractured replica. $PDI(\%) = [(x - a)/(b - a)] \times 100$, where x, a, and b represent, respectively, the particle density of a membrane under the conditions to be examined, the particle density of a completely fluid membrane showing random particle distribution, and the particle density of a maximally aggregated membrane at very low fluidity.

Recently, Kitajima and Thompson (1977a) extended this technique to observation of other membranes and found that the temperature causing the first appearance of particle-free regions differs from one membrane to another, thereby being indicative of physical change of each membrane. For example, the aggregation onset temperatures of outer and inner alveolar membrane, plasma membranes, endoplasmic reticulum, and mitochondrial membrane of 39.5°C-grown cells are 33°, 24°, 9°, 24°, and 21°C, respectively. The onset temperatures of *Tetrahymena* cells grown at different temperatures are plotted in Figure 7. The onset temperature for each particular membrane varies depending on the cell growth temperature. The ciliary membrane does not induce any particle aggregation at the temperatures tested.

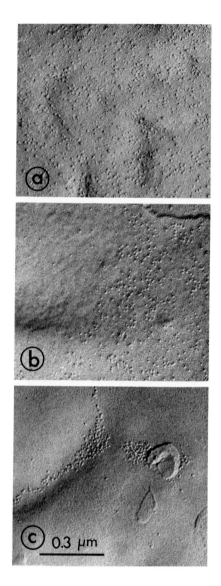

Figure 6.   Distribution pattern of membrane-intercalated particles at different temperatures in the outer alveolar membrane (PF face) of *T. pyriformis* NT-I. (a) 39.5°C; (b) 27°C; (c) 15°C.

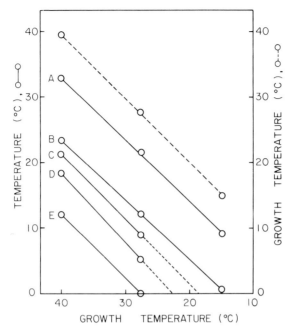

Figure 7. Temperatures for emergence of particle-free regions in several membranes of cells grown at 39.5°, 27°, and 15°C. (A) Outer alveolar membrane; (B) endoplasmic reticulum; (C) outer mitochondrial membrane; (D) nuclear envelope; (E) vacuolar membrane. (From Kitajima and Thompson, 1977a, reproduced by permission.)

V. MEMBRANE BIOSYNTHESIS AND ASSEMBLY

Although considerable progress has been made in elucidating the structure and organization of membranes, the biosynthetic processes in membrane biogenesis, e.g., site of synthesis, transfer and assembly of the synthesized molecules, and turnover of membrane constituents, are still poorly understood. Current research seems to be especially directed toward finding out how the membrane building blocks are assembled. Protozoan cells have proved to be suitable model systems for studying the membrane biogenesis, since we can label the unicellular culture with radioactive precursors of protein or lipid and then isolate various different subcellular membrane fractions from the same batch. Actually, *Acanthamoeba* and *Tetrahymena* have been employed. In this section, we shall describe principally intracellular movement of lipids in terms of membrane formation in *Tetrahymena*.

A. Growing Cells

1. Nonsynchronized Cells

Chlapowski and Band (1971) have followed the movement of ^{3}H-labeled glycerol phospholipids in various cell fractions of *Acanthamoeba palestinensis*. From the changes in specific activity of the labeled phospholipids, an ordered sequence of incorporation of newly synthesized lipids was observed: nuclei and endoplasmic reticulum > Golgi membranes > collapsed vesicles > plasma membrane. These observations can provide evidence that membrane phospholipids are synthesized in rough-surfaced endoplasmic reticulum and nuclei and are subsequently transferred to the Golgi membrane to be integrated into the collapsed membrane vesicles, precursors of the plasma membrane.

Using *T. pyriformis,* Nozawa and Thompson (1971b) also compared the rates at which various membranes incorporate newly synthesized lipids, and obtained results similar to these reported for *Acanthamoeba*. The logarithmic phase cells were pulse-labeled with [^{14}C]palmitic acid, [^{14}C]acetate, or [^{3}H]hexadecyl glycerol. For example, the incorporation pattern of [1-^{14}C]palmitate into phospholipids of various membrane fraction is illustrated in Figure 8. Immediately after [^{14}C]palmitate labeling, the highest specific radioactivity is found in the postmicrosomal supernatant, with the microsomal fraction being slightly less. On the other hand, the cilia and ciliary supernatant show the very slow incorporation, and require almost 6 hr to reach the level of specific activity equivalent to the other membrane fractions. These observations indicate the following sequence of new lipid movement to the several membrane destinations in the cell: endoplasmic reticulum → mitochondria and pellicles → cilia. The distribution of radioactivity in the three major phospholipids reveals a considerably delayed labeling of the phosphonolipid. However, after 6 hr all these phospholipids are equally radioactive in the cilia and the microsomes (Table III), suggesting the involvement of phospholipid exchange between different membranes. The similarity of the labeling pattern of [^{14}C]palmitate, [^{14}C]acetate, and even [^{3}H]chimyl alcohol provides further confidence that we are observing the natural pathways of lipid synthesis, and that intracellular movement of phospholipid molecules does occur.

2. Synchronized Cells

To obtain further information regarding the dynamic movement of membrane constituents during the cell cycle, synchronized cells have been examined. Baugh and Thompson (1975) have shown changes in the specific radioactivity of phospholipids of several cell fractions isolated

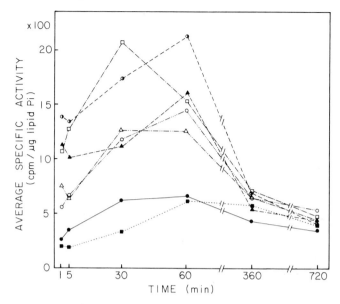

Figure 8. Incorporation of [1-¹⁴C]palmitic acid into phospholipids of various membranes in *T. pyriformis* WH-14. Microsomes, □; postmicrosomal supernatant, ◑; whole cells, ▲; mitochondria, ○; pellicles, △; ciliary supernatant, ●; cilia, ■. (From Nozawa and Thompson, 1971b, reproduced by permission.)

Table III   Relative Labeling of Ciliary Lipids and Microsomal Lipids after Short and Long Incubation with [¹⁴C]Palmitate[a]

Lipid fraction	Ratio of specific radioactivity in microsomes/specific radioactivity in cilia	
	1 min	36 min
2-Aminoethylphosphonolipids	7.9	1.0
Phosphatidylcholines	15.0	1.4
Phosphatidylethanolamines	4.5	1.2

[a] Data reprinted by permission from Nozawa and Thompson (1971b).

from synchronized cells at different times during the cell cycle. It is clearly indicated that the degree to which newly synthesized phospholipids are incorporated into certain membrane fractions depends significantly upon what stage of the division cycle the cell is in. Each membrane fraction has its own characteristic labeling pattern. Mitochondria and microsomes show peaks of labeling prior to the initiation of cytokinesis, whereas cilia are most rapidly labeled during furrow formation. The pellicles do not show significant variations during the cell cycle. The distribution of radioactivity among the major phospholipids classes at various times during the division cycle does not change dramatically during the cycle.

B. Nongrowing Cells

Cells which are not undergoing cell division do not require net biosynthesis of membrane materials, i.e., phospholipids and proteins. Actually, *Tetrahymena* cells have been studied in several nongrowing states, i.e., stationary phase cells, starving cells, and metabolically inhibited cells. The distribution rates of radioactive phospholipids to various membranes was measured in all these cells.

1. Stationary Phase Cells

Under the standard growth conditions we use, cells cease dividing after approximately 60–70 hr of growth at a cell density of about 2×10^6 cells/ml. After this time no further increase in phospholipid occurs, although triglycerides accumulate. Fatty acid synthesis is drastically inhibited. The incorporation of [^{14}C]palmitate into phospholipids was compared in various membrane fractions, and was found in general to be similar to that of the growing cells—rapid and slow in cytoplasmic membranes and surface membranes, respectively (Nozawa and Thompson, 1972). This observation is consistent with the concept that there is the same kind of rapid transfer (or exchange) of phospholipids between membranes as found with the growing cells.

2. Starving Cells

In a nonnutritive inorganic medium, cell division and fatty acid synthesis ceased within several hours, but the cells were still able to swim normally and remained viable for several days (Nozawa and Thompson, 1972). However, during the prolonged period of starvation, the cell volume was reduced to approximately one-fifth of the normal value. The tetrahymanol content was markedly increased in all membrane fractions as shown in Table IV, because there are no enzymes to degrade it.

Table IV Distribution of Tetrahymanol in Several Membrane
Fractions of *T. pyriformis* WH-14[a]

| Membrane fractions | Molar ratio (tetrahymanol : lipid phosphorus) | | Starved cell, tetrahymanol as a percentage of log phase cell tetrahymanol |
	Logarithmic phase cells	93-hr-starved cells	
Whole cells	0.065	0.177	270
Cilia	0.320	0.482	150
Microsomes	0.037	0.094	250
Postmicrosomal supernatant	0.017	0.046	270

[a] Data reprinted by permission from Thompson *et al.* (1972).

The general tendency of [14C]palmitate labeling pattern in various membrane fractions of starving cells showed a similarity to that of the rapidly growing cells. The microsomes and the postmicrosomal supernatant showed much higher specific activities than the other membrane fractions. With the passage of time, radioactive lipids were transferred from microsomes to other destinations in the cell, and after 12 hr all cell fractions were almost evenly labeled (Nozawa and Thompson, 1972).

3. Metabolically Inhibited Cells

The intracellular redistribution of lipids was examined using cells in which lipid synthesis had been inhibited by use of the lipid lowering agents, which are thought to be specific inhibitors of acetyl-CoA carboxylase. Tetralylphenoxyisobutyric acid (TPIA) and chlorophenoxyisobutyric acid (CPIB or clofibrate) strongly inhibited the incorporation of [14C]acetate into lipids of *Tetrahymena* (Nozawa and Thompson, 1972; Nozawa, 1973). The decrease in synthesis caused by these drugs was observed fairly uniformly in all lipids, except that the radioactivity level in triglycerides was much higher than that found in growing cells, with almost equal amounts in phospholipids and triglycerides.

On the other hand, logarithmic phase cells labeled with [14C]palmitate and then treated with TPIA (0.2 m*M*) for 3.5 hr prior to isolation of membrane fractions revealed a distribution pattern of radioactivity in various membranes which was fairly similar to that observed in the control cells. Thus *de novo* synthesis of fatty acids is not prerequisite for the lipid movement throughout the cell.

C. Transfer and Assembly of Membrane Lipids

In general, there are several ways of transferring lipids in biological membranes, with two of these being of prime importance. They are the unidirectional and the bidirectional transfer system. The former involves a net transfer to membranes incapable of biosynthesis, e.g., mitochondria, and the latter, called "exchange," involves simultaneous transfer of probably the same phospholipids between two different membranes. This exchange mechanism was comprehensively reviewed by Wirtz (1974). As described above, the rapid movement of intact phospholipid molecules clearly seems to occur between various membranes in *Tetrahymena*. It was rather surprising that even in the nongrowing cells in which no net lipid synthesis occurred phospholipids were transferred from one membrane to another at nearly the same rate as in the control, growing cells (Nozawa and Thompson, 1971b). This result suggests that the lipid exchange process is even more important than the net transfer in maintaining the proper membrane lipid composition. A tentative scheme for the mobility of membrane lipids within the *Tetrahymena* cell was proposed as illustrated in Figure 9 (Nozawa and Thompson, 1971b). Phospholipids travel on a vehicle, carrier, or exchange protein, such as that found in rat

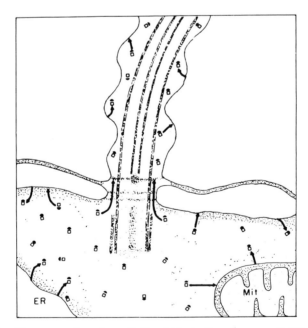

Figure 9. Movement of membrane lipids within the *Tetrahymena* cell. ER, endoplasmic reticulum; Mit, mitochondria; △□, lipid–carrier protein complex. (From Nozawa and Thompson, 1971b, reproduced by permission.)

liver and other cells (Wirtz, 1974). Previous results which showed a considerable delay in incorporation of newly synthesized phospholipids into the surface membranes, especially the ciliary membrane, were thought to be due to its physical separation from other cytoplasmic membranes. However, recently Iida *et al.* (1978) have observed by the use of a spin-labeling technique that another regulating factor, membrane fluidity, plays an important role in determining the incorporation of lipids into a target membrane. A great advantage of this technique is that there is no need to reisolate the incubated membranes because the transfer rate of lipids could be estimated by a decrease in exchange broadening in the ESR spectra. As demonstrated in Figure 10, spin-labeled lecithin was found to transfer from microsomes to other membranes at a rate decreasing in the following order: microsome > pellicles > cilia. Increased fluidity of the acceptor membrane induced enhancement of lipid insertion. This hypothesis was further supported by the experimental result that the labeled lipid was transferred into the high fluidity pellicles from 15°C-grown cells much more rapidly than into pellicles of lower fluidity from 39.5°C-grown cells.

Although it was clearly shown that lipids were rapidly transferred between different membranes, a question now arose about how the cell

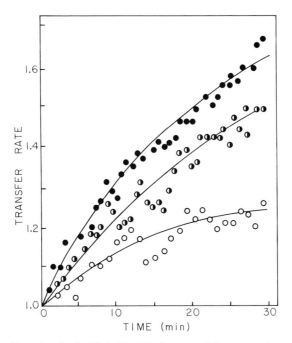

Figure 10. Transfer of spin-labeled lecithin between different membranes in *T. pyriformis* NT-I cell. Microsomes, ●; pellicles, ◑; cilia, ○. (From Iida *et al.*, 1978, reproduced by permission.)

accomplishes the apparently selective transfer of certain specific lipids such as phosphonolipids and tetrahymanol to membrane sites remote from their origin. Two mechanisms are available for inducing such lipid localization: the specificity in the transferring system and the specificity in the selection at the acceptor membrane. On the basis of the distribution pattern of lipid radioactivity in various membranes, the latter mechanism was found to be operating in *Tetrahymena* (Thompson *et al.,* 1971). A good example of experiments to support this theory is presented in Figure 11, in which cells were pulse-labeled with [³H]chimyl alcohol, a precursor of phosphonolipids. Although after a 5-min incorporation the radioactivity was essentially the same in most membranes, with the passage of time the relative radioactivity distribution changed strikingly, so that this labeling pattern nearly reflected the mass composition of the lipids. These findings indicate that all cellular membranes receive a similar assortment of lipids from the sites of synthesis or from other membranes, and that the lipids are then reshuffled and exposed to lipolytic enzymes to produce the final characteristic lipid composition.

D. Coordination (Association) of Lipid and Protein Assembly

Several models can be envisaged for membrane biosynthesis. However, there are, in principle, two major hypotheses for membrane assembly: a random and separate incorporation of newly made lipids and proteins into the membrane (uncoupled assembly) and an insertion of the two

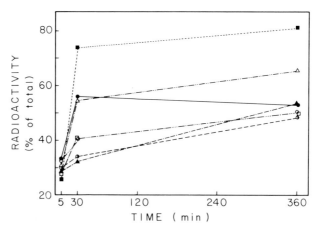

Figure 11. Incorporation of [8,9-³H]chimyl alcohol into phosphonolipids of various membranes in *T. pyriformis* WH-14. Cilia, ■; pellicles, △; ciliary supernatant, ●; microsomes, □; mitochondria, ○; postmicrosomal supernatant, ◑; whole cells, ▲. (Reprinted with permission from G. A. Thompson *et al.,* 1971, *Biochemistry* **10,** 4441–4447. Copyright 1971 by the American Chemical Society.)

components preassembled into a tight association (coupled assembly). The former mechanism has been supported by some investigators (Mindich, 1970; Kahane and Razin, 1969; Benjamins *et al.*, 1971), and the latter mechanism has been supported by others (Beattie, 1969; Wilson and Fox, 1971). To test which mechanism is more likely with *Tetrahymena*, we have compared the rates at which the radioactivity was incorporated into membrane lipids under conditions where protein or lipid synthesis was selectively inhibited (Nozawa and Thompson, 1972). Although protein synthesis was blocked almost completely by addition of cycloheximide, [^{14}C]palmitate was incorporated into lipids at nearly the normal rate, and the lipids were rapidly disseminated throughout the cell. After several hours a lowered rate of palmitate utilization was observed, indicating a decrease in the demand for new lipids. These and other data provide us with evidence that the lipid movement within the cell is not significantly dependent upon protein synthesis. More direct evidence arguing against the tight coupling of lipid and protein synthesis has been presented by Subbaiah and Thompson (1974). They compared the relative changes of lipid radioactivity and protein radioactivity in isolated cell fractions over the 12-hr period required for both components to complete their initial intracellular deployment. The findings that these two components exhibited kinetic differences offered clear evidence that lipids and proteins need not be transported as stable complexes within the cell.

VI. ALTERATIONS OF MEMBRANE LIPIDS CAUSED BY ENVIRONMENTAL CHANGES; MODIFICATION OF MEMBRANE LIPID

There is increasing evidence that a variety of microorganisms can alter their membrane lipid composition according to growth conditions, e.g., nutrition, temperature, light, aeration, and metabolic inhibitors. Since *Tetrahymena* cells have a proven ability to modify their lipid composition in response to changed growth environments, this cell is a most useful system for studying mechanisms of lipid modification as well as the interrelationship between lipid composition and membrane functions. In this section, we shall describe several methods which induce *Tetrahymena* cells to alter their membrane lipid composition.

A. Supplementation

1. Ergosterol

Ferguson *et al.* (1971) have shown that the ergosterol supplementation to the growth medium induces replacement of tetrahymanol by that sterol and causes small significant compensatory changes in the fatty acid composition of *Tetrahymena* cells. Recently, Ferguson *et al.* and one of our laboratories examined the effects of replacing tetrahymanol with ergosterol upon lipid composition (Ferguson *et al.*, 1975; Nozawa *et al.*, 1975; Kasai *et al.*, 1977). Although the former group found no significant change in the phospholipid class composition, the latter discovered considerable alterations in the proportions of certain phospholipids in some membranes, as shown in Table V. This discrepancy might result from the difference in strains used between the two groups. In ergosterol-replaced membranes of pellicles and microsomes, there is a marked increase of phosphatidylethanolamine (PE) with a compensating decrease of phosphatidylcholine (PC). The level of 2-aminoethylphosphonolipid (AEPL) does not change in pellicles but does to a significant extent in microsomes. On the other hand, both groups, using different strains, have noted that profound alterations in fatty acyl chain composition were induced by the tetrahymanol replacement with ergosterol. The overall tendency is a shortening of the fatty acyl chain length and a lowering in the degree of fatty acyl unsaturation. As shown in Table VI, despite some variations between different membranes, the level of myristic, palmitic, and palmitoleic acids generally increases at the expense of oleic and γ-linolenic acids. Similar changes in the lipid composition were observed to occur in cells supplemented with β-sitosterol.

It is reasonable to expect that such pronounced lipid alteration may affect the physical properties of membranes. Data from electron spin resonance indicated that cilia, pellicles, and microsomes isolated from

Table V Alterations Induced by Ergosterol Replacement in Phospholipid Composition of Pellicles and Microsomes from *T. pyriformis* WH-14 Cells[a]

Phospholipids	Whole cells		Pellicles		Microsomes	
	Ergosterol-replaced	Native	Ergosterol-replaced	Native	Ergosterol-replaced	Native
Phosphatidyl-cholines	26.2	30.4	22.5	26.9	30.4	35.6
Phosphatidyl-ethanolamines	37.9	32.9	36.8	31.6	40.0	32.6
2-Aminoethyl-phosphonolipids	23.4	21.7	28.7	29.5	20.7	24.4
Cardiolipin	5.0	6.0	3.2	2.8	3.4	2.6
Lysophosphatidyl-cholines	3.3	3.5	5.1	4.5	2.9	3.5

[a] Data reprinted by permission from Kasai *et al.* (1977).

Table VI Distribution of Major Fatty Acids in Various Membrane Fractions from Ergosterol-Replaced Cells of *T. pyriformis* WH-14[a]

Fatty acids	Cilia		Pellicles		Mitochondria		Microsomes	
	Ergosterol-replaced	Native	Ergosterol-replaced	Native	Ergosterol-replaced	Native	Ergosterol-replaced	Native
$C_{14:0}$	8.2	3.5	11.0	7.6	7.5	5.1	9.5	7.7
$C_{16:0}$	10.1	10.1	15.7	12.9	11.5	8.9	12.3	12.2
$C_{16:1,\Delta^9}$	8.2	5.9	10.1	8.2	10.3	7.6	13.3	9.8
$C_{18:0}$	3.8	4.5	2.1	2.9	2.4	3.0	1.9	2.0
$C_{18:1,\Delta^9}$	6.4	8.7	6.1	9.7	4.9	8.5	6.1	11.8
$C_{18:2,\Delta^{6,11}}$	3.6	3.5	3.2	1.8	1.8	1.5	2.9	2.0
$C_{18:2,\Delta^{9,12}}$	11.6	15.3	11.9	13.2	18.7	19.9	14.0	13.6
$C_{18:3,\Delta^{6,9,12}}$	34.4	41.0	23.3	30.1	31.9	37.5	26.0	31.3

[a] Data reprinted by permission from Kasai *et al.* (1977).

ergosterol-supplemented cells are less fluid than those from unsupplemented cells (Figure 12). This decreased membrane fluidity induced by ergosterol replacement was also revealed by the fluorescence polarization using 1,6-diphenyl-1,3,5-hexatriene (Figure 13). When 34°C-grown cells were examined by freeze–fracture electron microscopy, the onset temperature of outer alveolar membranes at which the membrane particles first began to aggregate (probably due to phase separation) was lower than 15°C for ergosterol-replaced cells, as compared with 22°C for native, tetrahymanol-containing cells (Thompson and Nozawa, 1977).

2. 1-O-Hexadecyl Glycerol (Chimyl Alcohol)

Tetrahymena cells contain a high level of glyceryl ether in phospholipids (Table I). Therefore, the effects on the lipid composition of membranes of feeding 1-O-hexadecylglycerol, chimyl alcohol, which is precursor of glyceryl ether-containing phospholipids were examined (Fukushima *et al.*, 1976b). When the cell density reached 1×10^5 cells/ml, a sonicated chimyl alcohol dispersion was added to the growth medium at a concentration 2 mg/200 ml of culture. Subsequently, when the cell density became 2×10^5 and 4×10^5 cells/ml, 4 and 8 mg of chimyl alcohol, respectively, were added as supplements to the culture. The glyceryl ether content in the membrane phospholipids from cells grown in the presence and absence of chimyl alcohol (Table VII) indicates that the glyceryl ether content of pellicle phospholipids was greatly increased by feeding: 32.7% for the native and 40.7% for the fed membranes. A smaller increase in glyceryl ether phospholipid was observed in microsomes. It was also found that

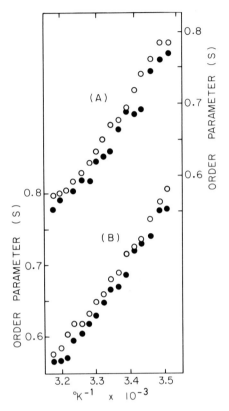

Figure 12. Order parameters versus temperature of ergosterol-replaced membranes from *T. pyriformis* WH-14 cell. (A) Microsomes; (B) pellicles. Ergosterol-replaced, ○; native, ●. (From Kasai *et al.*, 1977, reproduced by permission.)

incorporated chimyl alcohol was principally inserted into phosphatidyl-choline and 2-aminoethylphophonolipid but not into phosphatidyl-ethanolamine. In phosphatidylcholine, the molar content of glyceryl ether rose to 71.2% of the total phosphatidylcholine. Associated with this increased glyceryl ether content were marked alterations in both the phospholipid head group and fatty acyl chain composition. Table VIII shows that although no change in percentage of phosphatidylcholine was noted in pellicles of ergosterol-fed cells, there was a large increase in 2-amino-ethylphosphonolipid with a corresponding decrease in phosphatidyl-ethanolamine. In contrast, as for microsomes, small changes were found in the microsomal phospholipid composition, namely, a slight drop in phosphatidylethanolamine accompanied by a small increase in phos-phatidylcholine plus 2-aminoethylphosphonolipid.

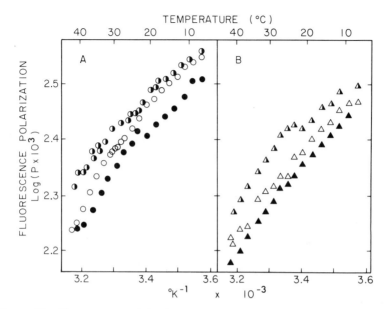

Figure 13. Fluorescence polarization versus temperature of lipid-manipulated membranes from *T. pyriformis* NT-I. (A) Pellicles; (B) microsomes. Native, ○△; chimyl alcohol-fed, ●▲; ergosterol-replaced, ◐▲. (From Shimonaka *et al.,* 1978, reproduced by permission.)

As depicted in Table IX, a remarkable increase in palmitic acid together with a slight elevation in the polyunsaturates and linoleic and γ-linolenic acids were noted in both membranes, thus preventing a noticeable change in the ratio of total unsaturated to saturated fatty acid content. The oleic acid content was found to be much lower in hexadecylglycerol-fed micro-

Table VII Glyceryl Ether Content in Native and Hexadecylglycerol-Fed *T. pyriformis* NT-1 Membrane Phospholipids[a]

Membrane fractions	Glyceryl ether (moles/100 moles of lipid phosphorus)	
	Native	Hexadecylglycerol-fed
Whole cells	28.8	33.0
Pellicles	32.7	40.7
Microsomes	33.1	39.3

[a] Data reprinted by permission from Fukushima *et al.* (1976b).

Table VIII Alterations Induced by Hexadecylglycerol Feeding in Phospholipid Composition in Pellicles and Microsomes from *T. pyriformis* NT-1[a]

Membrane fractions	2-Aminoethylphos-phonolipids	Phosphatidyl ethanolamines	Lysophosphatidyl cholines	Phosphatidyl cholines	Cardiolipin
Pellicles					
Native	21.9	46.8	2.6	25.0	1.0
Hexadecylglycerol-fed	30.5	37.1	2.7	25.5	1.3
Microsomes					
Native	20.9	40.9	1.5	33.2	1.9
Hexadecylglycerol-fed	22.6	36.1	1.9	35.5	2.1

[a] Data reprinted by permission from Fukushima *et al.* (1976b).

Table IX Major Fatty Acid Composition of Total Lipids from Native and Hexadecylglycerol-Fed *T. pyriformis* NT-I Membranes[a]

Fatty acids	Pellicles		Microsomes	
	Native	Hexadecylglycerol-fed	Native	Hexadecylglycerol-fed
$C_{12:0}$	2.7	2.6	0.8	0.6
$C_{14:0}$	10.4	6.9	7.8	5.5
$C_{16:0}$	16.9	21.0	12.4	19.9
$C_{16:1,\Delta^9}$	9.6	10.7	11.1	13.5
$C_{18:0}$	2.7	2.8	1.6	1.8
$C_{18:1,\Delta^9}$	11.9	9.4	14.4	4.5
$C_{18:2,\Delta^{6,11}}$	3.0	3.3	5.6	3.1
$C_{18:2,\Delta^{9,12}}$	11.9	14.2	14.5	16.7
$C_{18:3,\Delta^{6,9,12}}$	19.4	21.5	21.4	23.6

[a] Data reprinted by permission from Fukushima *et al.* (1976b).

somes than in native microsomes. The alteration in phospholipid fatty acid composition may well have been caused to some degree by the large supply of palmitic acid produced through cleavage of the chimyl alcohol ether bond. This process is known to destroy a significant proportion of the fed chimy alcohol (Kapoulas *et al.,* 1969).

Arrhenius plots of the membrane fluidity as estimated by the fluorescence polarization technique using 1,6-diphenyl-1,3,5-hexatriene are given in Figure 13. The pellicles and microsomes isolated from hexadecylglycerol-supplemented cells have greater fluidities than membranes isolated from native, unsupplemented cells. Enhancement of fluidity was especially noticeable below approximately 21°C in pellicles.

3. Fatty Acids

Lees and Korn (1966) have demonstrated that *Tetrahymena,* when grown in the presence of 11,14-eicosadienoate, 8,11,14-eicosatrienoate, 11-eicosenoate, or 11-octadecenoate, incorporated the fatty acids into their neutral and phospholipids, and that despite profound changes in fatty acid composition, the cells were normal in growth rate, appearance, and motility.

We have examined the influence of supplementation of large amounts of linoleic acid upon membrane lipid composition (Martin *et al.,* 1976). Cells were grown at 39.5°C to a density of 7.5×10^4 cells/ml. A sonified emulsion of 4.3 μmoles of linoleic acid in distilled water was added to the 200-ml culture (21.5 μM), which was allowed to incubate for desired periods of time. Table X presents the data showing a striking perturbation of phos-

Table X Major Fatty Acid Composition of Phospholipids from Membrane Fractions of $C_{18:2}$-Supplemented Cells of *T. pyriformis* NT-I[a]

Fatty acids	Unsupplemented cells			100 min following supplementation		
	Cilia	Pellicles	Microsomes	Cilia	Pellicles	Microsomes
$C_{14:0}$	6.7	8.9	7.3	5.2	12.1	7.8
$C_{16:0}$	16.8	16.4	13.2	20.9	17.8	14.4
$C_{16:1,\Delta^9}$	8.3	7.1	8.7	8.1	8.4	9.5
$C_{16:2,\Delta^7}$	2.3	4.0	4.0	3.0	2.9	2.7
$C_{18:0}$	6.1	3.0	2.2	5.2	2.0	1.6
$C_{18:1,\Delta^9}$	11.7	12.6	13.9	21.8	6.4	6.0
$C_{18:2,\Delta^{9,12}}$	8.1	10.5	13.0	14.8	23.3	29.2
$C_{18:3,\Delta^{6,9,12}}$	20.7	18.5	21.4	12.9	18.5	21.2

[a] Data reprinted with permission from R. Kasai *et al.* (1976), *Biochemistry* **15**, 5228–5233. Copyright 1976 by the American Chemical Society.

pholipid fatty acid composition in several membrane fractions. Linoleic acid increased greatly in relative amounts in all three membrane fractions at the expense of oleic acid. However, despite such large incorporation of linoleic acid, the concentrations of myristic and palmitic acids increased as compared with nonfed control cells. In contrast to the profound alteration of fatty acyl chain composition, analysis of phospholipids revealed little difference in head group distribution of membranes isolated from the fed and unfed cells. With regard to the membranes' physical properties, the distribution pattern of membrane particles as examined by freeze–fracture electron microscopy suggested that membranes of linoleic acid-supplemented cells are more fluid than those of control cells.

4. Choline Analogs

Methods were devised by Glaser *et al.* (1974) to alter significantly the phospholipid composition of LM cells by growing them in the presence of choline analogs. Then it was found that the analogs can be incorporated into phospholipids of rat hepatocytes (Åkesson, 1977) or embryo fibroblasts (Hale *et al.*, 1977), causing marked alterations in the polar head group composition. Although it might be expected in these cases that changes in the fatty acyl chain composition could compensate for changes in the polar head group, no significant changes were observed in the chain length and the degree of unsaturation of fatty acids. Therefore this method is a useful system for studying specifically the regulatory mechanism of the polar head group composition and also the role of the head group in membranes.

When *T. pyriformis* NT-I cells were grown in a medium containing choline analogs (8.1 mM) such as ethanolamine (HOCH$_2$CH$_2$NH$_2$), methylethanolamine (OHCH$_2$CH$_2$NHCH$_3$), or choline [HOCH$_2$CH$_2$N$^+$-(CH$_3$)$_3$], some differences were observed in the polar head group composition of phospholipids (Kasai and Nozawa, unpublished observations). The most striking alteration occurred upon methylethanolamine supplementation, in which case an unnatural phospholipid, phosphatidyl-mono-methylethanolamine, was found to accumulate up to 34% of the total phospholipids. The rise in this phospholipid was mainly compensated for by a decrease in phosphatidylethanolamine (43 → 13%). Supplementation with ethanolamine resulted in a slight increase in phosphatidyl-ethanolamine associated with a corresponding decrease in 2-amino-ethylphosphonolipid. In contrast, no consistent alteration in the phospholipid composition was seen in the choline-supplemented cells. The fatty acid compositions showed no significant differences when the cells were grown on any of the analogs.

5. Isovalerate

It was shown by Conner *et al.* (1974) that isovalerate-supplemented cells of *T. pyriformis* W revealed quantitative but not qualitative changes in fatty acyl chain composition. There was an increase in polar lipids that contained odd-numbered isofatty acids: C$_{13}$, C$_{15}$, C$_{17}$, C$_{19}$. This change was accompanied by a corresponding decrease in even-numbered normal acids. The saturated fatty acids in unsupplemented cells consist predominantly of myristic acid with smaller amounts of normal palmitic acid, whereas iso-C$_{17}$ is most abundant in isovalerate-supplemented cells. The unsaturated fatty acid content was not altered significantly by the supplementation. Recently, we have obtained an unusual strain of *Tetrahymena* which contains a very high level (up to 27%) of isofatty acids as compared with other strains—WH-14, W, and GL (Fukushima *et al.*, 1978). The two principal fatty acids iso-C$_{15:0}$ and iso-C$_{17:1}$ were present in greatest abundance in microsomes, and were measured at 12 and 22%, respectively. Since this strain is capable of synthesizing large amounts of isofatty acids, cells were incubated with isovaleric acid at various concentrations. As expected, there was a drastic increase in isofatty acids. Eleven hours after incubation with 4 mM isovalerate began, the level of isofatty acids rose to 74% of the total fatty acids: iso-C$_{13:0}$, 5.6%; iso-C$_{15:0}$, 24.8%; iso-C$_{17:0}$, 3.4%; and iso-C$_{17:1}$, 39.9%. The tremendous increase in these isofatty acids was concurrently associated with a marked decrease in even-numbered fatty acids—C$_{14:0}$, C$_{16:1}$, C$_{18:2}$, and γ-C$_{18:3}$. Therefore, it was postulated that isofatty acids can substitute for normal unsaturated fatty acids in maintaining proper membrane fluidity. No observation was obtained indicating a significant change in the phospholipid polar head group composition because of isovalerate supplementation.

B. Temperature

It is well known that poikilothermic organisms possess the uncanny ability to modify the fatty acyl composition of their membrane phospholipids in response to changes in the environmental temperature (Fulco, 1974; Cronan and Gelman, 1975). In general, an increase in the growth temperature results in the production of less unsaturated fatty acids and vice versa. It is almost certainly not merely fortuitous that this type of alteration in the fatty acid composition maintains precisely the optimal membrane fluidity at which a variety of membrane-associated functions can proceed effectively.

Erwin and Bloch (1963) observed that late log phase cells of *T. pyriformis* MT II var. I showed an altered fatty acid composition when grown at different growth temperatures—25° and 35°C. At 25°C there was a higher level of unsaturated fatty acids, particularly palmitoleate and γ-linolenate, as compared with 35°C-grown cells, which contain larger amounts of saturated fatty acids, namely, myristic and palmitic acids. Wunderlich *et al.* (1973) have demonstrated changes in the fatty acid composition of total lipids after chilling *T. pyriformis* GL cells from 28° to 10°C, and also showed altered membrane structures by freeze–fracture electron microscopy. To investigate in more detail the adaptive modification of phospholipid acyl chain composition in various membranes, we grew *T. pyriformis* WH-14 cells at 25°C for 24 hr and then transferred them to shakers set at 15° or 34°C (Nozawa *et al.,* 1974). The percentage of the major saturated fatty acid, palmitic acid, increased with the increase in growth temperature while a concurrent decrease of palmitoleic acid occurred, thus producing a decline of the ratio of palmitoleate to palmitate in all membrane fractions: cilia, pellicles, mitochondria, and microsomes. In contrast, the levels of two principal polyunsaturated acids, linoleate and γ-linolenate, were not affected to a marked extent by changing temperature, except that the percentage of γ-linolenate in the cilia is tremendously high at 25°C and low both at 15° and 34°C. The oleic acid level remained fairly constant at 15° and 25°C but, unexpectedly, increased at 34°C. Similar results were recently obtained by Conner and Stewart using the same strain (Conner and Stewart, 1976).

In contrast to the marked changes in fatty acid constituents, other parameters of the lipid composition were not detectably altered, and the relative proportions of phospholipids were not affected significantly by changing the growth temperature. It is of interest to note that when electron spin resonance was used the membrane fluidities of cells incubated at different temperatures were found to be modified.

Recently, we discovered that another strain of *T. pyriformis*, tentatively designated NT-I, that was isolated from a hot spring, can adapt to temperatures as high as 41°C and can alter its membrane lipids in a much more pronounced fashion than can the WH-14 strain (Fukushima *et al.*, 1976a). Therefore, it was thought that this thermotolerant strain was a better cell system for studying the adaptative mechanism by which membrane lipids respond to environmental temperature changes. The cells were grown isothermally at 15°, 24°, and 39.5°C, and the generation time at these temperatures was 16, 4, and 3 hr, respectively. Table XI shows the profound changes observed in the fatty acid composition of phospholipids from the whole cell and its several subcellular fractions. The principal effects of decreasing temperature are an increase in linoleic and linolenic acid and a decrease in palmitic acid. The same trends are noted in the fatty acids of phospholipids from cilia, pellicles, and microsomes, producing a generally higher degree of unsaturation. For example, the ratios of total unsaturated to saturated fatty acids of microsomal phospholipids from cells grown at 15°, 24°, and 39.5°C are 4.51, 3.97, and 2.66, respectively.

It was also rather surprising to observe that there were striking differences in the phospholipid head group composition of cells grown at the three different temperatures. There was a marked increase in phosphatidylethanolamines (PE) as the growth temperature increased, and the relative concentration of phosphatidylcholines (PC) remained constant. The level of glyceryl ether-containing phospholipids dropped appreciably with increasing temperature; 27.1 and 20.9 mole% of these phospholipids were present in 15°C- and 39.5°C-grown cells, respectively. There was little change in the tetrahymanol content. A detailed analysis of phospholipids in three functionally different membrane fractions is presented in Table XII. The pellicles and microsomes followed the distribution pattern of the whole cell phospholipids, whereas the ciliary phospholipids differed in showing a marked alteration in phosphatidylcholines and in lyso compounds, probably lyso-PE and lyso-AEP. The PC level was fairly consistent regardless of growth temperature. It can therefore be speculated that such pronounced alterations in the membrane phospholipids play an important role in adapting the membrane fluidity to changed temperatures. The fluidities of major membrane fractions isolated from 15°C- and 39.5°C-grown cells were measured by electron spin resonance as a function of temperature. The membranes of cilia, pellicles, and microsomes from cells grown at 15°C were more fluid than those from cells grown at 39.5°C. An example of pellicles (Figure 14) illustrates that the order parameters (*S* values) of pellicles from 39.5°C-grown cells are

Table XI Fatty Acid Composition of Phospholipids of Several Membranes from *T. pyriformis* NT-I Cells Grown at Different Temperatures[a]

Fatty acids	Whole cells			Cilia			Pellicles			Microsomes		
	15°C	24°C	39.5°C	15°C	24°C	39.5°C	15°C	24°C	39.5°C	15°C	24°C	39.5°C
$C_{14:0}$	6.9	6.6	6.5	6.0	3.9	6.7	8.4	9.6	8.9	7.2	7.4	7.3
$C_{16:0}$	8.9	10.4	12.6	16.8	19.4	16.8	12.7	15.8	16.4	8.9	10.4	13.2
$C_{16:1,\Delta^9}$	8.7	8.9	8.7	14.4	7.7	8.3	9.9	7.4	7.1	8.8	9.8	8.7
$C_{16:2,\Delta}$	2.2	3.3	4.4	1.3	1.4	2.3	2.6	2.8	4.0	2.0	3.2	4.9
$C_{18:0}$	0.6	1.4	2.1	2.5	3.8	6.1	0.7	2.0	3.0	0.5	1.4	2.2
$C_{18:1,\Delta^9}$	9.6	7.4	10.6	6.9	8.0	11.7	7.4	8.5	12.6	8.1	7.1	13.9
$C_{18:2,\Delta^{6,11}}$	7.0	5.1	3.4	11.7	7.3	3.9	8.0	5.8	3.2	6.6	4.9	3.3
$C_{18:2,\Delta^{9,12}}$	20.2	18.2	14.5	9.0	9.7	8.1	16.2	13.8	10.5	19.5	17.7	13.0
$C_{18:3,\Delta^{6,9,12}}$	31.1	32.6	24.5	27.2	28.2	20.7	28.0	27.4	18.5	31.8	31.2	21.4

[a] Data reprinted by permission from Fukushima *et al.* (1976a).

Table XII Phospholipid Composition of Various Membranes from *T. pyriformis* NT-1 Cells Grown at Different Temperatures[a]

Lipid component	Cilia			Pellicles			Microsomes		
	15°C	24°C	39.5°C	15°C	24°C	39.5°C	15°C	24°C	39.5°C
Cardiolipin	1.3	0.4	1.0	1.3	0.3	1.7	2.7	4.8	1.9
2-Aminoethylphosphonolipids	40.0	37.0	36.4	30.8	32.9	19.1	22.9	20.8	14.7
Phosphatidylethanolamines	20.8	21.4	20.0	33.2	36.2	49.2	34.1	35.5	43.9
Lysophosphatidylethanolamines + lyso-2-aminoethylphosphonolipids	17.7	14.0	4.8	7.0	4.6	1.4	4.3	3.7	1.1
Phosphatidylcholines	15.8	18.1	30.0	18.9	21.4	23.5	30.0	27.6	32.4
Lysophosphatidylcholines	1.7	1.5	1.9	4.3	2.7	2.7	2.6	3.8	3.8
Tetrahymanol/lipid phosphorus (molar ratio)	0.137	0.260	0.178	0.064	0.090	0.098	0.053	0.055	0.078

[a] Data reprinted by permission from Fukushima *et al.* (1976a).

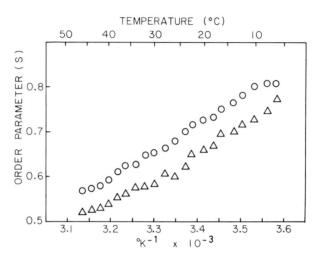

Figure 14. Order parameters versus temperature of pellicles from 39.5°C- and 15°C-grown cells. 39.5°C, ○; 15°C, △. (From Iida *et al.,* 1978, reproduced by permission.)

smaller than those of pellicles from 15°C-grown cells at all temperatures examined.

C. Drugs

1. Phenethyl Alcohol

Phenethyl alcohol (PEA) is known to affect the phospholipid and fatty acid metabolism in *Escherichia coli,* and therefore alters the phospholipid and its acyl chain composition of the cell membrane (Nunn, 1975, 1977). Recently, evidence was presented that PEA inhibits phospholipid synthesis primarily at the level of *sn*-glycerol-3-phosphate acyltransferase, and also that fatty acid synthesis would be inhibited as a secondary consequence of the altered phospholipid synthesis (Nunn, 1977).

When *Tetrahymena* cells were grown in the presence of 8 m*M* PEA, it was found that this drug induced a marked alteration in the relative proportions of both phospholipids and fatty acids in pellicles, mitochondria, and microsomes (Nozawa *et al.,* 1978). The general trend for these membranes was to increase the phosphatidylcholine content and to decrease concurrently the levels of ethanolamine-containing phospholipids (PE, AEP). An appreciable effect of PEA on the overall fatty acid composition of membrane phospholipids also was noted. PEA treatment resulted in higher unsaturation of fatty acids in the membrane phospholipids. The most pronounced changes were a drastic decrease in palmitoleic acid and an increase in linoleic and palmitic acids. Such effects on membrane lipid metabolism were found to be reversible. After the PEA-treated cells were transferred to fresh medium and incubated further the altered membrane lipid composition returned almost to the original composition. The physical properties of membranes were examined by freeze–fracture electron microscopy using the particle density index (PDI). The membranes from PEA-treated cells were presumed to have a greater fluidity as compared with control membranes.

2. Anesthetics

Recent data obtained by Nandini-Kishore *et al.* (1977) showed that the addition of a potent general anesthetic, methoxyflurane (3.0–4.35 m*M*), to *Tetrahymena* culture caused a significant change in the phospholipid fatty acid composition of whole cells; there was an increase in palmitate with an accompanying decrease in palmitoleate. Freeze–fracture electron microscopy confirmed that methoxyflurane had the strong fluidizing effect expected of this and other general anesthetics. The ability of *Tetrahymena* to change its lipid composition in response to membrane perturbation by anesthetics raises the question of whether such a phenomenon might occur during extended anesthesia of human subjects.

The above-mentioned study of anesthetic action prompted us to test ethyl alcohol, a compound which is much less efficient as an anesthetic but which is often present in the human body at physiologically active levels for extended periods of time. *Tetrahymena* responded to the presence of 0.35 *M* ethanol in a manner quite different from that noted for most other hydrophobic perturbing agents we have utilized (Nandini-Kishore *et al.*, 1978). The phospholipid fatty acid distribution was altered gradually but significantly, with the principal changes being a decrease in palmitoleate ($C_{16:1}$) and an increase in linoleate ($C_{18:2}$) (Table XIII). There was also an ethanol-induced decrease in the relative proportion of 2-aminoethyl-phosphonolipid from the control value of 16 to 6 mole% in 0.35 *M* ethanol, the difference being matched by an increase in phosphatidylethanolamine.

Fluorescence polarization measurements using the probe diphenyl-hexatriene disclosed that ethanol, like methoxyflurane, exerts a fluidizing influence on *Tetrahymena* membrane lipids (Nandini-Kishore *et al.*, 1979). Quite unexpectedly, however, long-term growth of the cells in ethanol-containing medium gave rise to membrane lipids which, when isolated and measured in the absence of ethanol, were consistently more fluid than equivalent lipids from control cells. On the basis of results gained from other fluidity-modifying factors tested, we might have anticipated a decrease in fluidity, thereby offsetting the influence of ethanol. At the present time the physiological significance of this finding is unclear, but it may be related to the fact that ethanol and other short-chain alcohols perturb the lipid bilayer selectively at a region near

Table XIII Fatty Acid Composition of Phospholipids from Control Cells and Cells of *T. pyriformis* NT-I in the Presence of Ethanol[a]

Fatty acids	Control	Days grown in presence of ethanol		
		2 days	4 days	8 days
$C_{14:0}$	12.8	9.2	8.7	8.5
$C_{15:0\,ai}$	1.4	5.3	3.9	3.9
$C_{16:0}$	11.9	13.0	14.5	16.0
$C_{16:1,\Delta^9}$	16.4	11.3	6.4	4.2
$C_{16:2,\Delta^7}$	6.7	3.8	2.5	0.8
$C_{18:0}$	0.8	2.3	2.5	2.7
$C_{18:1,\Delta^9}$	3.9	4.5	3.5	4.5
$C_{18:2,\Delta^{9,12}}$	13.7	19.1	22.2	24.9
$C_{18:3,\Delta^{6,9,12}}$	27.3	20.9	24.9	25.3

[a] Data reprinted by permission from Nandini-Kishore *et al.* (1979).

the carboxyl end of the fatty acyl groups (Lenaz *et al.*, 1976; MacDonald, 1978). It seems possible that this perturbation would trigger the observed compensatory alterations in the interior of the bilayer.

The response of *Tetrahymena* membranes to ethanol is in some respects similar to (but much greater than) that detected in mammals chronically exposed to intoxicating concentrations of ethanol for several weeks (Reitz *et al.*, 1973; Shorey *et al.*, 1978). Thus the potential value of *Tetrahymena* as a model system for the study of alcoholism is apparent.

3. Lipid Metabolism Inhibitors

The antibiotic cerulenin is a potent inhibitor of the condensing enzyme in the fatty acid synthetase complex, β-ketoacyl-acyl carrier protein synthetase, and 3-hydroxy-3-methyl-CoA synthetase (Omura, 1976). This drug also inhibits [^{14}C]acetate incorporation into lipid fractions in *Tetrahymena* cells (Frisch *et al.*, 1978; Kasai and Nozawa, unpublished observations), thereby resulting in lipid composition changes. There is a significant increase in phosphatidylethanolamine and a slight decrease in palmitoleic acid (Kasai and Nozawa, unpublished observations). Triparanol, 1-[*p*-(β-diethylaminoethoxy)phenyl]-1-(*p*-tolyl)-2-(*p*-chlorophenyl) ethanol, causes a reduction in tetrahymanol content by inhibition of squalene cyclization (Shorb *et al.*, 1965). Although little or no change was observed in phospholipid composition, a small decrease in palmitoleate and an increase in γ-linolenate were caused at 10 μg/ml triparanol (Kasai and Nozawa, unpublished observations). No significant alteration in lipid composition was caused by up to 3 m*M* CPIB (chlorophenoxyisobutyrate), even though the inhibitor markedly impaired the biosynthesis of fatty acid and tetrahymanol (Nozawa, 1973).

D. Cations

Tetrahymena is capable of acclimating to growth in high concentrations of cations. Whereas sudden exposure to $0.3\,M$ NaCl causes rapid death, gradually increasing the medium concentration of Na^+ over several weeks yields cells resistant to its toxic effects (Mattox and Thompson, unpublished observations). The phospholipid composition of Na^+-acclimated cells is significantly different from that of control cells, and the differences are most pronounced in the pellicles, whose outer membrane is directly exposed to the elevated Na^+ level in the medium. The largest change in phospholipid polar head group distribution is an increase in phosphatidylcholine in high Na^+ medium from 39 to 48%. This is offset by a decrease in cardiolipin and phosphonolipid.

The fatty acid composition of the phospholipids is also modified during acclimation to Na^+. The pattern of change is not as clearcut as the general increase of unsaturation accompanying reduced temperatures (see Section VI,B), but certain alterations are quite striking. For example, the ratio of $16:0/16:1$ in Na^+-acclimated pellicle phospholipids is 3.9 as compared with 0.9 in controls.

Lipid changes are also noted in cells adapted to grow in medium containing $0.1\,M$ $CaCl_2$ (Mattox and Thompson, unpublished observations). The effects are less striking than noted for Na^+, consisting, in the case of phospholipid fatty acids, of a general increase in saturated fatty acids coupled with a decrease in all unsaturated fatty acids so that the number of double bonds per 100 molecules is reduced by 10%.

Analysis of various membranes of high Ca^{2+}-adapted cells fixed for freeze–fracture electron microscope analysis indicated that Ca^{2+} exerts a physiologically important effect on certain structures. Most dramatically affected were the membranes of forming food vacuoles. The temperature of incipient phase separation in these membranes was elevated by 7°C in the Ca^{2+}-adapted cells. Changes in the physical properties of the nascent vacuolar membrane appear to be involved in the regulation of endocytosis by Ca^{2+}.

VII. MOLECULAR MECHANISMS OF MEMBRANE LIPID ADAPTATION TO TEMPERATURE CHANGES

Since Marr and Ingraham (1962) first found that *E. coli* adjusts the fatty acyl chain composition of its phospholipids in response to growth temperature, much information has accumulated concerning the adaptive changes of fatty acids in diverse microorganisms. However, the underlying molecular mechanism is not well understood although it has attracted the attention of numerous investigators. Most studies have been conducted with prokaryotes, and a number of mechanisms for temperature adaptation of bacterial fatty acyl composition have been proposed; fatty acid synthetase and acyltransferase systems have been implicated in *E. coli* (Cronan and Gelman, 1975) and the fatty acid desaturase system has been implicated in *Bacillus megaterium* (Fulco, 1974). On the other hand, less information is available regarding the temperature-mediated control mechanism in eukaryotes.

A. Adaptive Modulation of *Tetrahymena* Membrane Lipid Composition during Temperature Acclimation

After the shift of growth temperature of *T. pyriformis* NT-I from 39.5° to 15°C over 30-min period, changes in phospholipid, as well as fatty acyl chain composition, were observed, as demonstrated in Table XIV (Nozawa and Kasai, 1978). There was little significant alteration in the phospholipid class composition within 10 hr after the temperature shift, where no cell division occurred. However, a pronounced change was seen in the proportion of phospholipids thereafter; 2-amino-ethylphosphonolipids increased markedly at the expense of phosphatidylethanolamines, while the level of phosphatidylcholines was fairly consistent. The mechanism for such phospholipid adjustment remains to be clarified. In contrast, a rapid response to the temperature shift of the fatty acyl chain composition in phospholipids was noted. Palmitoleate began to increase immediately after the temperature shift and reached a maximum at 4 hr while the palmitate concentration was concurrently reduced (quick process). Therefore, the proportion of palmitoleate to palmitate increased during the first 4 hr after the shift, but declined thereafter with the passage of time. After this adjustment process, the concentration of palmitoleate continued to decrease with a corresponding increase in γ-linolenate (slow process). By these adjustment processes, the overall composition of phospholipid acyl chains gradually became similar to that of the 15°C isothermally grown cells and became almost identical to it 48 hr after the shift-down from 39.5 to 15°C.

It seems reasonable to think from these results that for the short-term adaptation within about 10 hr after the shift to 15°C the fatty acids may play a prime role for retailoring the perturbed membrane fluidity, and that changes in the phospholipid head groups which take in place as a consequent event after 10 hr may also be somehow involved in the adaptive processes for regulating membrane fluidity in *Tetrahymena*.

Table XIV Alteration of Lipid Composition during Temperature Acclimation in *Tetrahymena pyriformis* NT-I Cells[a]

	Isothermal 39.5°C	Time after shift from 39.5° to 15°C							Isothermal 15°C
		0 hr	2 hr	4 hr	6 hr	10 hr	24 hr	48 hr	
Phospholipid composition									
Phosphatidylethanolamines	44.9	—	—	46.5	—	46.7	36.0	32.4	25.6
2-Aminoethylphosphonolipids	17.4	—	—	14.9	—	16.5	26.3	30.2	29.0
Phosphatidylcholines	30.7	—	—	32.5	—	28.7	29.2	28.0	27.2
Fatty acid composition of phospholipids									
$C_{14:0}$	11.0	11.9	8.3	8.4	8.0	7.3	8.4	7.9	7.4
$C_{16:0}$	14.6	14.8	10.7	9.3	8.3	7.9	8.9	8.3	7.3
$C_{16:1,\Delta^9}$	14.3	17.4	18.1	19.5	17.8	15.7	13.3	11.3	12.0
$C_{16:2,\Delta^7}$	8.4	9.7	10.1	10.0	9.6	9.7	7.6	6.1	4.7
$C_{18:1,\Delta^9}$	3.8	1.9	3.7	3.8	5.3	6.0	7.2	7.9	6.9
$C_{18:2,\Delta^{6,11}}$	1.8	2.4	2.8	2.9	3.1	3.8	4.3	4.8	7.6
$C_{18:2,\Delta^{9,12}}$	12.0	9.3	9.5	9.2	9.5	10.2	13.4	17.1	16.3
$C_{18:3,\Delta^{6,9,12}}$	21.4	20.6	22.2	22.4	23.0	25.7	25.6	27.9	28.3
16:1/16:0	0.98	1.18	1.69	2.10	2.14	1.99	1.49	1.36	1.64

[a] Data reprinted by permission from Nozawa and Kasai (1978).

B. Temperature-Induced Alterations in Membrane Physical States

It was shown by many freeze–fracture electron microscopy studies from Wunderlich's laboratory (Speth and Wunderlich, 1973; Wunderlich *et al.*, 1973) that *T. pyriformis* GL cells exhibit changes in the distribution of membrane-intercalated particles when cooled from 28° to 10° or 5°C. Moreover, these workers drew the conclusion that the structural recovery of perturbed membranes in the shifted cells is accomplished by modifying the membrane phospholipid fatty acids. We have confirmed their observations using cells of the thermotolerant NT-I strain chilled from 39.5° to 15°C. As described above, the NT-I cells underwent a marked modification of their lipid composition following the temperature shift-down, and one would expect therefore that the membrane structures would recover from the effects of chilling and return to the random distribution of membrane particles observed in control cells before chilling. Figure 15 presents the distribution pattern of membrane-intercalated particles (110–130

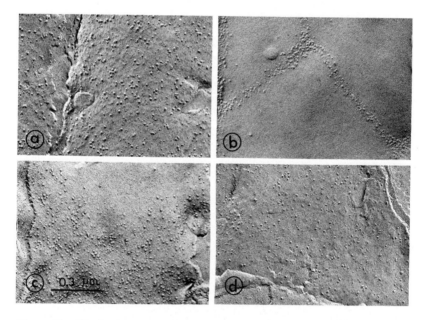

Figure 15. Distribution pattern after shift (39.5° → 15°C) of membrane-intercalated particles in the outer alveolar membrane (PF face) of *T. pyriformis* NT-I. (a) 39.5°C; (b) 0 hr; (c) 6 hr; (d) 11 hr. (From Nozawa and Kasai, 1978, reproduced by permission.)

Å diameter) in the outer alveolar membranes of the fractured pellicle membrane at different intervals after the shift of 15°C. The membrane particles began to move within the plane of the membrane by lateral diffusion, forming highly aggregated beltlike clusters upon cooling to 15°C. Six hours after the shift, many particles appeared to be randomly distributed once again, but some smooth areas devoid of particles still remained. However, 11 hr after chilling, the reorientation of membrane particles was almost complete, as inferred from the totally random, homogenous distribution. These thermotropic lateral separations are believed to reflect a phase transition from the liquid crystalline to the gel state in the membrane lipid bilayer. It is of importance to note that cells regain the ability to divide when the membrane particles are again randomly distributed.

Martin and Thompson (1978) have recently confirmed the freeze–fracture observations by a fluorescence polarization technique. As is clearly illustrated in Figure 16, the physical state of microsomes isolated at different intervals after the shift from 39.5 to 15°C revealed the trend of increasing fluidity in parallel with an increasing unsaturation index of phospholipid fatty acids. During the first 4 hr after the shift (quick adaption), where was a drastic rise in the fluidity, which might be brought about by enhanced conversion of palmitate into palmitoleate, leading to greater availability of palmitoleate for incorporation into microsomal phospholipids.

C. Mechanisms of Adaptive Control

As described in the Section VII,B, modifications of the phospholipid head group composition may be involved in accomplishing the temperature-induced homeoviscous adjustment of membrane fluidity. However, no information is available regarding the mechanisms responsible for such phospholipid modulations. Therefore, we will mainly discuss here the regulatory mechanisms for altering the acyl chain composition during temperature acclimation in *T. pyriformis* NT-I cells. Unlike certain bacteria, such as *E. coli*, in which two enzymes, fatty acid synthetase and acyltransferase, have been proposed to operate (Cronan and Gelman, 1975), *Tetrahymena* calls on its fatty acid desaturase systems to readjust its membrane acyl chain composition.

The apparent control of desaturase activity by molecular oxygen tension, as reported to occur in plant tissues (Harris and James, 1969; Brown

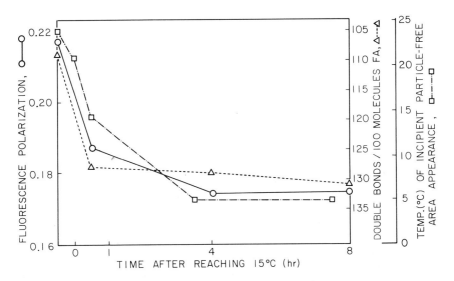

Figure 16. Properties of microsomal membranes during temperature acclimation. Temperature of particle-free appearance, □; fluorescence polarization, ○; number of double bonds in phospholipid fatty acids, △. (Reprinted with permission from C. E. Martin and G. A. Thompson, 1978, *Biochemistry* **17**, 3581–3586. Copyright 1978 by the American Chemical Society.)

and Rose, 1969), has been ruled out in *Tetrahymena* (Skriver and Thompson, 1976). On the other hand, our results (Martin *et al.*, 1976; Kasai *et al.*, 1976) have provided evidence that the desaturase activities may be regulated by the physical state of the membrane environment. To test the effects of membrane fluidity upon desaturase activities, cells were incubated with linoleic and γ-linolenic acids prior to a temperature shift. The exogenously added fatty acids were rapidly incorporated into membrane phospholipids, thereby inducing the pronounced changes in fatty acyl composition shown in Table XV. In the early acclimation phase the most striking change in membrane fatty acid occurs in microsomes, and is followed by a net movement to other membranes. Such rapid incorporation of fatty acids into microsomes was reflected in freeze–fracture replicas, indicating the fluidizing effect of unsaturated fatty acid supplementation. Figure 17 demonstrates the particle density index (PDI) curves for supplemented and unsupplemented cells during acclimation to 15°C. Although both kinds of acclimating cells were found to increase in fluidity with time, the fed cells were consistently more fluid than the unfed cells.

To measure the activity of fatty acid desaturase during acclimation, the incorporation of [¹⁴C]acetate into various fatty acids of phospholipids was

Table XV Distribution of Major Fatty Acids in Membranes of $C_{18:2}$-plus $C_{18:3}$-Fed Cells 1 and 4 Hours after Shifting from 39.5° to 15°C[a]

Fatty acids	Cilia 39.5°C control	1 hr	4 hr	Pellicles 39.5°C control	1 hr	4 hr	Microsomes 39.5°C control	1 hr	4 hr
$C_{14:0}$	6.7	5.9	5.9	8.9	11.5	8.9	7.3	8.6	5.6
$C_{16:0}$	16.8	17.4	15.7	16.4	16.2	12.8	13.2	12.0	8.0
$C_{16:1,\Delta^9}$	8.3	12.8	7.9	7.1	8.7	7.4	8.7	9.0	6.7
$C_{16:2,\Delta^?}$	2.3	2.2	2.2	4.0	2.0	2.3	4.9	3.0	2.1
$C_{18:1,\Delta^9}$	11.7	15.0	13.6	12.6	4.9	4.7	13.9	4.8	4.3
$C_{18:2,\Delta^{9,12}}$	8.1	11.4	11.8	10.5	14.0	17.4	13.0	16.2	18.7
$C_{18:3,\Delta^{6,9,12}}$	20.7	20.7	29.8	18.5	31.7	37.1	21.4	37.3	48.0

[a] Data reprinted with permission from C. E. Martin *et al.* (1976), *Biochemistry* **15**, 5218–5227. Copyright 1976 by the American Chemical Society.

followed. Thirty-two percent of the fatty acid radioactivity of isothermal 39.5°C cells was found in $C_{14:0}$ and $C_{16:0}$ after 2 hr, whereas only 12% was in these saturated fatty acids of cells shifted to 15°C. Unfed cells also showed a slightly higher rate of desaturation (mainly into $C_{16:1}$) than cells grown isothermally at 15°C during the first 3 hr after the shift (Figure

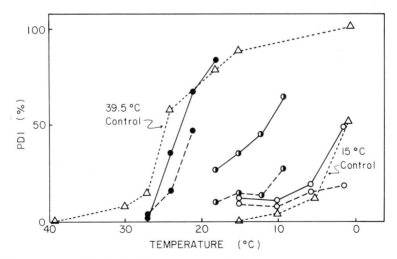

Figure 17. Particle density index (PDI) versus fixation temperature of the outer alveolar membrane of *T. pyriformis* NT-I cells supplemented with $C_{18:2}$ plus $C_{18:3}$. Control, △; 0.5 hr after shift (39.5° → 15°C), ●; 4 hr, ◗; 8 hr, ○. Solid line, unfed; thick broken line, fed. (Reprinted with permission from C. E. Martin *et al.*, 1976, *Biochemistry* **15**, 5218–5227. Copyright 1976 by the American Chemical Society.)

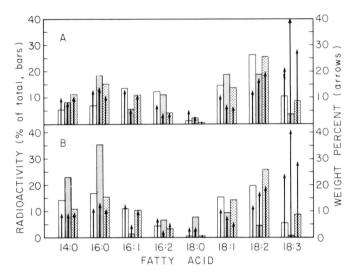

Figure 18. [1-¹⁴C]Acetate incorporation into phospholipid fatty acids in *T. pyriformis* NT-I cells supplemented with $C_{18:2}$ plus $C_{18:3}$. (A) Three hours after the shift (39.5° → 15°C); (B) 6 hr. Left-hand bar, unfed temperature-shifted; middle bar, $C_{18:2}$ plus $C_{18:3}$ fed temperature-shifted; right-hand bar, unfed 15°C-grown. (Reprinted with permission from C. E. Martin *et al.*, 1976, *Biochemistry* **15**, 5218–5227. Copyright 1976 by the American Chemical Society.)

18A). The more fluid membranes of cells supplemented with $C_{18:2}$ and $C_{18:3}$ revealed a decreased desaturation of fatty acids newly synthesized *de novo*. This result was observed in $C_{16:1}$, $C_{18:2}$, and $C_{18:3}$ (Figures 18A and 18B). These results indicate that increasing desaturation is correlated with decreasing membrane fluidity. Further evidence supporting the hypothesis of fluidity-dependent control of desaturase was obtained from results of several experiments. Kitajima and Thompson (1977b) have observed that incorporation of monomethoxy derivatives of stearic acid caused an increase in membrane fluidity, and also that the fluidizing effect resulted in a marked reduction in fatty acid desaturase activity similar to that caused by supplements of polyunsaturated fatty acids. Furthermore, they demonstrated that the desaturase activity was also depressed in cells incubated with a general anesthetic, methoxyflurane, which was shown to have a fluidizing effect (Nandini-Kishore *et al.*, 1977). Whereas data from the above experiments would seem adequate to support the conclusion that augmented membrane fluidity induces a reduction of fatty acid desaturase activity, there is another piece of evidence confirming this thesis. That is, the membrane-rigidifying effect induces a significant increase in desaturase activity even under isothermal conditions. Since membranes from cells supplemented with ergosterol or β-sitosterol have been known to be less fluid than those of control cells (Thompson and Nozawa, 1977; Kasai *et al.*, 1977; Shimonaka *et al.*, 1978), activities of fatty acid desaturation were compared in both cell systems. Thus, the ergosterol-replaced cells were found to show a higher rate of [^{14}C]palmitate to [^{14}C]palmitoleate conversion as compared with the control cells. Several experimental results, summarized in Table XVI, strongly implicate a close correlation of the fluidity with desaturase activity in endoplasmic reticulum. However, detailed information about how various desaturase enzymes are regulated by the membrane fluidity has to wait for future extensive work, including studies of the electron transport system in *Tetrahymena* microsomes. It is worth noting the contrasting evidence that no distinct correlation exists between stearyl-CoA desaturase activity and membrane fluidity in microsomes (Holloway and Holloway, 1977).

In addition to the fluidity-regulating theory, we have recently observed findings which would be strongly indicative of changes in the concentration of desaturase enzymes occurring during temperature acclimation (Nozawa and Kasai, 1978), which have been observed with *B. megaterium* (Fulco, 1974; Fujii and Fulco, 1977). In particular, the conversion rate of palmitate to palmitoleate, a key process for quick adaptation, was shown to be highest 2 hr after a shift from 39.5° to 15°C, and to be even higher than that found in the 15°C isothermally grown cells. However, addition of cycloheximide prior to the shift prevented a significant increase in desaturation activity of palmitate. This hypothesis of temperature-mediated induction of desaturases was just recently confirmed by direct assay of palmitoyl-CoA desaturase in microsomes isolated at different intervals after a shift to 15°C (H. Fukushima and co-workers, unpublished

Table XVI Effects of Membrane Fluidity on Fatty Acid Desaturation[a]

Type of experiment	Fluidity of endoplasmic reticulum	Effects on enzymatic desaturation
(A) Shifting 39.5° to 15°C	Suboptimal	Higher than 15°C controls
(B) Feeding $C_{18:2}$ plus $C_{18:3}$ before shifting from 39.5 to 15°C	Optimal—superoptimal(?)	Lower than 15°C controls
(C) Feeding $C_{18:2}$ to 39.5°C	Superoptimal	Lower than 39.5°C controls
(D) Shifting 15°C cells to 39.5°C	Superoptimal	Lower than 39.5°C controls
(E) Anesthetics treatment before shifting from 39.5 to 15°C	Superoptimal	Lower than 15°C controls
(F) Feeding ergosterol before shifting from 39.5 to 15°C	Suboptimal	Higher than 15°C controls

[a] Data from Martin *et al.* (1976), Nandini-Kishore *et al.* (1977), and Kasai *et al.* (1977).

observations). Therefore, it is conceivable that the content of palmitoyl-CoA desaturase might also be regulated by an effect of membrane fluidity on the rate of desaturase enzyme biosynthesis.

VIII. DYNAMIC BIOLOGICAL ACTIVITY IN *TETRAHYMENA* MEMBRANES

Abundant evidence is accumulating to support the concept that a variety of biological phenomena taking place in membranes are closely associated with the physical state of membrane lipids (Singer, 1974). The *Tetrahymena* cell is thought to be a highly suitable model system for understanding the nature of membrane-associated functions because it has highly developed, readily isolable organelles whose lipid composition can be manipulated with ease (Thompson and Nozawa, 1977). Several examples are described in this chapter.

A. Membrane-Bound Enzymes

1. Adenylate and Guanylate Cyclases

A body of evidence has accumulated which implicates cyclic AMP and cyclic GMP as regulators of many biological functions in the cell. Specifically, cell growth is known to be sensitively controlled by these cyclic nucleotides in a wide variety of cells (Pastan *et al.*, 1975), even in *Tetrahymena* (Wolfe, 1973; Dickinson *et al.*, 1976). It was somewhat surprising to find a fairly high activity of adenylate cyclase in a free-living unicellular eukaryote, *T. pyriformis,* which is apparently not exposed to the direct influence of hormones under its normal growth condition. However, the enzyme was observed to respond to epinephrine, and its stimulation by this hormone could be abolished by the β-adrenergic blocking agent, propranolol, but not by an α-adrenergic blocker (Rozensweig and Kindler, 1972). The membrane-bound cyclase could be partially extracted by washing the membrane fraction with 0.25 M sucrose. The enzyme thus dissociated was still responsive to epinephrine, whereas the cyclase preparation treated with Triton X-100 was no longer stimulated by epinephrine (Kassis and Kindler, 1975). We have recently shown a predominant enrichment of adenylate cyclase activity in the surface membrane (pellicle) as compared with the other membrane fractions in *T. pyriformis* NT-I (Shimonaka and Nozawa, 1977). Since this enzyme is tightly associated with the pellicle membrane, it seems likely that alterations in membrane structure or composition may drastically change the properties of this enzyme. Indeed, the specific activity of adenylate cyclase in the ergosterol-replaced pellicle was much lower than that observed in the native pellicle. Also, in the Arrhenius plots a marked change in activity occurred at about 22°C (Figure 19), whereas there was a sharp discontinuity at 28°C in the native pellicle containing tetrahymanol (Shimonaka and Nozawa, unpublished observations). These two temperatures are coincident with the onset temperatures of membrane particle aggregation (phase separation) in the native and the sterol-replaced pellicle. This may indicate that the activity of adenylate cyclase is dependent upon the physical state of the *Tetrahymena* surface membrane.

In addition to adenylate cyclase mentioned above, guanylate cyclase was found in *Tetrahymena* (Gray *et al.,* 1977), and was localized in its surface membrane (Nakazawa, 1979). Whereas guanylate cyclases from microorganisms and mammalian tissue have been reported to require Mn^{2+} as the sole metal cofactor with Mg^{2+} as a very poor substitute, the cyclase of *Tetrahymena* surface membranes was observed to prefer Mg^{2+} to Mn^{2+} as a cofactor. Further work on these two cyclases bound to the surface membrane is in progress.

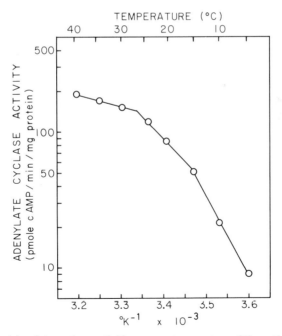

Figure 19. Adenylate cyclase activities versus temperature of *T. pyriformis* NT-I pellicles. (From Shimonaka and Nozawa, 1977, reproduced by permission.)

2. Glucose-6-Phosphatase

Glucose-6-phosphatase is known to be a microsome-bound enzyme, and is frequently used as a representative marker enzyme. Wunderlich and Ronai (1975) have found using freeze–fracture electron microscopy that *T. pyriformis* GL cells grown at 18°C developed particle-free areas in their microsomal membranes at 12°C, the same temperature at which the activity of glucose-6-phosphatase of smooth microsomes underwent a sudden reduction in rate. They have added more detailed data (Wunderlich *et al.*, 1975) which support evidence that the structural changes correlate with alterations in membrane functions. Thus the smooth microsomes isolated from 28°C-grown cells showed discontinuities at 17°C in Arrhenius plots of glucose-6-phosphatase (Figure 20), fluorescence intensity of 8-anilino-1-naphthalenesulfonate, separation of the outer hyperfine extrema $(2T_{\parallel})$ of 5-doxylstearate, and in the phase partition of 4-doxyldecane. To see whether microsomal lipids undergo a phase transition, liquid crystalline to crystalline, the extracted lipids were examined by proton nuclear magnetic resonance and also by differential scanning calorimetry. The absence of detectable changes in the transition tempera-

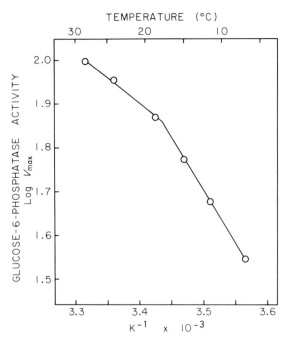

Figure 20. Glucose-6-phosphatase activities versus temperature of *T. pyriformis* GL microsomes. (From Ronai and Wunderlich, 1975, reproduced by permission.)

ture suggested that thermotropic membrane alterations in *T. pyriformis* may not be induced by free lipid phase transition, but rather by cluster formation of "quasi-crystalline" lipids.

3. Fatty Acid Desaturases

As described above (Section VII,C), fatty acid desaturases play crucial roles in adjusting the fatty acyl composition of membrane phospholipids in response to environmental temperature changes. Activities of these enzymes which are bound to microsomes were suggested to be highly dependent upon the physical state of membrane lipids. Thus, we have postulated a membrane fluidity-controlling theory for thermal regulation of desaturase activity (Martin *et al.*, 1976; Kasai *et al.*, 1976; Kitajima and Thompson, 1977b; Nandini-Kishore *et al.*, 1977). Recently, palmitoyl-CoA desaturase in microsomes from 39.5°C-grown cells, probably a key enzyme for the quick temperature adaptation, was shown to reveal a clear discontinuity at 30°C and a slight bend around 15°C in the Arrhenius plots of its activity (Fukushima *et al.*, 1977). Similarily, two breaks in the Arrhenius plots of stearyl-CoA desaturase activity was observed at 26° and 15°C. The higher temperatures at which marked transitions occur in the activation energy are roughly correlated with the temperature for emergence of membrane particle-devoid areas in endoplasmic reticulum. However, such a rearrangement of membrane particles is thought to be a disappearance, rather than an aggregation, as observed in other membranes, of membrane particles, indicating vertical displacement of membrane proteins caused by thermotropic changes in fluidity, as was demonstrated in lymphoid cells (Wunderlich *et al.*, 1974a) and erythrocytes (Borochov and Shinitzky, 1976). Thus membrane proteins may be squeezed out by perpendicular movement toward the outside and become more exposed on the exterior surface. Such dynamic protein movement may be put forward as a potential explanation for thermal regulation mechanism of fatty acid desaturases of *Tetrahymena*; increased lipid rigidity increases the exposure of the desaturases.

4. Adenosine Triphosphatase

In *T. pyriformis*, cilia are known to possess axonemal ATPase, a 14 S and a 30 S dynein (Gibbons, 1966). The surface membrane (pellicle) also has fairly high activity of ATPase which has not yet been characterized (Kawai and Nozawa, unpublished observations). In Arrhenius plots of 39°C-grown cell preparations this pellicle-bound ATPase shows a break at about 27°C. However, treatment of the membrane with the local anesthetic, dibucaine, or the inhalation anesthetic, methoxyflurane, was observed to inhibit and lower the transition temperature of the ATPase activity to 21°C (Saeki and Nozawa, unpublished observations). In the anesthetic-treated membranes, no distinct breaks due to phase transition were observed at 32°C, where a discontinuity was seen in the control, untreated membranes, as inferred by fluorescence polarization using 1,6-diphenyl-1,3,5-hexatriene. Instead, there was a break at 20°C in both anesthetic-treated membranes, with methoxyflurane exerting an especially abrupt change in activation energy at 20°C. These results may indicate that the anesthetic-induced alterations in the physical properties of the pellicular membrane might lead to inhibition of ATPase activity bound to the membrane.

Recently a Ca^{2+}-activated ATPase with a molecular weight of 89,000 was isolated from the cytosol of *T. pyriformis* (Chua *et al.*, 1977). The increase in the cytosol ATPase activity in stationary growth phase suggests some role in cell division. It is of great interest to note that this specific enzyme is different from Ca^{2+}-ATPase found in nuclei, mitochondria, and microsomes.

5. 5'-Nucleotidase

5'-Nucleotidase has been suggested to behave as an "ecto-enzyme" that is located in the outer surface of the plasma membrane (DePierre and Karnovsky, 1973). The Arrhenius plot of 5'-nucleotidase activity in the pellicular membrane shows no change in activation energy (Figure 21), suggesting that this enzyme may not be closely associated with physical state of membrane lipid (Kawai and Nozawa, unpublished observations).

6. Tetrahymanol-Synthesizing Enzymes

Whereas we have described previously that exogenous sterols such as ergosterol and cholesterol inhibit tetrahymanol biosynthesis and can be inserted into membranes, the precise mechanisms for this inhibition remain obscure. However, an explanation has been advanced that exogenous sterols cause structural alterations in membrane physical properties, thus inducing conformational changes capable of lowering the activity of membrane-bound enzymes that involve tetrahymanol biosynthesis (Beedle *et al.*, 1974).

7. Acid Phosphatase

Membrane-bound acid phosphatases have been reported from a variety of organisms. Recently, a particulate-bound form of acid phosphatase was partially purified from *Tetrahymena* (Williams and Juo, 1976). Since the acid phosphatase was activated by treatment with Triton X-100 or Nonidet P-40 and aggregated upon removal of the detergent and mem-

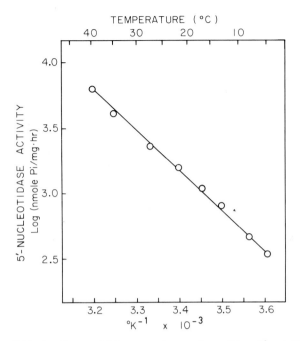

Figure 21. 5′-Nucleotidase activities versus temperature of *T. pyriformis* NT-I pellicles. (Kawai and Nozawa, unpublished observations.)

brane lipids, it seems that this enzyme is an integral membrane protein which might require lipids for effective function.

B. Mobilization of Cell Constituents

There are a couple of examples which indicate that the membrane fluidity might exert considerable influence on mobilization of cell components.

1. Transfer (Exchange) of Membrane Lipids

The transfer of lipids between various organelle membranes was examined with *Tetrahymena* using spin labeling (Iida *et al.*, 1978). The spin-labeled lecithin was observed to transfer from microsomes to other membranes at a greater rate in the order, cilia > pellicles > microsomes (Figure 10). Taking into consideration that the membrane fluidity increased in this order, the physical state of the acceptor membrane is thought to act as an effective regulator for membrane lipid assembly. Furthermore, such a concept was supported by another experiment. The transfer rate of lecithin from the spin-labeled microsomes to other organelles of 15°C-grown cells which are more fluid was found to be faster than that to the less fluid membranes of 39.5°C-grown cells.

2. Nucleocytoplasmic RNA Transport

The nuclear envelope is thought to be the critical site for regulation of nucleocytoplasmic exchanges of macromolecules, including ribonucleic acid. Furthermore, it is known that nucleocytoplasmic RNA transport is thermodependent; it decreases at low temperature (Horisberger and Amos, 1970). Wunderlich and co-workers have demonstrated by extensive work using the freeze–fracture electron microscope that nuclear membranes indeed undergo reversible, thermotropic structural alterations (Wunderlich et al., 1973, 1974a). When cells were chilled from the optimal growth temperature to lower temperatures, particle-depleted smooth areas appeared, accompanied by the decreased frequency of nuclear pore complexes. In addition to such membrane structure changes, they have shown data which clearly indicate that the higher the incubation temperature, the faster migration of RNA molecules from nucleus to cytoplasm occurs (Wunderlich et al., 1974b). In the Arrhenius plot of the nucleus to cytoplasm transport of RNA, cells grown at 23° and 18°C reveal breaks at 15° and 12°C, respectively (Nägel and Wunderlich, 1977). Figure 22 depicts the findings with 23°C-grown cells, demonstrating an abrupt change in activation energy at about 15°C. Accordingly, it should be noted that at those temperatures where the transition of RNA transport occurs, the particle-devoid smooth regions begin to emerge in nuclear membranes. On the basis of these results, they postulated a hypothesis that the thermotropic lipid phase transition may control the switch, "on" and "off" for RNA transport, by converting nuclear pore complexes from the "open" to the "closed" state and vice versa.

C. Membrane Fusion-Involving Cytokinesis

1. Endocytosis

There is a large body of evidence indicating that insoluble particles induce formation of food vacuoles in Tetrahymena and that this process may be the major site of nutrient uptake into the cell, as recently reviewed in detail by Nilsson (1976). Although the precise mechanisms by which endocytic vacuoles are formed remain obscure, several events are known to occur: (1) assembly of precursor vesicles, (2) ingestion of particles, (3) pinching off, and (4) fusion with lysosomes. Whereas the prexisting plasma membrane has been known in many cells to be used as bulk material for vacuole formation (Silverstein et al., 1977), an alternate

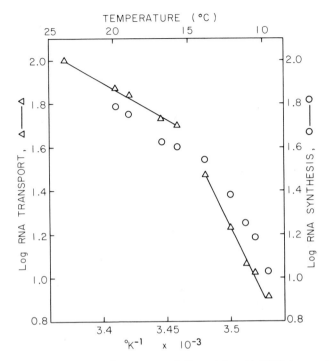

Figure 22. Nucleocytoplasmic transport of ribonucleic acid. RNA transport, △; RNA synthesis, ○. (From Nägel and Wunderlich, 1977, reproduced by permission.)

source of vacuolar membrane material, endoplasmic reticulum, was suggested from the radioactive labeling pattern of isolated vacuoles (Weidenbach and Thompson, 1974). Quite recently, Kitajima and Thompson (1977c) followed a series of events occurring during endocytosis by use of freeze–fracture electron microscopy and presented suggestive evidence that the appropriate physical state of the vacuole membranes or their precursor vesicles may be essential for continued vacuole formation. With *Tetrahymena*, there is, up to now, no direct evidence to support the concept that interior membrane is later cycled back for reuse, but such membrane recycling probably happens in this cell, as it has been reported to do in *Paramecium* (Allen, 1976).

2. Exocytosis

It is well known that *Tetrahymena* and *Paramecium* have exocytic mucocysts and trichocysts, respectively, which are involved in secretion in these protozoan cells. The mucocyst is a unit membrane-surrounded structure in the shape of an elongated sac, and, it is suggested, orginates from the endoplasmic reticulum (Tokuyasu and Scherbaum, 1965). The maturing mucocysts containing mucoprotein crystals approach the specialized fusion sites of the plasma membrane and consequently discharge their contents outside the cell. Satir and co-workers (1972, 1973) have observed by freeze–fracture electron microscopy a sequence of events demonstrating that a specific membrane array, the "fusion rosette," consisting of a ring of nine particles surrounding a central particle, is formed at the plasma membrane site where fusion will take place. Such similar membrane structures have been demonstrated at the fusion site of trichocyst discharge in *Paramecium* (Hausmann and Allen, 1976; Plattner, 1976). Although little is known about the trigger mechanism for membrane fusion and discharge, one could speculate that Ca^{2+} ion may be involved in it.

It has been proposed that the mucocyst membrane is irreversibly merged with the plasma membrane following fusion, therefore contributing appreciably to membrane growth (Satir *et al.*, 1972). It is tempting to think, based on calculation of membrane area, that surface membrane biogenesis can be accounted for by insertion of preexisting mucocyst membrane. However, some recent evidence from studies of *Paramecium* has led to an alternative conclusion that exocytosis involves retrieval of the secretory vesicle membranes rather than their insertion into the cell membrane for membrane growth (Plattner, 1976).

3. Conjugation

Another membrane fusion process in *Tetrahymena* is conjugation, which is one of the representative phenomena of cell surface recognition. Actually, three major events have been known to occur: (1) "co-stimulation" of random collison of two complementary types, (2) "pairing," and fusion (Frisch and Loyter, 1977). They also have reported that the lectin concanavalin A inhibits the induction of the conjugation process when added to the mixture of mating types and can even dissociate attached pairs. Therefore, it may be possible that specific surface glycoproteins participate in the conjugation process at the recognition stage that concanavalin A binds directly to them, thus inhibiting conjugation. With regard to membrane structure, there appear certain regions completely devoid of membrane particles prior to the terminal stage, fusion. More recently, the involvement of *de novo* lipid synthesis in these membrane phenomena, cell recognition and fusion, was implied (Frisch *et al.*, 1978). Indeed, cells otherwise competent to mate cannot initiate the conjugation sequence when incubated with cerulenin, a potential inhibitor of lipid synthesis. During the conjugation the ratio of saturated to unsaturated fatty acids increases from 0.30 to 0.45, and one would therefore expect that the membrane may become less fluid. However, if this is the case, the conjugation mechanism may not be simply explained by the widely accepted rationale that the fluid membrane regions are capable of undergoing fusion (Lucy, 1970).

IX. EPILOGUE

We have described several topics related to *Tetrahymena* membranes. Since this cell possesses numerous advantages for studying membrane biology, it is not unreasonable for us to believe that *Tetrahymena* is a sort of a treasury from which continued interest and extensive work will bring about much presently unanticipated knowledge of molecular cell biology, behavior, recognition, and even memory.

REFERENCES

Åkesson, B. (1977). *Biochem. Biophys. Res. Commun.* **76**, 93–99.

Allen, R. D. (1976). *Cytobiologie* **12**, 254–261.

Baugh, L. C., and Thompson, G. A. (1975). *Exp. Cell Res.* **94**, 111–121.

Beattie, D. S. (1969). *J. Membr. Biol.* **1**, 383–401.

Beedle, A. S., Mundy, K. A., and Wilton, D. C. (1974). *Biochem. J.* **142**, 57–64.

Benjamins, J. A., Herschkowitz, N., Robinson, J., and McKhann, G. M. (1971). *J. Neurochem.* **18**, 729–738.

Berger, H., Jones, P., and Hanahan, D. (1972). *Biochim. Biophys. Acta* **260**, 617–629.

Borochov, H., and Shinitzky, M. (1976). *Proc. Natl. Acad. Sci. U.S.A* **73**, 4526–4530.

Brown, C. M., and Rose, A. H. (1969). *J. Bacteriol.* **99**, 371–378.

Carter, H. E., and Gaver, R. C. (1967). *Biochem. Biophys. Res. Commun.* **29**, 886–891.

Child, F. M. (1959). *Exp. Cell Res.* **18**, 258–267.

Chlapowski, F. J., and Band, R. N. (1971). *J. Cell Biol.* **50**, 634–651.

Chua, B., Elson, C., and Shrago, E. (1977). *J. Biol. Chem.* **252**, 7548–7554.

Conner, R. L., and Stewart, B. Y. (1976). *J. Protozool.* **23**, 193–196.

Conner, R. L., Koo, K.-E., and Landrey, J. R. (1974). *Lipids* **9**, 554–559.

Cronan, J. E., and Gelman, E. P. (1975). *Bacteriol. Rev.* **39**, 232–256.

Curgy, J. J., Ledoigt, G., Stevens, B. J., and André, J. (1974). *J. Cell Biol.* **60**, 628–640.

DePierre, J. W., and Karnovsky, M. L. (1973). *J. Cell Biol.* **56**, 275–303.

Dickinson, J. R., Gravis, M. G., and Swoboda, B. E. P. (1976). *FEBS Lett.* **65**, 152–154.

Elgsaeter, A., and Branton, D. (1974). *J. Cell Biol.* **63**, 1018–1030.

Elliott, A. M., ed. (1973). "Biology of *Tetrahymena*." Dowden, Hutchinson & Ross, Stroudsburg, Pennsylvania.

Erwin, J. (1973). *In* "Lipids and Biomembranes of Eukaryotic Microorganisms" (J. Erwin, ed.), pp. 41–143. Academic Press, New York.

Erwin, J., and Bloch, K. (1963). *J. Biol. Chem.* **238**, 1618–1624.

Everhart, L. P. (1972). *Methods Cell Biol.* **5**, 219–288.

Ferguson, K. A., Conner, R. L., and Mallory, F. B. (1971). *Arch. Biochem. Biophys.* **144**, 448–450.

Ferguson, K. A., Davis, F. M., Conner, R. L., Landrey, J. R., and Mallory, F. B. (1975). *J. Biol. Chem.* **250**, 6998–7005.

Franke, W. W., Kartenbeck, J., Zentgraf, H., Sheer, V., and Falk, H. (1971). *J. Cell Biol.* **51**, 881–888.

Frisch, A., and Loyter, A. (1977). *Exp. Cell Res.* **110**, 337–346.

Frisch, A., Loyter, A., Levy, R., and Goldberg, I. (1978). *Biochim. Biophys. Acta* **506**, 18–29.

Fujii, D. K., and Fulco, A. J. (1977). *J. Biol. Chem.* **252**, 3660–3670.

Fukushima, H., Martin, C. E., Iida, H., Kitajima, Y., Thompson, G. A., and Nozawa, Y. (1976a). *Biochim. Biophys. Acta* **431**, 165–179.

Fukushima, H., Watanabe, T., and Nozawa, Y. (1976b). *Biochim. Biophys. Acta* **436**, 249–259.

Fukushima, H., Nagao, S., Okano, Y., and Nozawa, Y. (1977). *Biochim. Biophys. Acta* **488**, 442–453.

Fukushima, H., Kasai, R., Akimori, N., and Nozawa, Y. (1978). *Jpn. J. Exp. Med.* **48**, 373–380.

Fulco, A. J. (1974). *Annu. Rev. Biochem.* **43**, 215–241.

Gibbons, I. R. (1965). *J. Cell Biol.* **25**, 400–402.

Gibbons, I. R. (1966). *J. Biol. Chem.* **241**, 5590–5596.

Glaser, M., Ferguson, K. A., and Vagelos, P. R. (1974). *Proc. Natl. Acad. Sci. U.S.A.* **71**, 4072–4076.

Gorovsky, M. A. (1970). *J. Cell Biol.* **47**, 619–630.

Gray, N. C. C., Dickinson, J. R., and Swoboda, B. E. P. (1977). *FEBS Lett.* **81**, 311–314.

Hale, A. H., Pessin, J. F., Palmer, F., Weber, M. J., and Glaser, M. (1977). *J. Biol. Chem.* **252**, 6190–6200.

Harris, P., and James, A. T. (1969). *Biochim. Biophys. Acta* **187**, 13–18.

Hartman, H., Moss, P., and Gurney, T. (1972). *J. Cell Biol.* **55**, 107a.

Hausmann, K., and Allen, R. D. (1976). *J. Cell Biol.* **69**, 313–326.

Hill, D. L. (1972). *In* "Biochemistry and Physiology of *Tetrahymena*." Academic Press, New York.

Holloway, C. T., and Holloway, P. W. (1977). *Lipids* **12**, 1025–1031.

Holz, G. G., and Conner, R. L. (1973). *In* "Biology of *Tetrahymena*" (A. M. Elliott, ed.), pp. 99–122. Dowden, Hutchison & Ross, Stroudsburg, Pennsylvania.

Horisberger, M., and Amos, H. (1970). *Exp. Cell Res.* **62**, 467–470.

Iida, H., Maeda, T., Ohki, K., Nozawa, Y., and Ohnishi, S. (1978). *Biochim. Biophys. Acta* **508**, 55–64.

Jonah, M., and Erwin, J. A. (1971). *Biochim. Biophys. Acta* **231**, 80–92.

Kahane, I., and Razin, S. (1969). *Biochim. Biophys. Acta* **183**, 79–89.

Kapoulas, V. M., Thompson, G. A., and Hanahan, D. J. (1969). *Biochim. Biophys. Acta* **176**, 237–249.

Kasai, R., Kitajima, Y., Martin, C. E., Nozawa, Y., Skriver, L., and Thompson, G. A. (1976). *Biochemistry* **15**, 5228–5233.

Kasai, R., Sekiya, T., Okano, Y., Nagao, S., Ohki, K., Ohnishi, S., and Nozawa, Y. (1977). *Maku (Membrane)* **2**, 301–312.

Kassis, S., and Kindler, S. H. (1975). *Biochim. Biophys. Acta* **391**, 513–516.

Kennedy, E. K., and Thompson, G. A. (1970). *Science* **168**, 989–991.

Kitajima, Y., and Thompson, G. A. (1977a). *J. Cell Biol.* **72**, 744–755.

Kitajima, Y., and Thompson, G. A. (1977b). *Biochim. Biophys. Acta* **468**, 73–80.

Kitajima, Y., and Thompson, G. A. (1977c). *J. Cell Biol.* **75**, 436–445.
Koroly, M. J., and Conner, R. L. (1976). *J. Biol. Chem.* **251**, 7588–7592.
Krishnan, K. S., and Balaram, P. (1975). *FEBS Lett.* **60**, 419–421.
Lees, A. M., and Korn, E. D. (1966). *Biochemistry* **5**, 1475–1481.
Lenaz, G., Bertoli, E., Curatola, G., Mazzanti, L., and Bigi, L. (1976). *Arch. Biochem. Biophys.* **172**, 278–288.
Liang, C. R., and Rosenberg, H. (1966). *Biochim. Biophys. Acta* **125**, 295–305.
Lloyd, D., Brightwell, R., Venables, S. E., Roach, G. I., and Turner, G. (1971). *J. Gen. Microbiol.* **65**, 209–223.
Lucy, J. A. (1970). *Nature (London)* **227**, 815–817.
MacDonald, A. G. (1978). *Biochim. Biophys. Acta* **507**, 26–37.
Marr, A. G., and Ingraham, J. L. (1962). *J. Bacteriol.* **84**, 1260–1267.
Martin, C. E., and Thompson, G. A. (1978). *Biochemistry* **17**, 3581–3586.
Martin, C. E., Hiramitsu, K., Kitajima, Y., Nozawa, Y., Skriver, L., and Thompson, G. A. (1976). *Biochemistry* **15**, 5218–5227.
Mindich, L. (1970). *J. Mol. Biol.* **49**, 433–439.
Müller, M., Hogg, J. F., and de Duve, C. (1968). *J. Biol. Chem.* **243**, 5385–5395.
Nägel, W. C., and Wunderlich, F. (1977). *J. Membr. Biol.* **32**, 151–164.
Nakazawa, K., Shimonaka, H., Nagao, S., and Nozawa, Y. (1979). *J. Biochem.* (in press).
Nandini-Kishore, S. G., Kitajima, Y., and Thompson, G. A. (1977). *Biochim. Biophys. Acta* **471**, 157–161.
Nandini-Kishore, S. G., Matlox, S. M., Martin, C. E., and Thompson, G. A. (1979). *Biochim. Biophys. Acta* **551**, 315–327.
Nilsson, J. R. (1976). *C. R. Trav. Lab. Carlsberg* **40**, 215–355.
Nozawa, Y. (1973). *J. Biochem. (Tokyo)* **74**, 1157–1163.
Nozawa, Y. (1975). *Methods Cell Biol.* **10**, 105–133.
Nozawa, Y. (1977). *In* "Methods in Biochemical Experiments: Biomembranes" (Jap. Biochem. Soc., ed.), Vol. 15, pp. 245–249. Tokyo Kagakudojin, Tokyo.
Nozawa, Y., and Kasai, R. (1978). *Biochim. Biophys. Acta* **529**, 54–66.
Nozawa, Y., Kasai, R., Sakiya, T. (1979). *Biochim. Biophys. Acta* **552**, 38–52.
Nozawa, Y., and Thompson, G. A. (1971a). *J. Cell Biol.* **49**, 712–721.
Nozawa, Y., and Thompson, G. A. (1971b). *J. Cell Biol.* **49**, 722–729.
Nozawa, Y., and Thompson, G. A. (1972). *Biochim. Biophys. Acta* **282**, 93–104.
Nozawa, Y., Fukushima, H., and Iida, H. (1973). *Biochim. Biophys. Acta* **318**, 335–344.
Nozawa, Y., Iida, H., Fukushima, H., Ohki, K., and Ohnishi, S. (1974). *Biochim. Biophys. Acta* **367**, 134–147.
Nozawa, Y., Fukushima, H., and Iida, H. (1975). *Biochim. Biophys. Acta* **406**, 248–263.
Nunn, W. D. (1975). *Biochim. Biophys. Acta* **380**, 403–413.
Nunn, W. D. (1977). *Biochemistry* **16**, 1077–1081.
Nwanze, C. E. A., Hanks, R., Dodd, G., and Howarth, O. (1977). *Chem. Phys. Lipids* **18**, 267–273.
Ojakian, G. K., and Satir, P. (1974). *Proc. Natl. Acad. Sci. U.S.A.* **71**, 2052–2056.
Oldfield, E., and Chapman, D. (1972). *FEBS Lett.* **23**, 285–297.
Omura, S. (1976). *Bacteriol. Rev.* **40**, 681–697.
Pastan, I. H., Johnson, G. S., and Anderson, W. B. (1975). *Annu. Rev. Biochem.* **44**, 491–522.
Plattner, H. (1976). *Exp. Cell Res.* **103**, 431–435.
Reitz, R. C., Helsabeck, E., and Mason, D. P. (1973). *Lipids* **8**, 80–84.
Renaud, F. L., Rowe, A. J., and Gibbons, I. R. (1968). *J. Cell Biol.* **36**, 79–90.
Ronai, A., and Wunderlich, F. (1975). *J. Membr. Biol.* **24**, 381–399.

Rosenbaum, J. L., and Carlson, K. (1969). *J. Cell Biol.* **40,** 415–425.
Rozensweig, Z., and Kindler, S. H. (1972). *FEBS Lett.* **25,** 221–223.
Satir, B., Schooley, C., and Satir, P. (1972). *Nature (London)* **235,** 53–54.
Satir, B., Schooley, C., and Satir, P. (1973). *J. Cell Biol.* **56,** 153–176.
Seaman, G. R. (1960). *Exp. Cell Res.* **21,** 292–302.
Shimonaka, H., and Nozawa, Y. (1977). *Cell Struct. Funct.* **2,** 81–89.
Shimonaka, H., Fukushima, H., Kawai, K., Nagao, S., Okano, Y., and Nozawa, Y. (1978). *Experientia* **34,** 586–587.
Shinitzky, M., and Inbar, M. (1974). *J. Mol. Biol.* **71,** 2128–2130.
Shorb, M. S., Dunlap, B. E., and Pollard, W. O. (1965). *Proc. Soc. Exp. Biol. Med.* **118,** 1140–1145.
Shorey, R. A., Denim, C., and Thompson, G. A. (1978). In preparation.
Silverstein, S. C., Steinman, R. M., and Cohn, Z. A. (1977). *Annu. Rev. Biochem.* **46,** 669–722.
Singer, S. J. (1974). *Annu. Rev. Biochem.* **43,** 805–833.
Singer, S. J., and Nicolson, G. L. (1972). *Science* **175,** 720–731.
Skriver, L., and Thompson, G. A. (1976). *Biochim. Biophys. Acta* **431,** 180–188.
Smith, J. D., and Law, J. H. (1970). *Biochemistry* **9,** 2152–2157.
Smith, J. D., Snyder, W. R., and Law, J. H. (1970). *Biochem. Biophys. Res. Commun.* **39,** 1163–1169.
Speth, V., and Wunderlich, F. (1973). *Biochim. Biophys. Acta* **291,** 621–628.
Subbaiah, P. V., and Thompson, G. A. (1974). *J. Biol. Chem.* **249,** 1302–1310.
Taketomi, T. (1961). *Z. Allg. Mikrobiol.* **1,** 331–340.
Thompson, G. A. (1967). *Biochemistry* **6,** 2015–2022.
Thompson, G. A., and Nozawa, Y. (1971). *Annu. Rev. Microbiol.* **26,** 249–278.
Thompson, G. A., and Nozawa, Y. (1977). *Biochim. Biophys. Acta* **472,** 55–92.
Thompson, G. A., Bambery, R. J., and Nozawa, Y. (1971). *Biochemistry* **10,** 4441–4447.
Thompson, G. A., Bambery, R. J., and Nozawa, Y. (1972). *Biochim. Biophys. Acta* **260,** 630–638.
Thompson, G. A., Baugh, L. C., and Walker, L. F. (1974). *J. Cell Biol.* **61,** 253–257.
Tokuyasu, K., and Scherbaum, H. O. (1965). *J. Cell Biol.* **27,** 67–81.
Watson, M. R., and Hopkins, J. M. (1962). *Exp. Cell Res.* **28,** 280–295.
Weidenbach, A., and Thompson, G. A. (1974). *J. Protozool.* **21,** 745–751.
Williams, J. T., and Juo, P.-S. (1976). *Biochim. Biophys. Acta* **422,** 120–126.
Wilson, G., and Fox, C. F. (1971). *J. Mol. Biol.* **55,** 49–60.
Wirtz, K. W. A. (1974). *Biochim. Biophys. Acta* **344,** 95–117.
Wolfe, J. (1973). *J. Cell. Physiol.* **82,** 39–48.
Wunderlich, F., and Ronai, A. (1975). *FEBS Lett.* **55,** 237–241.
Wunderlich, F., and Speth, V. (1972). *J. Ultrastruct. Res.* **41,** 258–269.
Wunderlich, F., Speth, V., Batz, W., and Kleinig, H. (1973). *Biochim. Biophys. Acta* **298,** 39–49.
Wunderlich, F., Wallach, D. F. H., Speth, V., and Fisher, H. (1974a). *Biochim. Biophys. Acta* **373,** 34–43.
Wunderlich, F., Batz, W., Speth, V., and Wallach, D. F. H. (1974b). *J. Cell Biol.* **61,** 633–640.
Wunderlich, F., Ronai, A., Speth, V., Seelig, J., and Blume, A. (1975). *Biochemistry* **14,** 3730–3735.

Phagotrophy in *Tetrahymena* 8

JYTTE R. NILSSON

I. INTRODUCTION

The holotrich ciliate *Tetrahymena* is widely used in biological research. It may be grown in the absence of other organisms, has a short generation time, and grows to high cell densities, which makes it particularly useful for biochemical analysis. In nature, this freshwater ciliate is a suspension feeder, ingesting bacteria and organic detritus. In the laboratory, how-

BIOCHEMISTRY AND PHYSIOLOGY OF PROTOZOA
SECOND EDITION, VOL. 2

ever, *Tetrahymena* is almost exclusively kept axenically in fluid medium which may be chemically defined.

Tetrahymena has a well-developed oral structure. Phagotrophy occurs by formation of food vacuoles at the base of the funnel-shaped buccal cavity. Food vacuoles are formed during uptake both of particulate matter and of fluid. Such uptake via phagocytosis and pinocytosis, respectively, may be described under the common term of endocytosis (Holter, 1959, 1961; de Duve, 1963, 1967). Endocytic uptake in *Tetrahymena* may occur by ways other than by formation of food vacuoles, as will be discussed in Section II,H.

Tetrahymena has a pellicle composed of the cell membrane and of a system of flat membrane-bound alveoli (Pitelka, 1961). Adjacent to each somatic cilium the cell membrane penetrates the layer of alveoli and forms an indentation, the parasomal sac (Pitelka, 1961; Allen, 1967). Furthermore, mucocysts are positioned below the cell membrane between adjacent alveoli (Pitelka, 1961; Tokuyasu and Scherbaum, 1965; Allen, 1967; Williams and Luft, 1968; Wunderlich and Speth, 1972; Hausmann, 1972, 1978; Satir *et al.*, 1972, 1973; Satir, 1974). This pellicle structure is somewhat modified in the oral region.

Phagotrophic uptake leads to digestion of the ingested food, release of metabolites to the cytoplasm, and elimination of indigestible residues. These topics have been discussed previously for *Tetrahymena* and other protozoa (Mast, 1947; Seaman, 1955; Kitching, 1956, 1957; Carasso *et al.*, 1964; Holz, 1964, 1973; Conner, 1967; Müller, 1967; Kidder, 1967; Chapman-Andresen, 1973, 1977; Chapman-Andresen and Müller, 1974; Blum and Rothstein, 1975; Nilsson, 1976, 1977b; Rasmussen, 1976; Gebauer, 1977; Allen, 1978; Rasmussen and Ricketts, 1979). This discussion will be kept in general terms without indicating the particular strain or species of *Tetrahymena* studied, unless such a distinction is necessary for comprehension of the event. Primarily *Tetrahymena pyriformis* has been used in the studies cited; strains of this species have recently been elevated to species (Nanney and McCoy, 1976).

II. INGESTION (ENDOCYTOSIS)

Tetrahymena can utilize bacteria, crude organic media, e.g., proteose–peptone, and chemically defined media. In all cases, food vacuoles are formed in the oral region. The size of the newly formed food vacuole is largely independent of the type of food ingested. Cell growth is influenced by the nature of the ingested nutrients since it is dependent on the rate of utilization of the food. Hence the nutritional value of a particular medium is reflected in the rate of cell multiplication (Kidder, 1967; Holz, 1964, 1973), and *Tetrahymena* is an important tool for assaying the nutritional value of proteins (Hutner *et al.*, 1972, 1973, 1977; Frank *et al.*, 1975; Baker *et al.*, 1978).

Tetrahymena ingests inert particulate matter as readily as bacteria or other edible particles. A wide range of different particles has been used as markers for food vacuoles in *Tetrahymena* (Harding, 1937; Furgason, 1940; Seaman, 1961a; Mueller *et al.*, 1965; Cox, 1967; Nachtweg and Dickinson, 1967; Chapman-Andresen and Nilsson, 1968; Elliott and Clemmons, 1966; Hildebrandt and Duspiva, 1969; Rasmussen and Kludt, 1970; Ricketts, 1970, 1971a; Rasmussen and Modeweg-Hansen, 1973; Wolfe, 1973; Rothstein and Blum, 1973, 1974c; Ricketts and Rappitt, 1975a; Batz and Wunderlich, 1976). These particles are incorporated into all food vacuoles formed in their presence (Chapman-Andresen and Nilsson, 1968) provided sufficient particles are present.

Phagotrophy in *Tetrahymena* is a discontinuous process, because no food vacuoles are formed during cell division (Nachtweg and Dickinson, 1967; Chapman-Andresen and Nilsson, 1968). The cessation of phagotrophy lasts for about 30 min and food vacuole formation commences about 5 min after cell separation (Chapman-Andresen and Nilsson, 1968; Nilsson, 1972, 1976). This fact has been used to develop techniques for selecting cells that give synchronous populations by separation of non-feeding cells from feeding ones, either by passing the cells through a magnetic field after ingestion of iron particles (Hildebrandt and Duspiva, 1969), or through a density gradient after ingestion of tantalum particles (Wolfe, 1973). The duration of the exposure to particles is critical in these techniques and the final yield of cells is about 10% when an exponentially multiplying population is used.

Data on the efficiency with which *Tetrahymena* removes bacteria and particles from the medium have been tabulated (Table I). The volume of suspension medium cleared by 10^6 cells per hour (clearance factor) is large, apart from a single case (Seaman, 1961a), thus indicating that *Tetrahymena* is an efficient suspension feeder. The efficiency depends on the number of particles available per cell (Harding, 1937; Cox, 1967; Curds and Cockburn, 1968) and on the size of the particles. The clearance factor is relatively low at high concentrations of particles and extremely high at a fairly low, but sufficient, concentration of particles (Table I, examples 1 and 6); the apparent clearance factor is relatively low when particle concentration is insufficient to label all food vacuoles formed during the exposure. This last point may be illustrated by the last example in Table I: After 45 min each cell has removed 186 particles (approximately the number available per cell), whereas during the first 10 min each cell has re-

Table I Removal of Bacteria and Particles by *Tetrahymena*

Particle	Particles ingested per cell	Exposure (min)	Particles/ml	Cells/ml	Particles available/cell	Volume (ml) cleared by 10^6 cells/hr	References
Bacteria	10,000	60	6×10^8	2×10^4	3×10^4	17	Harding (1937)
Trypan blue	—	60	—	10^6	—	0.04	Seaman (1961a)
Bacteria	—	60	—	10^6	—	0.04	Seaman (1961a)
India ink	—	60	—	10^6	—	7.5–10	Cox (1967)
Bacteria	—	60	—	10^6	—	3–15	Curds and Cockburn (1968)
Carmine (1 μm)	100	10	6×10^6	1.5×10^4	4×10^2	17^a	Nilsson (unpublished results)
Dimethylbenzanthracene	2.5×10^4	15	2×10^{11}	10^6	2×10^5	0.5	Rothstein and Blum (1974c)
(0.1 μm)	9.7×10^4	45	2×10^{11}	10^6	2×10^5	0.7	Rothstein and Blum (1940)
India ink	—	180	—	2×10^5	—	3.5	Rasmussen et al. (1975)
Polyvinyltoluene latex	186	45	10^8	$3–5 \times 10^5$	$3.3–2 \times 10^2$	2.5	Batz and Wunderlich (1976)
(2.02 μm)	110	10	10^8	$3–5 \times 10^5$	$3.3–2 \times 10^2$	6.6	Batz and Wunderlich (1976)

[a] Ten-minute value.

moved 110 particles (about half the particles available per cell); the latter value represents a higher rate of clearance than does the former value, although both values derive from a single experiment (Batz and Wunderlich, 1976, Figure 1). Extension of such an experiment would reveal a final value close to that found at 45 min, and the calculated clearance factor would be low. Since a low clearance factor may be encountered at both low and high concentrations of particles, determination of this factor is only meaningful when correlated with the observation that particles are actually incorporated into food vacuoles throughout the experiment.

Particles have been proposed as an essential, mechanical stimulus for food vacuole formation (Mueller *et al.*, 1965) and few particles may suffice to trigger the mechanism (Nilsson, 1972). Normal proteose–peptone medium always contains some precipitate derived from heat sterilization; this precipitate is ingested and appears in black food vacuoles, resembling india ink. Furthermore, when such particles are not visible, *in vivo* observations reveal that single particles are swept into food vacuoles in formation (Nilsson, 1972). The correlation between the presence of particles and rapid cell multiplication of *Tetrahymena* was reported by Rasmussen and Kludt (1970); as opposed to cells in the normal proteose–peptone medium, cells in sterile-filtered (particle-free) proteose–peptone medium multiplied slowly and contained few food vacuoles. Addition of particles to the latter medium resulted in rapid cell multiplication and the presence of food vacuoles in the cells. This evidence supports the hypothesis that particles are essential for food vacuole formation.

Various digestible components induce food vacuole formation in *Tetrahymena* and the effect is dependent on the concentration of the solutes. Ricketts (1972b) reports that protein, polypeptides, and RNA are highly effective inducers of particle uptake, whereas glutamate, amino acid mixtures, polysaccharides, and glucose are moderately effective; furthermore, sodium β-glycerophosphate has slight effect and sodium acetate is ineffective.

After this general introduction to ingestion, the structural and physiological implications of the process will be discussed in the following sections.

A. Morphology

The structural aspects of ingestion imply a consideration of the components of the oral apparatus and of the detailed events in formation of a food vacuole.

The oral structure of *Tetrahymena* has been studied extensively (Furgason, 1940; Corliss, 1953; Metz and Westfall, 1954; Miller and Stone, 1963;

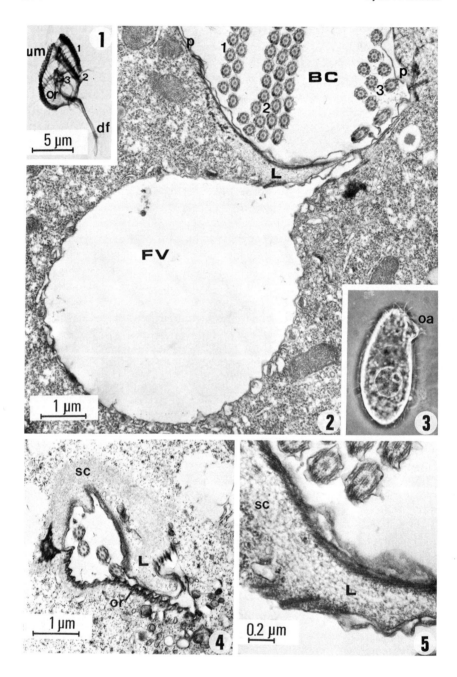

Nilsson and Williams, 1966; Elliott and Clemmons, 1966; Williams and Luft, 1968; Wolfe, 1970; Buhse *et al.,* 1970, 1973; Forer *et al.,* 1970; Frankel and Williams, 1973; Sattler and Straehelin, 1974, 1976; Nilsson, 1976). The oral apparatus is composed of four ciliary structures: the undulating membrane bordering the right and posterior margins of the oral overture, and the three adoral zone membranelles situated on the left wall of the buccal cavity. The cytostome lies at the base of the funnel-shaped buccal cavity. The kinetosomes of the ciliary components in the oral region are interconnected by microtubular and filamentous material and the structure may be isolated as an entity (Figure 1). Two structures composed primarily of microtubules extend from the oral membranelles (Figure 6, top half). First, the oral ribs extend from the undulating membrane and terminate just behind the cytostome; these oral ribs together with the modified pellicle structure form the right buccal cavity wall. Second, the oral deep fiber (Figure 1) extends from the three adoral zone membranelles far into the cytoplasm (Figures 9 and 10) after passing underneath the oral lip, a protrusion of the left buccal cavity wall (Figures 2, 4, 5). The region around the cytostome is composed of a filamentous network (Figures 2, 4, 5), the so-called specialized cytoplasm (Nilsson and Williams, 1966). The pellicle extends only a short distance beyond the cytostome into the cytopharynx, below which only the cell membrane shields the cytoplasm from the environment.

During phagotrophy the lively beating of the oral membranelles produces a current, together with the general swimming activity of *Tetrahymena,* whereby food is swept into the buccal cavity for accumulation in the forming food vacuole. The precise mode by which this interaction occurs is largely unexplored. However, an investigation along the lines of that elucidating the feeding mechanism in *Vorticella* (Sleigh and Barlow,

---

Figure 1. Isolated oral apparatus of *Tetrahymena* showing kinetosomes of the adoral zone membranelles (labeled 1, 2, 3) and the undulating membrane (um). Of the fibrillar components the oral ribs (or) and the oral deep fiber (df) are indicated. (From Nilsson, 1976, reproduced by permission.)

Figure 2. Section through buccal cavity (BC) showing cilia of the adoral zone membranelles (labeled 1, 2, 3) and the forming food vacuole (FV); note protruding lip structure (L). Pellicle structure (p) is present in the buccal cavity, but absent beyond the cytostome.

Figure 3. Light micrograph of *Tetrahymena* illustrating position of the oral region (oa). (From Nilsson, 1976, reproduced by permission.)

Figure 4. Cross-section of cytostome showing the oral ribs (or) and the cytostomal lip (L). Note specialized cytoplasm (sc) and numerous small vesicles (about 0.2 μm in diameter). (From Nilsson and Williams, 1966, reproduced by permission.)

Figure 5. Enlargement of the cytostomal lip (L) shown in Figure 2. Note the filamentous network of the specialized cytoplasm (sc).

1976) would add to the understanding of the mechanism operating in *Tetrahymena*.

In model systems, such as the ameba and mammalian phagocytes, endocytosis occurs by invagination of the cell membrane (Karnowsky, 1962; Chapman-Andresen, 1973, 1977; Allison, 1973; Stockem, 1977), although recent evidence suggests that a concomitant incorporation of new membrane occurs during this invagination (Vicker, 1977; Ryter and de Chastellier, 1977). In ciliates which have a more or less fixed cell surface area, because of the presence of a pellicle, endocytosis cannot occur by an invagination of the cell membrane. In this case, new membrane must be incorporated into the cytopharyngeal membrane to provide the limiting membrane for a new food vacuole.

Formation of a food vacuole in *Tetrahymena* may be divided into four stages as outlined in Figures 6A–6D (Nilsson, 1977b): (1) formation of the limiting membrane of the food vacuole. After detachment of a food vacuole only a small tubelike structure is visible at the cytostome (Figure 6D). This structure expands gradually (Figure 6A) as new food vacuolar membrane forms (Figure 6B); this stage lasts about 5–10 sec. (2) The "filling up" stage. The fully expanded vacuole (Figure 6B) is slightly oval in shape during accumulation of nutrients; this stage varies in duration, averaging about 20 sec. (3) The "closing off" of the vacuole. This event is correlated with vivid motion of the vacuole, ending in an anterior movement (pathway 1 in Figure 6C); this stage is brief, lasting 1–2 sec. (4) Movement of the vacuole away from the cytostome. After complete detachment and concomitant rounding, the vacuole moves posteriorly (pathway 2 in Figure 6D) into the cell, following an invisible structure (pathway 3 in Figure 6D), presumably the oral deep fiber. The entire cycle of formation of a food vacuole varies from 20 sec to 1 min even in individual cells (Nilsson, 1977b).

In connection with the formation of food vacuoles the following questions may be asked: (1) What is the source of the food vacuolar membrane? This question will be discussed in Section II,B. (2) How is the sealing off of the vacuolar membrane mediated? In phagocytes an actin–myosin system is likely to be involved (Stossel, 1977) and in *Tetrahymena* a contraction of the cytostome may be responsible, although the contractile, actinlike nature of the "specialized cytoplasm" has not yet been verified (Nilsson, 1976, 1977a). Relevant to these questions is also the process of membrane fusion (see Section II,B). (3) How are the different components of the oral structure (apart from the membranellar beating) involved in the process? The initial paths taken by the complete, rounded vacuole indicate an involvement of the oral ribs, a rigid structure, and of

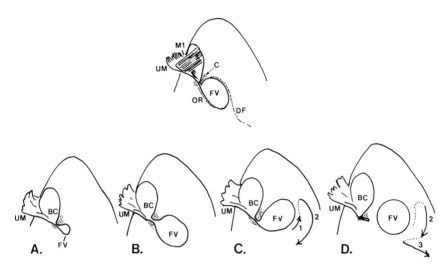

Figure 6. Oral region of *Tetrahymena* during food vacuole formation. Upper diagram: parts of the oral structures involved in formation of food vacuoles are shown. Undulating membrane (UM), first membranelle (M1), cytostome (C) marked by the oral lip (dotted area); and the forming food vacuole (FV) supported posteriorly by the oral ribs (OR) and anteriorly by the oral deep fiber (DF). Lower diagrams: different stages in formation of a food vacuole (FV); of the buccal cavity (BC) structures only the undulating membrane (UM) is shown. (A) Stage 1: gradual growth of the limiting membrane of the open food vacuole (of short duration). (B) Stage 2: "filling up" of the fully expanded, slightly oval vacuole (of long duration). (C) Stage 3: "closing off" of the vacuole (of brief duration) is correlated with an anterior motion (path 1). (D) Stage 4: initial path (2, 3) of the completed, now rounded vacuole leaving the cytostomal region. Only a small, tubelike structure remains at the cytostome. (From Nilsson, 1977b, reproduced by permission.)

the oral deep fiber, a more flexible structure (Nilsson, 1977b). The last part of the process could be a sliding motion along the oral deep fiber, which would explain the finding that newly formed food vacuoles take the same initial path away from the cytostome (Chapman-Andresen and Nilsson, 1968; Nilsson, 1972, 1976; Holz, 1973). Moreover, in electron micrographs the oral deep fiber may be seen as a ribbon of microtubules adjacent to the young food vacuole (Figures 9–11).

Newly formed food vacuoles in *Tetrahymena* are uniform in size (diameter, 5 μm) when fluid growth media or small particles are ingested (Nilsson, 1972), and their size may be related to the diameter of the cytostome. During ingestion of larger particles, however, the resulting food vacuoles may have a larger diameter than normal. Ingestion of 2-μm latex particles results in formation of uniformly sized food vacuoles about

7 μm in diameter (Ricketts, 1972a), whereas ingestion of baker's yeast results in formation of food vacuoles ranging from 5 to 15 μm in diameter in individual cells (Nilsson, 1977b). *Tetrahymena* shows little initial uptake of yeast but adapts within 30 min (Ricketts, 1970; Nilsson, 1977b); naturally, more membrane is incorporated into the limiting membrane of larger food vacuoles.

The factors governing the "closing off" of a food vacuole remain to be discovered. The presumed contraction of the cytostome could be a rhythmic event; however, this hypothesis is unlikely since *in vivo* observations have revealed that the second stage in vacuole formation, as discussed above, varies in duration in individual cells. Another possibility is that repletion of the vacuole, with, e.g., particles, triggers the mechanism; although this appears plausible it is not evident in the case of uptake of fluid medium (see, however, Section II,G). If the latter explanation represents the mode by which the "closing off" process is triggered, then the variable diameter of the yeast-containing vacuoles could be explained on the basis that passage of large particles through the cytostome might prevent the conjectured contraction, i.e., the cytostome becomes choked, and this results in a forced formation of a larger vacuole.

There are many questions about the mechanisms involved in the ingestion process that are to be resolved before our understanding is complete.

B. Food Vacuolar Membrane

The open food vacuole at the cytostome in *Tetrahymena* has membrane continuity with the cell membrane (Nilsson and Williams, 1966; Elliott and Clemmons, 1966), and hence the food vacuolar membrane must have properties in common with the cell membrane (Nilsson, 1976).

Since the limiting membrane of the food vacuole cannot arise by membrane invagination in *Tetrahymena,* as discussed earlier (Section II,A), new membrane must be incorporated in the cytopharyngeal membrane during phagotrophy. This membrane may be newly synthesized; however, a recycling of membrane may also occur, as has been proposed for other cells (McKanna, 1973a; Allen, 1974, 1978; Duncan and Praten, 1977; Tulkens *et al.,* 1977). This incorporation may occur by fusion of small vesicles with the cell membrane, a process believed to require common properties of the involved membranes (Lucy, 1969; Morré *et al.,* 1971; Poste and Allison, 1973; Ahkong *et al.,* 1975; Franke and Kartenbeck, 1976; Allen, 1978).

Small vesicles are present in the oral region of *Tetrahymena* (Nilsson and Williams, 1966; Batz and Wunderlich, 1976; Nilsson, 1976), and their membranes resemble the cell membrane and the food vacuolar membrane, because they are thicker than the membranes of the endoplasmic reticulum, nuclear envelope, contractile vacuole, mitochondria, and peroxisomes. The cell membrane shows polarity as the outer bilayer is more electron dense than the bilayer facing the cytoplasm (Wunderlich and Speth, 1972; Nilsson and Coleman, 1977); the same polarity is seen in the food vacuolar membrane where the electron-dense bilayer faces the lumen of the vacuole (Nilsson and Coleman, 1977). Membrane differentiation is generally believed to involve passage through the Golgi complex, whereby the endoplasmic reticulum membrane is converted into the cell membrane type; this flow is unidirectional (Morré *et al.*, 1971; Cook, 1973; Whaley, 1975; Franke and Kartenbeck, 1976). The Golgi complexes in *Tetrahymena* are inconspicuous (Franke *et al.*, 1971; Franke and Eckert, 1971; Franke and Kartenbeck, 1976; Nilsson, 1976) but they resemble those found in *Paramecium* (Estève, 1970). The possible relationship between vesicles found in the oral region of *Tetrahymena* is shown in Figure 7 (Nilsson, 1976). Relevant to the question of the origin of the food vacuolar membrane are membrane-renewal vesicles derived from the Golgi complex (type 2 and possibly type 1) and disk-shaped recycling-membrane vesicles derived from food vacuoles, either as they decrease in size (Section III,A) or when defecated (Section IV,A). Golgi-derived membrane-renewal vesicles have been observed in amebas (Daniels, 1964; Stockem, 1969, 1977; Wise and Flickinger, 1970). Furthermore, in ciliates vesicles have been proposed as the source of membrane for food vacuoles (McKanna, 1973a,b; Allen, 1974, 1978; Bradbury, 1974).

As reviewed elsewhere in this volume (Nozawa and Thompson, Chapter 7), much work has been done on the biochemical and physical characterization of membranes in *Tetrahymena*. Initially, membranes from isolated food vacuoles (formed during a 10-min period) were found to resemble the membrane of the microsomal fraction in lipid composition more than that of the cell membrane fraction represented by the cilia (Weidenbach and Thompson, 1974). A possible explanation for the apparent contradiction with the cytological evidence could be either that the microsome fraction contains the above-mentioned small membrane-renewal vesicles and Golgi complexes, or that cilia have a modified cell membrane (Nilsson, 1976). Further characterization of the membranes by freeze–fracture studies, in part with determination of the phase-separation temperature describing membrane fluidity (Wunderlich and Speth, 1972; Sekiya *et al.*, 1975; Kitajima and Thompson, 1977a,b), has revealed important information. First, the membrane of the disk-shaped vesicles in the oral region (Figure 7, type 2) is similar to that of the forming food vacuole both in distribution of membrane-intercalated particles and in phase-

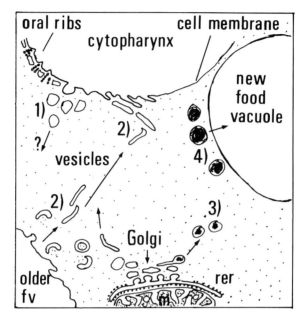

Figure 7. Possible relationships among four types of vesicles in *Tetrahymena*. The membranes of these vesicles resemble the cell membrane and the food vacuolar membrane in thickness and staining intensity; the possible flow of vesicles is indicated by arrows. Type 1: spherical vesicles with no visible contents may represent micropinocytic vesicles (?) formed in the cytopharyngeal region, or they may represent membrane-renewal vesicles derived from the Golgi complex. Type 2: disk-shaped, flattened vesicles may represent recycling-membrane vesicles for formation of the limiting membrane of new food vacuoles; they may derive from older food vacuoles (fv) as they decrease in size or from collapsing membranes of debris vacuoles. Type 3: spherical vesicles with electron-dense contents may represent primary lysosomes derived from the Golgi complex. Type 4: larger vesicles with laminated contents may be small secondary lysosomes (recycling acid hydrolases), possibly neutral red granules. The Golgi complex (Golgi) is a differentiated area of mitochondria (m)-associated endoplasmic reticulum (rer) and its role in vesicle formation is indicated. (From Nilsson, 1976, reproduced by permission.)

separation temperature (Kitajima and Thompson, 1977b). Second, the food vacuolar membrane exhibits close similarity to the cell membrane in the distribution of intercalated particles (Sekiya *et al.*, 1975) and in fluidity properties (Kitajima and Thompson, 1977a,b) (however, see also Section III,C), whereas the cilia membrane differs from the cell membrane in both respects (Wunderlich and Speth, 1972; Kitajima and Thompson, 1977a).

C. Rate of Phagocytosis

The rate of ingestion may be measured by the uptake of particles during a defined period. The results may be expressed as the average number of

labeled food vacuoles formed per cell or as a maximum number of vacuoles formed by a single cell. The prerequisite for such determinations is
that the concentration of particles is adequate to permit labeling of all food
vacuoles formed during the exposure (Section II,A).

Conflicting data have been reported on the capacity of *Tetrahymena* to
form food vacuoles. The time required to form one food vacuole has
been reported as 3 min (Nachtweg and Dickinson, 1967; Rasmussen
and Modeweg-Hansen, 1973), 2 min (Seaman, 1955), less than 1 min
(Chapman-Andersen and Nilsson, 1968; Nilsson, 1972), and about 20 sec
(Nilsson, 1976, 1977b); the latter value has also been observed *in vivo*
during uptake of fluid medium (Nilsson, 1977b). These determinations are
based either on the average value for a population or on the value of
individual cells. The capacity to form food vacuoles varies during the cell
cycle; thus the highest rate of ingestion is seen just prior to cell division
(Nilsson, 1976). No food vacuoles are formed during cell division (Section
II).

Addition of particles induces food vacuole formation in *Tetrahymena*.
About 30% of the vacuoles formed during a 10-min period are formed
during the first minute (Nilsson, 1972). A similar response in feeding activity is seen in the amebas, where about 60% of the uptake during a 4-hr
period occured during the first hour (Salt, 1961). A second exposure of
Tetrahymena to particles will not result in the formation of the same number of vacuoles, unless the two exposures are spaced at an interval of 1 hr
(Nilsson, 1972). The demonstration of a required recovery period to obtain a full second response of phagocytosis indicates that the process is
membrane limited (Nilsson, 1972), an interpretation which is in agreement
with the reported membrane-limited pinocytosis in amebas (Chapman-
Andresen, 1961). In *Tetrahymena* a cessation in particle uptake is also seen
after about 1 hr during continuous exposures (Ricketts, 1971b; Rothstein
and Blum, 1974b), although here the phenomenon was ascribed to a limitation in available lysosomes (Ricketts, 1971b).

D. Energy Dependence of Phagocytosis

Phagotrophic uptake in *Tetrahymena* is affected by low temperature
(Nilsson, 1972) and by the addition of dinitrophenol, an inhibitor of oxidative phosphorylation (Chapman-Andresen and Nilsson, 1968; Nilsson,
1976), indicating that phagocytosis is an energy-requiring process.

An energy-dependent uptake should be reflected in an increased uptake
of oxygen by cells induced to form food vacuoles. Such an increased rate
has been demonstrated during uptake of heat-killed bacteria (Burmeister,
1971) and during uptake of inert particles (Skriver and Nilsson, 1978).
Further experimental proof was obtained in the latter study, where no

increase in oxygen consumption was found when particles were added to a *Tetrahymena* mutant (*Tetrahymena thermophila* NP-1 (Orias and Pollock, 1975)) at the temperature where the cells are incapable of forming food vacuoles, or to heat-synchronized *T. pyriformis* during the synchronous cell division where food vacuole formation has ceased. The amount of energy *Tetrahymena* expends on formation of one food vacuole has been calculated to represent 6×10^{-13} mole ATP (Skriver and Nilsson, 1978).

That phagotrophic uptake in *Tetrahymena* is an energy-dependent process is in agreement with findings on endocytosis in general. Phagocytosis and certain types of pinocytosis in other protozoa and mammalian phagocytes have been found to be energy requiring (Karnowsky, 1962; Chapman-Andresen, 1967a, 1973, 1977; Allison and Davis, 1974; Aaronson, 1974).

E. Interference with Phagocytosis

Formation of food vacuoles in *Tetrahymena* is sensitive to changes in the cellular environment. The process is influenced by the temperature (Nilsson, 1972) and the pH of the medium (Mills, 1931; Nilsson, 1976), and it may be disturbed by handling of the cells (Nilsson, 1972). Some understanding of the mechanisms involved in the formation of food vacuoles may be gained from studies on the effects of various compounds added to the cells in their normal medium and in the presence of particles; an interference may be measured as a stimulated or an inhibited uptake of particles. A compound may often cause stimulation at low concentration and inhibition at high concentration. A wide range of compounds interfere with phagotrophy in *Tetrahymena,* as will be summarized below.

The presence of cations influences phagocytosis in *Tetrahymena*. Divalent cations have different effects; thus calcium (Nilsson, 1972, 1976; Brutkowska *et al.,* 1977), strontium (Nilsson, 1976), and lead (Nilsson, 1978) inhibit food vacuole formation, whereas magnesium has no effect (Nilsson, 1976) and copper stimulates the process (Nilsson, unpublished observations). Monovalent cations such as sodium and potassium stimulate the rate of phagocytosis (Nilsson, 1976). Furthermore, depending on the composition of the medium, presumably with respect to the content of divalent ions, the chelating reagent EDTA (ethylenediaminetetraacetic acid) also affects the rate of particle uptake (Nilsson, 1976).

Pharmacological compounds have been tested for their effect on food vacuole formation. Serotonin, caffeine, and dibutyryl cyclic AMP increase the rate of particle uptake, catecholamines do not affect the process, and the catecholamine antagonists, dichloroisoproterenol, desmethylimipramine, reserpine, and phenoxybenzamine, inhibit particle ingestion (Rothstein and Blum, 1974b). Moreover, animal hormones may stimulate phagocytosis, although histamine is most effective (Csaba and Lantos, 1973, 1975).

Inhibitors of protein synthesis have marked effects on food vacuole formation. Cycloheximide and puromycin at concentrations sufficient to block protein synthesis cause almost complete cessation of phagotrophy about 15 min after addition; chloramphenicol has only a slight effect on the process (Ricketts and Rappitt, 1975b).

Cryoprotective reagents may interfere with the rate of phagocytosis. Dimethyl sulfoxide (DMSO) and glycerol affect food vacuole formation in a similar manner; at low concentration they stimulate the formation and at high concentration, within the range used in cryoprotective work, they inhibit the process (Nilsson, 1974, 1976). The additional effects of the two reagents on the cells are, however, very different. DMSO is often employed as a solvent for water-insoluble compounds applied to living cells.

Drugs believed to affect specific cellular functions have also been tested. Cytochalasin B inhibits food vacuole formation (Nilsson *et al.*, 1973; Rothstein and Blum, 1974b; Hoffmann *et al.*, 1974; Ricketts and Rappitt, 1975c; Nilsson, 1977a) and the effect is readily reversible on removal of the drug. Vinblastine (Rothstein and Blum, 1974b) and colchicine (Rothstein and Blum, 1974b; Nilsson, 1976) have only moderate inhibitory effects on ingestion. On the basis of the assumed subcellular influence of these drugs (e.g., Allison and Davis, 1974) it may be concluded that microfilaments play a more important role than microtubules in the formation of food vacuoles.

Finally, the effect of ionic detergents has been tested. An anionic and a cationic detergent influence particle uptake in *Tetrahymena* similarly; at low concentration a slightly increased rate is seen and at high concentration a marked inhibition occurs (Brutkowska and Mehr, 1976).

This heterogeneous group of compounds has little else in common other than their interference with food vacuole formation. Moreover, this effect may represent only one of several effects of the compound (see Nilsson, 1976). Inhibition of phagotrophy leads to starvation, which again will be reflected in a low rate of cell multiplication; such an effect is, however, not observed in the presence of cytochalasin B (Nilsson, 1977a; see also Section II,H). Some of the compounds also affect pinocytosis and/or phagocytosis in other cells. Thus, sodium and potassium induce pinocytosis in amebas (Chapman-Andresen, 1958, 1962, 1973, 1977; Cooper, 1968) and calcium inhibits the process (Cooper, 1968); furthermore, cytochalasin B inhibits phagocytosis in various cells (Malawista *et al.*, 1971; Zigmond and Hirsch, 1972; Allison, 1973; Klaus, 1973; Allison and Davis, 1974; Hausmann and Eisenbarth, 1977).

A possible common action of the compounds may be an interference with the properties of the cell membrane. Such an interference may alter the functioning of the cell membrane, whereby the mechanism of food vacuole formation also may be altered, since the membrane events are important factors in this process (Sections II,B, and III,C). Moreover, membrane functioning is correlated with the lateral movement of intercalated particles within the lipid bilayer. Redistribution of the intramembranous particles may be induced by some of the factors or compounds which interfere with formation of food vacuoles in *Tetrahymena,* such as a change in pH (Pinto da Silva, 1972), change in temperature (Speth and Wunderlich, 1973; Wunderlich *et al.,* 1973b; Kitajima and Thompson, 1977a,b), exposure to DMSO or glycerol (McIntyre *et al.,* 1974), and exposure to colchicine (Wunderlich *et al.,* 1973a).

Tetrahymena has a "cell coat" which is pronounced on the cell membrane of cells from stationary phase cultures (Nilsson and Behnke, 1971); its function is unknown. The cell coat resembles that found in other ciliates (Wyroba and Przelecka, 1973; Tolloczka, 1975, 1976; Hausmann and Mocikat, 1976); in *Paramecium* its role in membrane functioning has been indicated, since treatment with hydrolytic enzymes reduces the thickness of the coat and inhibits food vacuole formation (Tolloczka, 1975).

F. Uptake of Dissolved Nutrients

In the laboratory *Tetrahymena* is grown mainly axenically in fluid medium. Essentially two types of media are used, namely, a proteose–peptone solution enriched with liver or yeast extract and inorganic salts (see Plesner *et al.,* 1964) and a chemically defined medium containing amino acids, nucleosides, vitamins, and inorganic salts (Kidder and Dewey, 1951, 1957). The question is how these dissolved nutrients enter *Tetrahymena.* In either case growing cells contain food vacuoles (Holz, 1964, 1973).

Uptake of proteose–peptone medium must occur largely by phagotrophy. This medium contains peptides of various lengths, some amino acids, and other low molecular weight compounds. The major part of this medium must be hydrolyzed prior to utilization. This assumption is supported by the finding that sterile-filtered (particle-free) proteose–peptone medium does not support appreciable cell multiplication unless particulate material is added to induce food vacuole formation (Rasmussen and Kludt, 1970; see also Section II). Furthermore, addition of fluid medium to starved synchronized cells results in an increased rate of oxygen consumption only during cell stages where food vacuoles are formed (Hamburger and Zeuthen, 1971). These studies indicate that phagotrophic uptake of proteose–peptone medium is essential for rapid cell growth. However, under special experimental conditions cell multiplication has been reported to occur in the absence of food vacuole formation (Rasmussen, 1973; see also Section II,H).

By uptake of the chemically defined medium the formation of food vacuoles could conceivably be avoided. *Tetrahymena* grows well in both autoclaved and sterile-filtered (particle-free) defined medium, although the rate of cell multiplication is lower than that found in proteose–peptone medium (e.g., Rasmussen and Kludt, 1970). However, little cell multiplication occurs in the defined medium if the trace metals are omitted (Rasmussen and Kludt, 1970), a finding which may indicate that these metals form particulate matter enough to stimulate food vacuole formation in the complete medium. Addition of particles also improves the rate of cell multiplication in the defined medium (Rasmussen and Modeweg-Hansen, 1973); however, even in this case the rate of cell multiplication does not exceed that found in proteose–peptone medium. Hence the chemically defined medium with all the essential building blocks for synthesis is not utilized more rapidly than when hydrolysis must precede utilization. The explanation may be either that the release of metabolites occurs only at a certain maturation stage of the food vacuole, independently of the nature of the contents, or that the rate of synthesis is the factor that prevents excessive cell growth. Support for the latter hypothesis is found in the observation that labeled amino acids or nucleosides are incorporated into macromolecules within minutes after their addition to the medium.

The conclusion is that food vacuole formation is essential for rapid cell proliferation in fluid medium.

G. Involvement of Mucus

Phagotrophy is generally accepted to be the only route of nutrient entry in *Tetrahymena* feeding on bacteria. However, when uptake of fluid medium is considered it becomes a much-debated question (Seaman, 1955, 1961a; Holz, 1964, 1973; Chapman-Andresen and Nilsson, 1968; Nilsson, 1972, 1976; Rasmussen, 1973, 1974, 1976; Rasmussen and Modeweg-Hansen, 1973; Hoffmann *et al.,* 1974).

The controversy largely originates in the number of food vacuoles a cell is capable of forming during a generation time. This number cannot be determined directly but has to be estimated from the rate of food vacuole formation (Section II,C) using either the average or the maximum rate. In either case the total volume ingested per generation time is insufficient to provide enough nutrients for a cell doubling if only medium as such is taken up (see Nilsson, 1976). Assuming that unconcentrated fluid medium is enclosed in food vacuoles, a cell has to form five times as many vacuoles as a cell ingesting bacteria to obtain the same amount of dry weight per unit time. Nevertheless, cells feeding on fluid medium or bacteria may have identical rates of food vacuole formation and of cell proliferation (Kidder, 1941; Harding, 1937). Since cells in fluid medium obtain adequate nutrients despite the relatively low number of food vacuoles, about 60, formed during a generation time, two possible explanations for this nutrient uptake have been suggested: (1) Nutrients become concentrated prior to ingestion in food vacuoles (Holz, 1964, 1973; Chapman-Andresen and Nilsson, 1968; Nilsson, 1972, 1976), or (2) the majority of nutrients enter *Tetrahymena* directly through the "cell surface" (Seaman, 1955, 1961a; Rasmussen, 1973, 1974, 1976). The latter possibility will be discussed in Section II,H.

If all nutrients of the fluid medium enter *Tetrahymena* by phagotrophy, calculations indicate that they must be concentrated by a factor of 10–20 times prior to enclosure in food vacuoles (Nilsson, 1976). The capacity of *Tetrahymena* to concentrate albumin by a factor of 20–50 times the amount present in the medium has been demonstrated (Ricketts and Rappitt, 1975c). To explain this phenomenon, attention may be focused on the mucous coat of amebas. This material plays a significant role in endocytosis (Chapman-Andresen, 1962, 1972, 1973, 1977) and is capable of concentrating proteins and inorganic cations by a factor of 10–50 times from the external medium (Schumaker, 1958; Brandt, 1962; Chapman-Andresen and Holter, 1964; Hendil, 1971; Allen and Winzler, 1973). Visually, such a binding capacity of the mucous coat is seen with the dye Alcian blue (Chapman-Andresen, 1962, 1973) prior to uptake of the dye–mucus complex by pinocytosis. Addition of Alcian blue to *Tetrahymena* causes extrusion of mucocysts and the dye becomes adsorbed to this material, which is then ingested (Nilsson, 1972); this finding indicates that mucocyst material has binding properties in common with the mucous coat of amebas. Binding of Alcian blue to extruding mucocyst material in *Tetrahymena* has also been demonstrated electron microscopically (Nilsson and Behnke, 1971); furthermore, not only Alcian blue but also proteose–peptone medium becomes adsorbed to extruded mucocyst material (Nilsson, 1976). These findings indicate that mucocyst material may play a role in the feeding of *Tetrahymena;* however, the concentration factor involved is unknown.

Unextruded mucocyst material has a paracrystalline structure, whereas the extruded material has a loose netlike structure (Hausmann, 1972, 1978; Nilsson, 1976). This configurational change could involve exposure of free binding sites for adsorption of solutes. Extruded mucocysts are spherical, about 2 μm in diameter, and the total volume of the approximately 1000 mucocysts, which *Tetrahymena* contains, corresponds to about 33% of that of a young daughter cell or to the volume of 65 food vacuoles. Mucocysts are abundant in the oral region (Williams and Luft, 1968; Nilsson, 1976) and secretory activity of the oral region has been indicated (Mills, 1931; Jahn *et al.*, 1965). Mucocysts are extruded in response to mechanical or chemical stimuli; furthermore, large amounts of mucocyst material are extruded when exposed to toxic substances, low pH, or osmotic shocks, and it may form an entire sheath around the organism (Bresslau, 1921, 1923, 1924; Mills, 1931; Tokuyasu and Scherbaum, 1965; Nilsson, 1972; Tiedtke, 1976). This reaction is also seen on exposure to diluted immune serum (Robertson, 1939a,b; Harrison, 1955; Watson *et al.*, 1964; Alexander, 1967) or to undiluted nonimmune serum (Brenner, 1973, 1974; Brenner *et al.*, 1975, 1976); in the former case, Alexander (1968) proposed that the antigen responsible for the reaction is the mucocyst protein, whereas in the latter case an osmotic shock may have been involved. This drastic response of the cells is undoubtedly a protective reaction; the mucocyst sheath shields the organism from the stimulus. When Alcian blue produces the response the dye becomes adsorbed to the sheath, within which *Tetrahymena* is found in clear solution. If the organisms survive the treatment, they leave the sheath, which may then be ingested (Nilsson, 1972). The "large bodies" found several hours after exposure to nonimmune serum (Brenner, 1973, 1974) are likely to have been formed in the same manner. This behavior agrees well with the observation that *Tetrahymena* preferentially ingests fluffy material when that is present (Nilsson, 1978).

If mucocyst material plays a general role in feeding of *Tetrahymena*, it may be difficult to visualize how such a mechanism could operate in practice. Mucocysts extruded from the general cell surface stand less chance of reaching the buccal cavity than those near the oral region, although *Tetrahymena* often reverses when feeding. However, little discrimination is seen during ingestion of, for example, Alcian blue–mucus complexes; material originating from one cell may be ingested by a different cell. Mucocyst material might possibly act as "opsonins" for *Tetrahymena* in a manner similar to that described for phagocytes (see Stossel, 1975); in uptake of particles it could cause agglutination (Mills, 1931; Harrison, 1955) in addition to labeling them for ingestion. Such a hypothesis is in accordance with the general finding that a certain cell density (10,000–20,000 cells/ml) is necessary for rapid multiplication of *Tetrahymena;* it is not unlikely that an efficient utilization of extruded mucocyst material would occur under these conditions.

Extruded mucocyst material is the likely candidate responsible for the high capacity of *Tetrahymena* to concentrate solutes from the medium. This material is rich in acidic residues and contains two lipoproteins and one protein (Alexander, 1968); the presence of mucopolysaccharide is indicated by the staining with Alcian blue (Nilsson, 1972). The adsorption capacity of the mucocyst material appears to be less specific than that of the mucous coat of amebas with respect to the charge of the adsorbed solutes (see Chapman-Andresen, 1977).

H. Uptake without Food Vacuole Formation

Under certain conditions cell multiplication may be induced in *Tetrahymena* without formation of food vacuoles. Recent reports (Rasmussen, 1973, 1974, 1976) have thrown new life into the concept that nutrients from the proteose–peptone medium enter the cells directly through the cell surface (Lwoff, 1923) (see also Section II,G). In 1941 Kidder stated it seems highly improbable that protein molecules could be absorbed through the pellicle and that if no proteolytic enzymes are released to the medium, dissolved nutrients must enter *Tetrahymena* by food vacuole formation. However, secretion of some hydrolases occurs (Müller, 1967; see also Section IV,C).

The question of nutrient uptake through the general "cell surface" must be based on assumptions about uptake of macromolecules and uptake of low molecular weight compounds thought to be readily utilizable in synthesis. Furthermore, consideration must be paid to the complex structure of the pellicle (Section I).

Uptake of macromolecules must involve endocytosis because hydrolysis is necessary. Endocytosis at the cell surface requires direct access from the cell membrane to the cytoplasm. The sites of the parasomal sacs (Section I) fulfill this requirement and their possible role in uptake of molecules has been proposed (Allen, 1967). If these sites play any substantial role in uptake, the rate of vesicle formation must be formidable since formation of one vesicle at each of the approximately 1000 sites in *Tetrahymena* would result in uptake of a volume corresponding only to 6% of that of one food vacuole; furthermore, this formation would require membrane corresponding to 1.6 times that of one food vacuole (Nilsson, 1972). Moreover, uptake at the parasomal sacs has not yet been demonstrated and the possible rate of vesicle formation is unknown.

Uptake of low molecular weight compounds could conceivably occur through the indentation of the cell membrane at the parasomal sacs, provided special "carriers" for such transport were present in the membrane. Little exact knowledge exists on the mechanism involved in carrier-mediated transport, and uptake of some amino acids involves endocytosis (Conner, 1967; Gordon, 1973; Bronk and Leese, 1974). Furthermore, compounds which inhibit the possible carrier-mediated transport of neutral amino acids in *Tetrahymena* (see Dunham and Kropp, 1973, p. 195) are also inhibitors of food vacuole formation.

Another explanation for cell growth in the absence of food vacuole formation could be pinocytic uptake at the cell membrane in the oral region. Structural evidence suggestive of this mode of uptake by formation of small vacuoles (diameter, 1 μm) is seen in cells that multiply during inhibition of phagocytosis by cytochalasin B in proteose–peptone medium (Nilsson, 1977a); these vacuoles are not seen in control cells. However, the rate of formation of the small vacuoles is unknown. The generation time of the drug-treated cells is prolonged by a factor of 1.4 above that of control cells, indicating that uptake of nutrients by formation of the small vacuoles is less efficient than uptake by formation of food vacuoles. In agreement with this assumption, cytochalasin B-treated cells concentrate labeled albumin from the medium at a rate corresponding to half of that of untreated cells (Ricketts and Rappitt, 1975c).

Cell multiplication in the absence of food vacuole formation was first induced by addition of high concentrations of nucleosides and glucose to *Tetrahymena* in sterile-filtered (particle-free) proteose–peptone medium (Rasmussen, 1973). The rate of cell multiplication in this medium was half of that of cells in particle-supplemented sterile-filtered medium. Clearly nucleosides and glucose cannot alone support cell growth in proteose–peptone medium; thus, an endocytic uptake must have occurred either at the parasomal sacs or, alternatively, at the cytopharyngeal membrane, possibly induced by the high concentration of the supplement. An uptake directly through the membrane of the open food vacuole (Rasmussen, 1974) appears problematic since only low molecular weight compounds could possibly enter here. High concentrations of nucleosides and glucose were also found to induce cell multiplication during inhibition of phagocytosis by cytochalasin B in the normal proteose–peptone medium (Hoffmann *et al.*, 1974). However, as indicated above, in this case nutrients are likely to have entered by pinocytosis at the cytopharyngeal membrane (Nilsson, 1977a).

Another approach to the problem of cell growth in the absence of phagotrophy has been made in studies using the wild type and the mutant NP-1 of *T. thermophila*. The mutant cells form normal functional oral structures at 28°–30°C, but form defective ones at 37°C (Orias and Pollock, 1975); at the high temperature, cell multiplication and food vacuole formation cease. However, addition of high concentrations of nucleosides and glucose induced cell multiplication of the mutant at 37°C in the proteose–peptone medium; the generation time was 1.75 times that of the wild type (Rasmussen and Orias, 1975). Furthermore, high concentrations of heavy metal salts and vitamins added to the chemically defined medium also permitted growth of the mutant at 37°C (Rasmussen and Orias, 1975). Further studies revealed that the vitamin supplement could be replaced by a high concentration of calcium folinate alone (Orias and Rasmussen, 1977) and that the heavy metal salts could be reduced to iron and copper alone (Rasmussen and Orias, 1976). On the basis of these investigations a dual capacity for nutrient uptake in *Tetrahymena* was proposed as an oral uptake system representing food vacuole formation and a surface uptake system (Rasmussen, 1976; Rasmussen and Orias, 1976; Orias and Rasmussen, 1976, 1977).

In conclusion, nutrients may enter *Tetrahymena* by routes other than that of food vacuole formation under special experimental conditions. The efficiency of such uptake is, however, less than that of formation of food vacuoles. Moreover, it remains to be established whether such uptake systems operate under normal growth conditions.

III. DIGESTION

Digestion commences after transfer to the food vacuole of hydrolytic enzymes that break down the ingested food to metabolites, which are then transferred to the cytoplasm for utilization. All organelles involved in the ingestion–digestion–egestion process may be described as the "vacuolar apparatus" (de Duve, 1967), much of which comprises the lysosomal system. Originally, the lysosome was determined biochemically as a membrane-bound cell organelle containing acid hydrolases (de Duve, 1959).

Components of the vacuolar apparatus have been divided into three groups of vacuoles (de Duve, 1967): (1) *The phagosome* (endocytic vacuole) containing material of exogenous origin destined to be digested, but no digestive enzymes, i.e., the newly formed food vacuole. (2) *The primary lysosome* containing newly synthesized acid hydrolases which have not yet been involved in digestion. (3) *The secondary lysosome* containing hydrolases, which are in the process of digesting (the digestive vacuole), which are present with residues of previous digestive events (the debris vacuole), and which have been involved in digesting (vesicles containing recycling hydrolases). This group of vacuoles is heterogeneous with respect to amounts of substrate; furthermore, in *Tetrahymena* the first two types of vacuoles are large in size, whereas the last type is small, as is the primary lysosome. Another type of secondary lysosome is the cytolysome, the digestive stage of autophagic vacuoles; however, since these vacuoles have contents of endogenous origin this aspect of digestion is beyond the scope of this review. The components of the vacuolar apparatus in *Tetrahymena* are outlined in Figure 8.

The dynamics of the vacuolar apparatus involves exchange between the different groups of vacuoles. This exchange is to a large extent based on membrane fusion either between the vacuoles or between the vacuoles and the cell membrane; the capacity of the membranes to fuse suggests that they have properties in common (see Section II,B). Another aspect of membrane fusion is that appropriate mechanical and energy-dependent factors may be required to bring the two membrane segments in close apposition to one another (de Duve, 1967).

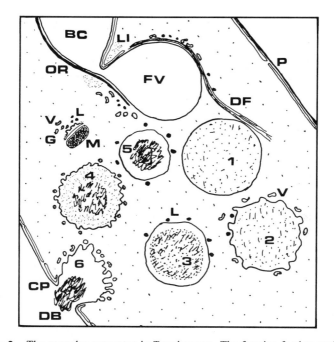

Figure 8. The vacuolar apparatus in *Tetrahymena*. The forming food vacuole (FV) at the cytostome, defined by the oral lip (LI), is open to the buccal cavity (BC). The newly formed food vacuole (1) is spherical and leaves the cytostomal region along the oral deep fiber (DF). The vacuole decreases in size (2), recycling membrane vesicles (V) become detached, and the vacuolar contents become concentrated. Lysosomes (L) fuse with the vacuole and 0.5 hr after formation the "halo" stage (3) is seen. After about 1 hr (4) finely granulated material is removed by micropinocytosis, the resorption stage. The debris vacuole (5) surrounded by lysosomes indicates removal of hydrolases for recycling. Membrane-renewal vesicles and primary lysosomes are formed at the Golgi complex (G) adjacent to a mitochondrion (M). Pellicle (P) composed of the cell membrane and underlying alveoli; oral ribs (OR).

In the following sections the structural aspects of digestion in *Tetrahymena* will be described (Section III,A), the literature on the lysosomal enzymes will be reviewed (Section III,B), the intravacuolar changes during digestion will be discussed (Section III,C), and the duration of the digestive events will be commented on (Section III,D).

A. Morphology

The structural aspects of digestion involve considerations of the food vacuole throughout the process and of the enzyme-bearing organelles—primary and small secondary lysosomes.

Before the food vacuole (phagosome) gives a positive reaction for acid phosphatase it undergoes structural changes (Figure 8). Initially, after detachment from the cytostome the limiting membrane is smooth in outline (Müller and Röhlich, 1961; Elliott and Clemmons, 1966; Nilsson, 1976) and the ingested material is evenly distributed within the vacuole. However, the vacuole soon decreases in size and its limiting membrane becomes irregular in outline. Small vacuoles, which become recycling membrane material, are pinched off from the phagosome, and as a result the ingested contents become concentrated. Little decrease is seen after ingestion of inert particles whereas a marked decrease is seen after ingestion of fluid medium (Nilsson, 1972). During this stage fine fibrous material is seen on the cytoplasmic side of the outpocketings of the food vacuolar membrane; this material is probably involved in the pinching off process (Nilsson, 1976). During the first 5–10 min after formation of a food vacuole, the density of the membrane-intercalated particles increases twofold (Kitajima and Thompson, 1977b).

Acid phosphatase activity has been demonstrated in food vacuoles 10 min after their formation (Nilsson, 1976) or somewhat later (Mueller *et al.*, 1965; Elliott and Clemmons, 1966). Food vacuoles receive hydrolases irrespective of whether inert or digestible particles have been ingested; however, only in the latter case are structural changes in the contents observable. Thus, after ingestion of bacteria a disintegration of their structure is evident in older vacuoles (Elliott and Clemmons, 1966). Half an hour after ingestion the partially disintegrated bacteria are found in the center of the vacuole, separated from the limiting membrane by a clear space, the so-called halo stage (Elliott and Clemmons, 1966). This stage is also seen after ingestion of proteose–peptone medium (Nilsson, 1976); typically such vacuoles are surrounded by rough endoplasmic reticulum (Müller and Röhlich, 1961; Elliott and Clemmons, 1966; Nilsson, 1970b) although the implications of the phenomenon are not understood. About 1 hr after formation the food vacuolar membrane exhibits micropinocytic activity (Müller and Röhlich, 1961; Elliott and Clemmons, 1966; Nilsson, 1976); finely granulated material from the digestive vacuole becomes enclosed in vesicles. This stage resembles the resorption stage in amebas (Mercer, 1959; Chapman-Andresen and Nilsson, 1967) and the process may represent the mode by which metabolites are transferred to the cytoplasm. Fusion of digestive vacuoles is frequently seen (Elliott and Clemmons, 1966; Nilsson, 1970a,b), although there is little evidence for a subdivision of the food vacuole as is regularly seen in amebas (Mast and Hahnert, 1935; Chapman-Andresen and Nilsson, 1967; Stockem and Stiemerling, 1976; Stockem, 1977). Acid phosphatase activity diminishes in old food vacuoles (Elliott and Clemmons, 1966; Elliott and Kennedy, 1973). From formation to defecation the diameter of the food vacuole may decrease from 5 to 3 μm, which means a decrease in the original volume and surface area to 22 and 35%, respectively.

The primary lysosomes, or small secondary lysosomes, are involved in the transfer of hydrolases to the food vacuole (Müller, 1971). The primary lysosomes presumably derive from the Golgi complex (Figure 7), as described for *Paramecium* (Estève, 1970); the hydrolases are synthesized on the rough endoplasmic reticulum and the passage through the Golgi complex probably involves package of the enzymes and transformation of the limiting membrane, as is believed to be the case in general (Cohn, 1968; de Duve, 1969; Estève, 1970; Cook, 1973; Whaley, 1975; Novikoff, 1976; Stockem, 1977). The enzyme complement of the primary lysosomes may vary (de Duve, 1969; Holtzman, 1976). The small secondary lysosomes may derive from the pinching off from old digestive vacuoles whereby the hydrolases become recycled (de Duve, 1969; Estève, 1970; Holtzman, 1976; Lloyd, 1977). These small lysosomes are probably identical to the neutral red granules, about 0.3 μm in diameter, seen in *Tetrahymena* (Elliott and Kennedy, 1973; Nilsson, 1977b). Neutral red granules are now generally accepted as representing lysosomes (Rosenbaum and Wittner, 1962; Chapman-Andresen, 1967b; de Duve, 1969; Allison and Young, 1969; Holtzman, 1976; Allen, 1978); their role in digestive events in protozoa has long been suggested (Prowazek, 1898; Nirenstein, 1905; Fortner, 1933).

Identification of the fine structure of the primary lysosome in *Tetrahymena* is a matter of controversy. The organelle first identified as the primary lysosome (Elliott, 1965) was later suggested to represent the peroxisome (Baudhuin and Müller, 1967; Williams and Luft, 1968; Müller, 1969; Nilsson, 1970a), as was demonstrated by a positive reaction for catalase (Stelly *et al.,* 1975); however, recently Kolb-Bachofen (1977) has reidentified the peroxisome as the primary lysosome. This controversy is not readily explained and further investigation is required, especially since the peroxisome does not show the expected membrane similarity with the vacuolar apparatus in general. Another interpretation of the primary and small secondary lysosomes is given in Figure 7 as the types 3 and 4 vesicles, respectively (Nilsson, 1976). Both these vesicles have membranes corresponding in structure to those of food vacuoles.

B. Lysosomal Enzymes

The lysosomal enzymes in *Tetrahymena* have been studied extensively (see Müller, 1967). They are all hydrolases with an acidic pH optimum. The enzymes typically show a latency in activity, a property which indicates that they are membrane bound (Müller *et al.*, 1966). An interesting aspect is that some hydrolases are secreted into the medium (Section IV,C).

The marker enzyme for lysosomes is acid phosphatase. This enzyme has been demonstrated cytochemically in small structures (i.e., in primary and small secondary lysosomes) and in digestive vacuoles (large secondary lysosomes) at both the light and electron microscope levels (Seaman, 1961b; Klamer and Fennell, 1963; Müller *et al.*, 1963; Allen *et al.*, 1963; Mueller *et al.*, 1965; Elliott and Clemmons, 1966; Nilsson, 1970a; Ricketts, 1970; Kolb-Bachofen, 1977). Biochemically, this enzyme has been studied extensively (Müller, 1967, 1971, 1972; Lazarus and Scherbaum, 1967, 1968; Lee, 1970; Ricketts, 1970, 1971a; Allen and Weremiuk, 1971; Lloyd *et al.*, 1971; Poole *et al.*, 1971; Rothstein and Blum, 1973, 1974a,b; Borden *et al.*, 1973; Ricketts and Rappitt, 1974, 1975a; Blum, 1975, 1976). Digestive vacuoles may not account for more than 34% of the acid phosphatase activity in *Tetrahymena* since the latency of the enzyme averaged 66% in broken cell preparations containing no intact food vacuoles (Ricketts and Rappitt, 1974).

Several hydrolases have been identified in *Tetrahymena*. These are acid phosphatase, ribonuclease, deoxyribonuclease, proteinase, amylase (Müller *et al.*, 1966), α-glucosidase, β-glucosidase, β-N-acetylglycosaminidase (Müller, 1971), β-N-acetylhexoseaminidase, β-galactosidase, and α-mannosidase (Blum, 1976). Several isoenzymes of the hydrolases have been found (Allen *et al.*, 1963; Klamer and Fennell, 1963; Allen and Weremiuk, 1971; Borden *et al.*, 1973; Nielsen and Andronis, 1975; Blum, 1976; Levy *et al.*, 1976).

In addition to digestive vacuoles *Tetrahymena* contains at least two populations of lysosomes (Müller, 1971, 1972; Blum, 1975, 1976) or possibly three or four populations (Morgan *et al.*, 1973; Ricketts and Rappitt, 1974; Rothstein and Blum, 1974c; Blum, 1976). This subdivision is based on the distribution of enzyme activity in different fractions of the homogenate. The two major populations have been designated low- and high-density lysosomes (Müller, 1971, 1972). The low-density lysosomes are particularly rich in proteinase and to a lesser extent in nucleases and phosphatase, whereas the high-density lysosomes are characterized by a high concentration of glucosidases (Müller, 1971). The low-density lysosomes are involved in intracellular digestion after fusion with food vacuoles, whereas the high-density lysosomes secrete their contents into the medium (Müller, 1972; see also Section IV,C).

The activity of acid hydrolases alters with the physiological state of the organisms. The low-density lysosome peak is distinct in growing cells but is markedly reduced in cells in stationary growth phase (Blum, 1976). Accordingly, the activity of the various enzymes varies during the culture cycle (Klamer and Fennell, 1963; Lazarus and Scherbaum, 1967, 1968; Lee, 1970; Lloyd *et al.,* 1971; Rothstein and Blum, 1974a; Blum and Rothstein, 1975); this variation is correlated with a varying latency of the hydrolases (Lee, 1970). Starvation causes a marked decrease in the intracellular content of acid phosphatase (Levy and Elliott, 1968) and only 42% of the original specific activity remains after the first 19 hr (Lloyd *et al.,* 1971).

Induction of endocytosis in starved cells results in an adaptive increase in the activity of acid phosphatase after 1 hr, reaching a maximum after 2.5 hr (Ricketts, 1970, 1971a). This adaptive response occurs only after ingestion of digestible material (Ricketts, 1971a; Ricketts and Rappitt, 1975a), which indicates that new synthesis is involved.

C. Intravacuolar Changes

Food vacuoles in protozoa have long been known to be acidic during digestion (Nirenstein, 1905; Fortner, 1933; Mast, 1947; Kitching, 1956). After ingestion of indicator dyes the studies revealed an initial alkaline phase followed by the acidic phase and a final increase in pH before defecation (see Kitching, 1956).

The intravacuolar pH value has been determined in *Tetrahymena* after ingestion of heat-killed yeast stained with indicator dyes (Nilsson, 1977b). The pH in the forming and newly formed food vacuole (phagosome) is unchanged from that of the medium (pH 7.3). Within 5 min after formation of the vacuole, the pH may decrease to 6.0–5.5, and 10 min later to pH 4.5. A maximum depression to pH 4.0–3.5 is reached within 2 hr; finally, an increase in pH is seen prior to defecation. These pH changes are in agreement with those observed in phagocytic vacuoles in mammalian phagocytes (Jensen and Bainton, 1973). The hydrolases in *Tetrahymena* have pH optima between 3.4 and 5.6 (Müller *et al.,* 1966; Müller, 1971),

and they are assumed to operate at the same pH values within the digestive vacuole as is also indicated by the pH determinations. The physicochemical factors responsible for changing the pH and for maintenance of a
low pH value within the digestive vacuole are not fully understood; some
possible mechanisms have been discussed recently (Jensen and Bainton,
1973; Coffey and de Duve, 1968; Holtzman, 1976).

Changes occur in the limiting membrane of the food vacuole during
digestion. Physical properties of cellular membranes may be measured by
determination of the temperature needed to induce aggregation of intramembranous particles (Speth and Wunderlich, 1973; Wunderlich *et al.*,
1975; Kitajima and Thompson, 1977a,b). In the study by Kitajima and
Thompson (1977b) the first sign of phase separation occurred at 9°C in the
membrane of the forming food vacuole, at 14°C in the completed phagosome, and at 12°C in older vacuoles after fusion with lysosomes. These
findings indicate that the properties of the membrane of the food vacuole
are altered during digestion; however, it is unknown whether these
changes are correlated with an altered permeability or whether they are
induced by the altered pH within the vacuoles; changes in pH induce
particle aggregation in cellular membranes (Pinto da Silva, 1972). An altered distribution of the intramembranous particles in the food vacuolar
membrane is also reflected at the normal growth temperature, and thus the
freeze–etch pattern of the membrane of the mature food vacuole differs
from that of the forming vacuole. Furthermore, smooth particle-free areas
in the vacuolar membrane may be the initial sign of fusion with lysosomes
(Batz and Wunderlich, 1976). Similar particle-free regions in the membrane of phagocytic vacuoles in leukocytes were observed after fusion
with lysosomes (Moore *et al.*, 1978). However, in *Tetrahymena* aggregation of intramembranous particles has also been suggested as representing
sites of membrane fusion (Kitajima and Thompson, 1977b). It is not clear
whether the distribution of membrane-intercalated particles will be the
same during the pinching off of membrane from the vacuole as during the
fusion with a lysosome. In food vacuoles in *Paramecium* the distribution
and the number of intramembranous particles also alter during digestion
and three different stages have been determined (Allen, 1977, 1978).

D. Duration of Digestion

Duration of the digestive events in *Tetrahymena* may depend on the
physiological state of the cells and on the nature of the ingested material.
The question may be studied in different ways.

One approach is to determine the rate at which food vacuoles disappear
in cells transferred from a nutrient to a nonnutrient medium. After inges-

tion of bacteria the number of food vacuoles decreased to 79, 31, 7, and 2% after 1, 2, 3, and 4 hr, respectively (Doyle and Harding, 1937). After ingestion of proteose–peptone medium the number of food vacuoles decreased to 40 and 5% after 1 and 4 hr, respectively (Chapman-Andresen and Nilsson, 1968). Cells starved for more than 4 hr contain a small number of—two to four—vacuoles; their presence is undoubtedly due to autophagic events (Nilsson, 1970b).

Another approach is to determine the rate of removal of labeled food vacuoles in continuously feeding cells. After ingestion of dimethylbenzanthracene particles for 1 hr the cells released 30–40% of the particles during the following hour (Rothstein and Blum, 1974b). Furthermore, after a 10-min ingestion period of carmine particles the cells eliminated about 25 and 50% of the labeled food vacuoles after 1 and 2 hr, respectively (Nilsson, 1977b).

Although defecation may be observed as early as 0.5 hr after ingestion, the process is most frequent about 2 hr after the initial uptake (Doyle and Harding, 1937; Rothstein and Blum, 1974b, 1977; Ricketts and Rappitt, 1976; Batz and Wunderlich, 1976; Nilsson, 1976, 1977b). Hence the duration of the digestive events in *Tetrahymena* is on the order of 2 hr.

IV. EGESTION

Indigestible residues retained within old food vacuoles are egested through the cytoproct. Silver impregnation reveals the cytoproct in *Tetrahymena* as a slitlike structure posteriorly on the oral kinety (Furgason, 1940). Within the debris vacuole the residues are packed into a sphere which is egested as an entity (Mueller *et al.*, 1965; Elliott and Clemmons, 1966); however, several debris vacuoles may fuse prior to egestion, and thus several defecation balls may be eliminated in one process (Nilsson, 1976).

The amount of debris left over from digestion varies with the nature of the ingested material. Little debris usually results from ingestion of fluid medium, whereas much residue remains after ingestion of yeast or inert particles (Nilsson, 1977b). The process of egestion is conveniently observed in slightly compressed cells around 2 hr after uptake of inert particles.

During egestion the pellicle becomes distorted in the cytoproct region, but normal cell shape is restored after extrusion of the defecation balls (Blum and Greenside, 1976). The debris vacuole enters into position at the cytoproct and remains here for a few seconds; the actual extrusion of the defecation ball takes a few milliseconds (Blum and Greenside, 1976). The

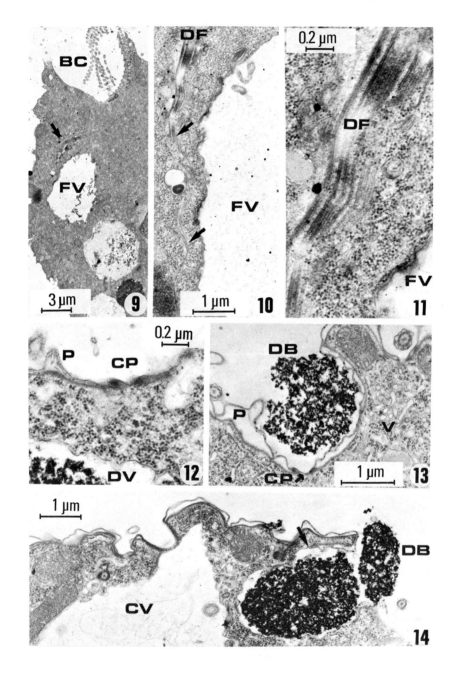

limiting membrane of the debris vacuole remains within the cell since defecation balls are not membrane bound (Elliott and Clemmons, 1966). After egestion of carmine-containing vacuoles single carmine particles may be found enclosed in small vacuoles at the cytoproct; this observation indicates that the particles have become trapped within the collapsing membrane of the debris vacuole during defecation (Nilsson, 1977b).

Defecation balls are rarely observed to be reingested (Cox, 1967). The explanation for this phenomenon could be either that the defecation balls are too large in size to permit reingestion or that they are "labeled" in some manner which prohibits ingestion (Batz and Wunderlich, 1976). Although the latter explanation appears attractive, it may not be the right one since *Tetrahymena* shows little discrimination in uptake of particulate material. Moreover, in cases where defecation balls fragment, as may be seen with carmine particles, the smaller fragments may eventually be reingested.

Egestion is inhibited by inhibitors of protein synthesis (Ricketts and Rappitt, 1975b), whereas two catecholamine antagonists enhance the rate of egestion (Rothstein and Blum, 1974b). Since ingestion is inhibited in both cases, the findings indicate that ingestion and egestion are uncoupled processes.

A. Morphology

The fine structure of the *Tetrahymena* cytoproct in cross section is rather inconspicuous (Figure 12). As described by Wolf and Allen (1974),

---

Figure 9. Section through *Tetrahymena* showing the buccal cavity (BC) (viewed from inside the cell) and food vacuoles of various ages. The newly formed vacuole (FV) is positioned next to the oral deep fiber (arrow); see also Figures 10 and 11.

Figure 10. Enlargement of the young food vacuole (FV) labeled in Figure 9. The oral deep fiber (DF) extends along (arrows) the vacuole, hence indicating that the structure guides the vacuole away from the cytostomal region.

Figure 11. Higher magnification of the oral deep fiber revealing its microtubular nature (enlargement of upper left corner of Figure 10).

Figure 12. Cross section of the cytoproct (CP). The electron-dense material marks the two cytoproct lips. On either side of the cytoproct typical pellicle structure (P) is evident. A debris vacuole (DV) adjacent to the cytoproct.

Figure 13. An extruded defecation ball (DB) shown in an indentation of the pellicle (P). The inconspicuous cytoproct (CP) is closed. Several small vesicles (V) in the cytoproct region may represent recycling membrane, derived from collapsed debris vacuolar membrane.

Figure 14. Section through the open cytoproct. One defecation ball (DB) has been extruded and another one is still present within the cell. The arrow marks the transition from pellicle to the membrane of the debris vacuole. The contractile vacuole (CV) with one of its two pores is shown.

electron-dense material borders the long narrow cytoproct and coats the margins of the adjacent alveoli; from these "cytoproct lips" numerous microtubules pass an unknown distance into the cytoplasm. Old food vacuoles near the cytoproct are draped with these microtubules, a finding which suggests a possible mechanism for guidance of the debris vacuole toward the cytoproct (Wolf and Allen, 1974).

The cytoproct opens when the membrane of the debris vacuole fuses with the cell membrane between the cytoproct lips (Wolf and Allen, 1974). During this stage the pellicle bordering the cytoproct indents (Figure 14), and thus a tangential section through a partially extruded defecation ball may show this surrounded by pellicle-coated cytoplasm (Figure 13). The indentation may be about 1 μm deep (Figure 14) and the transition from the pellicle structure to the limiting membrane of the debris vacuole is marked by the electron-dense material of the cytoproct lip.

The cytoproct closes when the now empty debris vacuole collapses, and as a result of endocytosis the limiting membrane breaks down to small vesicles (Wolf and Allen, 1974). This hypothesis is supported by the presence of small vesicles near the closing cytoproct (Figure 13). The ultimate fate of these small vesicles is probably a recycling to the cytostome, where after fusion they become part of the limiting membrane of a new food vacuole, as proposed for *Paramecium* (Allen and Wolf, 1974). The cytoproct region in *Tetrahymena* is less complex than that found in *Paramecium* (Estève, 1969; Allen and Wolf, 1974; Allen, 1978).

B. Sequence of Defecation

The time of egestion of residues from digestive vacuoles may be determined by the nature of the ingested material, by the sequential order of formation of food vacuoles, or by an incidental contact with the cytoproct region.

The nature of the ingested material has no apparent effect on the time of defecation. Irrespective of whether inert or digestible particles are ingested, the rate of defecation is high about 2 hr after uptake is initiated (see Section III,D).

If egestion is a temporal process based on the sequence in which the food vacuoles are formed, then the debris from a first vacuole must be defecated before that of a second vacuole. Ricketts and Rappitt (1976) presented data in support of this hypothesis. However, other investigators have found that labeled vacuoles may be retained for 4 hr or more in feeding cells (Doyle and Harding, 1937; Rothstein and Blum, 1974b, 1977; Nilsson, 1977b); thus, egestion cannot be a sequential process dependent on the time of formation of the food vacuole.

Most evidence points toward egestion as a random process. This assumption is also supported because defecation of labeled material may be observed as early as 0.5 hr after ingestion (Chapman-Andresen and Nilsson, 1968; Nilsson, 1977b). However, if egestion occurs completely at random so that a vacuole in any stage of digestion may stand a chance of being egested, then the result would be a low utilization of the egested material and a waste of hydrolytic enzymes. It seems reasonable to assume that some factor determines whether the digestive vacuole is ready to defecate its contents and that once this factor has become expressed then the capture by the cytoproct structures is a random process. Since egestion involves fusion of the membranes of the defecation vacuole and the cytoproct, then the factor involved may be an altered property of the vacuolar membrane. As outlined in Section III,C, the properties of the food vacuolar membrane change during digestion; moreover, since membrane fusion is dependent on a similarity of the involved membranes, it is not unlikely that these membrane changes prohibit egestion. It would be interesting if the drugs that enhance the rate of egestion (Rothstein and Blum, 1974b; see also Section IV,C) could cause premature alterations in the food vacuolar membrane.

C. Secretion of Hydrolytic Enzymes

Tetrahymena releases hydrolases into the medium (see Müller, 1967). The phenomenon occurs in nutrient and in starvation media (Müller, 1972; Ricketts and Rappitt, 1974), and the amount of hydrolases secreted is similar in both cases. Growing cells replace the loss by synthesis whereas starved cells show a decrease in their intracellular contents of these enzymes (Müller, 1972). Cells starved for 5 hr release two-thirds of their α-glucosidase, β-glucosidase, β-N-acetylglycosaminidase, and amylase, about one-third of their deoxyribonuclease and phosphatase, and only a small fraction of their proteinase (Müller, 1972). The population of hydrolases released by *Tetrahymena* corresponds to that found in the heavy-density lysosomes (Müller, 1971; see Section III,B) and a progressive decrease in this type of lysosomes is seen during starvation (Müller, 1972). The amount of glucosidases released during a generation time equals that present initially in the cells (Müller, 1972). The hydrolases are secreted as several isoenzymes (Dickie and Liener, 1962; Blum, 1975, 1976; Levy *et al.,* 1976) which differ from the intracellular enzymes in thermostability and in pH optimum (Müller, 1967; Blum, 1976), a finding which indicates a modification of the enzymes during or after release (Blum, 1976; Rothstein and Blum, 1977).

The amount of hydrolases secreted by *Tetrahymena* varies with the age of the culture and with the composition of the medium (Rothstein and Blum, 1973, 1974a; Blum and Rothstein, 1975; Blum, 1975, 1976; Müller, 1975). Thus, rapidly growing cells secrete large amounts of α-glucosidase and ribonuclease but little acid phosphatase, whereas cells from stationary phase cultures release large amounts of α-glucosidase and little ribonuclease (Rothstein and Blum, 1974a).

The released hydrolases undoubtedly originate from lysosomes, and the question is how these enzymes are secreted. The release is unaltered by stimulated or inhibited food vacuole formation (Müller, 1972; Rothstein and Blum, 1973, 1974b; Ricketts and Rappitt, 1974), is inhibited by high osmolarity of the medium or by inhibitors of protein synthesis (Müller, 1975), and is increased during enhanced rate of egestion (Rothstein and Blum, 1974b). This last point indicates that the hydrolases are released through the cytoproct together with debris from digestive vacuoles (Rothstein and Blum, 1974b; Blum and Rothstein, 1975). This assumption is supported by the above-mentioned inhibited release by inhibitors of protein synthesis since these have been shown to inhibit egestion of ingested particles (Ricketts and Rappitt, 1975b). Some release could also occur from recycling-membrane vesicles, derived from secondary lysosomes, during their incorporation in the cytopharyngeal membrane (Section II,B) since they may also contain hydrolases. Lloyd (1977) has proposed that recycling-membrane vesicles have some of the contained hydrolases bound to the inner side of the limiting membrane and that when it fuses with the cell membrane the unbound enzymes are released to the medium; this hypothesis would also explain a selective release of enzymes. An alternative mode of release of hydrolases in *Tetrahymena* has been proposed to occur by extrusion of mucocysts (Kolb-Bachofen, 1977); however, the possible enzyme complement of mucocyst material has not been investigated biochemically.

The significance of the secretion of hydrolases by *Tetrahymena* is not clear. That the purpose is predigestion of nutrients appears unlikely, since the mechanism would be highly uneconomical in nature because the enzymes would be greatly diluted (Müller, 1967). Moreover, it does not agree with the unaltered release of enzymes during starvation. Future investigation may solve the problem.

V. CONCLUDING REMARKS

This discussion on phagotrophy in *Tetrahymena* has touched on many aspects of the process. Of special interest is the dynamics of the membrane events in the ingestion–digestion–egestion cycle involving membrane fusion in a controlled fashion. Much of the recent investigation bears on the importance of the structure and of the chemical composition of the membranes in *Tetrahymena*, although many points are open for future studies. There is no doubt that components other than the membrane itself play a part in the proper functioning of membranes. With the ever-growing interest in *Tetrahymena* as a research organism, such an understanding should emerge.

REFERENCES

Aaronson, S. (1974). *J. Gen. Microbiol.* **83**, 21–29.
Ahkong, Q. F., Fisher, D., Tampion, W., and Lucy, J. A. (1975). *Nature (London)* **253**, 194–195.
Alexander, J. B. (1967). *Trans. Am. Microsc. Soc.* **86**, 421–427.
Alexander, J. B. (1968). *Exp. Cell Res.* **49**, 425–440.
Allen, H. J., and Winzler, R. J. (1973). *In* "The Biology of Amoeba" (K. W. Jeon, ed.), pp. 451–466. Academic Press, New York.
Allen, R. D. (1967). *J. Protozool.* **14**, 533–565.
Allen, R. D. (1974). *J. Cell Biol.* **63**, 904–922.
Allen, R. D. (1977). *Abstr. Int. Congr. Protozool., 5th, New York* p. 235.
Allen, R. D. (1978). *In* "Membrane Fusion" (G. Poste and G. L. Nicolson, eds.), pp. 657–763. Elsevier/North-Holland, Amsterdam.
Allen, R. D., and Wolf, R. W. (1974). *J. Cell Sci.* **14**, 611–631.
Allen, S. L., and Weremiuk, S. L. (1971). *Biochem. Genet.* **5**, 119–133.
Allen, S. L., Misch, M. S., and Morrison, B. M. (1963). *J. Histochem. Cytochem.* **11**, 706–719.
Allison, A. C. (1973). *In* "Locomotion of Tissue Cells" (R. Porter and D. W. Fitzsimons, eds.), Ciba Foundation Symposium, No. 14, pp. 109–148. Assoc. Sci. Publ., Amsterdam.
Allison, A. C., and Davis, P. (1974). *Symp. Soc. Exp. Biol.* **28**, 419–446.
Allison, A. C., and Young, M. R. (1969). *In* "Lysosomes in Biology and Pathology" (J. T. Dingle and H. B. Fell, eds.), Vol. 2, pp. 600–628. North-Holland Publ., Amsterdam.
Baker, H., Frank, O., Rusoff, I. I., Morck, R. A., and Hutner, S. H. (1978). *Nutr. Rep. Int.* **17**, 525–536.
Batz, W., and Wunderlich, F. (1976). *Arch. Microbiol.* **109**, 215–220.
Baudhuin, P., and Müller, M. (1967). Unpublished results, cited in Müller (1969).
Blum, J. J. (1975). *J. Cell. Physiol.* **86**, 131–142.
Blum, J. J. (1976). *J. Cell. Physiol.* **89**, 457–472.
Blum, J. J., and Greenside, H. (1976). *J. Protozool.* **23**, 500–502.
Blum, J. J., and Rothstein, T. L. (1975). *In* "Lysosomes in Biology and Pathology" (J. T. Dingle and R. T. Dean, eds.), Vol. 4, pp. 33–45. North-Holland Publ., Amsterdam.
Borden, D., Whitt, G. S., and Nanney, D. L. (1973). *J. Protozool.* **20**, 693–700.
Bradbury, P. C. (1974). *Protistologica* **10**, 533–542.
Brandt, P. W. (1962). *Circulation* **26**, 1075–1091.
Brenner, L. J. (1973). *Experientia* **29**, 347–348.
Brenner, L. J. (1974). *J. Protozool.* **21**, 417–418.

Brenner, L. J., Osborne, D. G., and Schumaker, B. L. (1975). *Annu. Proc. Electron Microsc. Soc. Am., 33rd, Las Vegas, Nev.* pp. 694–695.
Brenner, L. J., Osborne, D. G., and Schumaker, B. L. (1976). *J. Invertebr. Pathol.* **28,** 47–55.
Bresslau, E. (1921). *Naturwissenschaften* **9,** 57–62.
Bresslau, E. (1923). *Zentralbl. Bakteriol. Parasitenkd. Infektionskr. Abt. I.* **89,** 87–90.
Bresslau, E. (1924). *Verh. Dtsch. Zool. Ges.* **29,** 91–95.
Bronk, J. R., and Leese, H. J. (1974). *Symp. Soc. Exp. Biol.* **28,** 283–304.
Brutkowska, M., and Mehr, K. (1976). *Acta Protozool.* **15,** 66–76.
Brutkowska, M., Kubalski, A., and Kurdybacha, J. (1977). *Acta Protozool.* **16,** 195–200.
Buhse, H. E., Jr., Corliss, J. O., and Holsen, R. C., Jr. (1970). *Trans. Am. Microsc. Soc.* **89,** 328–336.
Buhse, H. E., Jr., Stamler, S. J., and Corliss, J. O. (1973). *Trans. Am. Microsc. Soc.* **92,** 95–105.
Burmeister, J. (1971). *Z. Allg. Mikrobiol.* **11,** 275–282.
Carasso, N., Favard, P., and Goldfischer, S. (1964). *J. Microsc. (Paris)* **3,** 297–322.
Chapman-Andresen, C. (1958). *C. R. Trav. Lab. Carlsberg, Ser. Chim.* **31,** 77–92.
Chapman-Andresen, C. (1961). *Proc. Int. Congr. Protozool., 1st, Prague* pp. 267–270.
Chapman-Andresen, C. (1962). *C. R. Trav. Lab. Carlsberg* **33,** 73–264.
Chapman-Andresen, C. (1967a). *Protoplasma* **63,** 103–105.
Chapman-Andresen, C. (1967b). *C. R. Trav. Lab. Carlsberg* **36,** 161–187.
Chapman-Andresen, C. (1972). *J. Protozool.* **19,** 225–231.
Chapman-Andresen, C. (1973). In "The Biology of Amoeba" (K. W. Jeon, ed.), pp. 319–348. Academic Press, New York.
Chapman-Andresen, C. (1977). *Physiol. Rev.* **57,** 371–385.
Chapman-Andresen, C., and Holter, H. (1964). *C. R. Trav. Lab. Carlsberg* **34,** 211–226.
Chapman-Andresen, C., and Müller, M. (1974). In "Actualités Protozoologiques" (P. de Puytorac and J. Grain, eds.), Vol. 1, pp. 265–277. Poul Couty, Clermont-Ferrand, France.
Chapman-Andresen, C., and Nilsson, J. R. (1967). *C. R. Trav. Lab. Carlsberg* **36,** 189–207.
Chapman-Andresen, C., and Nilsson, J. R. (1968). *C. R. Trav. Lab. Carlsberg* **36,** 405–432.
Coffey, J. W., and de Duve, C. (1968). *J. Biol. Chem.* **243,** 3255–3263.
Cohn, Z. A. (1968). *Excerpta Med. Int. Congr. Ser.* No. 166, p. 6.
Conner, R. L. (1967). In "Chemical Zoology" (G. W. Kidder, ed.), Vol. 1, pp. 309–350. Academic Press, New York.
Cook, M. W. (1973). In "Lysosomes in Biology and Pathology" (J. T. Dingle, ed.), Vol. 3, pp. 237–277. North-Holland Publ., Amsterdam.
Cooper, B. A. (1968). *C. R. Trav. Lab. Carlsberg* **36,** 385–403.
Corliss, J. O. (1953). *Parasitology* **43,** 49–87.
Cox, F. E. G. (1967). *Trans. Am. Microsc. Soc.* **86,** 261–267.
Csaba, G., and Lantos, T. (1973). *Cytobiologie* **7,** 361–365.
Csaba, G., and Lantos, T. (1975). *Acta Protozool.* **13,** 409–413.
Curds, C. R., and Cockburn, A. (1968). *J. Gen. Microbiol.* **54,** 343–358.
Daniels, E. W. (1964). *Z. Zellforsch. Mikrosk. Anat.* **64,** 38–51.
de Duve, C. (1959). In "Subcellular Particles" (T. Hayashi, ed.), pp. 128–159. Ronald Press, New York.
de Duve, C. (1963). In "Lysosomes" (A. V. S. de Reuck and M. P. Cameron, eds.), Ciba Foundation Symposium, pp. 1–31. Churchill, London.
de Duve, C. (1967). *Protoplasma* **63,** 95–98.

de Duve, C. (1969). *In* "Lysosomes in Biology and Pathology" (J. T. Dingle and H. B. Fell, eds.), Vol. 1, pp. 3–40. North-Holland Publ., Amsterdam.

Dickie, N., and Liener, I. E. (1962). *Biochim. Biophys. Acta* **64**, 41–51.

Doyle, W. L., and Harding, J. P. (1937). *J. Exp. Biol.* **14**, 462–469.

Duncan, R., and Praten, M. K. (1977). *J. Theor. Biol.* **66**, 727–735.

Dunham, P. B., and Kropp, D. L. (1973). *In* "Biology of *Tetrahymena*" (A. M. Elliott, ed.), pp. 165–198. Dowden, Hutchinson & Ross, Stroudsburg, Pennsylvania.

Elliott, A. M. (1965). *Science* **149**, 640–641.

Elliott, A. M., and Clemmons, G. L. (1966). *J. Protozool.* **13**, 311–323.

Elliott, A. M., and Kennedy, J. R. (1973). *In* "Biology of *Tetrahymena*" (A. M. Elliott, ed.), pp. 57–87. Dowden, Hutchinson & Ross, Stroudsburg, Pennsylvania.

Estève, J.-C. (1969). *C. R. Acad. Sci., Ser. D* **268**, 1508–1510.

Estève, J.-C. (1970). *J. Protozool.* **17**, 24–35.

Forer, A., Nilsson, J. R., and Zeuthen, E. (1970). *C. R. Trav. Lab. Carlsberg* **38**, 67–86.

Fortner, H. (1933). *Arch. Protistenkd.* **81**, 19–56.

Frank, O., Baker, H., Hutner, S. H., Rusoff, I. I., and Morck, R. A. (1975). *In* "Protein Nutritional Quality of Foods and Feeds" (M. Friedman, ed.), Part 1, pp. 203–209. Dekker, New York.

Franke, W. W., and Eckert, W. A. (1971). *Z. Zellforsch. Mikrosk. Anat.* **122**, 244–253.

Franke, W. W., and Kartenbeck, J. (1976). *In* "Progress in Differentiation Research" (N. Müller-Bérat, C. Rosenfeld, D. Tarin, and D. Viza, eds.), pp. 213–243. North-Holland Publ., Amsterdam.

Franke, W. W., Eckert, W. A., and Krien, S. (1971). *Z. Zellforsch. Mikrosk. Anat.* **119**, 577–604.

Frankel, J., and Williams, N. E. (1973). *In* "Biology of *Tetrahymena*" (A. M. Elliott, ed.), pp. 375–409. Dowden, Hutchinson & Ross, Stroudsburg, Pennsylvania.

Furgason, W. H. (1940). *Arch. Protistenkd.* **94**, 224–266.

Gebauer, H.-J. (1977). *Protistologica* **13**, 535–548.

Gordon, A. H. (1973). *In* "Lysosomes in Biology and Pathology" (J. T. Dingle, ed.), Vol. 3, pp. 89–137. North-Holland Publ., Amsterdam.

Hamburger, K., and Zeuthen, E. (1971). *C. R. Trav. Lab. Carlsberg* **38**, 145–161.

Harding, J. P. (1937). *J. Exp. Biol.* **14**, 422–430.

Harrison, J. A. (1955). *In* "Biological Specificity and Growth" (E. G. Butler, ed.), pp. 141–156. Princeton Univ. Press, Princeton, New Jersey.

Hausmann, K. (1972). *Protistologica* **8**, 401–412.

Hausmann, K. (1978). *Int. Rev. Cytol.* **52**, 197–276.

Hausmann, K., and Eisenbarth, A. (1977). *Protoplasma* **92**, 269–279.

Hausmann, K., and Mocikat, K.-H. (1976). *Cytobiologie* **13**, 469–475.

Hendil, K. B. (1971). *C. R. Trav. Lab. Carlsberg* **38**, 187–211.

Hildebrandt, A., and Duspiva, F. (1969). *Z. Naturforsch., Teil B* **24**, 747–750.

Hoffmann, E. K., Rasmussen, L., and Zeuthen, E. (1974). *J. Cell Sci.* **15**, 403–406.

Holter, H. (1959). *Int. Rev. Cytol.* **8**, 481–504.

Holter, H. (1961). *In* "Biological Structure and Function" (T. W. Goodwin and O. Lindberg, eds.), Vol. 1, pp. 157–168. Academic Press, New York.

Holtzman, E. (1976). "Lysosomes: A Survey." Springer-Verlag, Berlin and New York.

Holz, G. G., Jr. (1964). *In* "Biochemistry and Physiology of Protozoa" (S. H. Hutner, ed.), Vol. 3, pp. 199–242. Academic Press, New York.

Holz, G. G., Jr. (1973). *In* "Biology of *Tetrahymena*" (A. M. Elliott, ed.), pp. 89–98. Dowden, Hutchinson & Ross, Stroudsburg, Pennsylvania.

Hutner, S. H., Baker, H., Frank, O., and Cox, D. (1972). *In* "Biology of Nutrition" (R. N. Fiennes, ed.), pp. 85–177. Pergamon, Oxford.

Hutner, S. H., Baker, H., Frank, O., and Cox, D. (1973). *In* "Biology of *Tetrahymena*" (A. M. Elliott, ed.), pp. 411–433. Dowden, Hutchinson & Ross, Stroudsburg, Pennsylvania.

Hutner, S. H., Baker, H., Frank, O., and Cox, D. (1977). *Global Impacts Appl. Microbiol., 4th Int. Congr., Sao Paulo, 1973* pp. 51–57.

Jahn, T. L., Bovee, E. C., Dauber, M., Winet, H., and Brown, M. (1965). *Ann. N.Y. Acad. Sci.* **118**, 912–920.

Jensen, M. S., and Bainton, D. F. (1973). *J. Cell Biol.* **56**, 379–388.

Karnowsky, M. L. (1962). *Physiol. Rev.* **42**, 143–168.

Kidder, G. W. (1941). *Biol. Bull. (Woods Hole, Mass.)* **80**, 50–68.

Kidder, G. W. (1967). *In* "Chemical Zoology" (G. W. Kidder, ed.), Vol. 1, pp. 93–159. Academic Press, New York.

Kidder, G. W., and Dewey, V. C. (1951). *In* "Biochemistry and Physiology of Protozoa" (A. Lwoff, ed.), Vol. I, pp. 323–400. Academic Press, New York.

Kidder, G. W., and Dewey, V. C. (1957). *Arch. Biochem. Biophys.* **66**, 486–492.

Kitajima, Y., and Thompson, G. A., Jr. (1977a). *J. Cell Biol.* **72**, 744–755.

Kitajima, Y., and Thompson, G. A., Jr. (1977b). *J. Cell Biol.* **75**, 436–445.

Kitching, J. A. (1956). *Protoplasmatologia* **3(D)**, 1–54.

Kitching, J. A. (1957). *Symp. Soc. Gen. Microbiol.* **7**, 259–286.

Klamer, B., and Fennell, R. A. (1963). *Exp. Cell Res.* **29**, 166–175.

Klaus, G. G. (1973). *Exp. Cell Res.* **79**, 73–78.

Kolb-Bachofen, V. (1977). *Cytobiologie* **15**, 135–144.

Lazarus, L. H., and Scherbaum, O. H. (1967). *Life Sci.* **6**, 2401–2407.

Lazarus, L. H., and Scherbaum, O. H. (1968). *J. Cell Biol.* **36**, 415–418.

Lee, D. (1970). *J. Cell. Physiol.* **76**, 17–22.

Levy, M. R., and Elliott, A. M. (1968). *J. Protozool.* **15**, 208–222.

Levy, M. R., Sisskin, E. E., and McConkey, C. L. (1976). *Arch. Biochem. Biophys.* **172**, 634–647.

Lloyd, D., Brightwell, R., Venables, S. E., Roach, G. I., and Turner, G. (1971). *J. Gen. Microbiol.* **65**, 209–223.

Lloyd, J. B. (1977). *Biochem. J.* **164**, 281–282.

Lucy, J. A. (1969). *In* "Lysosomes in Biology and Pathology" (J. T. Dingle and H. B. Fell, eds.), Vol. 2, pp. 313–341. North-Holland Publ., Amsterdam.

Lwoff, M. A. (1923). *C. R. Acad. Sci.* **176**, 928–930.

McIntyre, J. A., Gilula, N. B., and Karnowsky, M. J. (1974). *J. Cell Biol.* **60**, 192–203.

McKanna, J. A. (1973a). *J. Cell Sci.* **13**, 663–675.

McKanna, J. A. (1973b). *J. Cell Sci.* **13**, 677–686.

Malawista, S. E., Gee, J. B. L., and Bensch, K. G. (1971). *Yale J. Biol. Med.* **44**, 286–300.

Mast, S. O. (1947). *Biol. Bull. (Woods Hole, Mass.)* **92**, 31–72.

Mast, S. O., and Hahnert, W. F. (1935). *Physiol. Zool.* **8**, 255–272.

Mercer, E. H. (1959). *Proc. R. Soc. London, Ser. B* **150**, 216–232.

Metz, C. B., and Westfall, J. A. (1954). *Biol. Bull. (Woods Hole, Mass.)* **107**, 106–122.

Miller, O. L., Jr., and Stone, G. E. (1963). *J. Protozool.* **10**, 280–288.

Mills, S. M. (1931). *J. Exp. Biol.* **8**, 17–29.

Moore, P. L., Bank, H. L., Brissie, N. T., and Spicer, S. S. (1978). *J. Cell Biol.* **76**, 158–174.

Morgan, N. A., Howells, L., Cartledge, T. G., and Lloyd, D. (1973). *In* "Methodological

Developments in Biochemistry'' (E. Reid, ed.), Vol. 3, pp. 219–232. Longman Group, London.

Morré, D. J., Keenan, T. W., and Mollenhauer, H. H. (1971). *Adv. Cytopharmacol.* **1**, 159–182.

Müller, M. (1967). *In* "Chemical Zoology" (G. W. Kidder, ed.), Vol. 1, pp. 351–380. Academic Press, New York.

Müller, M. (1969). *Ann. N.Y. Acad. Sci.* **168**, 292–301.

Müller, M. (1971). *Acta Biol. Acad. Sci. Hung.* **22**, 179–186.

Müller, M. (1972). *J. Cell Biol.* **52**, 478–487.

Müller, M. (1975). *J. Protozool.* **22**, 21A.

Müller, M., and Röhlich, P. (1961). *Acta Morphol. Acad. Sci. Hung.* **10**, 297–305.

Müller, M., Röhlich, P., Tóth, J., and Törö, I. (1963). *In* "Lysosomes" (A. V. S. de Reuck and M. P. Cameron, eds.), Ciba Foundation Symposium, pp. 201–216. Churchill, London.

Mueller, M., Röhlich, P., and Törö I. (1965). *J. Protozool.* **12**, 27–34.

Müller, M., Baudhuin, P., and C. de Duve (1966). *J. Cell. Physiol.* **68**, 165–176.

Nachtweg, D. S., and Dickinson, W. J. (1967). *Exp. Cell Res.* **47**, 581–595.

Nanney, D. L., and McCoy, J. W. (1976). *Trans. Am. Microsc. Soc.* **95**, 664–682.

Nielsen, P. J., and Andronis, P. T. (1975). *J. Protozool.* **22**, 185–187.

Nilsson, J. R. (1970a). *C. R. Trav. Lab. Carlsberg* **38**, 87–106.

Nilsson, J. R. (1970b). *C. R. Trav. Lab. Carlsberg* **38**, 107–121.

Nilsson, J. R. (1972). *C. R. Trav. Lab. Carlsberg* **39**, 83–110.

Nilsson, J. R. (1974). *J. Cell Sci.* **16**, 39–47.

Nilsson, J. R. (1976). *C. R. Trav. Lab. Carlsberg* **40**, 215–355.

Nilsson, J. R. (1977a). *J. Cell Sci.* **27**, 115–126.

Nilsson, J. R. (1977b). *J. Protozool.* **24**, 502–507.

Nilsson, J. R. (1978). *Protoplasma* **95**, 163–173.

Nilsson, J. R., and Behnke, O. (1971). *J. Ultrastruct. Res.* **36**, 542–545.

Nilsson, J. R., and Coleman, J. R. (1977). *J. Cell Sci.* **24**, 311–325.

Nilsson, J. R., and Williams, N. E. (1966). *C. R. Trav. Lab. Carlsberg* **35**, 119–141.

Nilsson, J. R., Ricketts, T. R., and Zeuthen, E. (1973). *Exp. Cell Res.* **79**, 456–459.

Nirenstein, E. (1905). *Z. Allg. Physiol.* **5**, 435–510.

Novikoff, A. B. (1976). *Proc. Natl. Acad. Sci. U.S.A.* **73**, 2781–2787.

Orias, E., and Pollock, N. A. (1975). *Exp. Cell Res.* **90**, 345–356.

Orias, E., and Rasmussen, L. (1976). *Exp. Cell Res.* **102**, 127–137.

Orias, E., and Rasmussen, L. (1977). *J. Protozool.* **24**, 507–511.

Pinto da Silva, P. (1972). *J. Cell Biol.* **53**, 777–787.

Pitelka, D. R. (1961). *J. Protozool.* **8**, 75–89.

Plesner, P., Rasmussen, L., and Zeuthen, E. (1964). *In* "Synchrony in Cell Division and Growth" (E. Zeuthen, ed.), pp. 543–563. Wiley (Interscience), New York.

Poole, R. K., Nicholl, W. G., Turner, G., Roach, G. I., and Lloyd, D. (1971). *J. Gen. Microbiol.* **67**, 161–173.

Poste, G., and Allison, A. C. (1973). *Biochim. Biophys. Acta* **300**, 421–465.

Prowazek, S. (1898). *Z. Wiss. Zool.* **63**, 187–194.

Rasmussen, L. (1973). *Exp. Cell Res.* **82**, 192–196.

Rasmussen, L. (1974). *Nature (London)* **250**, 157–158.

Rasmussen, L. (1976). *Carlsberg Res. Common.* **41**, 143–167.

Rasmussen, L., and Kludt, T. A. (1970). *Exp. Cell Res.* **59**, 457–463.

Rasmussen, L., and Modeweg-Hansen, L. (1973). *J. Cell Sci.* **12**, 275–286.

Rasmussen, L., and Orias, E. (1975). *Science* **190**, 464–465.
Rasmussen, L., and Orias, E. (1976). *Carlsberg Res. Commun.* **41**, 81–90.
Rasmussen, L., and Ricketts, T. R. (1979). *In* "Protozoological Actualities—1977" (S. H. Hutner, ed.). Allen Press, Lawrence, Kansas. In press.
Rasmussen, L., Buhse, H. E., Jr., and Groh, K. (1975). *J. Protozool.* **22**, 110–111.
Ricketts, T. R. (1970). *Protoplasma* **71**, 127–137.
Ricketts, T. R. (1971a). *Exp. Cell Res.* **66**, 49–58.
Ricketts, T. R. (1971b). *Protoplasma* **73**, 387–396.
Ricketts, T. R. (1972a). *Arch. Microbiol.* **81**, 344–349.
Ricketts, T. R. (1972b). *J. Protozool.* **19**, 373–375.
Ricketts, T. R., and Rappitt, A. F. (1974). *Arch. Microbiol.* **98**, 115–126.
Ricketts, T. R., and Rappitt, A. F. (1975a). *Protoplasma* **85**, 119–125.
Ricketts, T. R., and Rappitt, A. F. (1975b). *Arch. Microbiol.* **102**, 1–8.
Ricketts, T. R., and Rappitt, A. F. (1975c). *Protoplasma* **86**, 321–337.
Ricketts, T. R., and Rappitt, A. F. (1976). *Protoplasma* **87**, 221–236.
Robertson, M. (1939a). *J. Pathol. Bacteriol.* **48**, 305–322.
Robertson, M. (1939b). *J. Pathol. Bacteriol.* **48**, 323–338.
Rosenbaum, R. M., and Wittner, M. (1962). *Arch. Protistenkd.* **106**, 223–240.
Rothstein, T. L., and Blum, J. J. (1973). *J. Cell Biol.* **57**, 630–641.
Rothstein, T. L., and Blum, J. J. (1974a). *J. Protozool.* **21**, 163–168.
Rothstein, T. L., and Blum, J. J. (1974b). *J. Cell Biol.* **62**, 844–859.
Rothstein, T. L., and Blum, J. J. (1974c). *Exp. Cell Res.* **87**, 168–174.
Rothstein, T. L., and Blum, J. J. (1977). *Abstracts Int. Congr. Protozool., 5th, New York* p. 233.
Ryter, A., and de Chastellier, C. (1977). *J. Cell Biol.* **75**, 200–217.
Salt, G. W. (1961). *Exp. Cell Res.* **24**, 618–620.
Satir, B. (1974). *Symp. Soc. Exp. Biol.* **28**, 399–418.
Satir, B., Schooley, C., and Satir, P. (1972). *Nature (London)* **235**, 53–54.
Satir, B., Schooley, C., and Satir, P. (1973). *J. Cell Biol.* **56**, 153–176.
Sattler, C. A., and Staehelin, L. A. (1974). *J. Cell Biol.* **62**, 473–490.
Sattler, C. A., and Staehelin, L. A. (1976). *Tissue Cell* **8**, 1–18.
Schumaker, V. N. (1958). *Exp. Cell Res.* **15**, 314–331.
Seaman, G. R. (1955). *In* "Biochemistry and Physiology of Protozoa" (S. H. Hutner and A. Lwoff, eds.), Vol. 2, pp. 91–158. Academic Press, New York.
Seaman, G. R. (1961a). *J. Protozool.* **8**, 204–212.
Seaman, G. R. (1961b). *J. Biophys. Biochem. Cytol.* **1**, 243–245.
Sekiya, T., Kitajima, Y., and Nozawa, Y. (1975). *J. Electron Microsc.* **24**, 155–165.
Skriver, L., and Nilsson, J. R. (1978). *J. Gen. Microbiol.* **109**, 359–366.
Sleigh, M. A., and Barlow, D. (1976). *Trans. Am. Microsc. Soc.* **95**, 482–486.
Speth, V., and Wunderlich, F. (1973). *Biochim. Biophys. Acta* **291**, 621–628.
Stelly, N., Balmefrézol, M., and Adoutte, A. (1975). *J. Histochem. Cytochem.* **23**, 686–696.
Stockem, W. (1969). *Histochemie* **18**, 217–240.
Stockem, W. (1977). *In* "Mammalian Cell Membranes" (G. A. Jamieson and D. M. Robinson, eds.), Vol. 5, pp. 151–195. Butterworth, London.
Stockem, W., and Stiemerling, R. (1976). *Cytobiologie* **13**, 158–162.
Stossel, T. P. (1975). *Semin. Hematol.* **12**, 83–116.
Stossel, T. P. (1977). *Fed. Proc., Fed. Am. Soc. Exp. Biol.* **36**, 2181–2184.
Tiedtke, A. (1976). *Naturwissenschaften* **63**, 93.
Tokuyasu, K., and Scherbaum, O. H. (1965). *J. Cell Biol.* **27**, 67–81.
Tolloczko, B. (1975). *Acta Protozool.* **14**, 313–320.

Tolloczko, B. (1976). *Acta Protozool.* **15,** 359–366.
Tulkens, P., Schneider, Y.-J., and Trouet, A. (1977). *Biochem. Soc. Trans.* **5,** 1809–1815.
Vicker, M. G. (1977). *Exp. Cell Res.* **109,** 127–138.
Watson, M. R., Alexander, J. B., and Silvester, N. R. (1964). *Exp. Cell Res.* **33,** 112–129.
Weidenbach, A. L. S., and Thompson, G. A., Jr. (1974). *J. Protozool.* **21,** 745–751.
Whaley, W. G. (1975). "The Golgi Apparatus." Springer-Verlag, Berlin and New York.
Williams, N. E., and Luft, J. H. (1968). *J. Ultrastruct. Res.* **25,** 271–292.
Wise, G. E., and Flickinger, C. J. (1970). *J. Cell Biol.* **46,** 620–626.
Wolf, R. W., and Allen, R. D. (1974). *J. Protozool.* **21,** 425.
Wolfe, J. (1970). *J. Cell Sci.* **6,** 679–700.
Wolfe, J. (1973). *Exp. Cell Res.* **77,** 232–238.
Wunderlich, F., and Speth, V. (1972). *J. Ultrastruct. Res.* **41,** 258–269.
Wunderlich, F., Müller, R., and Speth, V. (1973a). *Science* **182,** 1136–1138.
Wunderlich, F., Speth, V., Batz, W., and Kleinig, H. (1973b). *Biochim. Biophys. Acta* **298,** 39–49.
Wunderlich, F., Ronai, A., Speth, V., Seelig, J., and Blume, A. (1975). *Biochemistry* **14,** 3730–3735.
Wyroba, E., and Przelecka, A. (1973). *Z. Zellforsch. Mikrosk. Anat.* **143,** 343–353.
Zigmond, S. H., and Hirsch, J. G. (1972). *Exp. Cell Res.* **73,** 383–393.

NOTE ADDED IN PROOF

The full paper by Wolf and Allen (1974) has now appeared: Allen, R. D., and Wolf, R. W. (1967). *J. Cell Sci.* **35,** 217–227.

Rumen Ciliate Protozoa

9

G. S. COLEMAN

I. INTRODUCTION

From the standpoint of a protozoon, the rumen is an unusual environment in that it is warm (39°C), anaerobic, rich in particulate matter that is more or less resistant to digestion, and usually deficient in soluble nutrients such as glucose and amino acids. The nutrition, physiology, and biochemistry of the rumen ciliates therefore reflect the demands placed on them by the environment, and they have adapted until they are intolerant of all but the smallest change in these conditions. There are two types of ciliate present, photomicrographs of which are shown in Figures 1 and 2. The first belong to the order Entodiniomorphida and are characterized by a firm pellicle often drawn out posteriorly into spines, the absence of cilia except on the peristome (Coleman and Hall, 1971) (Figures 1a and 1b) and in clearly defined bands elsewhere, and complex internal structure. The second (commonly called "holotrich") belong to the order Trichostomatida and family Isotrichidae and are represented by two genera *Isotricha* (Figure 2h) and *Dasytricha*. The body is uniformly ciliated with a mouth characteristically situated in different species.

BIOCHEMISTRY AND PHYSIOLOGY OF PROTOZOA
SECOND EDITION, VOL. 2

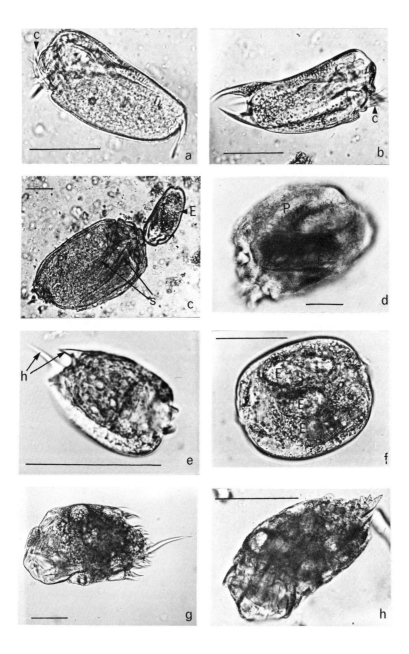

II. ENTODINIOMORPHID PROTOZOA

A. Cultivation *in Vitro*

Many of the Entodiniomorphid protozoa have now been cultivated *in vitro*, and there is no reason to believe that the others would not be culturable under the same conditions. Hungate (1942, 1943) first grew some of these protozoa for long periods *in vitro*, although this author and others, e.g., Sugden (1953), could not repeat his observations. The methods used by this author have been described in detail (Coleman, 1978a) and depend on inoculating individual protozoa into reduced buffered salts medium (principally potassium phosphate, pH 6.8) equilibrated with $N_2 : CO_2$ (95 : 5) or 100% CO_2 with the daily addition of starch and a little dried grass (for starch-utilizing protozoa) or powdered dried grass alone (for cellulose-utilizing protozoa, e.g., *Diplodinium pentacanthum*—Figure 2f). None of the protozoa have been grown in the absence of living bacteria, which probably keep the redox potential of the medium down. The success of the method depends on providing food for the protozoa in a form in which it cannot be readily utilized by the bacteria. The protozoa are therefore fed small amounts of particulate matter, such as starch grains, at daily intervals so that most of the food can be rapidly engulfed by the protozoa, leaving none available free in the medium for bacterial attack. Within these limits, the protozoal population density is proportional to the amount of starch added daily (Coleman, 1960, 1969a; Rahman et al., 1964).

The proportion of carbon dioxide in the gas phase with which the medium is equilibrated is also of great importance. Some protozoa such as *Entodinium caudatum* (Figure 1e) grow best in the presence of 5% CO_2, whereas others, e.g., *Entodinium simplex*, only grow in presence of 100% CO_2 (with appropriate dilution of the medium and addition of sodium

---

Figure 1. Photomicrographs of some rumen Entodiniomorphid protozoa. Bar marker, 50 μm. (a) *Epidinium ecaudatum caudatum* (form found in sheep). Note presence of cilia (c). (b) *Epidinium ecaudatum tricaudatum* (a species found in cattle). Note presence of cilia (c). (c) *Polyplastron multivesiculatum* (P) about to engulf an *Epidinium ecaudatum caudatum* (E) organism. Note the presence in *P. multivesiculatum* of two skeletal plates (s) which are actually composed of polysaccharide granules. (d) *Polyplastron multivesiculatum* (P) containing an engulfed *Epidinium ecaudatum caudatum* organism (E). (e) *Entodinium caudatum* taken directly from the rumen. Note the characteristic caudal spines (h) which are lost on growth in the absence of *E. bursa*. (f) *Entodinium bursa* containing the partially digested remains of at least three *E. caudatum* organisms (E). (g) *Ophryoscolex caudatus* taken directly from the rumen. (h) *Ophryoscolex caudatus* taken from a culture. Note that the main long caudal spine is much shorter.

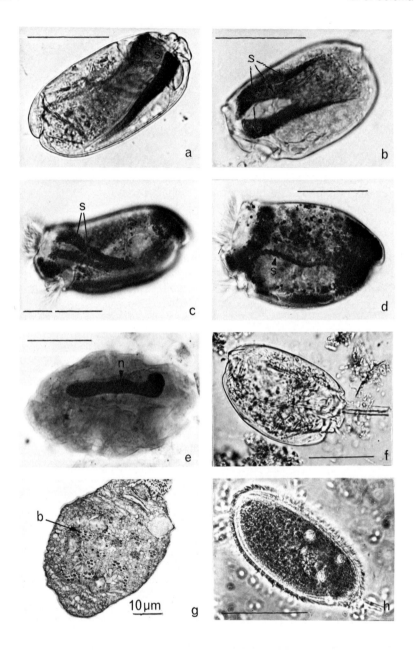

bicarbonate to keep the pH constant) (Coleman, 1960, 1969a, 1978a). However, this is not necessarily a constant property of one species. For example, different isolates of *Eudiplodinium maggii* grew best under either 5 or 100% CO_2 and one isolate of *D. pentacanthum* grew well in presence of 100% CO_2 on initial isolation, but after 2 months in culture only grew in presence of 5% CO_2 (Coleman *et al.*, 1976). The mean generation time of several protozoal species during logarithmic growth is shown in Table I. The minimum generation time for entodinia *in vivo* is 5.5 hr (determined from the number of division forms present, see Warner, 1962) but the average time over 24 hr is likely to be much lower and could be similar to those found *in vitro*.

1. Growth Requirements

During the initial isolation it is essential to add prepared fresh rumen fluid (10%, v/v, see Coleman, 1958) to the medium, probably to keep the redox potential down. Once protozoal cultures are established, the rumen fluid can be omitted with some protozoal species, without a decline in the numbers present (Coleman, 1978a; Coleman *et al.*, 1972, 1976). However, fresh or autoclaved rumen fluid was essential for the growth of *Entodinium caudatum* (Coleman, 1960), *E. simplex* (Coleman, 1969a), *Entodinium longinucleatum* (Owen and Coleman, 1976), *Entodinium bursa* (Coleman *et al.*, 1977), and *Ophryoscolex caudatus* (Coleman, 1978a). Tompkin *et al.* (1966) found that rumen fluid from animals containing a high population of entodinia supported larger numbers of entodinia *in vitro* than fluid from animals containing few entodinia. The growth of *E. caudatum* was stimulated by the soluble material in the rumen fluid (Coleman, 1960), whereas the growth of *E. simplex* was stimulated by the particulate matter (Coleman, 1969a). Some protozoal species, e.g., *E. longinucleatum* (Einszporn, 1961), mixed *Entodinium* spp., and *Diplodinium* (*Ano-*

Figure 2. (a–d) Photomicrographs of rumen Entodinimorphid protozoa stained with iodine to show the presence of skeletal plates (s). Bar marker, 50 μm. (a) *Ostracodinium obtusum bilobum*. Note the presence of a single large skeletal plate. (b) *Enoploplastron triloricatum*. Note the presence of three skeletal plates. (c) *Diploplastron affine*. The two skeletal plates are close together at the posterior end of the protozoon. (d) *Eudiplodinium maggii*. Note the single thin skeletal plate. (e) Photomicrograph of *Eudiplodinium maggii* stained with hematoxylin/eosin to show the macronucleus (n), which has a characteristically shaped anterior end. Bar marker, 50 μm. (f) Photomicrograph of the culture form of *Diplodinium pentacanthum* (no caudal spines) engulfing a cellulose fibre. Bar marker, 50 μm. (g) Electron micrograph of a section of *Entodinium caudatum* showing engulfed bacteria (*Escherichia coli*) (b). Bar marker, 10 μm. (h) Photomicrograph of *Isotricha intestinalis*. Bar marker, 50 μm.

Table I   Mean Generation Time of Rumen Entodiniomorphid Protozoa during Logarithmic Growth *In Vitro*

Protozoon	Mean generation time from growth curves (hr)	Generation time from uptake of mixed rumen bacteria[a] (hr)
Entodinium bursa	6	29
Entodinium caudatum	23	8.6
Entodinium longinucleatum	31	7.1
Entodinium simplex	38	—
Epidinium ecaudatum caudatum (ovine)	26	350
Diplodinium pentacanthum	8.5	—
Diploplastron affine	13	—
Metadinium medium	2[b]	—
Polyplastron multivesiculatum (+epidinia)	14	96

[a] Calculated from the rate of uptake of bacterial protein when the protozoa were incubated with 10^9 bacteria/ml and assuming that this was the only source of amino acids for protozoal protein synthesis (see Table II and Section II,C,4).

[b] Limited time—Naga and El-Shazly (1968).

plodinium) *denticulatum* (Michalowski, 1975) grow well in the presence of sterile undiluted rumen fluid.

Although all protozoal cultures contain bacteria which are engulfed by the protozoa as a source of food (see Section II,C), some information can be obtained on the possible food requirement of the protozoa from an examination of those food materials which will support growth. All the starch-utilizing protozoa except *E. caudatum* require ground whole wheat and will not grow on a purified material such as rice starch, although *E. simplex* will grow on rice starch if washed bran is also added (Coleman, 1969a). Because the stimulatory factor is heat sensitive it could be a substance such as thiamine. In contrast, Jarvis and Hungate (1968) found that protozoal extract stimulated the growth of *E. simplex*. Broad and Dawson (1976) showed that the rumen fluid and dried grass essential for the growth of *E. caudatum* could be replaced by β-sitosterol and that this protozoon also had a requirement for choline, which is present in commercial rice starch. The population density and division rate were lower in the purified medium. These results are not unexpected because soluble compounds are a less readily available source of these compounds than particulate materials which are engulfed (Coleman and Hall, 1969; Hall *et al.*, 1974) and digested in vesicles in the cytoplasm, presumably with release of compounds at a comparatively high concentration. Hino *et al.* (1973)

found that mixed entodinia required campesterol in addition to β-sitosterol. Wheat gluten is stimulatory to the growth of *E. longinucleatum* (Einszporn, 1961). Diethylstilbestrol increases ciliate population density *in vivo* (Ibrahim *et al.*, 1970) but this author has been unable to show a stimulatory effect *in vitro*.

Rumen protozoa cannot be grown for long periods in a conventional continuous culture apparatus (Rufener *et al.*, 1963). If, however, the fermentation vessel contains a large dead space (provided by a bag of food material with one to two times the volume of the free space), then at a dilution rate of 0.5–1.0 per day, populations of mixed protozoa of 10^3–4×10^4/ml can be maintained apparently indefinitely (Weller and Pilgrim, 1974; Czerkawski and Breckenridge, 1977). Little information is presently available on the performance of individual protozoal species under these conditions, except that *E. caudatum* is only present when starch is fed to the protozoa. It is of interest that *E. caudatum* cannot feed on starch grains when both protozoon and grains are present in gently stirred medium (Coleman, unpublished observations).

2. Requirement for Other Protozoa

Polyplastron multivesiculatum (Figures 1c and 1d) grows *in vitro* in the absence of other protozoa on a sodium chloride-rich medium (Coleman *et al.*, 1972) but will not grow on a potassium phosphate-rich medium except in the presence of certain other protozoa, preferably *Epidinium* spp. (Figures 1a and 1b). These are engulfed at a rate of up to 10 protozoa/ *P. multivesiculatum*/day: a minimum of one per day is essential for growth. Heat-killed epidinia are also engulfed, but *Entodinium* spp., although smaller, are not taken up and do not support growth. The reason for this selective uptake or rejection of certain species is not known but it is possible that *P. multivesiculatum* regards *Epidinium ecaudatum caudatum*, but not *Entodinium caudatum*, as a bacterium, because the former has bacteria attached to its pellicle whereas the latter does not (Coleman and Hall, 1974).

Entodinium bursa (Figure 1f) has an obligate requirement for the culture form of *E. cadatum* (i.e., the form lacking caudal spines) for growth *in vitro*; other *Entodinium* spp. will not support growth (Coleman *et al.*, 1977). Each *E. bursa* engulfs 2.2 *E. caudatum* per hour and divides every 6 hr. When *E. bursa* is grown in the presence of an excess of *E. caudatum* for 3 weeks, the latter gradually develops the caudal spines characteristic of the species *in vivo*. These spined protozoa are engulfed much less readily than the spineless form and support a lower population density of *E. bursa* (Coleman *et al.*, 1977). Since *E. caudatum* is engulfed posterior end first, the caudal spine must be an effective defence against uptake.

3. Morphology

Many Entodiniomorphid protozoa undergo a morphological change on cultivation *in vitro*. *Entodinium caudatum* (Figure 1e) loses its caudal spine after 3 months *in vitro* (Coleman, 1960) or *in vivo* (Eadie, 1967) in the absence of *E. bursa*. *Epidinium* spp. also tend to lose their caudal spines *in vitro* (Clarke, 1963), but this is not reversed in the presence of *Entodinium bursa* (Coleman, unpublished observations). *Polyplastron multivesiculatum,* which is 175 μm long *in vivo* (no epidinia present), is only 123 μm long during growth *in vitro* in absence of other protozoa but 205 μm long *in vitro* in the presence of epidinia (Coleman *et al.,* 1972). Although the *P. multivesiculatum* must increase in size to engulf an epidinium 118 μm long, the reason for the increase in size is unknown. The morphology of *Ophryoscolex caudatus* also changes markedly after 2–4 weeks in culture, as shown in Figures 1g and 1h: The reason is unknown.

4. Length of Time in Culture

The time that most Entodiniomorphid protozoa can be cultured *in vitro* is about 2 years [for a full list, see Coleman (1978a)], although there is considerable variation between species and isolates. This "loss of vigour" can be reversed by inoculating cultured protozoa into the defaunated rumen of sheep and reisolating the protozoa. In one experiment, *Epidinium ecaudatum caudatum* that had been cultured for 14 months was inoculated into a sheep and reisolated 6 days later. This reisolated strain survived for an additional 2 years, although the original culture lived for only another 6 months. The two exceptions to this limited life in culture are *Entodinium caudatum* ($>$18 years) and *E. simplex* [8.5 years (Coleman, 1960, 1969a)].

5. Monoxenic and Axenic Cultures

The problem of sterilizing the starch and dried grass used as substrates for *Entodinium* spp. has been solved by Hino and Kametaka (1977), who used radiation with ^{60}Co to sterilize the former and autoclaving under anaerobic conditions to sterilize the latter. By combined treatment with two antibiotics and a bacterial suspension, monoxenic cultures (with *Escherichia coli* or *Streptococcus bovis*) of *Entodinium caudatum* were maintained for over 2 months. Axenic cultures were also prepared by treatment with three antibiotics but the protozoal population density declined steadily and all the protozoa died after 3 weeks.

6. Reestablishment of Cultured Protozoa in Vivo

Entodinium caudatum grown for 10 years *in vitro* grew readily in the defaunated rumen of a sheep, apparently without a lag (Coleman and White, 1970). This result suggests that the physiology of this protozoon had not changed greatly during the period in culture.

B. Metabolism

1. Role of Bacteria

It is likely that all Entodiniomorphid protozoa contain bacteria in vesicles in their endoplasm when harvested from the rumen or culture media (Figure 2g) (Coleman and White, 1970; Coleman and Hall, 1974). All the protozoa taken directly from the rumen by Imai and Ogimoto (1978) had bacteria (principally *Streptococcus bovis* and *Ruminococcus albus*) attached to the pellicle, although with cultured species Coleman and Hall (1974) found bacteria on *Epidinium ecaudatum caudatum* but not on *Entodinium caudatum*. Each *E. caudatum* grown *in vitro* and harvested 16 hr after feeding contains approximately 16 *Klebsiella aerogenes*, 8 *Proteus mirabilis*, 1 *Bacteroides* spp., and 1 *Butyrivibrio* spp., the total number decreasing with increase in time since last feeding the protozoa (White, 1969). The *Klebsiella* and *Proteus* spp., which are present in protozoal cultures, are not found in *Entodinium caudatum* in the rumen and disappear when cultured protozoa are reinoculated into the rumen, to be replaced by an approximately equal number of typical rumen bacteria, e.g., *Veillonella* spp., *Butyrivibrio* spp., and *Streptococcus bovis* (Coleman and White, 1970). It is believed that, although most bacteria engulfed by the protozoa are killed and digested, some are comparatively resistant and remain alive in vesicles and that bacteria present are typical of the environment rather than the protozoon. With *Entodinium caudatum* grown *in vitro*, *P. mirabilis* is apparently resistant to digestion, whereas *K. aerogenes*, which is sensitive to digestion, produces a glucose-containing capsular polysaccharide which is comparatively resistant to attack (Coleman, 1967b, 1969b). The number of viable *K. aerogenes* present depends on a balance between the rate at which the bacterium can synthesize a capsule and the rate at which this can be digested by the protozoon's lytic enzymes (Coleman, 1975b).

Any investigation of the metabolism of suspensions of washed protozoa must take into account the bacteria closely associated with the protozoa. These bacteria have been shown by autoradiography to take up compounds that were free in the medium (Coleman and Hall, 1974). Although these bacteria can be killed by the prolonged action of antibiotics, the resultant protozoa are moribund (Coleman, 1962).

2. *Carbohydrates*

All rumen Entodiniomorphid protozoa studied, except possibly the smallest and including those considered to be cellulolytic, engulf starch grains very rapidly, the protozoon being completely filled with grains within a few minutes (Hungate, 1975). Many of these protozoa, but not *Entodinium* or *Diplodinium* spp., contain "skeletal plates," which are storage organelles composed of polysaccharide granules formed from products derived from ingested food particles. The skeletal plates have characteristic shapes in different genera (Figures 2a–2d).

a. Entodinium spp. When *E. caudatum* is incubated with carbohydrates, the endogenous rate of gas production is increased by rice starch but not by maltose or glucose, suggesting that this protozoon does not utilize free sugars (Abou Akkada and Howard, 1960). However, survival of polysaccharide-depleted *E. caudatum* is improved by some sugars as well as starch (Coleman, 1969b). It is now known that this protozoon hydrolyses starch to maltose and then to glucose (both of which appear in the pool and the medium) and that each product inhibits the enzyme that produced it (Coleman, 1969b). Maltose and glucose are both taken up by the protozoa, but because of this feedback inhibition the rate of starch breakdown is decreased and the actual intracellular concentration of sugars, and hence the rate of gas production, remains constant. Although external maltose is taken up more rapidly by *E. caudatum* than glucose, the rate of incorporation into protozoal polysaccharide is the same with both sugars, and the difference in overall rate can be accounted for because intracellular *Klebsiella aerogenes* incorporates maltose more than three times as fast as glucose. Little carbon from starch, glucose, or maltose is incorporated into protein (Coleman, 1969b).

Although *Entodinium caudatum* contains starch and maltose phosphorylases, the major path of starch utilization is by hydrolysis to glucose followed by phosphorylation to glucose 6-phosphate (Coleman, 1969b). There is no direct evidence for the further metabolism of this compound, although the labelling pattern when CO_2 carbon is incorporated into protozoal polysaccharide suggests the existence of a classical glycolysis pathway (Coleman, 1964a). The final products of starch fermentation are acetic and butyric acids with small amounts of lactic, propionic, and formic acids and carbon dioxide (Abou Akkada and Howard, 1960). Although *E. caudatum* only attacks starch components (Abou Akkada and Howard, 1960), other entodinia have wider degradative abilities (Bailey and Clarke, 1963). Bonhomme-Florentin (1975) found cellulolytic activity (against carboxymethylcellulose) in entodinia from the rumen, although cellulosic substrates will not maintain the life of *E. caudatum* (Coleman, 1960). Experiments (Coleman, unpublished data) on the digestion of phosphoric acid-regenerated cellulose showed that the activity of cell-free extracts of mixed entodinia (mostly *E. simplex*) was only 0.5% (on a per protozoon basis) of that of *Eremoplastron bovis,* which is a recognized cellulolytic protozoon.

Entodinium caudatum takes up particulate matter readily, but there is no evidence that soluble compounds are taken up by the formation of vesicles or that appreciable amounts of external medium are taken up with particles (Coleman, 1969b). Almost 40% of *E. caudatum* (probably the ectoplasm) is freely permeable to the external medium, and active uptake of sugars and amino acids probably represents uptake into the endoplasm. There is also a passive uptake of all low molecular weight compounds, and with glucose at least this is markedly increased by the inhibitor iodoacetate. It is suggested that this passive uptake might be diffusion of external medium into an organelle such as the contractile vacuole that, in the presence of an inhibitor of energy production, might be unable to contract and remain full of medium (Coleman, 1969b).

b. Epidinium spp. *Epidinium ecaudatum caudatum* (Figure 1a) differs from *Entodinium* spp. in that it will survive in culture in the presence of dried grass in the absence of added starch (Coleman *et al.,* 1972, 1976), and when isolated from a starch-fed sheep attacks a wider range of substrates (Bailey *et al.,* 1962; Bailey and Howard, 1963), including hemicelluloses (Bailey and Gaillard, 1965). *Epidinium ecaudatum caudatum* takes up glucose 100 times faster than *Entodinium caudatum* but utilizes both starch and glucose to only a very limited extent for the synthesis of protein (Coleman and Laurie, 1974a). There are probably two pools in *Epidinium ecaudatum caudatum,* glucose being taken up as such into the first pool and then phosphorylated to glucose 6-phosphate on transfer to the second pool. In this pool the glucose 6-phosphate is dephosphorylated and mixes with maltose and glucose derived from the hydrolysis of starch. Glucose from the second pool is utilized for the synthesis of polysaccharide, breakdown to carbon dioxide, and metabolism by bacteria (Coleman and Laurie, 1976).

Epidinium ecaudatum caudatum attaches itself to plant particles in the rumen (Bauchop and Clarke, 1976), and although the mechanism is not known it could be attraction to sugars released from the tissue as was found for holotrich protozoa (Orpin and Letcher, 1978).

c. Polyplastron multivesiculatum (Figures 1c and 1d). Like *Epidinium* spp., *P. multivesiculatum* is predominantly a starch-utilizing protozoon, although it has been grown at low population densities on grass alone

(Coleman *et al.*, 1972). When isolated from the rumen, *P. multivesiculatum* contains enzymes that hydrolyse starch, cellulose, pentosans, pectin, and a number of oligosaccharides (Abou Akkada *et al.*, 1963). Carbon from starch and glucose is incorporated into protozoal protein, and the latter can provide 10% of the protein required for the protozoon to divide once each day (Coleman and Laurie, 1977).

In contrast to *Epidinium* and *Entodinium* spp., *P. multivesiculatum* utilizes starch by phosphorolysis to glucose 1-phosphate and conversion to glucose 6-phosphate rather than by hydrolysis to glucose and phosphorylation with ATP. Free glucose was taken up as glucose 6-phosphate, which was then hydrolysed to glucose: The principal product of glucose uptake was protozoal polysaccharide (Coleman and Laurie, 1977).

d. Eudiplodinium maggii (Figures 2d and 2e). This protozoon grows better *in vitro* on dried grass than on starch, although *in vivo* numbers increase dramatically when concentrates are added to a hay ration (Coleman *et al.*, 1976). Washed protozoa utilized cellulose and to a lesser extent starch, but no soluble sugars, for survival. Cellulose is degraded to cellobiose and glucose, and fermented principally to acetic and butyric acids.

Although it has been known for many years that *E. maggii* digested and utilized cellulose for polysaccharide synthesis and that cell-free extracts produce reducing sugars from cellulose (Hungate, 1942, 1943), it was uncertain whether the initial digestion of cellulose was made by protozoal enzymes or enzymes from intracellular bacteria. Recently (Coleman, 1978b) it has been shown that the cellulase activity is not affected by incubation of intact protozoa with antibiotics and that at least 70% of the activity is of protozoal origin, although there is a bacterial enzyme in suspensions of protozoa grown *in vitro*. *Eudiplodinium maggii* utilizes cellulose, starch, and glucose for the synthesis of protein but only at a rate that would allow for division at less than 5% of the maximum rate.

e. Metadinium medium. This protozoon metabolizes starch, hemicellulose, xylan, and araban but not cellulose, although it grows for short periods on barley hay (Naga and El-Shazley, 1968). It may therefore be growing on other constituents of the hay than cellulose. The fermentation products were unusual because in addition to acetic acid, the major product (50%), approximately equal amounts of formic and butyric acids (20% of each) were formed.

f. Ophryoscolex spp. (Figures 1g and 1h). Little work has been done on this genus, although *O. caudatus* has an obligate requirement for starch for growth (Coleman, 1978a). Mah and Hungate (1965) showed an increased rate of fermentation in the presence of starch and pectin but no utilization of polygalacturonic acid.

3. Amino Acids and Proteins

The most important source of amino acids for protozoal protein synthesis is probably bacterial protein, which is taken up as living or dead intact bacteria (Coleman, 1975a). Although Einszporn (1961) and Bonhomme-Florentin (1974a,b) found that protein stimulated the growth of rumen protozoa, this has not been the experience of this author (Coleman, 1978a). However, particulate, but not soluble, proteins are taken up by Entodiniomorphid protozoa (Coleman, 1964b; Onodera and Kandatsu, 1970; Abou Akkada and Howard, 1962) and there is indirect evidence that *Epidinium ecaudatum caudatum* grown *in vitro* utilizes plant protein as an important source of amino acids for growth (Coleman and Laurie, 1974a). This protein could be taken up as intact chloroplasts, which are then digested in vesicles with the liberation of soluble protein (Hall *et al.*, 1974).

Rumen protozoa are proteolytic (Warner, 1956) and release amino acids into the medium. Although Clarke (1977) stated that ammonia is the end product of protozoal protein metabolism, it is more likely that ammonia only comes from amide groups (Abou Akkada and Howard, 1962). Suspensions of *Entodinium caudatum* incubated in the absence of substrate liberate nonprotein nitrogen including ammonia (Abou Akkada and Howard, 1962), and suspensions of *Epidinium ecaudatum caudatum* liberate protein and peptides (Coleman and Laurie, 1974b). With this latter protozoon, at least, it is likely that this is a generalized leakage of cell contents, possibly because of the trauma of the manipulations involved in preparing washed suspensions.

All Entodiniomorphid protozoa tested have a limited ability to take up amino acids from the medium, and as with other soluble compounds there is an active process predominant at low external concentrations which, with amino acids, is inhibited by amino acid analogues and a passive uptake predominant at high concentrations (Coleman, 1967a). With cultured *Entodinium caudatam,* the amino acids can be divided into two groups. Those of group 1 (e.g., alanine and aspartic acid) are taken up at only a quarter of the rate of those in group 2 (e.g., isoleucine and phenylalanine), the uptake of which is also increased another four times as the salt concentration in the medium is decreased to half that of normal. Although this is obviously a permeability effect, its significance is unknown.

The rate of uptake of amino acids also differs markedly between protozoal species even after allowing for their different sizes (Table II); *Polyplastron multivesiculatum* and *Eudiplodinium maggii* take up amino acids more than five times more rapidly per unit volume than *Entodinium* spp.

Table II Utilization of Free Amino Acids and Bacterial Protein for Protozoal Protein Synthesis

Protozoon	Volume[a]	Surface area[a]	Cytoplasmic protein (ng protein/hr)					
			Synthesis needed		From engulfed protozoa	From uptake of 18 free amino acids		Derived from 10^9 rumen bacteria/ml
			For division once a day	For maximum division rate		Without CAP[b]	With 80 μg CAP/ml	
Entodinium caudatum	1.0	1.0	0.023	0.022	—	0.0027	—	0.064
Entodinium longinucleatum	1.7	1.3	0.039	0.030	—	0.0017	0.0012	0.13
Entodinium bursa	3.8	2.3	0.17	0.68	0.12	0.154	0.116	0.14
Epidinium ecaudatum caudatum (ovine)	5.5	3.3	0.59	0.55	—	0.020	0.017	0.039
Eudiplodinium maggii	8.3	4.4	1.0	—	—	0.36	0.29	0.368
Polyplastron multivesiculatum								
– Epidinia	14.5	5.4	1.6	—	—	0.85	—	0.53
+ Epidinia	36	10.8	2.9	5.0	0.59	6.7	2.7	0.72

[a] Relative to that of *Entodinium caudatum*.
[b] CAP, chloramphenicol.

and *Epidinium ecaudatum caudatum.* Table II also shows the rate at which protein in the protozoal cytoplasm must be synthesized in order for the protozoon to divide once each day and the rate at which protein in this fraction was synthesized from 18 amino acids (present at 0.1 mM) assuming no competition between amino acids. It is apparent that if free amino acids were the only source of amino acids for protein synthesis then *Entodinium caudatum, E. longinucleatum,* and *Epidinium ecaudatum caudatum* could divide less than once every 10 days, whereas *Eudiplodinium maggii* and *Polyplastron multivesiculatum* could divide more frequently than once every 2 days.

The ^{14}C-labeled amino acids taken up by the protozoa are incorporated into protozoal protein and in *Entodinium caudatum* and *E. longinucleatum* there are no interconversions (Coleman, 1967a; Owen and Coleman, 1977). With *Polyplastron multivesiculatum* some metabolically related amino acids are slightly labelled and with *Epidinium ecaudatum caudatum* most amino acids are incorporated unchanged, although serine and aspartic acid are extensively metabolized (Coleman and Laurie, 1974a, 1977). As there is evidence (Coleman, 1975b, 1978b) for incorporation of added substrates by intracellular bacteria and subsequent digestion of the bacteria by the protozoa, it is possible for ^{14}C from external amino acids to reach protozoal protein after incorporation by intracellular bacteria. Some amino acids (e.g., glutamic acid and glycine) derived from the breakdown of bacterial protein or from the uptake of free amino acids are acetylated or formylated and appear in the pool or the medium in that form (Coleman, 1967a,b). A limited amount of the leucine, isoleucine, and valine taken up by *E. caudatum* is degraded to isovaleric acid, α-methylbutyric acid, and isobutyric acid, respectively, with the production of carbon dioxide from C1 (Coleman, 1967a,b; Wakita and Hoshino, 1975). Methionine can also be dethiomethylated by rumen protozoa but the process is probably not important (Merricks and Salisbury, 1976). Carbon from carbon dioxide and acetate is incorporated into alanine, aspartic acid, and glutamic acid in the protein of rumen entodinia (Harmeyer, 1967; Harmeyer and Hekimoglu, 1968), although Onodera and Kandatsu (1974) believe that net synthesis from acetate is small. Entodinia and other Entodiniomorphid protozoa can decarboxylate diaminopimelic acid, whether free or in *Escherichia coli* cell wall mucopeptide, with the production of lysine, but there is no evidence that the protozoa can synthesize diaminopimelic acid (Onodera and Kandatsu, 1974; Onodera *et al.,* 1974).

4. Purines and Pyrimidines

There is no evidence that Entodiniomorphid protozoa can synthesize purines or pyrimidines. *Entodinium caudatum* incorporates free adenine, guanine, and uracil at least but not thymine into protozoal nucleic acid, and has limited ability to interconvert adenine and guanine and a greater ability to convert uracil to cytosine (Coleman, 1968). *Entodinium caudatum* and *E. simplex* also deaminate adenine and guanine to hypoxanthine and xanthine, respectively, and also convert hypoxanthine into xanthine: These degradation products are found in the pool and the medium (Coleman, 1968, 1972). *Epidinium ecaudatum caudatum* and *Polyplastron multivesiculatum* degrade these metabolites to volatile material (Coleman and Laurie, 1974a, 1977). Uracil and thymine are metabolized to their dihydro derivatives. Although free purine and pyrimidine bases (except thymine) are taken up readily by Entodiniomorphid protozoa, nucleosides are taken up more rapidly and nucleotides faster still (Coleman, 1968, and unpublished observations). There is no evidence that these protozoa, except possibly *Eudiplodinium maggii,* can synthesize ribose or incorporate free ribose into nucleic acid, and this may be the reason for the preferential utilization of nucleosides and nucleotides over free bases.

As with amino acids for protein synthesis, the most important source of constituents for nucleic acid is probably engulfed bacteria. Bacterial nucleic acid is degraded to the nucleotide level and the nucleotides are then incorporated into protozoal nucleic acid with only a small exchange of phosphate (Coleman, 1968).

5. Choline and Ethanolamine

Entodinium caudatum cannot synthesize choline or ethanolamine, and has an absolute requirement for the former for growth (Broad and Dawson, 1976). Cultured *E. caudatum* take up ethanolamine much more slowly than choline, which is quantitatively converted into phosphatidylcholine in cell membranes (Broad and Dawson, 1975). Curiously, the rate of uptake of choline is not affected by the concomitant uptake of latex beads into vesicles when the rate of membrane synthesis might be expected to be high. Phosphatidylcholine is synthesized by the phosphorylcholine–CDP–choline pathway, the methylation or base-exchange pathways not being present. Phosphorylation of the choline after uptake is very rapid, the rate-limiting enzyme for the complete system being the choline phosphate cytidylyltransferase (Broad and Dawson, 1975; Bygrave and Dawson, 1976). A similar pathway operates for the synthesis of phosphatidylethanolamine. Although choline is rapidly incorporated into phosphatidylcholine, this and phosphatidylethanolamine turn over only slowly in intact protozoa, in contrast to phosphatidylinositol, which turns over very rapidly. Phosphatidylethanolamine is also converted into ceramide phosphorylethanolamine (Broad and Dawson, 1973, 1975).

6. Aminoethyl Phosphonate

Aminoethyl phosphonate, which occurs as the diglyceride and plasmalogen derivatives, was first isolated by Horiguchi and Kandatsu (1959) and is found in the rumen only in the protozoa. It is probably synthesized from a three-carbon glycolytic intermediate by *Entodinium caudatum,* where it occurs as the diglyceride derivative (Coleman *et al., 1971*).

7. N-(2-Hydroxyethyl)alanine

N-(2-Hydroxyethyl)alanine was first isolated esterified to phosphatidic acid from mixed rumen protozoa and from *Entodinium caudatum* by Kemp and Dawson (1969a,b). This compound is found only in the protozoa in the rumen, and the phosphatidyl derivative is synthesized in cultured *E. caudatum* from phosphatidylethanolamine. The *N*(1-carboxyethyl) grouping which substitutes on the amino group of the phosphatidylethanolamine is probably derived from a three-carbon glycolytic intermediate (Coleman *et al., 1971*).

8. Lipids

Most of the studies on the metabolism of lipids by rumen protozoa have been bedevilled by difficulties in interpreting the results obtained. Almost all have been made with protozoa (often mixed types) isolated from the rumen and contaminated with feed particles, which in turn are contaminated with bacteria.

a. Lipolysis. Latham *et al.* (1972) found that in cows on a high-roughage or a high-starch ration, 22 and 76%, respectively, of the lipolysis was carried out by the protozoa. However, the authors considered that "protozoal activity" in starch-fed animals could have been due to bacteria attached to starch grains that sedimented with the protozoa. On the basis of the action of penicillin, which was considered to remove bacterial activity specifically, Wright (1961) stated that 30–40% of the lipolytic activity in the rumen was due to protozoa. Clarke and Hawke (1970) found that most of the rumen lipolytic activity was present in the fraction that must have contained any protozoa present and that the activity was liberated into the supernatant on homogenization. This result could have been due to breakage of the protozoa or to removal of bacteria from plant particles. *Entodinium caudatum* at least can engulf oil droplets (Coleman and Hall, 1969) and is said to digest them (Warner, cited in Prins, 1977). It would therefore appear that Entodiniomorphid protozoa can hydrolyse lipids but the extent of the activity is uncertain.

b. Hydrogenation. There is also conflicting evidence on the ability of Entodiniomorphid protozoa to hydrogenate long-chain fatty acids. Wright

(1959) first showed that mixed protozoa (principally epidinia) hydrogenate linoleic acid and that, after engulfing chloroplasts, they rapidly convert triene to diene and monoene to stearic acid but hydrogenate diene to monoene more slowly. This view that Entodiniomorphid protozoa hydrogenate unsaturated fatty acids is supported by the findings (1) that on faunation of a defaunated animal, the proportion of oleic acid in plasma lipids rises and that of linoleic acid falls (Lough, 1968; Klopfenstein et al., 1966; Abaza et al., 1975); (2) that carefully washed Entodiniomorphid protozoa (but not holotrich protozoa) incubated with antibiotics reduce linoleic acid to stearic acid by intracellular reactions (Chalupa and Kutches, 1968); and (3) that washed Entodiniomorphid protozoa hydrogenate (as determined by a decrease in iodine number) free and combined (in plant oils) fatty acids (Abaza et al., 1975). In contrast, Dawson and Kemp (1969) found that the rumen organisms in a defaunated sheep could still hydrogenate fatty acids almost as rapidly as the original rumen contents and suggested that protozoa may not be important in biohydrogenation. However, it is possible that bacteria could have filled the ecological niche left by the protozoa, and carry out some of the same reactions. The situation is further complicated by the finding of Harfoot et al. (1973) that food particles are an important site for biohydrogenation and that protozoa showed little activity. If this is universally true then almost all experiments with protozoal suspensions could be suspect, because they are frequently contaminated with food particles. It is of interest that these protozoa can desaturate their own lipids (Emmanuel, 1974; Abaza et al., 1975).

Entodiniomorphid protozoa take up free long-chain fatty acids (Gutierrez et al., 1962; Chalupa and Kutches, 1968), and although there is no evidence for the mechanism, it is possible that the protozoa engulf fatty acid micelles in the same way as they take up other particulate matter.

It is likely that Entodiniomorphid protozoa can synthesize long-chain fatty acids from propionic and butyric acids and isoleucine (Emmanuel, 1974), but as the work was carried out with mixed Entodiniomorphid and holotrich protozoa, it is uncertain which species were involved.

9. Sterols

Entodinium caudatum requires β-sitosterol for growth (Broad and Dawson, 1976) and probably utilizes this and other sterols for the synthesis of stigmastanol and campestanol, which comprise 73–80% of the sterols present (Hino and Kametaka, 1975). The origin of the cholestanol (11–17% of the total) is unclear, although the amount present increases markedly in protozoa incubated in the absence of sterol and under conditions when the total sterol content is decreasing (Hino and Kametaka, 1975).

C. Metabolism of Bacteria

As mentioned above, bacteria are the most important single source of nitrogenous nutrients for rumen Entodiniomorphid protozoa. Bacteria are present *in vivo* at a population density of 10^9–10^{10}/ml and at this density most bacterial species are taken up progressively by washed suspensions of most protozoal species for at least a day (Figure 2g). Some protozoa digest some bacterial species readily, and with these combinations there is a progressive appearance of soluble bacterial constituents in the medium until after about 12 hr the rate of uptake and digestion of bacteria, as indicated by the release of soluble material, is almost equal.

1. Specificity for Different Bacterial Species

Most of the protozoa, except for *Entodinium caudatum*, show some dislike or preference for one or more bacterial species (Table III). There is no obvious explanation for this behaviour except that cultured *E. simplex* and *E. longinucleatum* prefer the bacteria present in their habitat *in vitro*, namely, *Klebsiella aerogenes* and *Proteus mirabilis* (Coleman, 1972; Owen and Coleman, 1977). All protozoal species engulf mixed rumen bacteria at a rate equal to or faster than those obtained with pure bacterial species (Coleman and Laurie, 1974a, 1977; Owen and Coleman, 1977; Coleman, unpublished observations).

2. Uptake by Individual Protozoal Species

The rate of uptake of bacteria increases with increasing bacterial population density until a limiting value is reached, and when the rate of uptake is plotted against bacterial population density a rectangular hyperbola is obtained (Coleman, 1964b). A number of parameters have been used to compare uptake by different protozoal species. The uptake at infinitely high bacterial population density [obtained from a reciprocal plot of rate against bacterial population density, see Coleman (1972)] probably measures a protozoon's ability to engulf bacteria without pursuit and depends on the size and shape of the bacterium. A few large bacteria will occupy the same volume as a large number of small bacteria. The uptake of bacteria can also be measured in terms of the volume of medium cleared of bacteria by the protozoa [for details of measurement, see Coleman (1972)]. Comparison between species is then made on the basis of the volume of medium cleared of bacteria from an infinitely dilute suspension. This probably measures the protozoon's ability to find, catch, and engulf bacteria from a large volume of medium and may depend on the attraction between the two organisms. It should, however, be independent of the size of the bacterium. Table IV shows a comparison of the rate of uptake

Table III Uptake of Bacterial Species by Entodiniomorphid Protozoa

Protozoon	Bacteria not engulfed	Bacteria poorly engulfed	Bacteria preferentially engulfed
Entodinium bursa	None	*Pseudomonas* sp.	None
Entodinium caudatum	None	None	None
Entodinium longinucleatum	*Bacteroides ruminicola*, *Pseudomonas* sp.	*Escherichia coli*, *Bacillus megaterium*	*Klebsiella aerogenes*, *Proteus mirabilis*
Entodinium simplex	None	None	*Klebsiella aerogenes*, *Proteus mirabilis*
Epidinium ecaudatum caudatum			
Ovine	None	*Klebsiella aerogenes*, *Streptococcus bovis*	*Proteus mirabilis*, *Pseudomonas aerugenosa*
Bovine	None	None	*Butyrivibrio fibrisolvens*, *Proteus mirabilis*
Epidinium ecaudatum tricaudatum	None	None	*Butyrivibrio fibrisolvens*, *Proteus mirabilis*
Eudiplodinium maggii	*Bacteroides ruminicola*, *Klebsiella aerogenes*	*Escherichia coli*	*Ruminococcus flavefaciens*, *Butyrivibrio fibrisolvens*, *Streptococcus bovis*
Polyplastron multivesiculatum			
− Epidinia	None	*Bacteroides ruminicola*	*Proteus mirabilis*
+ Epidinia	None	None	*Proteus mirabilis*

Table IV Uptake of *Butyrivibrio fibrisolvens* and *Escherichia coli* by Rumen Entodiniomorphid Protozoa

Protozoon	Volume	*Butyrivibrio fibrisolvens*				*Escherichia coli*			
		Bacteria engulfed/protozoon/hr		10^{-6} × volume cleared of bacteria/protozoon/hr		Bacteria engulfed/protozoon/h		10^{-6} × volume cleared of bacteria/protozoon/hr	
		Maximum	At 10^9 bacteria/ml	Maximum (μm^3)	At 10^9 bacteria/ml (μm^3)	Maximum	At 10^9 bacteria/ml	Maximum (μm^3)	At 10^9 bacteria/ml (μm^3)
Entodinium caudatum	1.0	4000	1400	2.3	1.5	3300	1300	2.9	1.8
Entodinium simplex	0.45	2800	1100	2.34	0.55	900	240	0.72	0.45
Entodinium longinucleatum	1.7	2200	780	1.14	0.76	400	147	0.68	0.125
Entodinium bursa	3.8	1090	695	0.71	0.82	237	114	0.23	0.12
Epidinium ecaudatum caudatum									
Ovine	5.5	1430	320	1.9	1.3	1870	770	1.2	0.70
Bovine	—	1700	1100	4.5	0.7	480	310	1.1	0.30
Epidinium ecaudatum tricaudatum	—	220	190	2.4	0.17	1100	210	0.30	0.19
Eudiplodinium maggii	8.3	12,600	2800	5.4	3.5	1620	740	1.6	0.70
Polyplastron multivesiculatum									
− Epidinia	14.5	7050	6550	100	6.0	15200	9600	30	9.3
+ Epidinia	36	32,000	8900	19	10	230,000	47,000	50	43

401

and medium clearance for two bacteria of similar size, one commonly (*Butyrivibrio fibrisolvens*) and one rarely (*Escherichia coli*) found in the rumen. At an infinitely high bacterial population density or at 10^9 bacteria/ml, *Escherichia coli* is taken up at less than 30% of the rate of *B. fibrisolvens* by *Entodinium bursa*, *E. longinucleatum*, *E. simplex*, *Epidinium ecaudatum caudatum* (bovine), and *Eudiplodinium maggii*, whereas with *Entodinium caudatum*, *Epidinium ecaudatum caudatum* (ovine), *Epidinium ecaudatum tricaudatum*, and *Polyplastron multivesiculatum*, *Escherichia coli* is engulfed at the same rate or more rapidly than *B. fibrisolvens*. When the parameter used is the volume of medium cleared of bacteria from an infinitely dilute suspension, all the protozoa except *Entodinium caudatum* and *Polyplastron multivesiculatum* (plus epidinia) clear *B. fibrisolvens* more efficiently than *Escherichia coli*.

Studies on the uptake of many bacterial species by one protozoon showed that, except for those bacteria which were not or only poorly engulfed (Table III), the volume of medium cleared of bacteria from an infinitely dilute suspension is similar for all bacterial species (Coleman, 1972; Coleman and Laurie, 1974a, 1977; Owen and Coleman, 1977). It is therefore considered likely that the volume of medium that passes in and out of the protozoal esophagus is constant for any one protozoal species and that variations between the uptake of different bacterial species depend on the efficiency with which the protozoon removes individual bacteria from the stream of medium.

Entodinium caudatum grown *in vitro* engulfs all particulate matter that is small enough to enter the esophagus, including live and dead bacteria, polystyrene latex beads, olive oil droplets, chloroplasts, and particles of palladium black (Coleman and Hall, 1969; Hall *et al.*, 1974). It is therefore not surprising that this protozoon shows no preference for any bacterial species and engulfs bacteria from a mixed suspension in the proportions in which they are present (Coleman, 1964b).

3. Digestion of Bacteria

The rates at which engulfed bacteria are killed by *Entodinium caudatum* and *E. simplex* vary greatly. The half-life for bacterial survival varies from a few minutes (*Escherichia coli*) to several hours (*Proteus mirabilis*) and is not related to gram reaction or cell shape. There is wide variation even among the Enterobacteriaceae (Coleman, 1967b, 1972). Amino acids from the bacterial protein are incorporated unchanged into protozoal protein or liberated free or as their N-acetyl or N-formyl derivatives into the medium. With *Entodinium caudatum* and *Escherichia coli* approximately equal amounts of bacterial carbon are present in the protozoa and in the medium after 24 hr.

Although different bacteria are digested at different rates as determined by studies on ^{14}C-labelled organisms, different subcellular structures of one bacterium show very variable degrees of resistance to digestion. The cell contents of *Escherichia coli* and *Klebsiella aerogenes* are digested rapidly when engulfed, as are the peptidoglycan and protein components of the cell envelope. In contrast, the lipopolysaccharide component of the envelope is very resistant to attack and is probably extruded via the anus (Coleman and Hall, 1972). Among gram-positive bacteria the cell walls of *Bacillus megaterium, Micrococcus lysodeikticus,* and *Bacillus subtilis* are rapidly attacked, suggesting the presence of a lysozymelike enzyme, whereas the walls of *Staphylococcus aureus* and *Streptococcus faecalis* are more resistant than the cell contents to attack.

4. Importance of Bacterial Protein for Protozoal Growth

As it is not possible to grow Entodiniomorphid protozoa without bacteria and complex food materials, no direct evidence can be obtained on the sources of amino acids for protozoal growth. However, indirect evidence can be obtained from the rates at which mixed rumen bacteria (at 10^9/ml) are engulfed. Since the protozoa have bacteria associated with them, only the rate of synthesis of cytoplasmic protein will be considered. When this criterion is used, *Entodinium caudatum* and *E. longinucleatum* incorporate sufficient material from bacteria to divide at least once each day, whereas *E. bursa, Epidinium ecaudatum caudatum, Eudiplodinium maggii,* and *P. multivesiculatum* must obtain amino acids from other sources to do so (Table II). *Entodinium bursa* and *Polyplastron multivesiculatum* may obtain this from the other protozoa which they engulf: At a maximum rate of uptake they could supply as much soluble protein as is supplied by the bacteria (Table II). *Epidinium ecaudatum caudatum* cannot obtain enough amino acids for division at the maximum rate, from free amino acids or bacteria, and there is speculation that this protozoon may utilize plant protein (Coleman and Laurie, 1974a).

III. HOLOTRICH PROTOZOA

A. Cultivation *in Vitro*

Although these protozoa are the easiest of all the rumen protozoa to maintain in culture for short periods (Sugden and Oxford, 1952; Purser and Weiser, 1963; Purser and Tompkin, 1965), they have proved difficult to culture for longer than a month. Holotrich protozoa rapidly convert extracellular sugars into reserve polysaccharide (Oxford, 1951) and in sheep fed once a day *Dasytricha ruminantium* only divides in the 5 hr before the next feed (Purser, 1961). It is likely that the failure to obtain division of protozoa *in vitro* or soon after feeding *in vivo* is due to an excess of storage polysaccharide. This prevents the protozoa from dividing and only when limited amounts of sucrose are added to cultures of *D. ruminantium* does satisfactory growth occur (Clarke and Hungate, 1966). The maximum division rate was once a day. *Isotricha* spp. have not yet been cultivated for long periods *in vitro*.

Holotrich protozoa can be grown as members of a mixed population in continuous culture (at populations of approximately 10^3/ml and dilution rates of 0.5–1.0 per day), apparently indefinitely, provided that there is a dead space in the fermentation vessel provided by a bag of food (hay) (Czerkawski and Breckenridge, 1977).

B. Metabolism

Because of the difficulties in culturing holotrich protozoa, all studies on the metabolism have had to be made with protozoa isolated from the rumen. At best these have been made with protozoa from animals containing only one protozoal species in the rumen (e.g., Prins and Van Hoven, 1977), whereas in others mixed *Isotricha* spp. or *D. ruminantium* have been prepared from normal rumen contents by differential centrifugation or filtration (Wallis and Coleman, 1967; Williams and Harfoot, 1976).

1. Carbohydrates

Dasytricha ruminantium and *Isotricha* spp. metabolize glucose, fructose, galactose, raffinose, and pectin (cellobiose and maltose in *D. ruminantium* only) with the production of hydrogen, carbon dioxide, acetic acid, butyric acid, lactic acid, and intracellular amylopectin (Heald and Oxford, 1953; Howard, 1959; Williams and Harfoot, 1976; Prins and Van Hoven, 1977; Van Hoven and Prins, 1977). *Isotricha* spp. can also rapidly engulf starch grains. For many years (see Hungate, 1966) it was thought that the rupturing of holotrich protozoa observed in the presence of excess glucose was due to uncontrolled amylopectin synthesis, but it is now believed that there is a maximum amount of amylopectin that can be synthesized and that rupture is due to low intracellular and environmental pH (Prins and Van Hoven, 1977; Van Hoven and Prins, 1977). Of the glucose incorporated by *D. ruminantium*, 3–5% of the carbon is incorporated into protein, and from the results of Holler and Harmeyer (1965) and Williams and Harfoot (1976) it can be calculated that under optimum conditions it could provide sufficient protein for the organism to divide every 13–22 hr.

Isotricha intestinalis (Figure 2h) and *Isotricha prostoma* both show chemotaxis to surfaces of particles that liberate sucrose, glucose, or fructose and attach themselves to that surface by a specialized organelle, provided that soluble protein is also present (Orpin and Letcher, 1978; Orpin and Hall, 1977). This phenomenon is of great importance *in vivo* because these protozoa become sequestered in the solid matter and they are specifically retained in the rumen (Weller and Pilgrim, 1974; Czerkawski and Breckenridge, 1978).

2. Proteins and Amino Acids

Several strains of rumen bacteria (Gutierrez, 1958) and *Eschericha coli* (Wallis and Coleman, 1967) are engulfed by *Isotricha* spp., but it is uncertain if the protozoa show any great specificity. After engulfment, *E. coli,* at least, is killed and digested to the extent that after 24 hr equal amounts of bacterial carbon are present in the protozoa and as low molecular weight compounds in the medium (Wallis and Coleman, 1967). As *Isotricha* spp. presumably engulf bacteria *in vivo,* the release of amino acids etc. by washed protozoa taken from the rumen (Heald and Oxford, 1953; Harmeyer, 1971b) may just represent the digestion of previously engulfed material. *Dasytricha ruminantium* also engulfs bacteria and may preferentially select certain small cocci (Gutierrez and Hungate, 1957).

Isotricha spp. take up ^{14}C-labeled amino acids from the medium but the rate is less than 10% of that found with the predatory form of *P. multivesiculatum,* which is of comparable size (Wallis and Coleman, 1967). However, Harmeyer (1971a), using the disappearance of amino acids from the medium as criterion, could detect no uptake by the protozoa.

Isotricha spp. and *D. ruminantium* incorporate ^{14}C from $^{14}CO_2$ and [^{14}C]acetate into their amino acids in protein (Harmeyer, 1965; Harmeyer and Hekimoglu, 1968). It is probable that these incorporations occur during synthesis of amino acids from glucose (Williams and Harfoot, 1976) and represent net synthesis rather than exchange.

3. Lipids

Isotricha spp. take up long-chain fatty acids (Gutierrez *et al.,* 1962; Williams *et al.,* 1963; Girard and Hawke, 1978), but there is a disagreement about whether these protozoa can hydrogenate unsaturated long-chain fatty acids. Gutierrez *et al.* (1962) and Williams *et al.* (1963) found that both *I. prostoma* and *I. intestinalis* converted oleic into stearic acid, whereas Katz and Keeney (1967) found little stearic acid in holotrich protozoa and considered that hydrogenation was unlikely to be important. However, this might not be significant, because Abaza *et al.* (1975) found that when protozoa were incubated alone desaturation could occur. Linoleic acid is not hydrogenated by washed isotrichs unless bacteria are present, and Girard and Hawke (1978) suggested that biohydrogenation may involve a symbiotic relationship between bacteria and protozoa.

There is little evidence on the ability of *Isotricha* spp. to hydrolyse lipids. Girard and Hawke (1978) found that these protozoa hydrolysed phosphatidylcholine only slowly unless bacteria were present and suggested that plant phospholipids in the food have to be broken down by plant or bacterial phospholipases before the fatty acids can be absorbed.

REFERENCES

Abaza, M. A., Abou Akkada, A. R., and El-Shazly, K. (1975). *J. Agric. Sci.* **85**, 135–143.
Abou Akkada, A. R., and Howard, B. H. (1960). *Biochem. J.* **76**, 445–451.
Abou Akkada, A. R., and Howard, B. H. (1962). *Biochem. J.* **82**, 313–320.
Abou Akkada, A. R., Eadie, J. M., and Howard, B. H. (1963). *Biochem. J.* **89**, 268–272.
Bailey, R. W., and Clarke, R. T. J. (1963). *Nature (London)* **198**, 787.
Bailey, R. W., and Gaillard, B. D. E. (1965). *Biochem. J.* **95**, 758–766.
Bailey, R. W., and Howard, B. H. (1963). *Biochem. J.* **86**, 446–452.
Bailey, R. W., Clarke, R. T. J., and Wright, D. E. (1962). *Biochem. J.* **83**, 517–523.
Bauchop, T., and Clarke, R. T. J. (1976). *Appl. Environ. Microbiol.* **32**, 417–422.
Bonhomme-Florentin, A. (1974a). *Ann. Sci. Nat., Zool. Biol. Anim.* **16**, 155–220.
Bonhomme-Florentin, A. (1974b). *Ann. Sci. Nat., Zool. Biol. Anim.* **16**, 221–283.
Bonhomme-Florentin, A. (1975). *J. Protozool.* **22**, 447–451.
Broad, T. E., and Dawson, R. M. C. (1973). *Biochem. J.* **134**, 659–662.
Broad, T. E., and Dawson, R. M. C. (1975). *Biochem. J.* **146**, 317–328.
Broad, T. E., and Dawson, R. M. C. (1976). *J. Gen. Microbiol.* **92**, 391–397.
Bygrave, F. L., and Dawson, R. M. C. (1976). *Biochem. J.* **160**, 481–490.
Chalupa, W., and Kutches, A. J. (1968). *J. Anim. Sci.* **27**, 1502–1508.
Clarke, R. T. J. (1963). *J. Gen. Microbiol.* **33**, 401–408.
Clarke, R. T. J. (1977). *In* "Microbial Ecology of the Gut" (R. T. J. Clarke and T. Bauchop, eds.), pp. 251–275. Academic Press, New York.
Clarke, R. T. J., and Hawke, J. C. (1970). *J. Sci. Food Agric.* **21**, 446–452.
Clarke, R. T. J., and Hungate, R. E. (1966). *Appl. Microbiol.* **14**, 340–345.
Coleman, G. S. (1958). *Nature (London)* **182**, 1104–1105.
Coleman, G. S. (1960). *J. Gen. Microbiol.* **22**, 555–563.
Coleman, G. S. (1962). *J. Gen. Microbiol.* **28**, 271–281.
Coleman, G. S. (1964a). *J. Gen. Microbiol.* **35**, 91–103.
Coleman, G. S. (1964b). *J. Gen. Microbiol.* **37**, 209–223.
Coleman, G. S. (1967a). *J. Gen. Microbiol.* **47**, 433–448.
Coleman, G. S. (1967b). *J. Gen. Microbiol.* **47**, 449–464.
Coleman, G. S. (1968). *J. Gen. Microbiol.* **54**, 83–96.
Coleman, G. S. (1969a). *J. Gen. Microbiol.* **57**, 81–90.
Coleman, G. S. (1969b). *J. Gen. Microbiol.* **57**, 303–332.
Coleman, G. S. (1972). *J. Gen. Microbiol.* **71**, 117–131.
Coleman, G. S. (1975a). *In* "Digestion and Metabolism in the Ruminant" (I. W. McDonald and A. C. I. Warner, eds.), pp. 149–164. Univ. of New England Publ. Unit, Armidale, Australia.
Coleman, G. S. (1975b). *Symp. Soc. Exp. Biol.* **29**, 533–558.

Coleman, G. S. (1978a). *In* "Methods of Cultivating Parasites in Vitro" (A. E. R. Taylor and J. R. Baker, eds.), pp. 39–54. Academic Press, New York.
Coleman, G. S. (1978b). *J. Gen. Microbiol.* **107**, 359–366.
Coleman, G. S., and Hall, F. J. (1969). *Tissue Cell* **1**, 607–618.
Coleman, G. S., and Hall, F. J. (1971). *Tissue Cell* **3**, 371–380.
Coleman, G. S., and Hall, F. J. (1972). *Tissue Cell* **4**, 37–48.
Coleman, G. S., and Hall, F. J. (1974). *J. Gen. Microbiol.* **85**, 265–273.
Coleman, G. S., and Laurie, J. I. (1974a). *J. Gen. Microbiol.* **85**, 244–256.
Coleman, G. S., and Laurie, J. I. (1974b). *J. Gen. Microbiol.* **85**, 257–264.
Coleman, G. S., and Laurie, J. I. (1976). *J. Gen. Microbiol.* **95**, 364–374.
Coleman, G. S., and Laurie, J. I. (1977). *J. Gen. Microbiol.* **98**, 29–37.
Coleman, G. S., and White, R. W. (1970). *J. Gen. Microbiol.* **62**, 265–266.
Coleman, G. S., Kemp, P., and Dawson, R. M. C. (1971). *Biochem. J.* **123**, 97–104.
Coleman, G. S., Davies, J. I., and Cash, M. A. (1972). *J. Gen. Microbiol.* **73**, 509–521.
Coleman, G. S., Laurie, J. I., Bailey, J. E., and Holdgate, S. A. (1976). *J. Gen. Microbiol.* **95**, 144–150.
Coleman, G. S., Laurie, J. I., and Bailey, J. E. (1977). *J. Gen. Microbiol.* **101**, 253–258.
Czerkawski, J. W., and Breckenridge, G. (1977). *Br. J. Nutr.* **38**, 371–384.
Czerkawski, J. W., and Breckenridge, G. (1978). *Proc. Nutr. Soc.* **37**, 70A.
Dawson, R. M. C., and Kemp, P. (1969). *Biochem. J.* **115**, 351–352.
Eadie, J. M. (1967). *J. Gen. Microbiol.* **49**, 175–194.
Einszporn, T. (1961). *Acta Parasitol. Pol.* **9**, 193–210.
Emmanuel, B. (1974). *Biochim. Biophys. Acta* **337**, 404–413.
Girard, V., and Hawke, J. C. (1978). *Biochim. Biophys. Acta* **528**, 17–27.
Gutierrez, J. (1958). *J. Protozool.* **5**, 122–126.
Gutierrez, J., and Hungate, R. E. (1957). *Science* **126**, 511.
Gutierrez, J., Williams, P. P., Davis, R. E., and Warwick, E. J. (1962). *Appl. Microbiol.* **10**, 548–551.
Hall, F. J., West, J., and Coleman, G. S. (1974). *Tissue Cell* **6**, 243–253.
Harfoot, C. G., Noble, R. C., and Moore, J. H. (1973). *Biochem. J.* **132**, 829–832.
Harmeyer, J. (1965). *Zentralbl. Veterinaermed., Reihe A* **12**, 9–17.
Harmeyer, J. (1967). *J. Protozool.* **14**, 376–378.
Harmeyer, J. (1971a). *Z. Tierphysiol., Tierernaehr. Futtermittelkd.* **28**, 65–75.
Harmeyer, J. (1971b). *Z. Tierphysiol., Tierernaehr. Futtermittelkd.* **28**, 75–85.
Harmeyer, J., and Hekimoglu, H. (1968). *Zentralbl. Veterinaermed., Reihe A* **15**, 242–254.
Heald, P. J., and Oxford, A. E. (1953). *Biochem. J.* **53**, 506–512.
Hino, T., and Kametaka, M. (1975). *Nippon Chikusan Gakkai-Ho* **46**, 693–705.
Hino, T., and Kametaka, M. (1977). *J. Gen. Appl. Microbiol.* **23**, 37–48.
Hino, T., Kametaka, M., and Kandatsu, M. (1973). *J. Gen. Appl. Microbiol.* **19**, 397–413.
Holler, H., and Harmeyer, J. (1965). *Zentralbl. Veterinaermed., Reihe A* **11**, 244–251.
Horiguchi, M., and Kandatsu, M. (1959). *Nature (London)* **184**, 901–902.
Howard, B. H. (1959). *Biochem. J.* **71**, 671–674.
Hungate, R. E. (1942). *Biol. Bull. (Woods Hole, Mass.)* **83**, 303–319.
Hungate, R. E. (1943). *Biol. Bull. (Woods Hole, Mass.)* **84**, 157–163.
Hungate, R. E. (1966). "The Rumen and its Microbes." Academic Press, New York.
Hungate, R. E. (1975). *Annu. Rev. Ecol.* **6**, 39–66.
Ibrahim, E. A., Ingalls, J. R., and Stanger, N. E. (1970). *Can. J. Anim. Sci.* **50**, 101–106.
Imai, S., and Ogimoto, K. (1978). *Jpn. J. Vet. Sci.* **40**, 9–19.
Jarvis, B. D. W., and Hungate, R. E. (1968). *Appl. Microbiol.* **16**, 1044–1052.
Katz, I., and Keeney, M. (1967). *Biochim. Biophys. Acta* **144**, 102–112.

Kemp, P., and Dawson, R. M. C. (1969a). *Biochem. J.* **113**, 555–558.
Kemp, P., and Dawson, R. M. C. (1969b). *Biochim. Biophys. Acta* **176**, 678–679.
Klopfenstein, J. J., Purser, D. B., and Tyznik, W. J. (1966). *J. Anim. Sci.* **25**, 765–773.
Latham, M. J., Storry, J. E., and Sharpe, M. E. (1972). *Appl. Microbiol.* **24**, 871–877.
Lough, A. K. (1968). *Proc. Nutr. Soc.* **27**, 30a.
Mah, R. A., and Hungate, R. E. (1965). *J. Protozool.* **12**, 131–136.
Merricks, D. L., and Salisbury, R. L. (1976). *J. Anim. Sci.* **42**, 955–959.
Michalowski, T. (1975). *J. Agric. Sci.* **85**, 151–158.
Naga, M. A., and El-Shazly, K. (1968). *J. Gen. Microbiol.* **53**, 305–315.
Onodera, R., and Kandatsu, M. (1970). *Nippon Chikusan Gakkai-Ho* **41**, 307–313.
Onodera, R., and Kandatsu, M. (1974). *Agric. Biol. Chem.* **38**, 913–920.
Onodera, R., Shinjo, T., and Kandatsu, M. (1974). *Agric. Biol. Chem.* **38**, 921–926.
Orpin, C. G., and Hall, F. J. (1977). *Proc. Soc. Gen. Microbiol.* **4**, 82–83.
Orpin, C. G., and Letcher, A. J. (1978). *J. Gen. Microbiol.* **106**, 33–40.
Owen, R. W., and Coleman, G. S. (1976). *J. Appl. Bacteriol.* **41**, 341–344.
Owen, R. W., and Coleman, G. S. (1977). *J. Appl. Bacteriol.* **43**, 67–74.
Oxford, A. E. (1951). *J. Gen. Microbiol.* **5**, 83–90.
Prins, R. A. (1977). *In* "Microbial Ecology of the Gut" (R. T. J. Clarke and T. Bauchop, eds.), pp. 73–183. Academic Press, New York.
Prins, R. A., and Van Hoven, W. (1977). *Protistologica* **13**, 549–556.
Purser, D. B. (1961). *Nature (London)* **190**, 831–832.
Purser, D. B., and Tompkin, R. B. (1965). *Life Sci.* **4**, 1493–1501.
Purser, D. B., and Weiser, H. H. (1963). *Nature (London)* **200**, 290.
Rahman, S. A., Purser, D. B., and Tyznik, W. J. (1964). *J. Protozool.* **11**, 51–55.
Rufener, W. H., Jr., Nelson, W. O., and Wolin, M. J. (1963). *Appl. Microbiol.* **11**, 196–201.
Sugden, B. (1953). *J. Gen. Microbiol.* **9**, 44–53.
Sugden, B., and Oxford, A. E. (1952). *J. Gen. Microbiol.* **7**, 145–153.
Tompkin, R. B., Purser, D. B., and Weiser, H. H. (1966). *J. Protozool.* **13**, 55–58.
Van Hoven, W., and Prins, R. A. (1977). *Protistologica* **13**, 599–606.
Wakita, M., and Hoshino, S. (1975). *J. Protozool.* **22**, 281–285.
Wallis, O. C., and Coleman, G. S. (1967). *J. Gen. Microbiol.* **49**, 315–323.
Warner, A. C. I. (1956). *J. Gen. Microbiol.* **14**, 749–762.
Warner, A. C. I. (1962). *J. Gen. Microbiol.* **28**, 129–146.
Weller, R. A., and Pilgrim, A. F. (1974). *Br. J. Nutr.* **32**, 341–351.
White, R. W. (1969). *J. Gen. Microbiol.* **56**, 403–408.
Williams, A. G., and Harfoot, C. G. (1976). *J. Gen. Microbiol.* **96**, 125–136.
Williams, P. P., Gutierrez, J., and Davis, R. E. (1963). *Appl. Microbiol.* **11**, 260–264.
Wright, D. E. (1959). *Nature (London)* **184**, 875–876.
Wright, D. E. (1961). *N.Z. J. Agric. Res.* **4**, 216–223.

Biological and Physiological Factors Affecting Pathogenicity of Trichomonads

10

B. M. HONIGBERG

I. INTRODUCTION

There are five species of trichomonads proved to contain pathogenic strains. These are *Trichomonas vaginalis* Donné, *Trichomonas gallinae* (Rivolta), *Tritrichomonas foetus* (Riedmüller), *Histomonas meleagridis* (Smith), and *Dientamoeba fragilis* Jepps & Dobell. Since *Tritrichomonas suis* (Gruby & Delafond) strains probably belong in *Tritrichomonas foetus* (see Honigberg, 1978a), they need not be considered separately from the latter species. Virulence of *Tetratrichomonas gallinarum* (Martin & Robertson) strains has not been conclusively demonstrated, and virtually nothing is known about their pathogenicity mechanisms (see Honigberg, 1978a). All adequately documented evidence militates against pathogenicity of any strain of *Pentatrichomonas hominis* (Davaine) (see Honigberg, 1978b).

BIOCHEMISTRY AND PHYSIOLOGY OF PROTOZOA
SECOND EDITION, VOL. 2

With the exception of *D. fragilis,* a rather atypical trichomonad, all the species with pathogenic strains inhabit sites other than the large intestine, which is thought to be the primary site of members of the order Trichomonadida. It seems, therefore, that there might be some connection between the migration of the flagellates from the large intestine and their acquisition of pathogenicity potential. Not all species, however, that inhabit secondary sites have pathogenic strains, e.g., *Trichomonas tenax* (O. F. Müller) (see Honigberg, 1978b).

Trichomonas vaginalis parasitizes the human urogenital tract; *Trichomonas gallinae* is found in the upper digestive tract, and some of its strains also parasitize other organs of a large variety of birds, but especially pigeons; *Tritrichomonas foetus* occurs in the genital passages of cattle; and *H. meleagridis* inhabits the ceca and liver of turkeys and chickens, but is typically more virulent in the former hosts. *Trichomonas vaginalis, Trichomonas gallinae,* and *Tritrichomonas foetus,* but not *H. meleagridis* and *D. fragilis,* can be grown in axenic culture.

Although ample evidence can be adduced for pathogenicity of *D. fragilis* [for references see Yang and Scholten (1977)], this organism will not be discussed here, because virtually nothing is known about its pathogenicity mechanisms or about the factors affecting its virulence. Furthermore, no attempt will be made to discuss here most of the descriptive aspects of the pathologic changes reported from natural and experimental infections involving *Trichomonas vaginalis, Trichomonas gallinae, Tritrichomonas foetus,* and the histomonad. For this information with regard to the three former species, the reader is referred to my recent reviews (Honigberg, 1978a,b); as far as *H. meleagridis* is concerned, the reviews by Lund (1969) or Reid (1967) can be consulted. In general, the pathologic manifestations will be mentioned only insofar as they help to understand pathogenicity mechanisms.

II. PATHOGENICITY AND CULTIVATION

A. Effect of Conditions of Cultivation on Pathogenicity

The highly pathogenic Jones' Barn (JB) strain of *Trichomonas gallinae* was reported to retain full virulence for nonimmune pigeons for between 17 and 21 weeks of axenic *in vitro* cultivation; thereafter, the parasites were no longer capable of causing a fatal disease in these avian hosts (Stabler *et al.,* 1964). On occasion, even infectivity for pigeons was lost by originally virulent or avirulent strains after prolonged cultivation in nonliving media (Honigberg, unpublished observations; Honigberg *et al.,* 1971).

Virulence could be restored by bird-to-bird passages at least for up to 28 weeks of cultivation. A very virulent strain of *H. meleagridis* retained its ability to cause a fatal infection in young chickens for an even shorter period, 8 weeks, of *in vitro* cultivation (Dwyer and Honigberg, 1970). For up to 15 weeks thereafter, virulence could be restored by passages in birds; ultimately it could not be restored.

Some light was shed on the factors affecting virulence of *Trichomonas gallinae* strains by the following observations of Stabler *et al.* (1964) and Honigberg *et al.* (1970): (1) More rapid virulence attenuation of the highly pathogenic JB strain was noted in the presence of antibiotics, i.e., penicillin and streptomycin, in the culture medium. The strain isolated in the medium containing these antibiotics lost its virulence between 7.5 and 9 weeks of *in vitro* cultivation. As will be pointed out later in this chapter, virulence is DNA- and RNA-dependent. Therefore, streptomycin could play a very important role in enhancing pathogenicity attenuation. (2) While *in vitro* cultivation in nonliving media resulted in virulence decrease in the pathogenic strain, stabilates of this strain stored in liquid nitrogen have retained full virulence for at least 18 years. These results suggest that cell division in nonliving media is responsible in some measure for the attenuation. That this hypothesis has merit was shown by maintaining the trichomonads in a medium without added carbohydrates. The flagellates, which underwent only a few divisions, remained fully virulent significantly longer than the trichomonads of the same isolate that were grown in conventional carbohydrate-supplemented media (23 versus 17 weeks of cultivation). (3) Trichomonads maintained for 1 year in the presence of chick liver cell cultures retained their virulence for pigeons, although in this system the division rate of the parasites was at least as high as of those grown in carbohydrate-supplemented media. No effect on the retention of virulence was exerted by the growth medium used for the cultivation of the chick liver cells. It seems that some unknown factor or factors present in chick liver cells prevent(s) the dilution of pathogenicity. (4) Parasites maintained for 1 year by intraperitoneal passages in mice seemed to be more pathogenic for pigeons than the trichomonads grown in nonliving media. Their virulence, however, was much below that of fully pathogenic isolates. On the other hand, their pathogenicity for mice had increased very significantly. Evidently different factors aid in preserving pathogenicity of *Trichomonas gallinae* for the natural avian and the experimental mammalian hosts.

It is suggested by all the foregoing results that some cytoplasmic factors (inclusions) may be instrumental in the maintenance of virulence of *Trichomonas gallinae* for pigeons. These factors appear to be diluted by division of the parasites in nonliving media, but not in the birds or in the presence of avian cell cultures; the environment of mammalian hosts seems much less favorable than that of birds for the maintenance of virulence for the latter. Evidently, up to a certain dilution the cytoplasmic factors can be restored to their effective levels by passages of the trichomonads in birds. Indeed, the entire situation resembles that known for the kappa particle of killer strains of *Paramecium aurelia* [for references, see Preer (1969) and Ball (1969)]. In view of this, we looked for kappalike units in the cytoplasm of virulent trichomonads with the aid of electron microscopy; no such inclusions have been found to date. However, a search for cytoplasmic factors, including viruses, that may be responsible for pathogenicity and its maintenance is continuing.

The situation is essentially similar with regard to *Trichomonas vaginalis*—prolonged *in vitro* cultivation results in pathogenicity decrease for mice inoculated intraperitoneally (Laan, 1966) or subcutaneously (Honigberg, 1961 and unpublished observations). However, the rate of pathogenicity attenuation appears to be lower than that observed for the avian trichomonad and histomonad. On the other hand, as reported by Dohnalová and Kulda (1975), cultivation does not seem to affect pathogenicity for mice of some strains of *Tritrichomonas foetus*. For still obscure reasons, other strains of this species actually appear to increase their pathogenicity for these rodent hosts after cultivation in nonliving media (Kulda and Honigberg, unpublished observations).

B. Relationship between Pathogenicity and *in Vitro* Growth Rate

Teras (1963) noted that the growth rate of freshly isolated *Trichomonas vaginalis* strains in axenic cultures was inversely proportional to their pathogenicity levels for mice. Furthermore, according to Laan (1966), fresh isolates of highly pathogenic strains had slower metabolism in axenic cultures than the less pathogenic parasites. In view of Laan's (1966) results, one may have some difficulty in understanding the statement published by Teras *et al.* (1977) that "A . . . correlation was established . . . between the virulence of the *T. vaginalis* strains and their hexokinase activity . . . " which was much higher in the virulent strains.

Conclusions similar to those of Teras (1963) with regard to the relationship of *T. vaginalis* strains for patients and mice and their generation times in axenic cultures could be drawn from the results of Newton *et al.* (1960) and Kulda *et al.* (1970). An alternative explanation of the inverse correlation of pathogenicity and growth rate was entertained, however, by Kulda and collaborators (1970). According to their records, all the slowly growing strains were isolated from women suffering from various diseases of the urogenital organs, but not in all instances could these diseases (e.g., adenocarcinoma of endometrial or ovarian origin) be related to the presence of *T. vaginalis*. In all cases there was much protein-rich exudate, and it seemed possible that the flagellates became dependent on some factors in the nutrient-rich environment, and, therefore, did not grow fast in the poorer environment of the culture media.

III. THE SUBCUTANEOUS MOUSE ASSAY FOR PATHOGENICITY

If axenic cultivation is possible, the presumably inherent pathogenicity levels of strains of various trichomonad species can be evaluated in systems which exclude most environmental factors known or assumed to affect pathogenicity expression, e.g., immunocompetence, hormonal balance, and bacterial flora, in the natural hosts. We have been employing with considerable success the subcutaneous mouse assay (Honigberg, 1961) for evaluating the inherent pathogenicity of freshly isolated *Trichomonas vaginalis* and *Trichomonas gallinae* strains grown in axenic cultures (Frost and Honigberg, 1962; Honigberg, 1961; Honigberg *et al.*, 1966, 1971; Kulda *et al.*, 1970, 1977; Stępkowski and Honigberg, 1972). The assay entails measurements of volumes of lesions produced in mice by subcutaneous inoculations of known numbers of axenically cultivated flagellates. Mean volumes of 6-day lesions not only reflect the relative pathogenicity of these parasites for mice but also have been found to be in essential agreement with pathogenicity estimates for the strains based on clinical and pathologic findings in experimental (*Trichomonas gallinae*) or natural infections (*Trichomonas vaginalis*) of natural hosts (Honigberg, 1961 and unpublished observations; Honigberg *et al.*, 1966, 1971; Kulda *et al.*, 1970, 1977; Stepkowski and Honigberg, 1972). The situation is often comparable also for *Tritrichomonas foetus* (Kulda and Honigberg, 1969); however, as was pointed out before, the bovine urogenital trichomonad strains differ from those of *Trichomonas gallinae* and *Trichomonas vaginalis* with regard to the effect of *in vitro* cultivation on pathogenicity for mice.

That volumes of lesions produced in mice reflect the severity of the pathologic processes following subcutaneous inoculation of various trichomonad strains is well documented by the histologic study of Frost and Honigberg (1962). Pathologic changes similar to those described from subcutaneous abcesses have been reported also from visceral and mesenteric lesions caused by *Trichomonas vaginalis* inoculated intraperitoneally into mice (Gobert *et al.*, 1969; Teras and Rõigas, 1966; for additional references, see these reports and Honigberg, 1978b). The intraperitoneal lesions, however, which are insusceptible to quantitative evaluation, cannot be correlated as accurately as the subcutaneous ones with the pathogenicity levels for natural hosts characteristic of a given parasite strain (see Kulda *et al.*, 1977; cf. Honigberg *et al.*, 1966; Kulda *et al.*, 1970; with Reardon *et al.*, 1961; Teras and Rõigas, 1966). In many, but not nearly in all, respects the cyto- and histopathologic changes described from mice inoculated with trichomonads by the subcutaneous and intraperitoneal routes resemble those reported from the urogenital passages of guinea pigs experimentally infected with *Trichomonas vaginalis* and from the passages of women harboring this parasite (Frost, 1967, 1975; Koss and Wolinska, 1959; for additional references, see these reports and Honigberg, 1978b). Such similarities exist also between the pathologic changes reported from mice receiving subcutaneous inoculations of the virulent JB strain of *Trichomonas gallinae* (Frost and Honigberg, 1962) and those described from the upper digestive tract and liver of pigeons infected with this strain (Perez-Mesa *et al.*, 1961). One cannot, however, expect perfect correlations between the abnormal alterations observed in the normal sites of natural or experimental hosts infected with pathogenic strains of trichomonads and the changes caused by these parasites in the subcutaneous or intraperitoneal sites of mice, where many of the environmental factors operating in the urogenital or digestive tract are absent.

IV. MOLECULAR BASIS OF PATHOGENICITY

The subcutaneous mouse assay was subsequently found useful in the analysis of the factors at the molecular level that affect the pathogenicity of *T. gallinae* strains. The strains employed in these studies were the very virulent JB strain, known to kill nonimmune pigeons in 8 days, on the average, and the Amherst (AG) strain which, originally avirulent, lost even its infectivity for these birds in the course of prolonged *in vitro* cultivation.

The general procedures used and the results of the virulence transformation experiments of Honigberg *et al.* (1971) are summarized in Table I. It is evident from the data presented in this table that virtually the same volumes of subcutaneous lesions were recorded from mice that received inoculations of untreated AG strain, this strain treated with either highly polymerized native DNA or high molecular weight RNA isolated from the virulent JB trichomonads, and in selfing experiments, in which the AG parasites were exposed to their own DNA–RNA mixture (in about a 1 : 10 proportion characteristic of trichomonads). The lesions resulting from in-

Table I Effects of DNA and RNA from the Virulent JB Strain on Pathogenicity of the Avirulent AG Strain of *Trichomonas gallinae*[a,b]

Recipient strain	Donor strain	Additives	No. of exposures	No. of replicate experiments	6-day mouse lesions (mm³)	
					n	Mean ($\pm$ SE)
AG	JB	DNA + RNA	2	9	302	27.6 ($\pm$ 0.35)
			3	1	36	31.2 ($\pm$ 1.85)
			5–6	3	113	39.6 ($\pm$ 1.38)
			12	1	31	137.5 ($\pm$28.89)
			17	3	80	199.3 ($\pm$16.27)
AG	JB	DNA	2	2	60	14.0 ($\pm$ 1.23)
		RNA	2	4	118	14.9 ($\pm$ 0.86)
	None		0	8	242	13.4 ($\pm$ 0.56)
AG	AG	DNA + RNA	3	1	36	13.4 ($\pm$ 1.41)
JB[c]	None		0	0	1	35	558.0 ($\pm$66.27)

[a] From Honigberg *et al* (1971), reproduced by permission of the American Society of Parasitologists.

[b] Experiments were controlled by the omission of nucleic acids and selfing.

[c] A fully virulent JB strain isolate which caused a fatal disease in *Trichomonas*-free, nonimmune pigeons.

jections of AG strain treated with a DNA–RNA mixture from the JB parasites were significantly larger than any of the aforementioned abscesses. It is evident also from the data given in Table I that further enhancement of the lesion size resulted from repeated exposures of the nonpathogenic AG trichomonads to the nucleic acid mixture isolated from the JB flagellates. More recent results (Honigberg, unpublished) suggested that lower molecular weight RNA, perhaps mRNA, may be responsible for the entire RNA effect. Additional experiments are being conducted to elucidate this point. It is not known whether RNA acts as a specific "priming" factor or if its effect is far less specific, simply affording protection to the DNA. The negative results of the "selfing" experiments in which the AG strain was exposed to its own DNA and RNA, as well as the possibility that low molecular weight RNA may be involved, militate against the second hypothesis. Of considerable significance was the finding that, although the subcutaneous lesions caused in mice by the "transformed" AG strain never reached the size of those resulting from inoculations of the JB trichomonads (199.3 versus 558.0 mm³) and no pathologic changes were noted in nonimmune pigeons infected with this strain, the transformed parasites were found capable of establishing a lasting infection in the birds; at least the infectivity component of pathogenicity was restored to the avirulent trichomonads. We intend to continue the transformation experiments using ^{14}C-labeled nucleic acids.

V. EFFECT ON VERTEBRATE CELL CULTURES AND ON HOST TISSUES

A. Cytological and Cytochemical Findings

Much can be learned about some aspects of pathogenicity mechanisms of trichomonads by studying vertebrate cell cultures infected with virulent and avirulent strains of these parasites (Farris and Honigberg, 1970; Honigberg *et al.*, 1964; Kulda and Honigberg, 1969; for additional references, see these reports and Honigberg, 1978a,b). Among various kinds of cell cultures tested, the primary chick liver cell cultures used routinely in our laboratory were found especially valuable for studying the cytopathologic (see the immediately preceding references) and cytochemical changes (Abraham and Honigberg, 1965; Sharma and Honigberg, 1966, 1967, 1969, 1971). Three types of cells are found in such cultures: epithelial cells, fibroblastlike cells (fibroblasts), and macrophages. Only a few pathologic changes can be noted in cultures inoculated with avirulent strains of *Trichomonas vaginalis, Trichomonas gallinae,* and *Tritrichomonas foetus,* although the mild strains of the human urogenital trichomonad appear to be somewhat more cytopathogenic than such strains of the other two species. Many largely nonspecific pathologic and cytochemical changes can be seen in the epithelial cells and fibroblasts in the infected cell cultures. Although the more pathogenic strains of each of the three species cause far greater changes in cell cultures than the mild strains, it is difficult to find differences among the effects exerted by these species. Actually, comparable cytopathologic and cytochemical changes have been reported from cell cultures infected with various living, protozoan and nonprotozoan, microbial agents or exposed to certain chemicals (for descriptive accounts and discussion, see Farris and Honigberg, 1970; Honigberg *et al.*, 1964; Kulda and Honigberg, 1969; Sharma and Honigberg, 1966, 1967, 1969, 1971).

Of interest is the observation of Conrad and Edward (1970) that hygromycin B, a compound said to be of value in the control of certain nematode parasites of chickens, when administered in feed to pigeons causes avirulent strains of *Trichomonas* (probably *T. gallinae*) to become invasive. The lesions reported from these latter birds are quite similar to those seen in infections with virulent strains of *Trichomonas gallinae*.

As far as the lack of specificity of tissue response is concerned, the basic response in the natural host to trichomonads is inflammation (e.g., Frost, 1967; Koss and Wolinska, 1959). Also, in subcutaneous abscesses of mice, beyond palisading of the parasites on the lesion margins, the general picture is that of a more or less severe inflammation, depending on the virulence level of a given strain (Frost and Honigberg, 1962). Histologic and cytologic as well as histochemical and cytochemical changes observed in laboratory animals experimentally infected with pathogenic strains of *T. vaginalis,* e.g., in vaginae of guinea pigs [for references, see Kazanowska (1966)] or cell cultures [for references, see Sharma and Honigberg (1971)], resemble those characteristic of preneoplastic or actual neoplastic conditions. There is, however, no incontestable evidence that any *T. vaginalis* strain is carcinogenic.

The lack of specificity of the host tissue responses has been demonstrated in the cytochemical studies of cell cultures infected with pathogenic strains of *T. vaginalis* (Sharma and Honigberg, 1966, 1967, 1969, 1971) and *T. gallinae* (Abraham and Honigberg, 1965). A few examples will be cited to illustrate this point.

Reduction and changes in distribution of glycogen observed in natural and experimental hosts of *T. gallinae* and *T. vaginalis* [for references, see Honigberg (1978a,b)], as well as in chick liver cell cultures exposed to pathogenic strains of these species (Abraham and Honigberg, 1965; Sharma and Honigberg, 1966), have been reported also from animals subjected to a variety of experimental procedures or harboring various parasites (for references, see Sharma and Honigberg, 1966). The same is true of the fatty changes noted in chick liver cell cultures infected with the aforementioned strains of the two trichomonad species. As in the *Trichomonas*-infected cells, the fatty changes are accompanied by glycogen losses in experimental animals subjected to a variety of surgical procedures or treated with hydrazine [for references, see Sharma and Honigberg (1966)]. Furthermore, both these changes have been reported from hosts infected with various viruses (Orsi *et al.,* 1957; Love, 1959) and with intracellular or blood protozoa (von Brand, 1973). Progressive depletion of proteins and decline in ATPase activity in chick liver cell cultures in the presence of pathogenic strains of the human urogenital and avian trichomonads has been found also in the liver of hosts infected with the Newcastle virus and from necrotic livers [for references, see Sharma and Honigberg (1967)]. The enhancement of malate dehydrogenase activity noted in chick liver cell cultures exposed to a pathogenic *T. vaginalis* strain is by no means characteristic of trichomonad or even protozoal infections. Indeed, similar increases in enzymic activity were found in cell cultures inoculated with various viruses, in neoplasms, and in tissues of rats treated with phenobarbital [for references, see Sharma and Honigberg (1969)]. Increase in the activity of nonspecific esterase reported from *T. vaginalis*-infected cell cultures was found in a number of pathologic conditions both in *in vivo* and *in vitro* systems [for references, see Sharma and Honigberg (1969)]. It has been suggested by Brodie *et al.* (1958) that this enzyme might be a detoxifying agent; however, other interpretations must also be entertained (Sharma and Honigberg, 1969). Decline in monoamine oxidase in chick liver cell cultures exposed to *T. vaginalis* has been observed previously in rats given pyrogallol or subjected to a variety of endocrinologic experiments [for references, see Sharma and Honigberg (1969)]. The changes in the activity of this latter enzyme appear to be related injuries of mitochondria, whose outer membranes constitute an important site of the oxidase (Schnaitman *et al.*, 1967). The diminished glucose-6-phosphatase activity is shared by liver cell cultures infected with a pathogenic strain of *T. vaginalis,* livers of rats treated with barbital or carbon tetrachloride, necrotic livers, kidneys from rats with experimentally induced hypertension, and with mammalian hepatomas and renal and prostatic carcinomas [for references, see Sharma and Honigberg (1971)]. There is evidence that this decline in enzymic activity reflects the structural changes in the endoplasmic reticulum (Jones and Fawcett, 1966; Mao *et al.,* 1966; Trump *et al.,* 1965). Similarly, the loss of phosphorylases from vertebrate cells in the presence of *T. vaginalis* has been reported also from a number of diseases and deficiencies, as well as from certain neoplastic cells [for references, see Sharma and Honigberg (1971)].

It is evident from all the aforementioned examples that in the preponderant majority of cases the cytopathologic and cytochemical changes found *in vivo* and in cell cultures can be caused by a variety of conditions, including malignancy, hormonal excess and deficiency, and poisons as well as by viral, bacterial, and protozoal (including trichomonad) infections. Evidently, therefore, such changes are largely nonspecific. In contrast to the foregoing changes, significant differences can be seen among the effects the pathogenic strains of the three trichomonad species exert on the macrophages and upon fibroblast division. There are also differences in percentages of the epithelial and fibroblast cells harboring trichomonads within their cytoplasm, and in the effects cell-free supernatant fluids from actively growing trichomonad cultures have on the cell cultures.

B. Mechanisms of Pathogenicity for Cell Cultures and for Mice

As far as the actual mechanisms whereby the three trichomonad species injure cell cultures, and perhaps also tissues in their hosts, are concerned, pathogenic *Trichomonas vaginalis* strains adhere closely to epithelial cells, especially macrophages and fibroblasts, which they destroy mostly from the outside; there seems to be only relatively limited phagocytic activity on the part of the macrophages (Farris and Honigberg, 1970). Virulent strains of *Trichomonas gallinae* are less likely to adhere to the vertebrate cells in culture (Honigberg *et al.*, 1964), and there is only minimal contact between pathogenic strains of *Tritrichomonas foetus* and the cell culture elements (Kulda and Honigberg, 1969). Both virulent and avirulent strains of all three species stimulate the macrophages, as reflected in the larger size and vacuolization of these phagocytes in cell cultures infected with trichomonads (Farris and Honigberg, 1970; Honigberg, unpublished observations; Honigberg *et al.*, 1964; Kulda and Honigberg, 1969). Especially in cell cultures infected with pathogenic strains, the macrophages tend to accumulate around islands of epithelial cells. Although relatively little phagocytic activity on the part of the macrophages was noted in cell cultures exposed to pathogenic strains of *Trichomonas vaginalis,* this activity was greatly enhanced in cultures infected with virulent and avirulent *Trichomonas gallinae* and *Tritrichomonas foetus* strains. As reported by Honigberg *et al.* (1964) and Kulda and Honigberg (1969), the virulent bovine urogenital trichomonads and *Trichomonas gallinae* multiply actively within and ultimately destroy the phagocytes. On the other hand, when engulfed the mild strains of all three trichomonad species are digested by the macrophages. The highest numbers of fibroblasts and epithelial cells with trichomonads lodged in the cytoplasm were noted in infections with virulent strains of *Trichomonas gallinae* (Honigberg *et al.*, 1964) and the lowest in those involving such strains of *Tritrichomonas foetus* (Kulda and Honigberg, 1969); pathogenic strains of *Trichomonas vaginalis* fell between the former two with regard to becoming intracellular, especially in the fibroblasts (Farris and Honigberg, 1970). These observations are essentially in agreement with those made in the natural hosts—many liver epithelial cells of a pigeon infected with the very virulent JB strain of *Trichomonas gallinae* were found to harbor this flagellate, and on rare occasions *Trichomonas vaginalis* was reported from cells of the cervix uteri (Frost *et al.*, 1961). As far as can be ascertained, there are no published records of *Tritrichomonas foetus* in cells other than phagocytes from infected cattle. With regard to *Trichomonas vaginalis,* Nielsen and Nielsen (1975), on the basis of their electron microscope observations, asserted that these parasites were never found within intact cells of the human female urogenital system. In light of our observations of tissues of natural and experimental hosts (Frost *et al.*, 1961) and of vertebrate cell cultures (Farris and Honigberg, 1970), the assertion of the Nielsens cannot be accepted without reservations. It must be remembered also that electron microscope investigations never include the examination of sufficient numbers of host cells to provide a firm basis for such an assertion. In their report, Nielsen and Nielsen (1975) described a very intimate relationship between the parasite and host cells, the areas of contact involving inter-digitation of cytoplasmic projections of trichomonads and host cells. This situation corresponds closely to that reported at the light microscope level by Farris and Honigberg (1970) from vertebrate cell cultures infected with a pathogenic *T. vaginalis* strain. In their discussion, however, the Nielsens suggested that, unlike the situation prevailing in cell cultures, e.g., Farris and Honigberg (1970), most of the damage inflicted by *T. vaginalis* on the host tissues depended on indirect effects mediated by some "toxic" sub-stances rather than by direct contact. This suggestion was based on the observations that "the clusters of *T. vaginalis* cells covered only a small part of the mucosa" and that "the local inflammatory reaction of the mucous membrane was generally the same whether or not *T. vaginalis* cells were present on the epithelial surface." On the basis of my observa-tions of lesions caused by pathogenic urogenital trichomonads of man in cell cultures, as well as of subcutaneous and visceral lesions caused by these parasites in mice, I am still inclined to support the view that both indirect, via diffusable toxins, and direct means, by contact, are responsi-ble for the damage the parasites inflict upon the host tissues.

Pathogenic strains of all three trichomonad species exert an inhibitory effect on division of fibroblasts in cell cultures. *Trichomonas gallinae* and *Tritrichomonas foetus* randomly inhibit all stages of mitosis (Honigberg, unpublished observations; Honigberg *et al.*, 1964; Kulda and Honigberg, 1969); however, *Trichomonas vaginalis* tends to inhibit fibroblasts during prophase (Farris and Honigberg, 1970). Supernatant fluids from cultures of the urogenital trichomonad of cattle appear to be about as inhibitory as the flagellates themselves; such fluids from cultures of the human urogeni-tal and the avian parasite are less inhibitory than the living parasites.

Since cytopathologic changes and inhibition of fibroblast mitosis have been found in cell cultures exposed to cell-free filtrates of cultures of pathogenic strains of all three trichomonad species whose effect on verte-brate cells *in vitro* has been investigated, it has been postulated that the flagellates can cause damage to vertebrate cells by producing toxic sub-stances (Farris and Honigberg, 1970; Honigberg *et al.*, 1964; Kulda and Honigberg, 1969; also see these papers for earlier pertinent reports). The substances produced by *Tritrichomonas foetus*, capable of causing about as extensive cytopathologic changes in the fibroblast and epithelial cells and about the same level of mitosis inhibition in the former cells as the living parasites, appear to be especially effective. The intensity of these effects of the bovine urogenital trichomonad on vertebrate cells is consis-tent with the observation that this organism has little tendency to remain in direct contact with the cell culture elements. The importance of trichomonad "toxins" in damaging cell cultures was first suggested by Hogue (1938, 1943), who ascribed to these substances all cytopathologic changes caused by *Tritrichomonas foetus* and *Trichomonas vaginalis*. It has become evident from subsequent studies that the importance of toxic substances in pathogenesis of trichomonad infection varies with the species. Direct contact appears to be more important in the case of *Trichomonas vaginalis* (Farris and Honigberg, 1970). There is much evidence, however, that "direct" action, involving contact, and "indirect" action, mediated by toxic products of trichomonads, may be involved in the pathogenic processes (for references, see Farris and Honigberg, 1970).

The nature of the toxic substances produced by trichomonads is not understood. The very complexity of the media renders chemical analysis of the supernatant fluid and of the substances most difficult. Müller and Saathoff (1972) demonstrated large quantities of neuraminidase in cultures of a freshly isolated *Tritrichomonas foetus* strain, and in another that had been maintained in culture for many years. The enzyme was found to act upon a wide spectrum of glycoproteins in human serum. The authors attempted to explain the possible role of neuraminidase in pathogenesis of the urogenital disease of cattle by direct effect of the enzyme on the trophoblast and placenta or by its involvement in immunologic changes leading to abortion and sterility. The involvement of neuraminidase in abortion and sterility is in need of additional experimental proof, and the same is true of the role hyaluronidase plays in pathogenesis associated with trichomonad infections. Both these enzymes appear about equally active in virulent and avirulent strains (Kulda and Zavadová, 1975; Honigberg, unpublished observations).

Most recently, Budilová and Kulda (1977) reported the virulence-enhancing effect of ferric ammonium citrate (FAC) on *Tritrichomonas foetus*. Mice inoculated intraperitoneally with 5×10^5 trichomonads of KV-1a strain were treated, also intraperitoneally, with 5 to 200 mg/kg body weight/day of FAC for 6 or more days. Mice receiving the trichomonads but not FAC and those given FAC but not the parasites served as experimental controls. The percentages of mice which became infected and those which succumbed to the infection were employed as pathogenicity indicators. The infectivity of the strain for mice was increased from 18 to over 90%, and virulence, as reflected in the mortality of the experimental animals, went from 13 to 85%. Higher dosages of FAC were needed to increase virulence than those required for infectivity enhancement. Evidently, as with bacteria (Weinberg, 1974, 1977), the availability of iron, in a form utilizable by the parasites, is important for the expression of pathogenicity of *Tritrichomonas foetus* for mice.

VI. RELATIONSHIP BETWEEN PATHOGENICITY AND
ANTIGENIC COMPOSITION

Much evidence has accumulated in support of the view that a high correlation exists between the antigenic composition and pathogenicity levels of trichomonads. This relationship was demonstrated clearly by gel diffusion and quantitative direct fluorescent antibody methods in *H. meleagridis* (Dwyer, 1971; Dwyer and Honigberg, 1972) and *Trichomonas gallinae* (Honigberg and Goldman, 1968; Stępkowski and Honigberg, 1972). As was mentioned before, in the course of *in vitro* cultivation by serial transfers, *H. meleagridis* tends to lose its virulence for turkeys and chickens within 9 weeks (Dwyer and Honigberg, 1970). It was noted that during cultivation in nonliving media the histomonads became progressively richer in antigens capable of stimulating antibody production in rabbits (Dwyer, 1971; Dwyer and Honigberg, 1972). It has been assumed on the basis of these data that the attenuated strains evoke a stronger immune response in the avian hosts than the more pathogenic ones, and that, therefore, the disease does not develop even though the parasites become established in the ceca. Indeed, it was demonstrated by Lund *et al.* (1966) that an *H. meleagridis* strain with virulence reduced during *in vitro* cultivation was capable of conferring a certain degree of protection on turkeys and chickens.

Equally convincing evidence for the relationships between the antigenic composition and pathogenicity of *T. gallinae* was adduced by Honigberg and Goldman (1968), with the aid of the quantitative direct fluorescent antibody method, and by Stępkowski and Honigberg (1972) in gel diffusion experiments. Since the results reported by the latter workers were based on the analysis of more strains than those of the former, they will be summarized below (Table II).

Stępkowski and Honigberg (1972) employed the following strains of *T. gallinae:* (a) the very virulent JB strain; (b) its descendant substrain JBC, attenuated by 1 year of cultivation *in vitro;* (c) a fresh isolate of a mild strain, SG; (d) a substrain, SGC, of this latter strain obtained by 1 year of *in vitro* growth; and (e) an originally avirulent strain, AG, maintained for many years in nonliving media. Several important observations were made and conclusions were drawn on the basis of this investigation (for the results and some conclusions, see Table II): (1) The virulent JB strain trichomonads have significantly less capacity than the primarily or secondarily avirulent strains to stimulate antibody production in rabbits. (2) The virulent parasites have either very small amounts of or incomplete antigens capable of combining with antibodies produced by rabbits and presumably also by the natural hosts against a variety of nonpathogenic

Table II Analysis of Group B Antigens in Five Strains or Substrains of *Trichomonas gallinae*[a,f]

Strain or substrain	Virulence	B1 Stimulates Ab[b] production	B1 Combines with Ab	B2 Stimulates Ab production	B2 Combines with Ab	B3 Stimulates Ab production	B3 Combines with Ab	B4 Stimulates Ab production	B4 Combines with Ab	B5 Stimulates Ab production	B5 Combines with Ab	Total content
JB	Very high	+++	+++	+p	+(+)p	-	++p	-	+p	-	-	B1; B2-tr[c] & h[d], p[e]; B3-h,p; B4-h,p
JBC	None	+++	+++	+++	+++	-	-(?)	-	-	-	-	B1; B2; B3
SG	None	+(+)p	+(+)p	-	-	++(+)	+++	-	+p	-	-	B1-r-p; B3
SGC	None	++p	++p	-	-	+++	+++	-	-	-	-	B1-r-p; B3
AG	None	-	-	-	-	-	-	+++	+++	+++	+++	B4; B5

[a] +++ and ++(+), strong; ++ and +(+), intermediate; +, weak stimulation of antibody production or reaction with antibodies.
[b] Ab, antibody.
[c] tr, Trace of an antigenic complex.
[d] h, Hapten.
[e] p, Partial stimulation of, or reaction with antibodies to some antigens of a complex.
[f] From Stępkowski and Honigberg (1972), reproduced with permission of the Society of Protozoologists.

strains. (3) Avirulent strains have many more antigenic components capable of evoking immune responses. This property of mild strains appears to be the same irrespective of whether they are fresh isolates, substrains of such isolates maintained *in vitro* for prolonged periods, or substrains of originally virulent strains that have been attenuated by being grown in culture for many months.

The foregoing findings with regard to *T. gallinae* strains and substrains could explain the observation that a nonimmune pigeon infected with an avirulent strain of this trichomonad and challenged some time later with a virulent strain develops neither symptoms nor any pathologic changes. However, birds infected simultaneously with a virulent and an avirulent strain develop a disease characteristic of the pathogenic trichomonads. Furthermore, the pathogenicity potential of the virulent member of the mixture is not impaired by its stay in immune birds—a transfer of the mixture to a nonimmune pigeon results in the development of symptoms and pathologic changes normally associated with this virulent strain.

As far as strains of *T. vaginalis* are concerned, several workers using agglutination were unable to observe antigenic changes in stocks maintained in nonliving media for prolonged periods (Kott and Adler, 1961; Laan, 1966; Teras, 1965, 1966; Teras and Tompel, 1963), nor were they able to correlate antigenic constitution and pathogenicity (e.g., Laan, 1966). As I pointed out previously (Honigberg, 1978b), "The results obtained by us with the aid of gel diffusion and fluorescent antibody methods in experiments involving *T. gallinae* (Honigberg and Goldman, 1968; Stępkowski and Honigberg, 1972) and *T. vaginalis* (B. M. Honigberg, unpublished data) throw doubts upon the alleged antigenic stability of trichomonads cultivated *in vitro* for prolonged periods." In light of the limited number (four) of antigens of *T. vaginalis* reported by the Estonian group of Professor Teras (e.g., Laan, 1966; Teras and Rõigas, 1977; Teras *et al.,* 1966; see these reports and Honigberg, 1978b, for additional references) as compared to the larger numbers of antigens found by other workers (for references, see Honigberg, 1978b), and most recently by Su-Lin (1977), I suggested (Honigberg, 1978b) that the Estonian investigators have dealt with "basic" antigenic types in well-defined geographic areas. Perhaps these "basic" antigens do not change when cultivated. Actually, Laan (1966), on the basis of his studies of the effects of cultivation upon pathogenicity and other physiologic attributes, concluded that the antigenic constitution is the only stable characteristic of *T. vaginalis* strains. Most recently, however, the members of the Estonian school (Rõigas *et al.,* 1977) indicated (in contradistinction to their previous assertions) ". . . that the antigenic properties of [*T. vaginalis* clones] had changed . . . in vitro. . . ." Admittedly, the correlation between antigenic constitution and pathogenicity levels of the human urogenital trichomonad strains appears not to be as clear as it is in those of *T. gallinae* and *H. meleagridis* (Honigberg, unpublished observations). Yet the recent results (Su-Lin, 1977) obtained by gel diffusion and immunoelectrophoretic methods suggest that the mild strains of *T. vaginalis,* like those of the avian trichomonad and histomonad, are richer in antigens than the more pathogenic ones. More research must be done before this relationship is fully understood.

In conclusion, it should be emphasized that although probably as much is known about the pathogenicity mechanisms and the factors affecting virulence of trichomonads as of any protozoal parasites, a great deal of research remains to be done before these mechanisms and the functional aspects of these factors are more fully understood.

ACKNOWLEDGMENT

Published and unpublished data obtained in the author's laboratory are the result of investigations supported by Research Grants AI 00742-1 to 23 from the National Institute of Allergy and Infectious Diseases, U. S. Public Health Service.

REFERENCES

Abraham, R., and Honigberg, B. M. (1965). *J. Parasitol.* **51,** 823–833.
Ball, G. H. (1969). *In* "Research in Protozoology" (T. T. Chen, ed.), Vol. 3, pp. 565–718. Academic Press, New York.
Brodie, B. B., Gilette, J. R., and La Du, B. N. (1958). *Annu. Rev. Biochem.* **27,** 427–454.
Budilová, M., and Kulda, J. (1977). *Collect. Abstr., Congr., Int. Partic., Trichomoniasis, Bratislva* p. 14.
Conrad, R. D., and Edward, A. G. (1970). *Avian Dis.* **14,** 599–605.
Dohnalová, M., and Kulda, J. (1975). *J. Protozool.* **22,** 61A.
Dwyer, D. M. (1971). *J. Protozool.* **18,** 372–377.
Dwyer, D. M., and Honigberg, B. M. (1970). *J. Parasitol.* **56,** 694–700.
Dwyer, D. M., and Honigberg, B. M. (1972). *Z. Parasitenkd.* **39,** 39–52.
Farris, V. K., and Honigberg, B. M. (1970). *J. Parasitol.* **56,** 849–882.
Frost, J. K. (1967). *In* "Novak's Gynecologic and Obstetric Pathology" (E. R. Novak and J. D. Woodruff, eds.), pp. 595–628. Saunders, Philadelphia, Pennsylvania.
Frost, J. K. (1975). *In* "Textbook of Gynecology" (E. R. Novak, G. S. Jones, and H. W. Jones, eds.), pp. 782–812. Williams & Wilkins, Baltimore, Maryland.
Frost, J. K., and Honigberg, B. M. (1962). *J. Parasitol.* **48,** 898–918.
Frost, J. K., Honigberg, B. M., and McLure, M. T. (1961). *J. Parasitol.* **47,** 302–303.
Gobert, J. G., Georges, P., Savel, G., Genet, P., and Piette, M. (1969). *Ann. Parasitol. Hum. Comp.* **44,** 687–696.
Hogue, M. J. (1938). *Am. J. Hyg.* **28,** 288–298.
Hogue, M. J. (1943). *Am. J. Hyg.* **37,** 142–152.

Honigberg, B. M. (1961). *J. Parasitol.* **47,** 545–571.
Honigberg, B. M. (1978a). *In* "Parasitic Protozoa" (J. P. Kreier, ed.), Vol. 2, pp. 163–273. Academic Press, New York.
Honigberg, B. M. (1978b). *In* "Parasitic Protozoa" (J. P. Kreier, ed.), Vol. 2, pp. 275–454. Academic Press, New York.
Honigberg, B. M., and Goldman, M. (1968). *J. Protozool.* **15,** 176–184.
Honigberg, B. M., Becker, R. DiM., Livingston, M. C., and McLure, M. T. (1964). *J. Protozool.* **11,** 447–465.
Honigberg, B. M., Livingston, M. C., and Frost, J. K. (1966). *Acta Cytol.* **10,** 353–361.
Honigberg, B. M., Stabler, R. M., Livingston, M. C., and Kulda, J. (1970). *J. Parasitol.* **56,** 701–708.
Honigberg, B. M., Livingston, M. C., and Stabler, R. M. (1971). *J. Parasitol.* **57,** 929–938.
Jones, A. L., and Fawcett, D. W. (1966). *J. Histochem. Cytochem.* **14,** 215–232.
Kazanowska, W. (1966). *Wiad. Parazytol.* **12,** 139–150.
Koss, L. G., and Wolinska, W. H. (1959). *Cancer (Philadelphia)* **12,** 1171–1193.
Kott, H., and Adler, S. (1961). *Trans. R. Soc. Trop. Med. Hyg.* **55,** 333–344.
Kulda, J., and Honigberg, B. M. (1969). *J. Protozool.* **16,** 479–495.
Kulda, J., and Závadová, H. (1975). *J. Protozool.* **22,** 65A–66A.
Kulda, J., Honigberg, B. M., Frost, J. K., and Hollander, D. H. (1970). *Am. J. Obstet. Gynecol.* **108,** 908–918.
Kulda, J., Zavadil, M., Vojtechovská, M., Donhalová, M., Kárasková, I., and Kuncová, E. (1977). *Collect. Abstr., Congr., Int. Partic. Trichomoniasis, Bratislava* p. 58.
Laan, I. (1966). *Wiad. Parazytol.* **12,** 173–182.
Love, R. (1959). *Ann. N.Y. Acad. Sci.* **81,** 101–117.
Lund, E. E. (1969). *Adv. Vet. Sci. Comp. Med.* **13,** 355–390.
Lund, E. E., Augustine, P. C., and Ellis, D. J. (1966). *Exp. Parasitol.* **18,** 403–407.
Mao, P., Nakao, K., and Angrist, A. (1966). *Cancer Res.* **26,** 955–973.
Müller, H. E., and Saathoff, M. (1972). *Zentralbl. Bakteriol., Parasitenkd., Infektionskr. Hyg., Abt. 1: Orig.* **222,** 275–279.
Newton, W. L., Reardon, L. V., and Di Leva, A. M. (1960). *Am. J. Trop. Med. Hyg.* **9,** 56–61.
Nielsen, M. H., and Nielsen, R. (1975). *Acta Pathol. Microbiol. Scand., Sect. B* **83,** 305–320.
Orsi, E. V., Love, R., and Koprowski, H. (1957). *Cancer Res.* **17,** 306–311.
Perez-Mesa, C., Stabler, R. M., and Berthrong, M. (1961). *Avian Dis.* **5,** 48–60.
Preer, J. R., Jr. (1969). *In* "Research in Protozoology" (T. T. Chen, ed.), Vol. 3, pp. 129–278. Academic Press, New York.
Reardon, L. V., Ashburn, L. L., and Jacobs, L. (1961). *J. Parasitol.* **47,** 527–532.
Reid, W. M. (1967). *Exp. Parasitol.* **21,** 249–275.
Rõigas, E., Kazakova, I., Mirme, E., Tompel, H., Palm, T., and Ellama, M. (1977). *Collect. Abstr., Congr., Int. Partic., Trichomoniasis, Bratislava* p. 89.
Schnaitman, C., Erwin, V. G., and Greenwalt, J. W. (1967). *J. Cell Biol.* **32,** 719–735.
Sharma, N. N., and Honigberg, B. M. (1966). *J. Parasitol.* **52,** 538–555.
Sharma, N. N., and Honigberg, B. M. (1967). *J. Protozool.* **14,** 126–140.
Sharma, N. N., and Honigberg, B. M. (1969). *J. Protozool.* **16,** 171–181.
Sharma, N. N., and Honigberg, B. M. (1971). *Int. J. Parasitol.* **1,** 67–83.
Stabler, R. M., Honigberg, B. M., and King, V. M. (1964). *J. Parasitol.* **50,** 36–41.
Stepkowski, S., and Honigberg, B. M. (1972). *J. Protozool.* **19,** 306–315.
Su-Lin, K.-E. (1977). Ph.D. Thesis, Grad. Sch., Univ. of Massachusetts, Amherst.
Teras, J. (1963). *Prog. Protozool., Proc. Int. Conf. Protozool., 1st, Prague, Czechoslovakia, 1961* pp. 572–576.

Teras, J. (1965). *Prog. Protozool., Proc. Int. Conf. Protozool., 2nd, London, England; Excerpta Med. Int. Congr. Ser.* No. 91, pp. 197–198.

Teras, J. (1966). *Wiad. Parazytol.* **12,** 357–363.

Teras, J., and Rõigas, E. (1966). *Wiad. Parazytol.* **12,** 161–172.

Teras, J., and Rõigas, E. (1977). *Collect. Abstr., Congr., Int. Partic., Trichomoniasis, Bratislava* p. 104.

Teras, J., and Tompel, H. (1963). *In* "Genito-Urinary Trichomoniasis" (K. S. Klenskii, ed.), Collect. Pap., pp. 43–50. Acad. Sci. Est. SSR, Tallinn. (In Russ., Engl. summ.)

Teras, J., Jaakmees, H., Nigesen, U., Rõigas, E., and Tompel, H. (1966). *Wiad. Parazytol.* **12,** 364–369.

Teras, J., Mirme, E., and Vokk, R. (1977). *Collect. Abstr., Congr., Int. Partic., Trichomoniasis, Bratislava* p. 109.

Trump, B. F., Goldblatt, P. J., and Stowell, R. E. (1965). *Lab. Invest.* **14,** 2000–2028.

von Brand, T. (1973). "Biochemistry of Parasites," 2nd ed. Academic Press, New York.

Weinberg, E. D. (1974). *Science* **184,** 952–956.

Weinberg, E. D., ed. (1977). "Microorganisms and Minerals." Dekker, New York and Basel.

Yang, J., and Scholten, T. (1977). *Am. J. Trop. Med. Hyg.* **26,** 16–22.

The Pathogenesis of African and American Trypanosomiasis

11

P. F. L. BOREHAM

I. INTRODUCTION

Why do people die of trypanosomiasis? This is a question often asked but so far it has not been studied extensively. The pathogenesis of a disease is the pathophysiological changes that occur during that disease. This may be defined as a study of the disturbances to normal physiology and the description of the mechanisms producing these functional abnormalities and the way in which they are expressed as symptoms and clinical signs (Lämmler, 1976). This definition demonstrates the interdisciplinary nature of the subject. The two approaches for investigating this subject are exemplified by the study of the pathogenesis of American and African trypanosomiasis. In the former case extensive pathological data are available from numerous autopsies carried out in many areas where the disease is found, so that speculation on the cellular and biochemical

BIOCHEMISTRY AND PHYSIOLOGY OF PROTOZOA
SECOND EDITION, VOL. 2

mechanisms can be drawn from a detailed knowledge of the anatomical defects. In the case of African trypanosomiasis many experimental studies have been undertaken to try to explain pathogenic processes, but our knowledge of human and cattle pathology and histopathology is rather sparse.

An understanding of the mechanisms of disease processes could lead to better methods of treatment by possibly inhibiting or reversing the important pathological changes or by alleviating some of the symptoms of the human disease. In addition, some of the side effects of specific trypanocidal chemotherapy may be prevented if an understanding of host–parasite–drug interactions is obtained.

II. AFRICAN TRYPANOSOMIASIS

A. Description of the Pathology

The pathology of African trypanosomiasis has been reviewed recently for both the human disease (Ormerod, 1970; Goodwin, 1970, 1974; Hutt and Wilks, 1971) and for the disease of other animals (Fiennes, 1970; Losos and Ikede, 1972). The African pathogenic trypanosomes are usually divided into two groups: (1) the *Trypanosoma brucei* subgroup, which includes those species pathogenic to man and widely distributed throughout the tissues and fluids of the body; and (2) *Trypanosoma vivax* and *Trypanosoma congolense,* which are confined to the blood and pathogenic to cattle (Losos and Ikede, 1972). This separation is not strictly correct because recent studies have demonstrated that *T. congolense* multiplies and develops extravascularly (Luckins and Gray, 1978). However, in terms of pathogenesis it is useful to retain this separation since the humoral group of trypanosomes causes an inflammatory degenerative disease with extensive necrosis whereas the hematic group causes a severe anemia, which is much more pronounced than that caused by *T. brucei* subgroup infections. In the early stages of *T. brucei* subgroup infections the chief lesions are found in the reticuloendothelial system and in the serous surfaces. In the later stages diffuse chronic meningoencephalitis is seen. Pathologically it is not easy to distinguish between the two human forms of the disease caused by *Trypanosoma rhodesiense* and *Trypanosoma gambiense,* although in the former case cardiac lesions are more frequent whereas in the latter the central nervous system manifestations are greater (Hutt and Wilks, 1971). There does not appear to be any significant pathological change in the solid tissues and organs of cattle trypanosomiasis (Losos and Ikede, 1972).

B. The Cardiovascular System

1. Anemia

Anemia is a consistent feature of human and animal trypanosomiases, although it is much more severe in animals infected with either *T. vivax* or *T. congolense*.

There is good agreement amongst different groups of workers on changes in the blood of animals experimentally infected with *T. brucei*. The major features of the anemia are decreases in erythrocyte count, hemoglobin concentration, packed cell volume, and half-life of the circulating erythrocytes. Polychromasia, anisocytosis, and basophilic stippling are present together with circulating normoblasts and spherocytes (Boreham, 1967; Jennings *et al.*, 1972, 1974; Jenkins *et al.*, 1974). Many red cells are crenated but rabbit cells show no increase in fragility in *T. brucei* infections (MacKenzie and Boreham, 1974), although this is not true for *T. congolense* infections of mice (Ikede *et al.*, 1977). These observations suggest that a young red cell population is present in the circulation and that it is the older cells which are being preferentially broken down. This conclusion is supported by examination of bone marrow smears which reveal no maturation defect. There is no increase in iron loss although hemosiderin is deposited in various organs, including the spleen and liver. The anemia in cattle caused by *T. congolense* and *T. vivax* has been variously described as normocytic and normochromic, with an absence of reticulocytes and an increased activity in the femoral bone marrow (Losos *et al.*, 1973; Mamo and Holmes, 1975), as normochromic and macrocytic accompanied by a leucopenia (Naylor, 1971), and as macrocytic in the acute stages and microcytic in chronic stages (Fiennes, 1970). Wellde *et al.* (1974) agree that it is a normocytic and normochromic anemia after a few weeks of infection but describe anisocytosis and polychromasia as the earliest changes, together with the presence of basophilic stippling and circulating normoblasts. The erythropoietic response is prominent during the sixth to ninth weeks of the infection. The macrocytic and normochromic nature of the disease has recently been confirmed for both infections by Maxie *et al.* (1976), who have shown an associated leucopenia.

The increase in plasma volume reported in *T. congolense* infections of cattle (Fiennes, 1970; Holmes and Mamo, 1975; Naylor, 1971), *T. brucei* infections of rabbits (Boreham, 1967), and *T. vivax* infections of sheep (Clarkson, 1968), will add to the apparent anemia. In infected animals there is an increased removal of red cells by the spleen (Holmes and Jennings, 1976; Ikede *et al.*, 1977).

Several early workers attributed the anemia in cattle to the inhibition of hemopoiesis (Hornby, 1952; Fiennes, 1954); however, most people now agree that the anemia is hemolytic and due, at least in part, to an immunological cause. If this is so, trypanosome antigen–antibody complexes are most probably formed on the surface of the erythrocytes, which, in the presence of complement, cause lysis. There are several pieces of evidence to support this idea. Herbert and Inglis (1973) demonstrated that antigens of *T. brucei* are adsorbed onto erythrocytes of infected mice by showing that, after inoculation of erthrocytes from infected animals into noninfected animals, immunity against homologous challenge occurred in the latter. Woodruff *et al.* (1973) showed that the red cells of patients with *T. rhodesiense* infections have complement on their surface whereas Woo and Kobayashi (1975) and Kobayashi *et al.* (1976) showed that *T. brucei* antigen is readily adsorbed onto normal red blood cells at 37°C and that erythrocytes taken from cattle infected with *T. congolense* contain immunoglobulin on their surface.

Kobayashi *et al.* (1976) were able to elute immunoglobulin from infected cells and in 16 of 74 eluates found IgM and IgG to be present by using the complement fixation, indirect hemagglutination, and indirect antiglobulin tests. In mice infected with *T. brucei,* suppression of the immune response by corticosteroids also reduces the anemia (Balber, 1974). Another interesting study supports the idea of an immune hemolytic mechanism. Multiple injections into rats of soluble *Trypanosoma evansi* antigen caused a marked fall in hematocrit plus an antibody response. Antigen–antibody complexes are probably formed in the blood stream, and are then removed from the circulation extravascularly (Assoku, 1975). It is not yet definitely known whether antigen and antibody are adsorbed onto the erythrocyte surface separately or as a complex.

Huan *et al.* (1975) have described a hemolytic factor with a molecular weight of approximately 10,000 which was isolated from living *T. brucei.* This factor was detected as early as 48 hr after infection and it was concluded that immune phenomena were not important in the acute phase of the anemia.

The hematological data for rabbits infected with *T. brucei* suggest that possibly a second mechanism exists in addition to immune hemolysis. The term microangiopathic hemolytic anemia was first introduced by Brain *et al.* (1962) to describe conditions where the destruction of red cells occurs secondarily to disease of the small blood vessels. It is found in a variety of disorders where pathological changes occur in small blood vessels, including microthrombi formation in capillaries and arteries, fibrinoid necrosis, necrotizing arteritis, and invasion of capillary walls by carcinoma. Many of these diseases are associated with diffuse intravascular coagulation

(DIC), and it is suggested that the microangiopathic hemolytic anemia is caused by the formation of microthrombi which trap and fragment red cells as they pass through the fibrin clot and so result in hemolysis (Hardisty and Weatherall, 1974). Features of microangiopathic hemolytic anemia are polychromasia, rouleaux formation, a high degree of crenation of red cells, and reticulocytosis but no autoagglutination of red cells (Dacie, 1967). The observations reported on the blood of rabbits infected with *T. brucei* are consistent with a microangiopathic hemolytic anemia (Boreham, 1974; Jenkins *et al.*, 1974).

2. The Fibrinolytic and Coagulation Systems

Platelets play an essential functional role in hemostasis. There is growing evidence that thrombocytopenia occurs in trypanosomiasis (Barrett-Connor *et al.*, 1973; Sadun *et al.*, 1973; Davis *et al.*, 1974; Robins-Browne *et al.*, 1975; Maxie *et al.*, 1976; Wellde *et al.*, 1978). Davis *et al.* (1974) have shown that *in vitro*, live, separated trypanosomes and extracts of trypanosomes are able to aggregate blood platelets taken from rats, rabbits, or humans. Greenwood and Whittle (1976a) have been unable to repeat these experiments and have suggested that the thrombocytopenia results from enhanced splenic trapping of platelets rather than aggregation.

Increases in fibrin–fibrinogen degradation products (FDP) are found in rabbits infected with *T. brucei* as a result of plasminogen activation, suggesting increased fibrinolysis and a state of DIC (Boreham and Facer, 1974). In addition, qualitative changes in rabbit plasma fibrinogen have been found; tests for fibrinogen B, heparin-precipitable fibrinogen, and cryofibrinogen are positive and indicative of microthrombi formation and DIC (Facer, 1974). FDP have also been reported in human disease (Greenwood and Whittle, 1976a; Robins-Browne and Schneider, 1977).

Few studies have been undertaken on components of the coagulation pathway. Trincão *et al.* (1953a) observed that in some cases of human sleeping sickness the prothrombin time was below normal levels and that this was corrected by administration of the anticoagulant ethyl bis-coumacetate (Tromexan, 600 mg). The partial thromboplastin time test with kaolin is often prolonged (Essien and Ikede, 1976). Boulton *et al.* (1974) suggest that there is increased production of many of the clotting proteins in *T. brucei* infections of the rabbit, in particular factors VIII and XII. A more detailed study has been undertaken by Robins-Browne and Schneider (1977) on five patients who had been recently infected in northern Botswana. Four of these patients showed clinical evidence of a bleeding disorder. This was attributed to the marked thrombocytopenia caused by platelet pooling and a reduction in platelet half-life. The concentrations of factors V and VIII and fibrinogen were generally elevated, suggesting rebound hypercoagulability. These authors also produced evidence that suramin may initially aggravate the coagulation defects, possibly by causing the formation of immune complexes which damage platelets and/or endothelial cells. It is, however, known that suramin inhibits the activation of the first component of complement, the action of thrombin on fibrinogen and plasminogen. It also interferes with the formation of kinins (Eisen and Loveday, 1973).

3. Plasma Constituents

The important changes in plasma proteins in trypanosomiasis infections are a decrease in serum albumin and an increase in globulins (Jenkins and Robertson, 1959; Clarkson, 1968; Goodwin and Guy, 1973; Wellde et al., 1974; Boreham et al., 1977). Fibrinogen concentrations in the plasma of rabbits, cattle, and man infected with T. brucei subgroup are raised (Trincão et al., 1953b; Boreham and Facer, 1974; Greenwood and Whittle, 1976a). Protein catabolism in mice and humans with trypanosomiasis is increased (Jennings et al., 1973; Robins-Browne and Schneider, 1977).

The serum total lipid concentration increases in rabbits infected with T. brucei by about 400%. This is almost entirely due to the triglyceride component, although there is a small increase in cholesterol but no change in phospholipid (Goodwin and Guy, 1973; Boreham et al., 1977). These findings are in contrast to observations made in cattle trypanosomiasis caused by either T. congolense or T. vivax, where there is a very marked decrease in all serum lipid components (Roberts, 1974).

Changes in the blood chemistry are given by Goodwin and Guy (1973) for rabbits infected with T. brucei, by Fiennes (1970) and Wellde et al. (1974) for cattle trypanosomiasis, and by Evens et al. (1963) for human infections. The presence of proteinuria with increases in serum urea and creatinine together with an increase in phosphate and a decrease in bicarbonate is suggestive of renal insufficiency in rabbits infected with T. brucei. The increase in aspartate transaminase and creatinine phosphokinase activity indicates necrosis of cells, especially muscle cells. The other striking feature in rabbits is a marked increase in serum pyruvate concentration. In T. congolense infections Wellde et al. (1974) found slight increases in serum urea nitrogen, whereas serum creatinine and glucose concentrations remained almost constant.

In terms of the pathogenesis of trypanosomiasis, one of the most important changes in plasma constituents is the increase in macroglobulins, which combined with the increase in plasma fibrinogen will certainly affect the flow properties of blood. A study has been undertaken to investigate changes in the rheological properties of blood of rabbits infected with T. brucei (Boreham, 1974; Facer, 1976; Boreham et al., 1977). It was found that the whole blood viscosity increased by 25% while serum and plasma viscosities increased by about 50%. Such increases mean that a greater force is required to circulate the blood around the body and also that blood will tend to pool in the microvessels with subsequent microthrombi formation. However, this effect may be counteracted to some extent by the hypotension which is present in infected animals, although it does imply that there will be a decrease in cardiac perfusion. There is also evidence of marked pooling of blood from histopathological studies in rabbits and cattle (Goodwin, 1971; Losos and Ikede, 1972).

4. Blood Pressure

Rabbits infected with *T. brucei* are hypotensive (Boreham and Wright, 1976a). The mean blood pressure of infected rabbits is 31/25 mm Hg, compared to 70/65 mm Hg in control rabbits. The heart rate is unchanged. This finding of gross hypotension in rabbits is a little surprising since recent reviews give no indication of such changes in human or cattle trypanosomiasis. However, Sicé (1937) did record hypotension as one of the pathological features of the human disease. Sicé made his observations in West Africa with patients infected with *T. gambiense,* but it is difficult to know whether the hypotension he recorded was due to the trypanosome infection or to concurrent disease. It is important to remember that control of blood pressure is complex, involving nervous as well as hormonal elements, and thus many compensatory mechanisms exist.

Some interesting studies have recently been carried out on blood pressure of rabbits. If whole, washed, separated trypanosomes are injected intravenously into a normal rabbit, there is no effect on the blood pressure. However, if the same experiment is repeated with a rabbit previously infected with *T. brucei,* there is a slow fall in blood pressure commencing 3–5 min after the injection. Similarly, if trypanosomes are incubated *in vitro* with immune serum and then washed to remove any blood components and injected intravenously into a noninfected rabbit, there is a similar, and often more dramatic, fall in blood pressure. In several of these experiments the animals died (Boreham and Wright, 1976a). These hypotensive responses can be prevented by using the kallikrein inhibitor aprotinin (Trasylol). These findings strongly suggest that the formation of immune complexes *in vivo* causes the activation of prekallikrein to form kallikrein, which itself has hypotensive properties and will also cause the release of hypotensive bradykinin from high molecular weight kininogen.

5. The Heart

Myocarditis is a well-known lesion in experimental and clinical trypanosomiasis (Koten and de Raadt, 1969; M. Murray *et al.,* 1974; Poltera *et al.,* 1976, 1977). Myocarditis may vary in intensity and the cellular

infiltrate contains morular cells. These cells are also found in the inflammatory infiltrate of the brain and contain immunoglobulin and free light
chains (Greenwood and Whittle, 1975). It is possible that the syndrome of
congestive cardiomyopathy described in the Cameroon may be caused by
trypanosomiasis since high antibody titres are commonly found and similar patients in Senegal responded well to trypanocidal therapy (Blackett
and Ngu, 1976). Changes in the electrocardiogram (ECG) during *T. brucei*
infections are complex. M. Murray *et al.* (1974) found that, while the S–T
segments of the ECG of the rat were isoelectric prior to infection, as early
as 1 week after inoculation, elevations were detected and by 3 weeks they
were very marked, suggesting cell destruction. Electrocardiographic studies have been undertaken in patients with *T. gambiense* and *T. rhodesiense*
infections (Bertrand *et al.*, 1967; Jones *et al.*, 1975). The major findings
were alterations in the T wave suggestive of a reversible ischemia and QT
abnormalities indicative of necrosis. The ischemia and necrosis of heart
tissue are probably secondary to other pathological changes. The ECG
gives no indication of a cardiotoxin in trypanosomiasis.

C. Inflammation

Inflammation is a complex series of independent reactions which begin
following sublethal injury to tissue and end with the permanent destruction of tissue or with complete healing. The inflammatory process is usually divided into four stages: increased vascular permeability, neutrophil
exudation, mononuclear cell exudation, and cellular proliferation and repair (Page, 1972). Edema is a common feature of inflammation and has
two major causes: sodium retention and hypoproteinemia (especially a
decrease in albumin). In the latter case, if the colloid osmotic pressure of
the plasma is diminished sufficiently, water will not be attracted from the
tissues into the blood and will tend to accumulate.

In the rabbit infected with *T. brucei* there is gross edema affecting all
tissues (Goodwin, 1970) (Figure 1). Since there is no sodium retention
(Goodwin and Guy, 1973), it is probable that the decrease in serum albumin and increase in vascular permeability are at least partly responsible for this edema. Ormerod (1970) sums up the pathology of human
trypanosomiasis as a chronic lymph-borne inflammation of glands and
vessels in the early stages, extending in the later stages to the nervous
system, especially where the spinal fluid is most abundant. Inflammatory
infiltration of cells is common especially around the meninges (Gallais and
Badier, 1952) (Figure 2). The majority of the inflammatory cells in the
rabbit are of the mononuclear series (Goodwin, 1970) known to be
phagocytic, especially on large particles. In man there are many lympho-

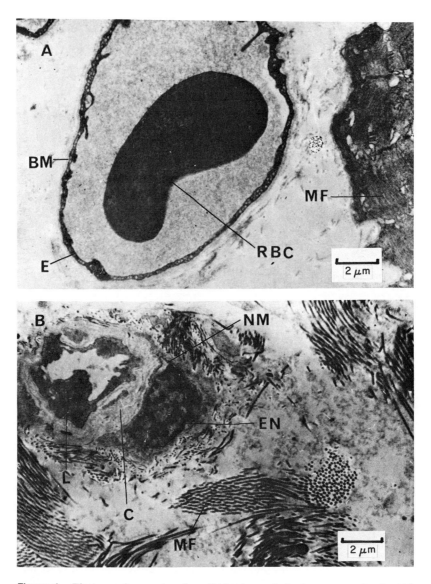

Figure 1. Electron micrographs of small blood vessels in the eye muscle of rabbits. Magnification: ×7500. (A) Noninfected rabbit showing normal structure of capillary BM, Basement membrane; E, endothelial lining; RBC, red blood cell; MF, muscle fibres. (B) Rabbit infected with *Trypanosoma brucei,* showing generalized edema of the endothelial cell. NM, Nuclear membrane; EN, endothelial cell nucleus; C, cytoplasm of endothelial cell; L, leukocyte migrating through vessel wall. The muscle fibres are forced apart by the gross edema. Stained with phosphotungstic acid.

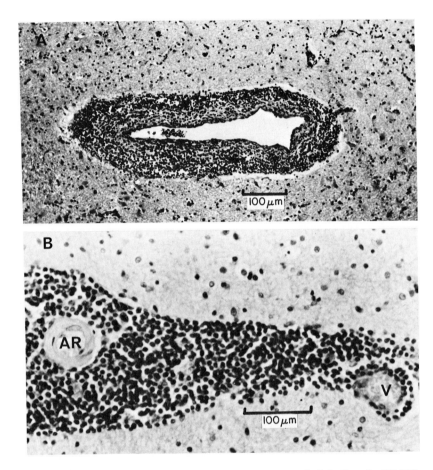

Figure 2. Lesions in the brain of human case of trypanosomiasis in Uganda. (A) Diffuse chronic encephalitis with perivascular cuffing. Magnification: × 160. (B) Chronic inflammation with mononuclear cells around arteriole (AR) and venule (V). Magnification: × 240. Stained with hematoxylin and eosin. (Courtesy of Dr. A. A. Poltera.)

cytes, plasma cells, and morular cells in the inflammatory infiltrations (Figure 3). In cattle infected with *T. congolense* or *T. vivax* the major inflammatory lesions occur in the brain and are a focal polioencephalomalacia. This probably occurs because of pressure necrosis caused by edema resulting from the accumulation of trypanosomes in the capillaries and venules (Losos *et al.*, 1973).

In a study of the function of the choroid plexus in rabbits infected with *T. rhodesiense* Ormerod and Segel (1973) showed that the inflammatory reaction probably had little effect on physiological function. A variable

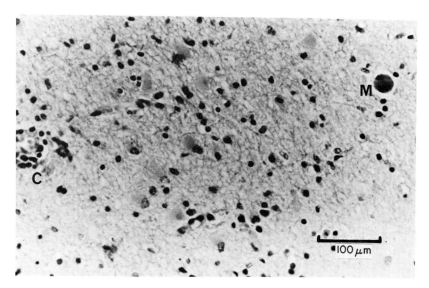

Figure 3. Perivascular infiltration around cerebral capillary (C) of human trypanosomiasis case from Uganda. A morular cell (M) is present. Magnification: × 240. Stained with hematoxylin and eosin. (Courtesy of Dr. A. A. Poltera.)

decrease in secretion of cerebrospinal fluid was observed, but this was probably due to a variable blocking of the blood vessels by parasites that occurs during the infection.

1. C-Reactive Protein (CRP)

In inflammatory conditions one of the serum constituents normally present in very low amounts increases significantly and reacts, in the presence of calcium ions, with somatic C polysaccharide of *Pneumococcus* (Tillett and Francis, 1930). This acute phase protein (CRP) appears in a variety of inflammatory conditions and is a good indication of the severity of the inflammatory response. In rabbits infected with either *T. congolense* or *T. brucei,* increased levels of CRP are found in the serum 1 or 2 weeks after infection. Thereafter, levels remain raised but below maximum concentrations (Thomasson *et al.,* 1973; Cook, 1979). The presence of CRP cannot be used as a diagnostic test for trypanosomiasis since it is found in many diseases such as malaria (Ree, 1971), but it could perhaps be a useful indication of the severity of illness.

2. Mediators of Inflammation

One of the most important pathological changes that occurs in *T. brucei* infections is the increase in vascular permeability (Boreham and Goodwin, 1967; Goodwin and Hook, 1968; Goodwin, 1970; Boreham, 1974). Such increases could be brought about in many ways, and several pharmacologically active substances have been implicated in trypanosomiasis. These include the plasma kinins (Boreham, 1968, 1970), fibrinogen degradation products (Boreham and Facer, 1974), the permeability factor of trypanosomes (Seed, 1969), and a variety of other pharmacologically active substances including 5-hydroxytryptamine and histamine (Goodwin, 1970, 1976; Boreham and Wright, 1976b).

The release of plasma kinins is known to occur early in the infection as the result of an antigen–antibody reaction (Boreham and Goodwin, 1970; Boreham and Wright, 1976b). Further studies recently carried out on the kallikrein–kinin system have given more information on pathogenic mechanisms. Urinary kallikrein levels in rabbits increase four- to eightfold during the infection, as measured by their esterase activity or biological activity on blood pressure (Wright and Boreham, 1977). Similar results are being found for plasma kallikrein (Boreham and Parry, 1979). It seems possible that plasma kallikrein is causing an increased blood flow to the glomerulus and an increase in glomerular filtration rate, which may be partly due to glomerular damage by immune complexes, resulting in the release of urinary kallikrein.

Goats infected with *T. vivax* show kinin release together with decreased concentrations of 5-hydroxytryptamine (Veenendaal *et al.*, 1976). *Trypanosoma vivax* antigen and specific antibody will cause release of 5-hydroxytryptamine from goat platelets *in vitro*, whereas antigen or antibody alone was unable to do so (Slots *et al.*, 1977).

3. Immune Complexes

Circulating immune complexes have been detected in human and experimental trypanosome infections (Fruit *et al.*, 1977; Lambert and Castro, 1977). In rabbits infected with *T. brucei* there is a fivefold increase in circulating IgG complexes with maximum amounts at the time of kallikrein activation (Boreham and Parry, unpublished observations). Deposits of complexes have been found in the renal glomeruli and it is possible that they may be important in causing glomerular damage (Lambert and Houba, 1974; Nagle *et al.*, 1974) as well as activating the kallikrein–kinin and complement systems.

4. Complement

Complement has several roles to play in disease processes, especially in the inflammatory reaction (Nelson, 1974). These are mainly through the chemotactic and anaphylatoxin activities of C3a and C5a, the adherence reactions of C3b, and the neutrophil-mobilizing factor, which is a cleavage product of C3 distinct from C3a.

The levels of C3 decrease in experimental infections (Nagle *et al.*, 1974; Lambert and Castro, 1977; Kobayashi and Tizard, 1976) and in man (Greenwood and Whittle, 1976b). The latter authors demonstrated decreases of factor B as well as C3 and C4, indicating activation of both the alternative and classical pathways of complement. Activation of the classical pathway would be expected since levels of IgM are high in this disease and it is known that one molecule of IgM in the presence of antibody can activate this pathway. Recent work has shown that trypanosomes themselves contain substances capable of activating complement (Musoke and Barbet, 1977; Nielsen and Sheppard, 1977).

5. Phagocytosis

Cremaster muscles of rabbits infected with *T. brucei* show a large number of phagocytic cells lining the venules and capillaries (Goodwin and Hook, 1968). In addition, many phagocytes have been observed attached to the endothelial lining of small blood vessels, often causing the occlusion or partial blockage of the vessel (Goodwin, 1971) (Figure 4). The majority of these cells are mononuclear cells which are presumably removing trypanosomal debris. Cook (1977) has studied *in vitro* the mechanisms of chemotaxis of peritoneal exudate cells using an adaptation of the Boyden technique. He has shown that immune complexes of trypanosomes and antibody have a significantly higher chemotactic index than trypanosomes or immune serum alone. This reaction is not complement dependent. During the process of phagocytosis many cells die, releasing catheptic enzymes which will contribute to the tissue necrosis. Ormerod (unpublished observations) has also shown that short, stumpy but not long, thin trypanosomes are rich in cathepsins.

Various factors present in trypanosome infections enhance the phagocytosis of the organisms. *In vitro* studies have shown that serum opsonic activity and cytophilic antibody activity both increase in *T. brucei* infection of the rabbits. Maximum activities are found 2 to 3 weeks after infection (Cook, 1977).

D. Toxins

Although several early workers postulated the presence of toxins (Laveran and Pettit, 1911; Novy *et al.*, 1917), injection of large numbers of trypanosomes intravenously or intracranially produced no clinical symptoms (Boreham, 1974). Renewed interest in this subject has occurred with

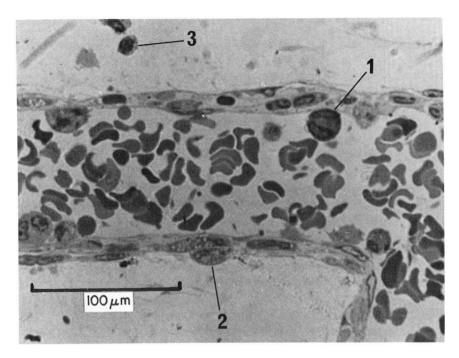

Figure 4. Mobilization of leukocytes in mesenteric vessel of rabbit infected with *T. brucei*. (1) Leucocyte attached to vascular endothelium; (2) leucocyte migrating through wall of blood vessel; and (3) leucocyte outside blood vessel in connective tissue. Stained with toluidene blue. Magnification: × 400. (Courtesy of H. Edeghere.)

the description of an active phospholipase A in autolysates of *T. congolense* (Tizard and Holmes, 1976; Tizard *et al.*, 1977). This enzyme will act on endogenous phosphatidylcholine to produce free fatty acids. Since free fatty acids can cause lysis of cells, immunosuppression, and thrombosis by destroying vascular endothelial cells, provoke myocarditis, and stimulate insulin secretion, it was postulated that this enzyme plays a central role in the pathogenesis of trypanosomiasis. This hypothesis has been extended by Assoku *et al.* (1977), who suggest that the major changes in the disease result from hypocomplementemia, elevated macroglobulins, and immunosuppression. Complement-activating factors have been demonstrated in trypanosomes (Nielsen and Sheppard, 1977; Musoke and Barbet, 1977) as well as mitogenic factors (Assoku and Tizard, 1978), and these together with the free fatty acids are responsible for the pathogenesis of the disease.

It is known that trypanosomes are able to metabolize tryptophan to tryptophol *in vitro* and *in vivo* (Stibbs and Seed, 1973, 1975). Lowered brain tryptophan would result in a reduction of 5-hydroxytryptamine synthesis and possibly could be responsible for the behavioural changes seen in the human disease. Pharmacological doses of tryptophol such as would be expected to be found by trypanosome metabolism of tryptophan are known to produce alterations in sleep and activity patterns (Jovet, 1969). Daily administration of tryptophol to voles and mice also depressed humoral antibody production but did not affect the cell-mediated immune response and could be involved in the immunosuppression seen in this disease (Ackerman and Seed, 1976).

E. Summary

As a result of recent studies, carried out mainly in the rabbit, a working hypothesis has been suggested to describe the main sequence of events leading to death from *T. brucei* subgroup infections. This hypothesis is certainly an oversimplification since a number of secondary changes will occur which will modify the disease process (Figure 5).

Infection of the host by the parasite stimulates the production of antibody, which removes the majority of the parasites by rendering them

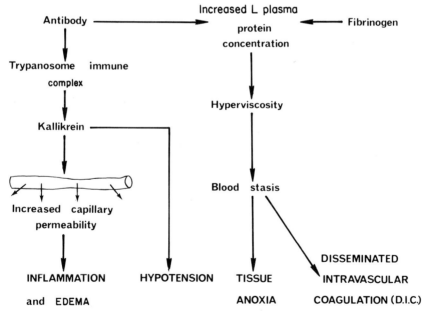

Figure 5. Diagram to show the main pathogenic mechanisms in rabbits infected with *Trypanosoma brucei*.

susceptible to phagocytosis. This will occur about 10 days after infection, but some parasites are not destroyed and form the new variant population (Vickerman, 1978). Soluble immune complexes formed at this time activate Hageman factor (Factor XII) with the subsequent production of kallikrein and kinin, which cause an increase in vascular permeability and, in the rabbit, hypotension. The increase in plasma proteins, especially globulins and fibrinogen, causes an increase in blood viscosity leading to stasis and microthrombi formation. As a consequence of these changes, ischemia and tissue necrosis occur. A combination of the leaky blood vessels and the decrease in serum albumin, which produces a lowered osmotic pressure within the circulation, results in edema that causes many of the pathological changes in the brain and is part of the inflammatory reaction.

The anemia in *T. brucei* infections is largely due to the formation of immune complexes on the surface of the erythrocytes but may be partly of microangiopathic origin resulting from microthrombi formation. These hemodynamic changes lead to a state of cardiovascular shock. In addition, there is a central depression of the immune response (P. K. Murray *et al.*, 1974) that will lead to secondary bacterial infections, which are often the final cause of death in trypanosomiasis. The possible role of phospholipase *in vivo* in the pathogenesis of trypanosomiasis has yet to be assessed.

The major pathological lesion in cattle trypanosomiasis is the anemia which appears to be hemolytic and is caused by an immunological reaction on the surface of the erythrocyte. The pathological changes occurring in the central nervous system are largely the result of edema and inflammatory changes.

III. CHAGAS' DISEASE

A. Description of the Pathology

Four distinct phases of Chagas' disease are generally recognized (World Health Organization, 1974) and have been described in detail in several recent reviews (Köberle, 1968, 1974; Andrade and Andrade, 1971; Marsden, 1971).

Infection of man by *Trypanosoma cruzi* is normally via the feces of triatomine bugs. The parasite usually enters the body through an abrasion of the skin or via mucosae but it is known that other methods of infection such as congenital infection and transmission via blood transfusion do occur, and these may be important in some circumstances (Zeledón, 1974).

The first stage, which lasts 1 to 3 weeks, is the incubation period when proliferation of amastigotes occurs within the cells with the appearance of trypomastigotes in the blood. The most common site of chagoma is the conjunctiva when infected bug feces are rubbed into the eye. This results in edema of the palpebral region (Romana sign) (Figure 6), which may possibly be due in part to the free histamine present in the bug feces (Harington, 1956). Marsden (unpublished observations), however, has been unable to demonstrate any role for histamine in Romana sign. This stage is followed by the second or acute phase of the disease, which may last for several months. It is characterized by chagoma, fever, hepatosplenomegaly, lymphodenopathy, and often tachycardia. The chagoma, or swelling at the site of infection, consists of an area of inflammation and necrosis of the skin and subcutaneous tissue with a mononu-

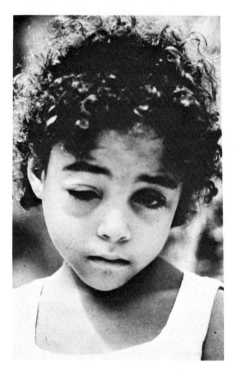

Figure 6. Patient with acute Chagas' disease, at São Felipe, Bahia State, Brazil, showing edema of the palpebral region (Romana sign). (Courtesy of Dr. M. A. Miles.)

clear cell infiltration (Pick, 1954). Multiplication and invasion of tissue cells elsewhere occur at this stage and pseudocysts are formed. Different strains of the parasite are known to infect different organs, for example, myotropic and reticulotropic strains are recognized (Bice and Zeledón, 1970; Melo and Brener, 1978).

Inflammatory reactions are marked in the acute disease, especially in the vicinity of ruptured pseudocysts. Pathological changes in the heart are usually not great in the acute disease, although there may be a moderate enlargement. The main findings are a decrease in contractility of the cardiac fibres resulting from the reduction of their distensibility following an increase in interstitial space caused by inflammation and edema (Anselmi and Moleiro, 1971). The electrocardiograms of patients in the acute phase of Chagas' disease may indicate some disturbance of cardiac conduction by a prolonged P–R interval and changes in the T wave. Approximately 95% of patients survive the acute phase. The major causes of death are encephalomyelitis or myocarditis (Köberle, 1968).

An asymptomatic acute phase may be common in children in some endemic areas when parasitaemia is found only very early in the infection (Marsden, unpublished observations).

The third stage is the intermediate or latent phase, when no clinical disease is seen and only a very few parasites are present. These may be extremely difficult to demonstrate. This phase may last for many years before developing into the chronic disease or it may last indefinitely. There is no evidence of self-cure in chagasic patients.

The final or chronic phase of the disease is characterized by dilatation of the hollow organs of the body such as the gastrointestinal tract and heart. The heart is enlarged and hypertrophied and shows evidence of vascular congestion with frequent mural thromboses (Andrade and Andrade, 1971). Cardiac arrhythmias and disturbances of cardiac conduction are common. The apical aneurism which occurs in Chagas' disease is of particular interest since it has been suggested that this may be used as a pathological diagnostic feature of the disease (Köberle, 1974). The aneurism results from dilatation and thinning of the apices of the ventricles and is most often seen on the left chamber. This lesion has been described from other parts of the heart such as the posterior or lateral wall of the ventricle (Köberle, 1968).

In a series of 1700 autopsies of patients with chronic Chagas' disease in Brazil, 90% showed cardiopathies, 20% megacolon, 18% megaesophagus, 7% bronchiectasis, and 5% other mega syndromes (Köberle, 1974). Thus, the major pathological findings for which mechanisms must be described are the cardiopathies and the enteromegalies together with the associated inflammatory changes.

B. Cardiopathy

In the acute phase of the disease the main histological changes in the heart are degenerative processes which occur in the myocardial fibres, the infiltration of mononuclear cells into the interstitial spaces, and the development of pseudocysts. However, in the chronic stage of the disease there is evidence of dilatation and hypertrophy together with fibrous tissue, inflammatory changes, and degeneration of the myocardial fibres (Figure 7) (Anselmi and Moleiro, 1974; Andrade, 1976).

There are two important pathological changes that occur in the heart. Firstly, as a result of the inflammatory reaction the interstitial spaces are increased and the capacity of oxygen to diffuse to the tissue is reduced (Anselmi et al., 1966). The contractility of the myocardial fibres is thus reduced, leading to a reduction in stroke volume and cardiac output. This occurs when the edema fluid attains 4–5% of the original weight of the heart (Cross et al., 1961). A second consequence of the inflammatory changes is that a compensatory dilatation of the cardiac chambers occurs to counteract the destruction of the contractile muscle fibres. Changes in the electrical conductivity of the heart in Chagas' disease are common and there appears to be a preferential involvement of the right bundle branch (Andrade et al., 1978). In some cases complete atrioventricular blockade

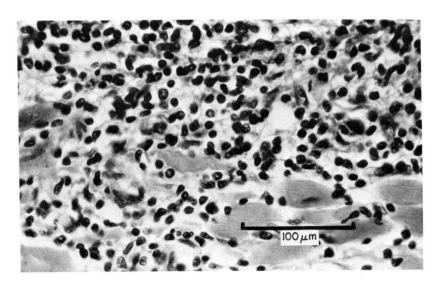

Figure 7. Chronic chagasic carditis in a fatal case showing intense inflammatory infiltration of mononuclear cells around the cardiac muscle fibres. Stained with hematoxylin and eosin. Magnification: × 400. (Courtesy of Dr. E. Reis Lopes.)

may be seen. These conduction changes are probably due to inflammatory infiltrations and tissue necrosis. One consequence of advanced cardiopathies will be a diminution of renal plasma flow and the possibility of renal infarct formation (Acquatella, 1969).

The second important pathological change is the destruction of ganglia (Figure 8). Examination of the intracardiac plexus has revealed a striking decrease in and sometimes complete absence of parasympathetic nerve cells (Köberle, 1957, 1959; Reis, 1966; Alcantara, 1970). The degree of sympathetic nerve involvement is less striking but is present, as shown by the decrease in stellate ganglion cells (Alcantara, 1970; Ferreira and Rossi, 1972). Studies on the degree of denervation required to produce pathological change have revealed a much greater sensitivity in the heart than other tissues. If 20% of heart ganglion cells are destroyed changes are seen but tolerance to denervation in the esophagus is about 80% (Köberle, 1974) (Figure 9).

The blood pressure in patients with Chagasic cardiomyopathy is low mainly because the pump is defective, although other cardiovascular changes may be contributory.

It has recently been demonstrated that rats infected with *T. cruzi* contain no detectable levels of noradrenaline in their hearts during the acute phase of the disease (Machado *et al.*, 1975) but that the concentration of catecholamines recovers during the chronic phase (Machado *et al.*, 1978). This implies functional degeneration and regeneration of the sympathetic

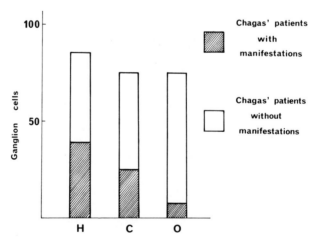

Figure 8. Reduction in the number of ganglion cells in the heart (H), colon (C), and esophagus (O) of patients with Chagas' disease with and without morphological manifestations. (Adapted from Köberle, 1974, by permission.)

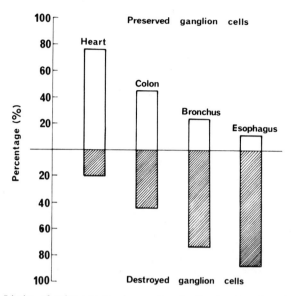

Figure 9. Limits of tolerance to denervation in the heart, colon, bronchus, and esophagus of patients infected with *Trypanosoma cruzi*. (From Köberle, 1974, reproduced by permission.)

nerves. Such nerve destruction would lead to loss of modulation of intrinsic cardiac nervous activity and contribute to pathogenesis of the disease. There is as yet no evidence for physical destruction of the adrenergic fibres or an impaired mechanism of amine synthesis or storage. The electron microscope studies of Tafuri (1970, 1976) provide some evidence in support of the first hypothesis. The most intensive degeneration of nerves was found in the vicinity of disintegrating parasites, suggesting the possibility that toxic products are liberated by the parasites or parasitized cells which are responsible for the inflammatory reaction and which may be involved in nerve destruction.

C. Enteromegaly

It is well known that the prevalence of enteromegaly varies considerably in different areas of South and Central America. It is likely that the reason for these pathological differences lies in the strain of parasite present. The very detailed and painstaking work of Köberle has shown a marked reduction in the number of ganglion cells in chagasic organs with hypertrophy and dilatation (Figure 8) (Köberle, 1968, 1974).

Peristalsis is controlled by the parasympathetic nerves arising from

Auerbach's plexuses, which are situated in the smooth muscle of the digestive tract. This plexus is destroyed in Chagas' disease (Tafuri, 1971). As a consequence, control of peristaltic waves is lost and solid residues accumulate and are retained in the intestine, which subsequently dilates. Thus aperistalsis is the cause of the enteromegalies seen in Chagas' disease, a conclusion which is supported by the demonstration of loss of esophageal motility in this disease (Brasil, 1955). It has recently been shown that the number of dense vesicles in Auerbach's plexus of the colon are reduced during the acute phase of *T. cruzi* infections of mice, and this is associated with a decrease in the potent hypotensive undecapeptide substance P (Almeida *et al.*, 1977). Noradrenaline is certainly present in these dense vesicles (Tranzer *et al.*, 1969) and possibly substance P is also stored in these granules. The release of pharmacologically active substances such as noradrenaline and substance P could be responsible for the changes in intestinal motility.

Mast cells which also contain biogenic amines have been shown to be present in increased numbers in the esophageal musculature (Pereira, 1972) and in the myocardium of the heart (Almeida *et al.*, 1975). The distribution of mast cells is not related to the inflammatory response and no real explanation for their increase has been offered, although they may be responsible for the sclerosing character of Chagas' myocarditis.

D. Other Pathological Findings

A number of other effects of denervation have been noted. These include a supersensitivity to pharmacologically active substances (Vieira and Godoy, 1963), a reduction in acid mucopolysaccharides in the digestive tract and bronchus (Britto-Costa *et al.*, 1973), decreased fat absorption (Meneghelli *et al.*, 1972), increased absorption of glucose (Meneghelli *et al.*, 1971), and carbohydrates (Meneghelli and Reis, 1967). Denervation does not lead to hypertension or prevent it from developing because of other causes (Soato *et al.*, 1974).

A number of authors have commented on pulmonary hemosiderosis in Chagas' disease (Giglio and Rossi, 1970; Rossi *et al.*, 1970; Siqueira and Ayala, 1972). This may be present in as many as 75% of cases. Köberle *et al.* (1969) suggested that destruction of the autonomic ganglia of the lungs in the chronic phase of the disease gives rise to disturbances of the intrapulmonary vasculature with pulmonary bleeding and subsequent hemosiderosis. Destruction of Purkinje cells in the cerebellum has also been described (Brandão and Zulian, 1966). Other workers have been unable to confirm these observations (Marsden, unpublished data).

In a series of 20 patients with Chagas' disease, 75% showed normal values in clotting tests (Guevara, 1969) and the author concludes that hypercoagulability does not occur in chronic Chagas' heart disease. Other hematological findings are a nonspecific mild anemia (9–12 gm of hemoglobin/100 ml of blood), a leucocytosis, and raised erythrocyte sedimentation rate (Marsden, 1971). Transient disturbances in liver function may also occur because of fatty infiltrations. During the acute phase of the disease *T. cruzi* is frequently found in the cerebrospinal fluid together with elevated albumin and leukocyte levels, which is suggestive of a mild meningitis or encephalitis (Hoff *et al.*, 1978).

Very few studies have been undertaken on plasma constituents in patients with Chagas' disease. The most comprehensive study of serum antibody concentrations was made by Lelchuk *et al.* (1970). They demonstrated that in the acute disease there were no changes in IgM, IgG, or IgA concentrations but that IgG was slightly raised in the chronic disease. Heterophile antibodies were present early in the acute phase. Fasting blood cholesterol, triglycerides, and phospholipids were all slightly lower in Chagas' disease patients (Santana *et al.*, 1969).

E. Possible Mechanisms of Ganglionic and Tissue Destruction

Inflammation seen in Chagas' disease appears to be an important mechanism in neuronal destruction. If the inflammatory response in mice is inhibited by cortisone, degenerative changes in the colon do not occur despite the presence of numerous parasites within and near Auerbach's plexus (Andrade and Andrade, 1966). Three major mechanisms have been postulated to explain ganglionic destruction: (1) the release of toxins, (2) an allergic mechanism, and (3) autoimmunity. The possibility of the release of a neurotoxin into tissues adjacent to pseudocysts when they rupture has been suggested (Köberle and Nador, 1955; Köberle, 1956). Evidence for the presence of such neurotoxins, which is not conclusive, has been reviewed by Köberle (1970). Lysosomes are present in *T. cruzi* (Vickerman and Preston, 1976) and it is possible that these may be responsible for the damage to the membranes of nerve cells. Alternatively, the protease enzyme recently isolated from *T. cruzi* may be involved (Bongertz and Hungerer, 1978). A toxin, a lipopolysaccharide, obtained from *T. cruzi* has been shown to be lethal to experimental animals. This toxin not only causes degeneration of vital organs but also induces hepatitis, myocarditis, and nephropathy (Seneca, 1969).

A number of authors have suggested a hypersensitivity mechanism to explain the pathogenesis of Chagas' disease (Muniz and de Azevedo,

1947; Torres, 1942). Necrotizing arteritis was demonstrated in some cases of megaesophagus (de Brito and Vasconcelos, 1959) but was not considered to play a significant role in plexus destruction.

The third major mechanism postulated to explain ganglion destruction is an autoimmune reaction (Jaffe, 1943). There are a number of reports of autoantibodies in Chagas' disease but the finding of Cossio *et al.* (1974) that 95% of patients have an antibody which reacts with human heart muscle has led to new ideas of possible pathogenic mechanisms. The possible pathological significance of this endocardial–vascular–interstitial factor (EVI) has recently been reviewed (Anonymous, 1977). The human antibody combined with isolated preparations of beating rat atrium, causing damage to the sarcolemma as demonstrated in the electron microscope (Sterin Borda *et al.,* 1976). Interestingly, this antibody also blocked the action of noradrenaline on this tissue. Lymphocytes appear to be involved in the pathogenic role of autoantibodies. *In vitro* studies have shown that lymphocytes sensitized with *T. cruzi* antigen are able to damage nonparasitized heart cells (Santos-Buch and Teixeira, 1974).

Although the main pathological features of Chagas' disease have been described we really know very little of how these changes are brought about.

ACKNOWLEDGMENTS

My sincere thanks are due to P. D. Marsden and M. A. Miles for help and advice in the preparation of the section on Chagas' disease.

REFERENCES

Ackerman, S. B., and Seed, J. R. (1976). *Experientia* **32,** 645–647.
Acquatella, H. (1969). *Rev. Venez. Sanid. Asist. Soc.* **34,** 327–380.
Alcantara, F. G. (1970). *Rev. Goiana Med.* **16,** 159–177.
Almeida, H. O., Pereira, F. E. L., and Tafuri, W. L. (1975). *Rev. Inst. Med. Trop. Sao Paulo* **17,** 5–9.
Almeida, H. O., Tafuri, W. L., Cunha-Melo, J. R., Freire-Maia, L., Raso, P., and Brener, Z. (1977). *Virchows Arch. A* **376,** 353–360.
Andrade, S. G., and Andrade, Z. A. (1966). *Rev. Inst. Med. Trop. Sao Paulo* **8,** 219–224.
Andrade, Z. A. (1976). *In* "American Trypanosomiasis Research," Pan Amer. Health Org., Sci. Publ. No. 318, pp. 146–151. Pan American Health Organization, Washington, D.C.
Andrade, Z. A., and Andrade, S. G. (1971). *In* "Pathology of Protozoal and Helminthic Diseases" (R. A. Marcial-Rojas, ed.), pp. 69–85. Williams & Wilkins, Baltimore, Maryland.
Andrade, Z. A., Andrade, S. G., Oliveira, G. B., and Alonso, D. R. (1978). *Am. Heart J.* **95,** 316–324.

Anonymous (1977). *Br. Med. J.* ii, 1243–1244.

Anselmi, A., and Moleiro, F. (1971). *Bull. W.H.O.* **44,** 659–665.

Anselmi, A., and Moleiro, F. (1974). *In* "Trypanosomiasis and Leishmaniasis with Special Reference to Chagas' Disease," Ciba Foundation Symposium, No. 20 (New Series), pp. 125–136. Elsevier, Amsterdam.

Anselmi, A., Pifano, F., Suarez, J. A., and Gurdiel, O. (1966). *Am. Heart J.* **72,** 469–481.

Assoku, R. K. G. (1975). *Int. J. Parasitol.* **5,** 137–145.

Assoku, R. K. G., and Tizard, I. R. (1978). *Experientia* **34,** 127–129.

Assoku, R. K. G., Tizard, I. R., and Nielsen, K. H. (1977). *Lancet* ii, 956–959.

Balber, A. E. (1974). *Exp. Parasitol.* **35,** 209–218.

Barrett-Connor, E., Ugoretz, R. J., and Braude, A. I. (1973). *Arch. Intern. Med.* **131,** 574–577.

Bertrand, E., Baudin, L., Vacher, P., Sentilhes, L., Ducasse, B., and Veyret, V. (1967). *Bull. Soc. Pathol. Exot.* **60,** 360–369.

Bice, D. E., and Zeledón, R. (1970). *J. Parasitol.* **56,** 663–670.

Blackett, K., and Ngu, J. L. (1976). *Br. Heart J.* **38,** 605–611.

Bongertz, V., and Hungerer, V. (1978). *Exp. Parasitol.* **45,** 8–18.

Boreham, P. F. L. (1967). *Trans. R. Soc. Trop. Med. Hyg.* **61,** 138.

Boreham, P. F. L. (1968). *Br. J. Pharmacol.* **32,** 493–504.

Boreham, P. F. L. (1970). *Trans. R. Soc. Trop. Med. Hyg.* **64,** 394–400.

Boreham, P. F. L. (1974). *Rev. Elev. Med. Vet. Pays Trop., Suppl.,* 279–282.

Boreham, P. F. L., and Facer, C. A. (1974). *Int. J. Parasitol.* **4,** 143–151.

Boreham, P. F. L., and Goodwin, L. G. (1967). *Int. Sci. Counc. Trypanosomiasis Res.* **11,** 83–84.

Boreham, P. F. L., and Goodwin, L. G. (1970). *In* "Bradykinin and Related Kinins" (F. Sicuteri, M. Rocha e Silva, and N. Back, eds.), pp. 539–542, Plenum, New York.

Boreham, P. F. L., and Parry, M. G. (1979). *In* "Current Concepts in Kinin Research" (G. L. Haberland, ed.). Pergamon, Oxford. In press.

Boreham, P. F. L., and Wright, I. G. (1976a). *Br. J. Pharmacol.* **58,** 137–139.

Boreham, P. F. L., and Wright, I. G. (1976b). *Prog. Med. Chem.* **13,** 159–204.

Boreham, P. F. L., Facer, C. A., and Wright, I. G. (1977). *Proc. Eur. Multicolloq. Parasitol., 2nd, Trogir, 1975* pp. 77–81.

Boulton, F. E., Jenkins, G. C., and Lloyd, M. J. (1974). *Trans. R. Soc. Trop. Med. Hyg.* **68,** 153.

Brain, M. C., Dacie, J. V., and Hourihane, C. O'B. (1962). *Br. J. Haematol.* **8,** 358–374.

Brandão, H. J. S., and Zulian, R. (1966). *Rev. Inst. Med. Trop. Sao Paulo* **8,** 281–286.

Brasil, A. (1955). *Rev. Bras. Gastroenterol.* **7,** 21–44.

Britto-Costa, R., Martins, N. E., Jr., and Mabtum, J. (1973). *Rev. Inst. Med. Trop. Sao Paulo* **15,** 227–234.

Clarkson, M. J. (1968). *J. Comp. Pathol.* **78,** 189–193.

Cook, R. M. (1977). Ph.D. Thesis, Univ. of London, London.

Cook, R. M. (1979). *Vet. Parasitol.* **5** (in press).

Cossio, P. M., Diez, C., Szarfman, A., Kreutzer, E., Candiolo, B., and Arana, R. M. (1974). *Circulation* **49,** 13–21.

Cross, C. E., Rieben, P. A., and Salisbury, P. F. (1961). *Am. J. Physiol.* **201,** 102–108.

Dacie, J. V. (1967). "The Haemolytic Anaemias Congenital and Acquired. Part III. Secondary or Symptomatic Haemolytic Anaemias," 2nd ed. Churchill, London.

Davis, C. E., Robbins, R. S., Weller, R. D., and Braude, A. I. (1974). *J. Clin. Invest.* **53,** 1359–1367.

de Brito, T., and Vasconcelos, E. (1959). *Rev. Inst. Med. Trop. Sao Paulo* **1,** 195–206.

Eisen, V., and Loveday, C. (1973). *Br. J. Pharmacol.* **49**, 678–687.

Essien, E. M., and Ikede, B. O. (1976). *Haemostasis* **5**, 341–347.

Evens, F., Niemegeers, C., and Charles, P. (1963). *Acad. R. Sci. Outre-Mer (Brussels), Cl. Sci. Nat. Med.* **14**, 1–179.

Facer, C. A. (1974). Ph.D. Thesis, Univ. of London, London.

Facer, C. A. (1976). *J. Comp. Pathol.* **86**, 393–407.

Ferreira, A., and Rossi, M. (1972). *Beitr. Pathol.* **145**, 213–220.

Fiennes, R. N. T.-W. (1954). *Vet. Rec.* **66**, 423–434.

Fiennes, R. N. T.-W. (1970). *In* "The African Trypanosomiases" (H. W. Mulligan, ed.), pp. 729–750. Allen & Unwin, London.

Fruit, J., Santoro, F., Afchain, D., Duvallet, G., and Capron, A. (1977). *Ann. Soc. Belge Med. Trop.* **57**, 257–266.

Gallais, P., and Badier, M. (1952). *Med. Trop. (Marseilles)* **12**, 633–675.

Giglio, J. R., and Rossi, M. A. (1970). *Hospital (Rio de Janeiro)* **78**, 363–368.

Goodwin, L. G. (1970). *Trans. R. Soc. Trop. Med. Hyg.* **64**, 797–812.

Goodwin, L. G. (1971). *Trans. R. Soc. Trop. Med. Hyg.* **65**, 82–88.

Goodwin, L. G. (1974). *In* "Trypanosomiasis and Leishmaniasis with Special Reference to Chagas' Disease," Ciba Foundation Symposium, No. 20 (New Series), pp. 108–124. Elsevier, Amsterdam.

Goodwin, L. G. (1976). *In* "Pathophysiology of Parasitic Infection" (E. J. L. Soulsby, ed.), pp. 161–170. Academic Press, New York.

Goodwin, L. G., and Guy, M. W. (1973). *Parasitology* **66**, 499–513.

Goodwin, L. G., and Hook, S. V. M. (1968). *Br. J. Pharmacol.* **32**, 505–513.

Greenwood, B. M., and Whittle, H. C. (1975). *Clin. Exp. Immunol.* **20**, 437–442.

Greenwood, B. M., and Whittle, H. C. (1976a). *Am. J. Trop. Med. Hyg.* **25**, 390–394.

Greenwood, B. M., and Whittle, H. C. (1976b). *Clin. Exp. Immunol.* **24**, 133–138.

Guevara, J. M. (1969). *Acta Med. Venez.* **16**, 342–345.

Harington, J. S. (1956). *Nature (London)* **178**, 268.

Hardisty, R. M., and Weatherall, D. J. (1974). "Blood and Its Disorders." Blackwell, Oxford.

Herbert, W. J., and Inglis, M. D. (1973). *Trans. R. Soc. Trop. Med. Hyg.* **67**, 268.

Hoff, R., Teixeira, R. S., Carvalho, J. S., and Mott, K. E. (1978). *N. Engl. J. Med.* **298**, 604–606.

Holmes, P. H., and Jennings, F. W. (1976). *In* "Pathophysiology of Parasitic Infection" (E. J. L. Soulsby, ed.), pp. 199–210. Academic Press, New York.

Holmes, P. H., and Mamo, E. (1975). *Trans. R. Soc. Trop. Med. Hyg.* **69**, 274.

Hornby, H. E. (1952). "Animal Trypanosomiasis in Eastern Africa, 1949." HM Stationery Off., London.

Huan, C. N., Webb, L., Lambert, P. H., and Miescher, P. A. (1975). *Schweiz. Med. Wochenschr.* **105**, 1582–1583.

Hutt, M. S. R., and Wilks, N. E. (1971). *In* "Pathology of Protozoal and Helminthic Diseases" (R. A. Marcial-Rojas, ed.), pp. 57–68. Williams & Wilkins, Baltimore, Maryland.

Ikede, B. O., Lule, M., and Terry, R. J. (1977). *Acta Trop.* **34**, 53–60.

Jaffe, R. (1943). *Rev. Sanid. Asist. Soc.* **8**, 85–93.

Jenkins, A. R., and Robertson, D. H. H. (1959). *Trans. R. Soc. Trop. Med. Hyg.* **53**, 524–533.

Jenkins, G. C., Forsberg, C. M., Brown, J. L., and Parr, C. W. (1974). *Trans R. Soc. Trop. Med. Hyg.* **68**, 154.

Jennings, F. W., Urquhart, G. M., and Murray, M. (1972). *Trans. R. Soc. Trop. Med. Hyg.* **66,** 342.

Jennings, F. W., Murray, P. K., Murray, M., and Urquhart, G. M. (1973). *Trans. R. Soc. Trop. Med. Hyg.* **67,** 277.

Jennings, F. W., Murray, P. K., Murray, M., and Uruqhart, G. M. (1974). *Res. Vet. Sci.* **16,** 70–76.

Jones, I. G., Lowenthal, M. N., and Buyst, H. (1975). *Trans. R. Soc. Trop. Med. Hyg.* **69,** 388–395.

Jovet, M. (1969). *Science* **163,** 32–41.

Kobayashi, A., and Tizard, I. R. (1976). *Tropenmed. Parasitol.* **27,** 411–417.

Kobayashi, A., Tizard, I. R., and Woo, P. T. K. (1976). *Am. J. Trop. Med. Hyg.* **25,** 401–406.

Köberle, F. (1956). *Rev. Goiana Med.* **2,** 101–110.

Köberle, F. (1957). *Virchows Arch. Pathol. Anat. Physiol.* **330,** 267–295.

Köberle, F. (1959). *Muench. Med. Wochenschr.* **101,** 1308–1310.

Köberle, F. (1968). *Adv. Parasitol.* **6,** 63–116.

Köberle, F. (1970). *Bull. W.H.O.* **42,** 739–743.

Köberle, F. (1974). *In* "Trypanosomiasis and Leishmaniasis with Special Reference to Chagas' Disease," Ciba Foundation Symposium, No. 20 (New Series), pp. 137–158. Elsevier, Amsterdam.

Köberle, F., and Nador, E. (1955). *Rev. Paul. Med.* **47,** 89–107.

Köberle, F., Oliveira, J. S. M., and Rossi, M. A. (1969). *Rev. Goiana Med.* **15,** 135–148.

Koten, J. W., and de Raadt, P. (1969). *Trans. R. Soc. Trop. Med. Hyg.* **63,** 485–489.

Lämmler, G. (1976). *In* "Pathophysiology of Parasitic Infection" (E. J. L. Soulsby, ed.), pp. xv–xvii. Academic Press, New York.

Lambert, P. H., and Castro, B. G. (1977). *Ann. Soc. Belge Med. Trop.* **57,** 267–269.

Lambert, P. H., and Houba, V. (1974). *In* "Progress in Immunology" (L. Brent and J. Holborow, eds.), Vol. 5, pp. 57–67. Elsevier, Amsterdam.

Laveran, A., and Pettit, A. (1911). *Bull. Soc. Pathol. Exot.* **4,** 42–45.

Lelchuk, R., Dalmasso, A. P., Inglesini, C. L., Alvarez, M., and Cerisola, J. A. (1970). *Clin. Exp. Immunol.* **6,** 547–555.

Losos, G. J., and Ikede, B. O. (1972). *Vet. Pathol.* **9,** Suppl., 1–71.

Losos, G. J., Paris, J., Wilson, A. J., and Dar, F. K. (1973). *Bull. Epizoot. Dis. Afr.* **21,** 239–248.

Luckins, A. G., and Gray, A. R. (1978). *Nature (London)* **272,** 613–614.

Machado, A. B. M., Machado, C. R. S., and Gomes, G. B. (1975). *Experientia* **31,** 1202–1203.

Machado, C. R. S., Machado, A. B. M., and Chiari, C. A. (1978). *Am. J. Trop. Med. Hyg.* **27,** 20–23.

MacKenzie, A. R., and Boreham, P. F. L. (1974). *Acta Trop.* **31,** 360–368.

Mamo, E., and Holmes, P. H. (1975). *Res. Vet. Sci.* **18,** 105–106.

Marsden, P. D. (1971). *Int. Rev. Trop. Med.* **4,** 97–121.

Maxie, M. G., Losos, G. J., and Tabel, H. (1976). *In* "Pathophysiology of Parasitic Infection" (E. J. L. Soulsby, ed.), pp. 183–198. Academic Press, New York.

Melo, R. C., and Brener, Z. (1978). *J. Parasitol.* **64,** 475–482.

Meneghelli, U. G., and Reis, L. C. F. (1967). *Rev. Assoc. Med. Bras.* **13,** 3–10.

Meneghelli, U. G., Padovan, W., Lima Filho, E. C., and Godoy, R. A. (1971). *Arq. Gastroenterol.* **8,** 109–118.

Meneghelli, U. G., Iazigi, N., Vieira, C. B., Padovan, W., and Godoy, R. A. (1972). *Rev. Goiana Med.* **18,** 75–90.

Muniz, J., and de Azevedo, A. P. (1947). *Hospital (Rio de Janeiro)* **32,** 165–188.
Murray, M., Murray, P. K., Jennings, F. W., Fisher, E. W., and Urquhart, G. M. (1974). *Res. Vet. Sci.* **16,** 77–84.
Murray, P. K., Jennings, F. W., Murray, M., and Urquhart, G. M. (1974). *Immunology* **27,** 825–840.
Musoke, A. J., and Barbet, A. F. (1977). *Nature (London)* **270,** 438–440.
Nagle, R. B., Ward, P. A., Lindsley, H. B., Sadun, E. H., Johnson, A. J., Berkaw, R. E., and Hildebrant, P. K. (1974). *Am. J. Trop. Med. Hyg.* **23,** 15–26.
Naylor, D. C. (1971). *Trop. Anim. Health Prod.* **3,** 159–168.
Nelson, R. A. (1974). *In* "The Inflammatory Process" (B. W. Zweifach, L. Grant, and R. T. McCluskey, eds.) Vol. 3, pp. 37–84. Academic Press, New York.
Nielsen, K., and Sheppard, J. (1977). *Experientia* **33,** 769–771.
Novy, F. G., De Kruif, P. H., and Novy, R. L. (1917). *J. Infect. Dis.* **20,** 499–535.
Ormerod, W. E. (1970). *In* "The African Trypanosomiases" (H. W. Mulligan, ed.), pp. 587–601. Allen & Unwin, London.
Ormerod, W. E., and Segal, M. B. (1973). *J. Trop. Med. Hyg.* **76,** 121–125.
Page, A. R. (1972). *In* "Pathophysiology Altered Regulatory Mechanisms in Disease" (E. D. Frohlich, ed.), pp. 671–682. Lippincott, Philadelphia, Pennsylvania.
Preira, F. E. L. (1972). *Rev. Inst. Med. Trop. Sao Paulo* **14,** 30–32.
Pick, F. (1954). *Acta Trop.* **11,** 105–138.
Poltera, A. A., Cox, J. N., and Owor, R. (1976). *Br. Heart J.* **38,** 827–837.
Poltera, A. A., Owor, R., and Cox, J. N. (1977). *Virchows Arch. A* **373,** 249–265.
Ree, G. H. (1971). *Trans. R. Soc. Trop. Med. Hyg.* **65,** 574–580.
Reis, E. L. (1966). *Hospital (Rio de Janeiro)* **70,** 1421–1433.
Roberts, C. J. (1974). *Trans. R. Soc. Trop. Med. Hyg.* **69,** 275.
Robins-Browne, R. M., and Schneider, J. (1977). *In* "Medicine in a Tropical Environment" (J. H. S. Gear, ed.), pp. 565–572. Publ. for S. Afr. Med. Res. Counc. by Balkema, Rotterdam.
Robins-Browne, R. M., Schneider, J., and Metz, J. (1975). *Am. J. Trop. Med. Hyg.* **24,** 226–231.
Rossi, M. A., Pessoa, J., and Mesquita, C. (1970). *Hospital (Rio de Janeiro)* **78,** 869–871.
Sadun, E. H., Johnson, A. J., Nagle, R. B., and Duxbury, R. E. (1973). *Am. J. Trop. Med. Hyg.* **22,** 323–330.
Santana, G., Valecillos, R. I., Puigbo, J. J., and Yepez, C. G. (1969). *Acta Med. Venez.* **16,** 346–348.
Santos-Buch, C. A., and Teixeira, A. R. L. (1974). *J. Exp. Med.* **140,** 38–53.
Seed, J. R. (1969). *Exp. Parasitol.* **26,** 214–223.
Seneca, H. (1969). *Trans. R. Soc. Trop. Med. Hyg.* **63,** 535–539.
Sicé, A. (1937). "La Trypanosomiase Humaine en Afrique Intertropicale." Vigot, Paris.
Siqueira, L. A., and Ayala, M. A. R. (1972). *Rev. Soc. Bras. Med. Trop.* **6,** 251–255.
Slots, J. M. M., van Miert, A. S. J. P. A. M., Akkerman, J. W. N., and de Gee, A. L. W. (1977). *Exp. Parasitol.* **43,** 211–219.
Soato, G. G., Vichi, F. L., Ruffino Netto, A., Machado, R. R., and Carvalho, D. S. (1974). *Rev. Paul. Med.* **84,** 121–123.
Sterin Borda, L., Cossio, P. M., Gimeno, M. F., Gimeno, A. L., Diez, C., Laguens, R. P., Meckert, P. C., and Arana, R. M. (1976). *Cardiovasc. Res.* **10,** 613–622.
Stibbs, H. H., and Seed, J. R. (1973). *Experientia* **29,** 1563–1565.
Stibbs, H. H., and Seed, J. R. (1975). *J. Infect. Dis.* **131,** 459–461.
Tafuri, W. L. (1970). *Am. J. Trop. Med. Hyg.* **19,** 405–417.
Tafuri, W. L. (1971). *Virchows Arch. A* **354,** 136–149.

Tafuri, W. L. (1976). *In* "American Trypanosomiasis Research" Pan Amer. Health Org., Sci. Pub. No. 318, pp. 152–161. Pan American Health Organization, Washington, D.C.

Thomasson, D. L., Mansfield, J. M., Doyle, R. J., and Wallace, J. H. (1973). *J. Parasitol.* **59,** 738–739.

Tillett, W. S., and Francis. T. J. R. (1930). *J. Exp. Med.* **52,** 561–571.

Tizard, I. R., and Holmes, W. L. (1976). *Experientia* **32,** 1533–1534.

Tizard, I. R., Nielsen, K., Mellors, A., and Assouku, R. K. (1977). *Lancet* **i,** 750–751.

Torres, M. C. (1942). *An. Acad. Bras. Cienc.* **14,** 1–6.

Tranzer, J. P., Thoenen, H., Cnipes, R., and Richards, J. (1969). *Prog. Brain Res.* **31,** 33–46.

Trincão, C., Franco, A., Gouveia, E., and Parreira, F. (1953a). *An. Inst. Med. Trop., Lisbon* **10,** 11–14.

Trincão, C., Parreira, F., Gouveia, E., and Franco, A. (1953b). *Gaz. Med. Port.* **6,** 121–122.

Veenendaal, G. H., and van Miert, A. S. J. P. A. M., van den Ingh, T. S. G. A. M., Schotman, A. J. H., and Zwart, D. (1976). *Res. Vet. Sci.* **21,** 271–279.

Vickerman, K. (1978). *Nature (London)* **273,** 613–617.

Vickerman, K., and Preston, T. M. (1976). *In* "Biology of the Kinetoplastida" (W. H. R. Lumsden and D. A. Evans, eds.), Vol. 1, pp. 35–130. Academic Press, New York.

Vieira, C. B., and Godoy, R. A. (1963). *Rev. Goiana Med.* **9,** 21–28.

Wellde, B., Lotzsch, R., Deindl, G., Sadun, E., Williams, J., and Warui, G. (1974). *Exp. Parasitol.* **36,** 6–19.

Wellde, B., Kovatch, R. M., Chumo, D. A., and Wykoff, D. E. (1978). *Exp. Parasitol.* **45,** 26–33.

Woo, P. T. K., and Kobayashi, A. (1975). *Ann. Soc. Belge Med. Trop.* **55,** 37–45.

Woodruff, A. W., Ziegler, J. L., Hathaway, A., and Gwata, T. (1973). *Trans. R. Soc. Trop. Med. Hyg.* **67,** 329–337.

World Health Organization (1974). *Bull. W.H.O.* **50,** 459–472.

Wright, I. G., and Boreham, P. F. L. (1977). *Biochem. Pharmacol.* **26,** 417–423.

Zeledón, R. (1974). *In* "Trypanosomiasis and Lieshmaniasis with Special Reference to Chagas' Disease." Ciba Foundation Symposium, No. 20 (New Series), pp. 51–76. Elsevier, Amsterdam.

Author Index

Numbers in italics refer to pages on which the complete references are listed.

A

Aaronson, S., 41, 42, 43, *57, 60, 62,* 352, *373*
Abaza, M. A., 398, 405, *406*
Abell, C. W., 168, *175*
Abou Akkada, A. R., 390, 392, 393, 398, 405, *406*
Abraham, R., 416, 417, *425*
Ackerman, S. B., 443, *452*
Acquatella, H., 448, *452*
Adegoke, J., 127, *144*
Adelstein, R. S., 202, *214*
Adler, S., 424, *426*
Adouette, A., 363, *378*
Afchain, D., 440, *454*
Afzelius, B. A., 166, *175*
Ahkong, Q. F., 348, *373*
Akesson, B., 308, *335*
Akimori, N., 309, *336*
Akkerman, J. W. N., 440, *456*
Albertini, D. F., 169, *178*
Alcantara, F., 185, 199, *214,* 448, *452*
Alexander, J. B., 357, 358, *373, 379*
Alexander, M., 16, *60*
Allaway, E., 36, *59*
Allen, C., 160, 162, 165, *175, 176, 178*
Allen, H. J., 356, *373*
Allen, R. D., 201, *214,* 333, 334, *335, 336,* 343, 348, 349, 358, 363, 366, 369, *373, 379*
Allen, S. L., 364, *373*
Allison, A. C., 346, 348, 352, 353, 363, *373, 377*

Almeida, H. O., 450, *452*
Alonso, D. R., 447, *452*
Alper, R. E., 10, 45, *60*
Alvarez, M., 451, *455*
Amos, H., 332, *336*
Amos, L. A., 154, *177*
Anderson, E., 10, *57*
Anderson, K., 159, *180*
Anderson, W. B., 326, *337*
Andrade, S. G., 444, 446, 447, 451, *452*
Andrade, Z. A., 444, 446, 447, 451, *452*
André, J., 281, *335*
Andronis, P. T., 364, *377*
Angrist, A., 418, *426*
Anonymous, 452, *453*
Anselmi, A., 446, 447, *453*
Antipa, G., 77, 78, 79, 80, 81, 82, 83, 86, 87, 88, 90, 92, 142, *144, 146, 147*
Aprille, J. R., 17, *57*
Arai, T., 157, *175*
Arana, R. M., 452, *453, 456*
Arce, C. A., 170, *175*
Argarana, C. E., 170, *175*
Arnold, C. G., 22, 54, *57, 60, 64*
Aschoff, J., 69, 142, *144*
Ashburn, L. L., 414, *426*
Ashman, D. F., 205, *214*
Ashworth, J. M., 200, 206, *214, 216*
Assoku, R. K. G., 432, 442, *453, 457*
Astrachan, L., 69, 105, *146*
Auclair, W., 165, *175*
Augustine, P. C., 422, *426*

459

Subject Index

A

A23187, *see* Ionophore A23187

Acanthamoeba palestinensis, membranes, 293, 294

Acetabularia
  biological clock, 69, 71, 113, 119–124
  in enucleate fraction, 120
  oxygen evolution, 120, 121

Acetabularia mediterranea, biological clock, 113, 119–124

Acetaldehyde, use by *Polytomella,* 33

Acetate
  incorporation in lipid, by *Tetrahymena,* 294, 316, 322, 324
  repression of glycolate pathway, in *Euglena,* 24
  substrate
    acetate flagellates, 12
    Astasia, 12
    Euglena, 12, 13, 26, 31, 32, 35, 36

Acetate flagellates
  cell cycle, 48, 49
  ecology, 6, 12
  glyoxylate cycle, 13, 44
  life cycle, 44, 45
  light effect, 26
  mitochondria, 18–23
  nutrition, 1, 6, 30–44
  peroxisome, 24, 25
  photosynthesis, 10–12

structure, 10
substrates
  acetate, 12
  fatty acid, short-chain, 13
  peroxisome, 24, 25
  pyruvic acid, 10
tolerance
  alcohols, 10
  fatty acids, 10

Acetate thiokinase, induction, synchronized *Euglena,* 57

Acetazolamide, *Gonyaulax,* biological clock, 117

Acetic acid, *see also* Acetate
  and rumen ciliates, 390, 392, 404, 405

Acetyl-CoA carboxylase, *Tetrahymena,* lipid synthesis, 297

N-Acetylglucosaminidase
  Euglena, 25
  Tetrahymena, 364, 371
  β-N-Hexoseaminidase, *Tetrahymena,* 364

Acid phosphatase
  Chlamydomonas, 55, 56
  Euglena, 25, 34, 36, 37
  Polytomella, 25
  Tetrahymena, 330–334

Acrasians, *see Dictyostelium*

Acrasin, 3, *see also* Adenosine 3′,5′-cyclic monophosphate; Chemoreception

Acriflavine

481

Atractyloside, *Euglena* mitochondria, 20, 23
Atrioventricular blockade, Chagas' disease, 447
Auerbach's plexus, destruction, Chagas' disease, 450, 451
Autoantibodies, and Chagas' disease, 452
Autogamy, *Paramecium,* 227, 238, 251, 256, 260–270
Autoimmunity hypothesis, Chagas' disease, 451, 452
Autonomic ganglia, destruction, Chagas' disease, 450
Autotrophy
    acetate flagellates, 2
    evolution, 5
Auxotrophy, evolution, 6
Avoiding reaction, *Paramecium,* 223, 224, 228, 229
Axoneme, proteins, 165, 172, 329
Axostyle
    microtubules, 164
    Saccinobaculus, 157
8-Azaguanine, *Astasia,* inhibition of cell cycle, 50
Azide
    Dictyostelium, calcium uptake, 204
    Polytomella, inhibition, propionate oxidation, 33

B

Bacillus megaterium
    lipid, adaptation to temperature, 318, 325
    and rumen ciliates, 400, 403
Bacillus subtilis, and rumen ciliates, 403
Bacteria, *see also* specific organism
    chemotactic behavior, 203, 205
    as food, phagotrophs, 340–342, 351, 355, 356, 362, 366
    infection, in trypanosomiasis, 444
    lipid, temperature adaptation, 317, 318
    and rumen ciliates, 383–386, 388, 389, 391–397, 399–403
Bacteroides ruminicola, and rumen ciliates, 400
Barium, and *Paramecium* mutants, 230–234, 239, 240, 246, 249–251, 253, 255
Basal bodies
    microtubules, 172–174
    proteins, 165

Basophilic stippling, trypanosomiasis, 431
Behavior, *Tetrahymena,* 335
Belousev–Zhabotinskii reaction, *see* Zhabotinskii–Belousev reaction
BHP, *see tert*-Butyl hydroperoxide
Bifurcations, limit cycle model of biological clock, 139–141
Biogenic amines, *see also* specific compound
    biological clock, 69, 88, 91, 92
    Chagas' disease, 450
Biological clock
    Acetabularia, 71, 113
        enucleate cell, 120
        model, 122–124
        oxygen evolution, 120, 121
        phase control in fragments, 121, 122
    Chlamydomonas, 69
    dynamic model, 133–142
    Escherichia coli, 69
    Euglena, 69, 72
        cell division, 92–96, 102–104
        enzyme activities, 92–102
        motility, 94
        nutritional effects, 97–99
        other rhythms, 96, 97
        sulfur compounds and photosynthetic mutants, 97, 99
    evolution, 2
    glossary, 142–144
    Gonyaulax, 104–119
        bright light and critical temperature, 118, 119
        change in particle size, 113–115
        chronotypic luciferase activity, 108
        heavy water, effect, 108, 109
        intracellular potassium, 115–117
        multiple circadian output, 105–107
        thylakoid spacing, chloroplast, 113
        transducing mechanism, 109
    Klebsiella aerogenes, 69
    ontogeny, 3
    Paramecium, 69
    Phaeodactylum, 69
    Physarum, 129–137
    protozoa, 67–149
    Tetrahymena
        biogenic amines, 69
        bioluminescence, 69
        cell division, 71–75
        glycogen metabolism, 87–89

Polytoma, 55
Polytomella, 53, 54
Euglena
 giant mitochondria, 53, 54
 influence of light, 29
 isolation
  Chlamydomonas, 22
  Euglena, 18, 20, 24, 35
  Polytomella, 21
 morphology, *Euglena,* 21, 34, 54
 ribosome, *Euglena,* 23
 Tetrahymena, membranes, 276–282, 284,
  285, 291, 293–296, 298, 300, 303, 310,
  314, 330, 349, 350, 361
 trichomonad pathogenicity, 418
Mitogens
 biological clock, 127
  Physarum, 129–133
  Tetrahymena, 127
 trypanosomiasis, 442
Mitosis
 Astasia and *Euglena,* 48, 49
 biological rhythm, 128
 Physarum, 129–138
 and cyclic nucleotide, 168
 and *Trichomonas* pathogenicity, 420
Mitotic inhibitors, 158, 159, 162–164
Mitotic spindle, and microtubule polymerization, 162–165, 168
MNNG, *see* N-methyl-N'-nitro-N-
 nitrosoguanidine
Monoamine oxidase, and trichomonad
 pathogenicity, 418
Mononuclear cells
 Chagas' disease, 445–447
 trypanosomiasis, 436, 441
Morphogenesis, *see also* Differentiation
 slime molds, 207, 208, 210
Mouse
 experimental host for trichomonads, 411,
  412, 418–421
 subcutaneous assay for trichomonad
  pathogenicity, 413, 414
MTOC, *see* Microtubule organizing center
Mucocyst, *Tetrahymena*
 membranes, 276, 334
 phagotrophy, 340, 356, 357, 372
Mucoid secretion, in feeding, 7
Mucopolysaccharide
 in esophagus and bronchi, Chagas' disease, 450

Tetrahymena, mucocyst, 358
Mucus, and phagotrophy, *Tetrahymena,*
 355–358
Mutation, *Paramecium* behavior, 221, 224,
 229, 262
 mutagenesis, 225, 226, 252
Myocarditis, trypanosomiasis, 435, 436, 442
 Chagas' disease, 446, 451
Myosin
 Dictyostelium, 202
 phagocyte, 346

N

NADH, NADPH diaphorase, *Euglena,* 19
NADH, NADPH oxidase
 Euglena, 19
 Polytoma, 21
NADH-Cytochrome *c* oxidoreductase, in-
 hibition by high oxygen, 39
NADH dehydrogenase
 activity, synchronized *Astasia,* 56
 inhibition, cycloheximide, synthesis by
  Polytomella, 57
NADH-Lipoyl dehydrogenase, *Euglena,* 19
Naegleria
 transformation, 3, 165
 tubulin, 165
Nalidixic acid, *Euglena* bleaching, 43
Neoplasm, and trichomonads, 417
Nephelostat, *see* Continuous culture
Nephropathy, *Trypanosoma cruzi* toxin, 451
Neoxanthin
 Chlamydomonas, 52
 Euglena, 40
Nervous system, and trypanosomiasis, 436
Neuraminidase, and trichomonad
 pathogenicity, 421
Neuroblastoma
 nontubulin protein, 161, 164, 166
 tubulin, 170
Neurospora, biological rhythm, 125
Neurotoxin, and Chagas' disease, 451
Neurotransmitters, and signal transduction,
 199, 207
Neutral lipid, *Tetrahymena,* 284, 286, 307
Neutral red granule, *Tetrahymena,* 363
Neutrophil exudation, trypanosomiasis, 436
Neutrophil-mobilizing factor, trypano-
 somiasis, 441